D0583701

Structural Acoustics and Vibration

Mechanical Models, Variational Formulations and Discretization

Structural Acoustics and Vibration

Mechanical Models
Variational Formulations and Discretization

ROGER OHAYON

Conservatoire National des
Arts et Métiers (CNAM)
Paris, France

CHRISTIAN SOIZE

Office National d'Etudes et de
Recherches Aérospatiales (ONERA)
Châtillon, France

ACADEMIC PRESS
San Diego London Boston
New York Sydney Tokyo Toronto

ACADEMIC PRESS LIMITED
24–28 Oval Road
LONDON NW1 7DX

U.S. Edition Published by
ACADEMIC PRESS
San Diego, CA 92101

This book is printed on acid free paper

Copyright © 1998 ACADEMIC PRESS LIMITED

All rights reserved

No part of this book may be reproduced or transmitted in any form or by any means, electronic or mechanical including photocopying, recording, or any information storage and retrieval system without permission in writing from the publisher.

A catalogue record for this book is available from the British Library

ISBN 0–12–524945–4

Printed and bound in Great Britain by MPG Books Ltd, Bodmin, Cornwall

TA
654
O35
1998

Contents

Preface

This book is devoted to mechanical models, variational formulations and discretization for calculating linear vibrations in the frequency domain of complex structures with arbitrary shape, coupled or not with external and internal acoustic fluids at rest. Such coupled systems are encountered in the area of internal and external noise prediction, reduction and control problems. The excitations can arise from different mechanisms such as mechanical forces applied to the structure, internal acoustic sources, external acoustic sources and external incident acoustic plane waves. These excitations can be deterministic or random. We are interested not only in the low-frequency domain for which modal analysis is suitable, but also in the medium-frequency domain for which additional mechanical modeling and appropriate solving methods are necessary. The main objective of the book is to present appropriate theoretical formulations, constructed so as to be directly applicable for developing computer codes for the numerical simulation of complex systems.

Paris, May 1997 — Roger Ohayon and Christian Soize

CHAPTER I

A Strategy for Structural-Acoustic Problems

1. Introduction

This book is devoted to mechanical models and appropriate numerical formulations for calculating the vibration of *complex structures with arbitrary shape* coupled with external and internal acoustic fluids. In practice, such coupled systems are encountered in the area of internal and external noise prediction, reduction and control problems. The external noise problem is also called the acoustic radiation of structures. Excitations can arise from different mechanisms such as mechanical forces applied to the structure, internal acoustic sources, external acoustic sources and external incident acoustic plane waves. These excitations can be deterministic or random. Concerning problems solved by analytical methods, such as rectangular thin plates, circular cylindrical thin shells, with particular acoustic and structural boundary conditions, we refer the reader to Junger and Feit, 1993; Bowman et al., 1969; Fahy, 1987; Jones, 1986.

In Section 2, we define a structural-acoustic master system.

In Section 3, we introduce the concept of fuzzy structure related to the structural complexity and we define a structural-acoustic fuzzy system.

Section 4 is devoted to the definition of the LF (low-frequency), MF (medium-frequency) and HF (high-frequency) ranges for structural-acoustic problems.

In Section 5, we present the theoretical and numerical strategy used for the LF and MF ranges.

In Section 6, we introduce the different types of prescribed deterministic and random excitations for a structural-acoustic system.

Finally, in Section 7, we indicate the general organization of the book.

2. Structural-Acoustic Master System

In general, a complex structure is composed of a main part called the *master structure*, defined as the "primary" structure accessible to conventional modeling, and a secondary part called the *fuzzy substructure* related to the structural complexity and including for example many equipment units attached to the master structure.

An internal (or an external) structural-acoustic master system is defined as a master structure coupled with an *internal acoustic fluid* (or an *external acoustic fluid*). The acoustic fluid can be a gas or a liquid. A master structure coupled with internal and external acoustic fluids is called a *structural-acoustic master system.*

In this book, mechanical modeling of a structural-acoustic master system is based on the use of the linearized equations of structural and fluid mechanics (acoustic fluids) around a state of equilibrium at rest. For a structural-acoustic problem, the fields of interest are the displacement, velocity and acceleration fields of the master structure, the pressure field in the internal fluid and the pressure field in the external fluid (near field and far field). Since the problem is linear, the structural-acoustic master system can be characterized by a Frequency Response Function (FRF) which allows the response to deterministic and random excitations to be calculated.

3. Concept of Fuzzy Structure and Structural-Acoustic Fuzzy System

A *fuzzy structure* is defined as a master structure coupled with a fuzzy substructure. A fuzzy substructure is defined as a structural complexity consisting of a large number of small secondary dynamical subsystems attached to the master structure. Each secondary subsystem is a discrete or continuous elastodynamic system, whose resonant frequencies lie in the frequency band of analysis considered (for instance, a large number of equipment units together can constitute a fuzzy substructure). More generally, a fuzzy substructure is a part of the structure that is not accessible to conventional modeling because the details of it are unknown, or are known with insufficient accuracy (this explains the choice of the word "fuzzy"). For this reason a statistical approach is proposed for modeling fuzzy substructures, by introducing a random boundary impedance operator related to the interfaces where the fuzzy substructure is connected to the master structure. This approach, called "fuzzy structure theory", was introduced by Soize, 1986. It should be noted that the term "fuzzy" has nothing to do with the mathematical theory concerning fuzzy sets and fuzzy logic. In the definition introduced in Section 2, if the master structure is replaced by a fuzzy structure, then we obtain a *structural-acoustic fuzzy system.*

4. LF, MF and HF Frequency Ranges

The different types of dynamical responses of a weakly dissipative complex structure leads to the definition of three frequency ranges of analysis. Fig. 1 is a qualitative diagram showing the frequency response function (FRF) of a component of the displacement in a fixed point of the structure.

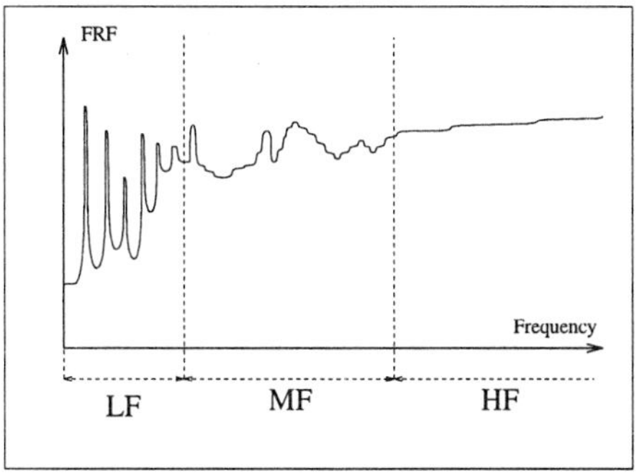

Fig. 1. Frequency ranges for a complex structure

The three frequency ranges are defined as follows.

- The *low-frequency range* (LF) is defined as the modal domain for which the associated conservative system has a small number of modes (low modal density).

- The *high-frequency range* (HF) is defined as the frequency band for which there is a uniform high modal density. Among the analysis methods for this band, we can mention the wave approach (see for instance Maidanik and Dickey, 1988 and 1994), the global statistical energy approach and the local energy approach. The most popular global statistical energy approach is the Statistical Energy Analysis. For its basic formulation, we refer the reader to Lyon, 1975; Maidanik, 1981; Lyon and DeJong, 1995; Soize, 1995a; Lesueur, 1988; Cremer et al., 1988; Crighton et al., 1992. For extensions to the case of a structure coupled with an external unbounded liquid, see David and Soize, 1994. Concerning the local energy approach, we can mention the power flow analysis based on continuous energy equations (Nefske and Sung, 1988).

- For complex systems, an intermediate frequency range called *medium-frequency range* (MF) appears. This MF range is defined as the intermediate frequency band for which the modal density exhibits large variations over the band. In addition, if there is a structural complexity related to the presence of a fuzzy substructure, this fuzzy substructure plays an important role in the dynamical behavior of the master structure. The fuzzy substructure induces an "apparent strong dissipation" in the master structure due to the power flow between the master structure and all the dynamical subsystems constituting the fuzzy substructure. Therefore, the frequency response functions of the master structure coupled with the fuzzy substructure have smooth variations which means that this system is not resonant in the MF range. The analysis is similar for internal structural-acoustic systems (master system or fuzzy system). The presence of an external acoustic fluid does not substantively modify the above qualitative analysis.

This book is devoted to LF and MF analyses for complex structures and complex structural-acoustic systems.

5. Strategy Used for the LF and MF Ranges

5.1. Structural-acoustic master systems in the LF and MF ranges

Concerning the internal structural-acoustic master system (master structure coupled with internal fluids), the LF and MF models are based on the symmetric variational formulation of the local equations including appropriate dissipative terms. The master structure is described by the displacement field and the internal fluids are described by scalar fields. An adapted formulation is presented for the LF range and the MF range. The internal structural-acoustic master system is discretized by the finite element method. The external fluid model is based on a symmetric variational formulation of an appropriate boundary integral representation which is valid for all values of the real frequencies and does not exhibit the nonphysical irregular frequencies (spurious frequencies) encountered in standard boundary integral representation. Discretization is carried out using the finite element method which leads to boundary elements. For the structural-acoustic master system, a linear symmetric matrix system is obtained whose dimension is the number of degrees of freedom of the discretized model of the internal structural-acoustic master system due to the use of a boundary integral formulation for the external acoustic fluid.

LF computation strategy. We construct a modal reduction procedure using the structural modes of the master structure *in vacuo* and the acoustic modes of the internal acoustic cavities. The reduced symmetric matrix equation obtained is solved frequency by frequency.

MF computation strategy. In the MF range, the dimension of the matrix equation resulting from finite element discretization is large because the mesh size must be adapted with respect to the wavelengths in the system. Consequently, a frequency-by-frequency solution is not practicable. The method proposed consists in solving the finite element matrix equation MF narrow band by MF narrow band.

5.2. Fuzzy structure in the MF range

In this case, the matrix equation obtained is the sum of the finite element matrix systems corresponding to the master structure with a frequency-dependent random symmetric matrix corresponding to the fuzzy substructure. An appropriate process is used for solving the linear random matrix equation.

5.3. Structural-acoustic systems in the MF range

In this case, the matrix equation obtained is the sum of the finite element matrix systems corresponding to the structural-acoustic master system with a frequency-dependent random symmetric matrix corresponding to the fuzzy substructure and, as above, an appropriate solver is used.

6. Excitations and Responses

For deterministic excitations such as mechanical forces applied to the master structure, internal acoustic sources, external acoustic sources and external incident acoustic plane waves, the responses are calculated using the frequency response functions constructed in the LF and MF ranges. Concerning random excitations, two cases are considered. The first case concerns discrete random excitations modeled by vector-valued stochastic processes. This is the case for random mechanical forces or for random aeroacoustic internal sources induced by machines. The second case concerns continuous random excitations corresponding to random wall pressure fields modeled by vector-valued random fields. This is the case for a turbulent boundary layer due to an external flow around a launcher, an aircraft, an automobile, a naval structure, a high-speed train or an external random field due to a spatial distribution of multipole sources such as complex aeroacoustic sources generated by a launcher engine. For time-stationary random excitations, spectral analysis of the linear filtering theory is used and, for nonstationary random excitations, appropriate methods are used. In all these random cases, the frequency response functions are used in the LF and MF ranges.

7. Organization of the Book

This book is organized in three main parts. Chapters III to IX are devoted to vibration of the master structures. Chapters X to XIV deal with vibration of structural-acoustic master systems and finally, in Chapter XV, fuzzy structures are introduced.

Chapter II is devoted to basic notions on variational formulations, the Ritz-Galerkin method and finite element discretization used throughout the book.

In Chapters III, IV and V, we introduce the structural modes, the damping model based on the use of the linear viscoelasticity theory and the frequency response function (FRF) for the master structure.

Chapter VI presents the master structure FRF calculation in the LF range using modal reduction and a frequency-by-frequency construction.

Chapters VII and VIII give the master structure FRF calculation in the MF range.

Chapter VII presents a narrow-frequency-band-by-narrow-frequency-band construction applied to the finite element discretization of the master structure and Chapter VIII introduces a reduced model in the MF range for which a frequency-by-frequency construction can be used.

Chapter IX gives the methods for constructing the response of the master structure submitted to deterministic and stationary or nonstationary random excitations in the LF and MF ranges.

Chapter X is devoted to the derivation of the linear acoustic equations introducing a damping model, a frequency-dependent wall acoustic impedance and appropriate terms for the internal fluid to ensure that the equations are correctly-stated at zero frequency.

Chapter XI introduces the acoustic modes of an internal cavity and presents the construction of the FRF for internal acoustic problems in the LF and MF ranges.

Chapter XII presents a symmetric boundary integral formulation for the external acoustic problem which is valid for all frequency values in the LF and MF ranges.

Chapter XIII describes the construction of the FRF calculation in the LF range for a structural-acoustic master system using a modal reduced symmetric matrix model and a frequency-by-frequency construction. The methods presented in Chapter IX can be applied directly for calculation of the response of a structural-acoustic master system submitted to deterministic or random excitations.

Chapter XIV deals with the construction of the FRF calculation in the MF range for a structural-acoustic master system using a narrow-frequency-band-by-narrow-frequency-band construction applied to the finite element symmetric matrix of the system. The methodology presented in Chapter VIII can be extended for constructing a reduced symmetric matrix model of the structural-acoustic master system in the MF range. The methods presented in Chapter IX can be applied directly for calculation of the response of a structural-acoustic master system submitted to deterministic or random excitations.

Chapter XV introduces the fuzzy structure theory in structural dynamics for the MF range. The basic tools introduced allow structural-acoustic fuzzy systems to be considered, i.e. structural-acoustic master systems coupled with fuzzy substructures corresponding to additional structural complexities.

Mathematical notations are given in the appendix, presenting the main mathematical tools used in the book and particularly elements concerning linear, antilinear, bilinear and sesquilinear forms on Hilbert space and Fourier transforms of functions.

A comprehensive list of references is given and is followed by a Subject Index.

It should be noted that the footnotes concern purely mathematical aspects which can be skipped in a first reading.

CHAPTER II

Basic Notions on Variational Formulations

1. Introduction

The objective of this chapter is to review the classical tools for the construction of variational formulations of boundary value problems and the corresponding discretization, which will be used throughout the book. In this chapter, these tools will be illustrated using a classical basic problem.

In Sections 2 and 3, we introduce the basic boundary value problem in a bounded medium and its strong solution.

Sections 4 and 5 give the method for constructing a variational formulation of the boundary value problem using the test function (weighted function) method. For that, we define the admissible function space of the problem and we use Green's formula. We introduce the weak solution of the boundary value problem which corresponds to the solution of the variational formulation. Finally the converse is established.

In Section 6, we introduce the linear operator equation associated with the variational formulation.

Section 7 reviews the Ritz-Galerkin and finite element methods which constitute the basic tools used in this book.

Section 8 is devoted to bibliographical comments.

2. Boundary Value Problem

Let Ω be a three-dimensional open bounded domain of $\mathbb{R}^3$ with a smooth boundary $\Gamma = \partial\Omega$. The external unit normal to Γ is denoted as $\mathbf{n}$ (see Fig. 1). We denote by $\mathbf{x} = (x_1, x_2, x_3)$ any point in Ω. Let g be a given real-valued field defined on Ω and G be a given real-valued field defined on Γ. For $k > 0$, we consider the following boundary value problem

$$-\nabla^2 u + k\, u = g \quad \text{in} \quad \Omega \quad , \tag{1}$$

with the Neumann boundary condition

$$\frac{\partial u}{\partial \mathbf{n}} = G \quad \text{on} \quad \Gamma \quad , \tag{2}$$

where u is a real-valued field defined on Ω. The Laplacian operator is defined by $\nabla^2 = \nabla \cdot \nabla$ in which ∇ denotes the gradient operator with respect to $\mathbf{x} = (x_1, x_2, x_3)$. The normal derivative $\partial u/\partial \mathbf{n}$ is defined by $\nabla u \cdot \mathbf{n}$.

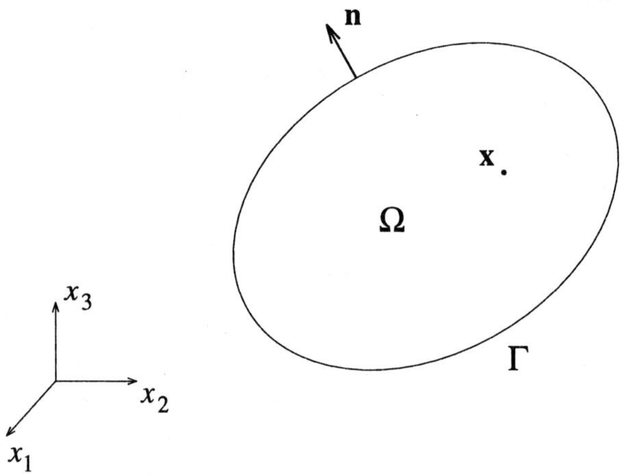

Fig. 1. Geometrical configuration

3. Strong Solution of the Boundary Value Problem

For sufficiently regular functions g and G, the boundary value problem defined by Eqs. (1) and (2) has a unique solution u called the *strong solution* [1].

---

[1] From a mathematical point of view, if $g \in L^2(\Omega)$ and $G \in H^{1/2}(\Gamma) \subset L^2(\Gamma)$, then strong solution u belongs to $H^2(\Omega)$. Since $u \in H^2(\Omega)$, then the trace u_Γ of u on Γ belongs to $H^{3/2}(\Gamma)$ and the trace $\partial u/\partial \mathbf{n}$ on Γ belongs to $H^{1/2}(\Gamma)$.

4. Variational Formulation and Weak Solution of the Boundary Value Problem

4.1. Variational formulation

We proceed by the test-function method (also called weighted-function method). In a first step, we introduce the *admissible function space* C of the problem constituted by the "sufficiently differentiable"[2] real-valued functions δu defined on Ω. In a second step, we consider strong solution u of the boundary value problem defined by Eqs. (1) and (2). Multiplying Eq. (1) by an arbitrary function $\delta u \in C$ and integrating over the domain Ω yields

$$-\int_\Omega \nabla^2 u \, \delta u \, d\mathbf{x} + k \int_\Omega u \, \delta u \, d\mathbf{x} = \int_\Omega g \, \delta u \, d\mathbf{x} \quad , \tag{3}$$

in which $d\mathbf{x} = dx_1 dx_2 dx_3$. We have Green's formula [3]

$$-\int_\Omega \nabla^2 u \, \delta u \, d\mathbf{x} = \int_\Omega \nabla u \cdot \nabla \delta u \; d\mathbf{x} - \int_{\partial\Omega} \frac{\partial u}{\partial \mathbf{n}} \, \delta u \, ds \quad , \tag{4}$$

in which ds is the surface element. Taking into account the Neumann condition defined by Eq. (2) and using Eq. (4), Eq. (3) can be written as

$$\int_\Omega \nabla u \cdot \nabla \delta u \; d\mathbf{x} + k \int_\Omega u \, \delta u \, d\mathbf{x} = \int_\Omega g \, \delta u \, d\mathbf{x} + \int_\Gamma G \, \delta u \, ds \quad , \quad \forall \, \delta u \in C \quad . \tag{5}$$

The variational formulation of the boundary value problem defined by Eqs. (1) and (2) is then stated as follows. For given g and G, find u in C such that Eq. (5) is satisfied, i.e.

$$a(u\,,\delta u) = f(\delta u) \quad , \quad \forall \, \delta u \in C \quad , \tag{6}$$

in which $a(u\,,\delta u)$ is the symmetric positive-definite bilinear form on $C \times C$ defined by

$$a(u\,,\delta u) = \int_\Omega \nabla u \cdot \nabla \delta u \; d\mathbf{x} + k \int_\Omega u \, \delta u \, d\mathbf{x} \quad , \tag{7}$$

and $f(\delta u)$ is the linear form on C defined by

$$f(\delta u) = \int_\Omega g \, \delta u \, d\mathbf{x} + \int_\Gamma G \, \delta u \, ds \quad . \tag{8}$$

---

[2] Space $C = H^1(\Omega)$.

[3] Green's formula is used with $u \in H^2(\Omega)$ and $\delta u \in H^1(\Omega)$. Since $H^{1/2}(\Gamma) \subset L^2(\Gamma)$, $(u,\delta u) \mapsto \int_\Gamma (\partial u/\partial \mathbf{n}) \, \delta u \, ds$ is the inner product in $L^2(\Gamma)$ of $\partial u/\partial \mathbf{n} \in H^{1/2}(\Gamma)$ with $\delta u \in H^{1/2}(\Gamma)$.

4.2. Weak solution of the boundary value problem

A *weak solution* of the boundary value problem defined by Eqs. (1) and (2) is a function $u \in \mathcal{C}$ satisfying Eq. (6). It should be noted that the strong solution of Eqs. (1) and (2) is a solution of Eq. (6), i.e. also a weak solution of Eqs. (1) and (2).

4.3. Existence and uniqueness of the weak solution of the boundary value problem

For sufficiently regular functions [4] g and G, the variational formulation defined by Eq. (6) has a unique solution [5] $u \in \mathcal{C}$.

5. Converse

Let u be the solution of Eq. (6) (weak solution of Eqs. (1) and (2)) and let us assume that u is sufficiently regular [6]. Using Eq. (6) and applying the Green formula defined by Eq. (4) yields

$$\int_\Omega (-\boldsymbol{\nabla}^2 u + k\, u - g)\, \delta u\, d\mathbf{x} + \int_\Gamma (\frac{\partial u}{\partial \mathbf{n}} - G)\, \delta u\, ds = 0 \quad , \quad \forall\, \delta u \in \mathcal{C} \quad . \tag{9}$$

To recover Eqs. (1) and (2) from Eq. (9), we proceed in two steps. First, we choose a test-function δu in the set of infinitely differentiable functions vanishing in a neighborhood [7] of Γ. Then Eq. (9) yields

$$\int_\Omega (-\boldsymbol{\nabla}^2 u + k\, u - g)\, \delta u\, d\mathbf{x} = 0 \quad . \tag{10}$$

From Eq. (10), we deduce that [8]

$$-\boldsymbol{\nabla}^2 u + k\, u = g \quad \text{in} \quad \Omega \quad . \tag{11}$$

---

4 For $g \in L^2(\Omega)$ and $G \in H^{1/2}(\Gamma)$, Eq. (6) has a unique solution $u \in \mathcal{C}$ in which $\mathcal{C} = H^1(\Omega)$. In effect, bilinear form $a(u\,, \delta u)$ is continuous on $\mathcal{C} \times \mathcal{C}$ and coercive ($\mathcal{C}-$ elliptic) on $\mathcal{C}$, and linear form $f(\delta u)$ is continuous on $\mathcal{C}$. Consequently, the Lax-Milgram theorem can be used. It should be noted that the weak solution exists for $g \in \mathcal{C}'$ (dual space of $\mathcal{C}$) and $G \in H^{-1/2}(\Gamma)$ (dual space of $H^{1/2}(\Gamma)$). In this case the strong solution does not exist in the sense defined in footnote 1 related to Section 3, but a weak solution exists.

5 We have the regularity property of the weak solution: for $g \in L^2(\Omega)$ and $G \in H^{1/2}(\Gamma)$, then weak solution u belongs to $H^2(\Omega)$. In this case, the weak solution of Eqs. (1) and (2) coincides with the strong solution of Eqs. (1) and (2).

6 It is assumed that $u \in H^2(\Omega)$ (see footnote 5).

7 More precisely, test-functions δu must be taken in the space $\mathcal{D}(\Omega)$ of the infinitely differentiable functions on Ω with compact support.

8 Since Eq. (10) holds for all $\delta u \in \mathcal{D}(\Omega)$, we deduce that $-\boldsymbol{\nabla}^2 u + k\, u - g = 0$ in the space $\mathcal{D}'(\Omega)$ of the generalized functions on Ω. Since $g \in L^2(\Omega)$, generalized derivative $-\boldsymbol{\nabla}^2 u$ of u is represented by a function in $L^2(\Omega)$. We can then consider Eq. (11) as an equality of functions belonging to $L^2(\Omega)$ and consequently, Eq. (11) is satisfied almost everywhere in Ω.

Substituting Eq. (11) into Eq. (9) yields

$$\int_{\Gamma} (\frac{\partial u}{\partial \mathbf{n}} - G)\, \delta u \, ds = 0 \quad , \quad \forall\, \delta u \in \mathcal{C} \quad . \tag{12}$$

We then deduce that [9]

$$\frac{\partial u}{\partial \mathbf{n}} = G \quad \text{on} \quad \Gamma \quad . \tag{13}$$

6. Associated Linear Operator Equation

Bilinear form $a(u\,,\delta u)$ given by Eq. (7) defines a linear operator **A** such that [10]

$$< \mathbf{A}\, u\,, \delta u >= a(u\,, \delta u) \quad , \tag{14}$$

and the element **f** such that [11]

$$< \mathbf{f}\,, \delta u >= f(\delta u) \quad . \tag{15}$$

Consequently, the linear operator equation corresponding to the variational formulation defined by Eq. (6) is written as [12]

$$\mathbf{A}\, u = \mathbf{f} \quad . \tag{16}$$

---

[9] Since $\delta u \in H^1(\Omega)$, the trace of δu on Γ belongs to $H^{1/2}(\Gamma) \subset L^2(\Omega)$. Since $u \in H^2(\Omega)$ (see footnote 6), $\partial u/\partial \mathbf{n}$ belongs to $H^{1/2}(\Gamma)$ and, since $G \in H^{1/2}(\Gamma)$, function $\partial u/\partial \mathbf{n} - G$ belongs to $H^{1/2}(\Gamma) \subset L^2(\Omega)$. Then, the left-hand side of Eq. (12) can be considered as the inner product in $L^2(\Gamma)$ and consequently Eq. (13) is satisfied almost everywhere on Γ.

[10] Since bilinear form $a(u,\delta u)$ is continuous on $\mathcal{C} \times \mathcal{C}$ (see footnote 4), $a(u,\delta u)$ defines a bounded linear operator (continuous operator) **A** from $\mathcal{C}$ into its dual space $\mathcal{C}'$. In Eq. (14), the angle brackets $<.\,,.>$ denote the duality product between $\mathcal{C}'$ and $\mathcal{C}$ (see Mathematical Notations in the appendix).

[11] Since linear form $f(\delta u)$ is continuous on $\mathcal{C}$ (see footnote 4), $f(\delta u)$ defines an element **f** belonging to the dual space $\mathcal{C}'$ of $\mathcal{C}$. In Eq. (15), the angle brackets $<.\,,.>$ denote the duality product between $\mathcal{C}'$ and $\mathcal{C}$ (see Mathematical Notations in the appendix).

[12] It should be noted that Eq. (16) is an equality in $\mathcal{C}'$.

7. Ritz-Galerkin Approximation and Finite Element Method

7.1. Ritz-Galerkin approximation

The Ritz-Galerkin method consists in constructing an approximate solution of the variational formulation defined by Eq. (6) in a subspace $\mathcal{C}_N \subset \mathcal{C}$ of finite dimension N. To construct $\mathcal{C}_N$, we consider a complete countable family $\{u_\alpha\}_{\alpha=1,2,\ldots}$ of elements in $\mathcal{C}$ such that $\mathcal{C}_N$ is spanned by $\{u_1,\ldots,u_N\}$. Any element $u^N \in \mathcal{C}_N \subset \mathcal{C}$ can then be written as

$$u^N = \sum_{\alpha=1}^{N} q_\alpha \, u_\alpha \quad , \tag{17}$$

in which $\mathbf{q} = (q_1,\ldots,q_N)$ is the $\mathbb{R}^N$-vector of components $\{q_\alpha\}$ called the *generalized coordinates*. A test function δu in subspace $\mathcal{C}_N$ can be written as

$$\delta u = \sum_{\beta=1}^{N} \delta q_\beta \, u_\beta \quad , \tag{18}$$

in which $\delta\mathbf{q} = (\delta q_1,\ldots,\delta q_N)$ is in $\mathbb{R}^N$. The restriction of the variational formulation defined by Eq. (6) to subspace $\mathcal{C}_N$ is obtained by substituting Eqs. (17) and (18) into Eq. (6),

$$\sum_{\alpha=1}^{N}\sum_{\beta=1}^{N} a(u_\alpha, u_\beta)\, q_\alpha \, \delta q_\beta = \sum_{\beta=1}^{N} f(u_\beta)\, \delta q_\beta \ , \ \ \forall\, \delta\mathbf{q} \in \mathbb{R}^N \ . \tag{19}$$

From Eq. (19), we deduce the linear matrix equation of dimension N

$$[\mathcal{A}]\,\mathbf{q} = \boldsymbol{\mathcal{F}} \quad , \tag{20}$$

in which $[\mathcal{A}]$ is an $(N \times N)$ matrix such that

$$[\mathcal{A}]_{\beta\alpha} = a(u_\alpha, u_\beta) \quad , \tag{21}$$

and $\boldsymbol{\mathcal{F}} = (\mathcal{F}_1,\ldots,\mathcal{F}_N) \in \mathbb{R}^N$ such that

$$\mathcal{F}_\beta = f(u_\beta) \quad . \tag{22}$$

Since bilinear form $a(u\,,\delta u)$ is symmetric positive definite,

$$a(u\,,\delta u) = a(\delta u\,,u) \quad , \tag{23}$$

$$a(u\,,u) > 0 \quad , \quad \forall u \neq 0 \quad , \tag{24}$$

we deduce that $[\mathcal{A}]$ is a symmetric positive-definite matrix and consequently, that matrix $[\mathcal{A}]$ is invertible.

7.2. Finite element method

Let us briefly recall that the finite element method is a powerful tool for constructing the basis used in the Ritz-Galerkin method described in Section 7.1. In the context of the finite element method, we denote the dimension of the finite approximation as n and the basis of $\mathcal{C}_n$ as $\{e_1, \ldots, e_n\}$. In this method, the generalized coordinates are denoted as U_j and coincide with the values $u^n(\mathbf{x}_j)$ of u^n at the n nodes $\mathbf{x}_j$ of a finite element mesh of domain $\overline{\Omega} = \Omega \cup \Gamma$ (see Fig. 2). Approximation u^n is defined by a low-order

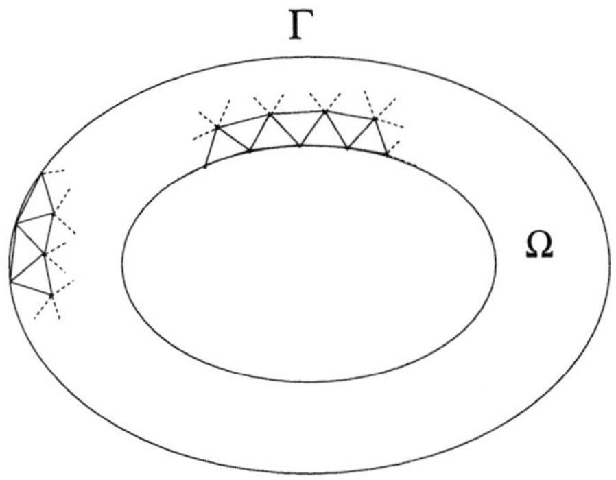

Fig. 2. Finite element mesh of domain $\overline{\Omega}$

polynomial interpolation from its nodal values. Eq. (17) can then be rewritten as

$$u^n = \sum_{j=1}^{n} U_j\, e_j \quad . \tag{25}$$

The values $U_j = u^n(\mathbf{x}_j), j = \{1, \ldots, n\}$, are called the *degrees of freedom* (DOFs). It should be noted that $e_j(\mathbf{x}_k) = \delta_{jk}$ (δ_{jk} denotes the Kronecker symbol) and that basis function $e_j(\mathbf{x})$ is equal to zero for all $\mathbf{x}$ belonging to the finite elements which are not connected to node $\mathbf{x}_j$ (this basis, defined on $\overline{\Omega}$, is called a local basis). For this particular basis related to the finite element method, Eq. (20) is rewritten as

$$[A]\,\mathbf{U} = \mathbf{F} \quad , \tag{26}$$

in which $\mathbf{U} = (U_1, \ldots, U_n) \in \mathbb{R}^n$, $[A]$ is an $(n \times n)$ symmetric positive-definite sparse matrix such that

$$[A]_{\ell j} = a(e_j, e_\ell) \quad , \tag{27}$$

and $\mathbf{F} = (F_1, \ldots, F_n) \in \mathbb{R}^n$ is such that

$$F_\ell = f(e_\ell) \quad . \tag{28}$$

Remark on the use of the linear operator equation. We emphasize that linear operator equation $\mathbf{A}\,u = \mathbf{f}$ introduced in Section 6 has a direct counterpart for its corresponding finite dimension approximation, i.e., $[\mathcal{A}]\,\mathbf{q} = \mathcal{F}$ for the Ritz-Galerkin method and $[A]\,\mathbf{U} = \mathbf{F}$ for the particular case of the finite element method. In addition, the algebraic properties of operator $\mathbf{A}$ hold for the matrices $[\mathcal{A}]$ and $[A]$ which are symmetric positive definite. This presentation will be used throughout the book in order to have similar expressions for the continuous cases and the corresponding finite dimension matrix equations.

7.3. Comments concerning convergence of the Ritz-Galerkin method

The Ritz-Galerkin method allows a finite dimension approximation u^N of solution u of Eq. (6) (weak solution of Eqs. (1) and (2)) to be constructed. The convergence of sequence $\{u^N\}_N$ to u is obtained in space $\mathcal{C}$ independently [13] of the regularity of solution u,

$$\lim_{N \to +\infty} \|u - u^N\|_{\mathcal{C}} = 0 \quad , \tag{29}$$

in which $\|\,.\,\|_{\mathcal{C}}$ denotes the norm in space $\mathcal{C}$. Let us assume that g belongs to $L^2(\Omega)$ (space of square integrable functions) and G belongs to a space $\mathcal{C}_\Gamma$ of regular [14] functions on Γ, such that trace $\partial u/\partial \mathbf{n}$ on Γ of the normal derivative of weak solution u belongs [15] to $\mathcal{C}_\Gamma$ (in this case the weak solution coincides with the strong solution, see Sections 4.2 and 4.3). In addition, we assume that basis functions $\{u_\alpha\}_\alpha$ are such that trace $\partial u^N/\partial \mathbf{n}$ on Γ of the normal derivative of u^N belongs to $\mathcal{C}_\Gamma$. In general, we have [16]

$$\lim_{N \to +\infty} \|\frac{\partial u^N}{\partial \mathbf{n}} - G\,\|_{\mathcal{C}'_\Gamma} \neq 0 \quad , \tag{30}$$

---

13 From footnote 5, $u \in H^2(\Omega) \subset H^1(\Omega)$, but we have only $\lim \|u - u^N\|_{H^1(\Omega)} = 0$ as $N \to +\infty$.

14 See footnote 1, $\mathcal{C}_\Gamma = H^{1/2}(\Gamma)$.

15 See footnote 5.

16 For all $v \in H^{1/2}(\Gamma) \hookrightarrow L^2(\Gamma) \hookrightarrow H^{-1/2}(\Gamma)$, we have $\|v\|_{H^{-1/2}(\Gamma)} \leq c\|v\|_{H^{1/2}(\Gamma)}$, $c > 0$.

in which C'_Γ denotes the dual space[17] of C_Γ. This means that the convergence of u^N in $\mathcal{C}$ (see Eq. (29)) does not imply the convergence of $\partial u^N/\partial \mathbf{n}$ to G in space C'_Γ. Consequently, the constructed approximation defined by Eq. (17) does not allow G to be retrieved on boundary Γ for any value of N. Nevertheless, since Eq. (29) implies the convergence of ∇u^N to ∇u in space $L^2(\Omega)$, then the first derivatives are convergent inside Ω, but not on boundary Γ (see Fig. 3).

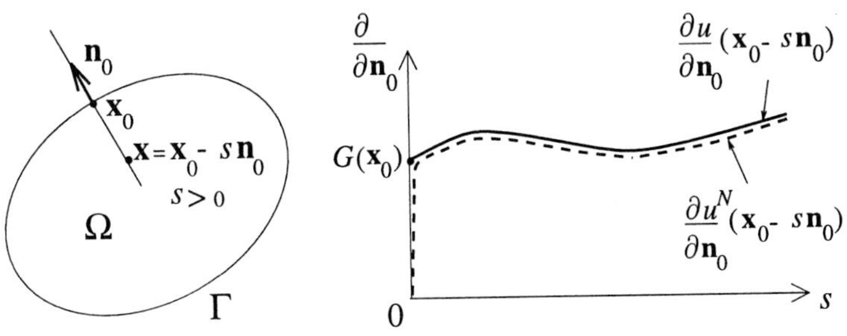

Fig. 3. Convergence of the normal derivative

8. Bibliographical Comments

Concerning the mathematical aspects of functional analysis using generalized functions, generalized derivatives, Sobolev spaces and trace theorems in the context of the variational formulations of boundary value problems, we refer the reader, for instance, to Brezis, 1987; Dautray and Lions, 1992; Reed and Simon, 1980. For a general presentation of variational methods for boundary value problems, see for instance Oden and Reddy, 1983; Dautray and Lions, 1992; Washizu, 1975. For more details on the Ritz-Galerkin method and variational principles, see Dautray and Lions, 1992; Finlayson, 1972; Mikhlin, 1964. Presentation of the finite element method and its applications in science and technology can be found in Zienkiewicz and Taylor, 1989; Hughes 1987; Bathe and Wilson, 1976; Ciarlet, 1979; Strang and Fix, 1973; Dautray and Lions, 1992; Raviart and Thomas, 1983; Oñate et al., 1991.

[17] Space $C_\Gamma = H^{1/2}(\Gamma)$ and space $C'_\Gamma = H^{-1/2}(\Gamma)$.

CHAPTER III

Linearized Vibrations of Conservative Structures and Structural Modes

1. Introduction

The objectives of this chapter are to give a synthesis of linear vibrations of conservative elastic structures and to introduce the fundamental notion of structural modes, the formulations appropriate to their calculation for complex structures and the corresponding numerical procedures. Among the structural vibration textbooks, we can mention Argyris and Mlejnek, 1991; Clough and Penzien, 1975; Landau and Lifchitz, 1992a; Meirovitch, 1980 and 1990; Roseau, 1980.

In Section 2, we introduce the equations and the boundary conditions in the time domain for linear vibrations of an undamped structure occupying a bounded domain.

In Section 3, we present the associated spectral problem which allows the introduction of structural modes.

Sections 4 and 5 deal with the variational formulation of the spectral problem and the corresponding linear operator equation.

In Sections 6 and 7, we introduce the basic properties of the spectral problem for a *fixed structure* (zero displacement on a part of the boundary) and for a *free structure*: countable number of positive eigenfrequencies, discussion of rigid body modes, orthogonality and completeness of the structural modes.

In Section 8, we discuss the cases of structures having symmetry properties: structures with one plane of symmetry, axisymmetric structures and structures with cyclic symmetries. We present the appropriate formulation allowing us to simplify the calculation of the structural modes.

Section 9 is concerned with the finite element discretization of the spectral problem and the corresponding generalized symmetric matrix eigenvalue problem. The finite element method is the most efficient method for computing the structural modes in the case of an arbitrary domain, boundary conditions and materials.

Finally, in Section 10 we briefly discuss substructuring techniques allowing a large eigenvalue problem to be replaced by several smaller eigenvalue problems posed on substructures.

2. Conservative Elastodynamic Boundary Value Problem in a Bounded Medium with Initial Cauchy Conditions

2.1. General equations, boundary conditions and initial conditions

The physical space $\mathbb{R}^3$ is referred to a Cartesian reference system $(\mathbf{i}, \mathbf{j}, \mathbf{k})$ and we denote the generic point of $\mathbb{R}^3$ as $\mathbf{x} = (x_1, x_2, x_3)$. Let Ω be a three-dimensional bounded connected domain of $\mathbb{R}^3$ occupied by the structure at equilibrium. It is assumed that boundary $\partial\Omega$ is smooth. The external unit normal to $\partial\Omega$ is denoted as $\mathbf{n}$ (see Fig. 1).

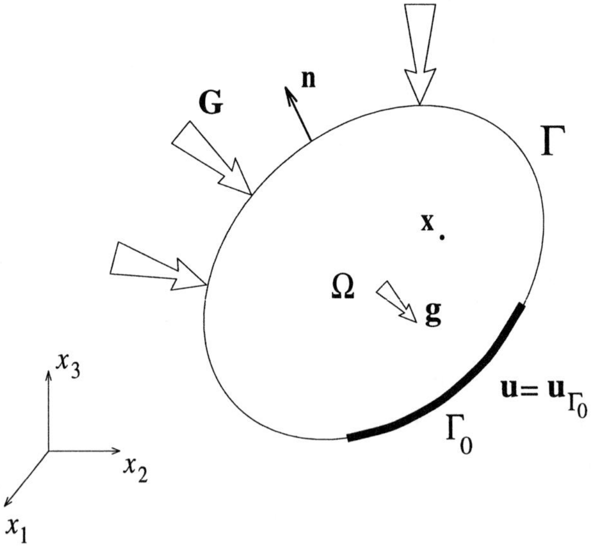

Fig. 1. Geometrical configuration

Let $\mathbf{u}(\mathbf{x}, t) = (u_1(\mathbf{x}, t), u_2(\mathbf{x}, t), u_3(\mathbf{x}, t))$ be the displacement at time t of a particle located at point $\mathbf{x}$ in Ω. For any function $a(\mathbf{x}, t)$, we use the notation $a_{,j}(\mathbf{x},t) = \partial a(\mathbf{x},t)/\partial x_j$, $\partial_t a(\mathbf{x},t) = \partial a(\mathbf{x},t)/\partial t$ and $\partial_t^2 a(\mathbf{x},t) = \partial^2 a(\mathbf{x},t)/\partial t^2$. We also use the classical convention for summations over repeated Latin indices, but not over Greek indices. In this chapter, we consider linear vibrations of a structure around a position of static equilibrium taken as reference configuration Ω (for the sake of brevity, we do not consider a reference configuration with prestresses: we use the natural state at rest). We operate within the framework of three-dimensional linearized elasticity theory and refer the reader to basic books (Truesdell, 1960; Fung, 1968; Malvern, 1969; Washizu, 1975; Abraham and Marsden, 1978; Germain, 1973 and 1986; Marsden and Hughes, 1983; Salençon, 1988; Ciarlet, 1988; Landau and Lifchitz, 1992a). The symmetric stress tensor σ_{ij} is related to the linearized symmetric strain tensor ε_{ij} by the constitutive equation

$$\sigma_{ij} = a_{ijkh}\,\varepsilon_{kh} \quad , \tag{1}$$

in which a_{ijkh} is the tensor of the elastic coefficients of the material having the usual properties of positive definiteness (for all second-order real symmetric tensors X_{ij}, we have $a_{ijkh}\,X_{kh}\,X_{ij} \geq c\,X_{ij}\,X_{ij}$ with $c > 0$) and symmetry ($a_{ijkh} = a_{jikh} = a_{ijhk} = a_{khij}$). Strain tensor ε_{ij} is related to displacement field $\mathbf{u}$ by

$$\varepsilon_{kh} = \frac{1}{2}(u_{k,h} + u_{h,k}) \quad . \tag{2}$$

Denoting as $\mathbf{g}(\mathbf{x},t) = (g_1(\mathbf{x},t), g_2(\mathbf{x},t), g_3(\mathbf{x},t))$ a given body force field applied in Ω, the elastodynamic equation is written as

$$\rho\,\partial_t^2 u_i(\mathbf{x},t) - \sigma_{ij,j}(\mathbf{x},t) = g_i(\mathbf{x},t) \quad \text{in} \quad \Omega \quad , \quad t > 0 \quad , \tag{3}$$

in which $\rho(\mathbf{x})$ is the mass density field defined on Ω. For prescribed displacements $\mathbf{u}_{\Gamma_0}(\mathbf{x},t)$ on the part Γ_0 of the boundary $\partial\Omega$ of Ω, we have the boundary condition

$$\mathbf{u}(\mathbf{x},t) = \mathbf{u}_{\Gamma_0}(\mathbf{x},t) \quad \text{on} \quad \Gamma_0 \quad . \tag{4}$$

For a given surface force field $\mathbf{G}(\mathbf{x},t) = (G_1(\mathbf{x},t), G_2(\mathbf{x},t), G_3(\mathbf{x},t))$ applied to the part $\Gamma = \partial\Omega\backslash\Gamma_0$, we have the boundary condition

$$\sigma_{ij}(\mathbf{x},t)\,n_j(\mathbf{x}) = G_i(\mathbf{x},t) \quad \text{on} \quad \Gamma \quad . \tag{5}$$

Finally, the above equations must be completed by the Cauchy initial conditions

$$\mathbf{u}(\mathbf{x},0) = \mathbf{u}_0(\mathbf{x}) \quad , \quad \partial_t\mathbf{u}(\mathbf{x},0) = \mathbf{v}_0(\mathbf{x}) \quad \text{in} \quad \Omega \quad , \tag{6}$$

in which $\mathbf{u}_0$ and $\mathbf{v}_0$ are given displacement and velocity fields respectively.

2.2. Boundary value problem in terms of displacement field u

In order to introduce the spectral problem in the next section, we consider the following time boundary value problem for which $\mathbf{u}_{\Gamma_0} = \mathbf{0}$ on Γ_0, $\mathbf{g} = \mathbf{0}$ in Ω and $\mathbf{G} = \mathbf{0}$ on Γ (see Fig. 2-a).

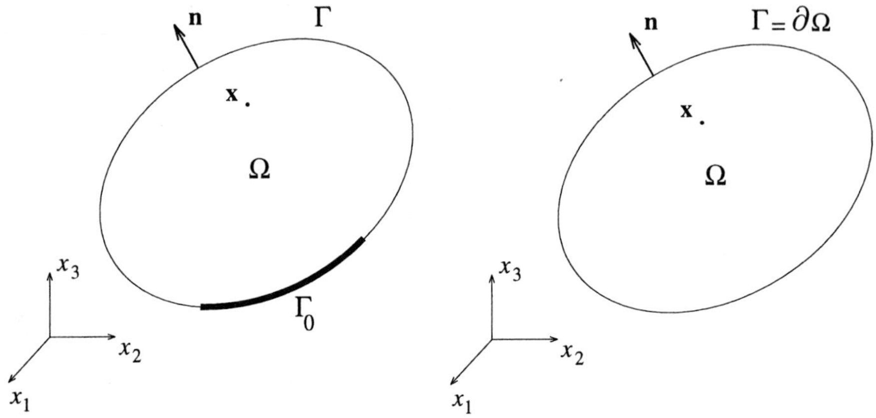

a- Structure fixed on Γ_0 b- Free structure

Fig. 2. Geometrical configuration for the spectral problem

Below, $\sigma_{ij}(\mathbf{u})$ denotes the stress tensor σ_{ij} as a function of $\mathbf{u}$ obtained by substituting ε_{kh} defined by Eq. (2) into the constitutive equation defined by Eq. (1), i.e.

$$\sigma_{ij}(\mathbf{u}) = a_{ijkh}\, \varepsilon_{kh}(\mathbf{u}) \quad . \tag{7}$$

Substituting Eq. (7) into Eq. (3) and from Eqs. (4) to (6), we deduce the boundary value problem in terms of $\mathbf{u}$:

$$\rho\, \partial_t^2 u_i - \sigma_{ij,j}(\mathbf{u}) = 0 \quad \text{in} \quad \Omega \quad , \quad t > 0 \quad , \tag{8}$$

$$\mathbf{u}(\mathbf{x}, t) = \mathbf{0} \quad \text{on} \quad \Gamma_0 \quad , \tag{9}$$

$$\sigma_{ij}(\mathbf{u})\, n_j = 0 \quad \text{on} \quad \Gamma \quad , \tag{10}$$

$$\mathbf{u}(\mathbf{x}, 0) = \mathbf{u}_0(\mathbf{x}) \quad , \quad \partial_t \mathbf{u}(\mathbf{x}, 0) = \mathbf{v}_0(\mathbf{x}) \quad \text{in} \quad \Omega \quad , \tag{11}$$

in which $\mathbf{u}_0$ and $\mathbf{v}_0$ are given initial conditions. Eqs. (8) to (11) correspond to a *structure fixed* on Γ_0. For a *free structure* ($\Gamma_0 = \emptyset$), Eq. (9) does not exist and Eq. (10) holds on the global boundary $\Gamma = \partial\Omega$ of domain Ω (see Fig. 2-b).

3. Associated Spectral Problem: Eigenfrequencies and Structural Modes

We consider all the solutions $\mathbf{u}(\mathbf{x}, t) = \mathbf{u}(\mathbf{x}) \exp(i\omega t)$ of the boundary value problem defined by Eqs. (8) to (10), where i is the pure imaginary complex number. The spectral problem consists in searching for $\{\omega, \mathbf{u}\}$, with $\mathbf{u} \neq \mathbf{0}$, satisfying the boundary value problem

$$-\omega^2 \rho u_i - \sigma_{ij,j}(\mathbf{u}) = 0 \quad \text{in} \quad \Omega \quad , \tag{12}$$

$$\mathbf{u} = \mathbf{0} \quad \text{on} \quad \Gamma_0 \quad , \tag{13}$$

$$\sigma_{ij}(\mathbf{u}) \, n_j = 0 \quad \text{on} \quad \Gamma \quad . \tag{14}$$

We will see in Section 6 that for a structure fixed on Γ_0, this spectral problem has a countable number of solutions $\{\omega_\alpha, \mathbf{u}_\alpha\}_{\alpha=1,2,\ldots}$, in which $\omega_\alpha > 0$ and $\mathbf{u}_\alpha$ are real vector fields. For a free structure ($\Gamma_0 = \emptyset$), Eq. (13) does not exist and Eq. (14) holds on the global boundary $\Gamma = \partial\Omega$ of domain Ω. In this case, we will see in Section 7 that $\omega_\alpha \geq 0$ and $\mathbf{u}_\alpha$ are real vector fields. For the two cases, the angular frequencies ω_α and the corresponding $\mathbf{u}_\alpha$ are called the *eigenfrequencies* (or natural frequencies) and the *structural modes* (or eigenmodes or mode shapes of vibrations) respectively.

4. Variational Formulation of the Spectral Problem

The variational formulation is constructed using the test-function method presented in Chapter II.

4.1. Construction of the variational formulation

First, we introduce the vector space $\mathcal{C}$ of "sufficiently differentiable" [1] functions $\delta\mathbf{u} = (\delta u_1, \delta u_2, \delta u_3)$ defined on Ω with values in $\mathbb{R}^3$. In a second step, multiplying Eq. (12) by an arbitrary function $\delta\mathbf{u} \in \mathcal{C}$ and integrating over domain Ω yields

$$-\omega^2 \int_\Omega \rho \, u_i \, \delta u_i \, d\mathbf{x} - \int_\Omega \sigma_{ij,j}(\mathbf{u}) \, \delta u_i \, d\mathbf{x} = 0 \quad . \tag{15}$$

---

[1] Space $\mathcal{C}$ is the Sobolev space $(H^1(\Omega))^3$.

The second term on the left-hand side of Eq. (15) is transformed using the following identity

$$-\int_\Omega \sigma_{ij,j}\,\delta u_i\,d\mathbf{x} = \int_\Omega \sigma_{ij}\,\delta u_{i,j}\,d\mathbf{x} - \int_\Omega (\sigma_{ij}\,\delta u_i)_{,j}\,d\mathbf{x} \quad . \tag{16}$$

The second integral on the right-hand side of Eq. (16) is transformed to an integral on $\partial\Omega$ using Stokes's formula

$$\int_\Omega (\sigma_{ij}\,\delta u_i)_{,j}\,d\mathbf{x} = \int_{\partial\Omega} \sigma_{ij}\,n_j\,\delta u_i\,ds \quad , \tag{17}$$

where ds is the surface element. From Eqs. (16) and (17), we deduce Green's formula

$$-\int_\Omega \sigma_{ij,j}\,\delta u_i\,d\mathbf{x} = \int_\Omega \sigma_{ij}\,\delta u_{i,j}\,d\mathbf{x} - \int_{\partial\Omega} \sigma_{ij}\,n_j\,\delta u_i\,ds \quad . \tag{18}$$

Substituting the left-hand side of Eq. (18) into Eq. (15) yields

$$-\omega^2\int_\Omega \rho\,u_i\,\delta u_i\,d\mathbf{x} + \int_\Omega \sigma_{ij}(\mathbf{u})\,\delta u_{i,j}\,d\mathbf{x} - \int_{\partial\Omega} \sigma_{ij}(\mathbf{u})\,n_j\,\delta u_i\,ds = 0 \quad . \tag{19}$$

In a third step, we transform the second term of the left-hand side of Eq. (19) in order to get a symmetric formulation in terms of **u** and $\delta\mathbf{u}$. From the symmetry of the stress tensor and the definition of the strain tensor given by Eq. (2), we deduce the equalities $\sigma_{ij}\,\delta u_{i,j} = \sigma_{ij}(\delta u_{i,j} + \delta u_{j,i})/2 = \sigma_{ij}\,\varepsilon_{ij}(\delta\mathbf{u})$. Substitution into Eq. (19) yields

$$-\omega^2\int_\Omega \rho\,u_i\,\delta u_i\,d\mathbf{x} + \int_\Omega \sigma_{ij}(\mathbf{u})\,\varepsilon_{ij}(\delta\mathbf{u})\,d\mathbf{x} - \int_{\partial\Omega} \sigma_{ij}(\mathbf{u})\,n_j\,\delta u_i\,ds = 0 \quad . \tag{20}$$

In a fourth step, we introduce the *admissible function space.* Two cases must be distinguished.

Structure fixed on Γ_0. In this case, $\Gamma_0 \neq \emptyset$ and the admissible function space must satisfy the constraint on **u** defined by Eq. (13) and consequently is the subspace $\mathcal{C}_0$ of $\mathcal{C}$ defined by [2]

$$\mathcal{C}_0 = \{\,\mathbf{u} \in \mathcal{C} \quad ; \quad \mathbf{u} = \mathbf{0} \quad \text{on} \quad \Gamma_0\,\} \quad . \tag{21}$$

2 Space $\mathcal{C}_0$ is the Sobolev space $H^1_{0,\Gamma_0}(\Omega)=\{\mathbf{u}\in(H^1(\Omega))^3\,,\mathbf{u}=0 \text{ on } \Gamma_0\}$.

Let $\delta\mathbf{u}$ be any test function in $\mathcal{C}_0 \subset \mathcal{C}$. Since $\delta\mathbf{u} = \mathbf{0}$ on Γ_0 and using Eq. (14), Eq. (20) yields

$$-\omega^2 \int_\Omega \rho\, u_i\, \delta u_i\, d\mathbf{x} + \int_\Omega \sigma_{ij}(\mathbf{u})\, \varepsilon_{ij}(\delta\mathbf{u})\, d\mathbf{x} = 0 \quad , \quad \forall\, \delta\mathbf{u} \in \mathcal{C}_0 \quad . \tag{22}$$

Free Structure. In this case, $\Gamma_0 = \emptyset$ and Eq. (13) does not exist. Therefore, the admissible function space of the problem is $\mathcal{C}$ (there is no constraint on $\mathbf{u}$). Let $\delta\mathbf{u}$ be any test function in $\mathcal{C}$. Due to Eq. (14), the integral over $\partial\Omega$ in Eq. (20) is equal to zero and Eq. (20) yields

$$-\omega^2 \int_\Omega \rho\, u_i\, \delta u_i\, d\mathbf{x} + \int_\Omega \sigma_{ij}(\mathbf{u})\, \varepsilon_{ij}(\delta\mathbf{u})\, d\mathbf{x} = 0 \quad , \quad \forall\, \delta\mathbf{u} \in \mathcal{C} \quad . \tag{23}$$

4.2. Variational formulation

Structure fixed on Γ_0. The variational formulation of the spectral problem defined by Eqs. (12), (13) and (14) is stated as follows. Find real $\omega^2 > 0$ and $\mathbf{u}$ in $\mathcal{C}_0$ such that

$$\int_\Omega \sigma_{ij}(\mathbf{u})\, \varepsilon_{ij}(\delta\mathbf{u})\, d\mathbf{x} = \omega^2 \int_\Omega \rho\; \mathbf{u}\cdot \delta\mathbf{u}\; d\mathbf{x} \quad , \quad \forall\, \delta\mathbf{u} \in \mathcal{C}_0 \quad . \tag{24}$$

Free structure. The variational formulation of the spectral problem defined by Eqs. (12) and (14) is stated as follows. Find real $\omega^2 \geq 0$ and $\mathbf{u}$ in $\mathcal{C}$ such that

$$\int_\Omega \sigma_{ij}(\mathbf{u})\, \varepsilon_{ij}(\delta\mathbf{u})\, d\mathbf{x} = \omega^2 \int_\Omega \rho\; \mathbf{u}\cdot \delta\mathbf{u}\; d\mathbf{x} \quad , \quad \forall\, \delta\mathbf{u} \in \mathcal{C} \quad . \tag{25}$$

5. Associated Linear Operators and Algebraic Properties

We proceed as in Chapter II in order to introduce a linear operator equation corresponding to the variational formulation defined by Eq. (24) or (25).

5.1. Mass operator

We introduce the bilinear form $m(\mathbf{u}\,,\delta\mathbf{u})$, called the structural mass bilinear form, defined by

$$m(\mathbf{u}\,,\delta\mathbf{u}) = \int_\Omega \rho\; \mathbf{u}\cdot \delta\mathbf{u}\; d\mathbf{x} \quad , \tag{26}$$

which is symmetric because

$$m(\mathbf{u}\,,\delta\mathbf{u}) = m(\delta\mathbf{u}\,,\mathbf{u}) \quad , \tag{27}$$

and which is positive definite because $\rho(\mathbf{x}) > 0$ (for all $\mathbf{u} \neq \mathbf{0}$, we have $m(\mathbf{u}\,,\mathbf{u}) > 0$). This result holds for the fixed and free cases. We then introduce the linear operator $\mathbf{M}$, called the *mass operator*, such that [3]

$$<\mathbf{M}\,\mathbf{u}\,,\delta\mathbf{u}>= m(\mathbf{u}\,,\delta\mathbf{u}) \quad . \tag{28}$$

5.2. Stiffness operator

Using Eq. (7), we introduce the bilinear form $k(\mathbf{u}\,,\delta\mathbf{u})$, called the structural stiffness bilinear form, defined by

$$\begin{aligned} k(\mathbf{u}\,,\delta\mathbf{u}) &= \int_\Omega \sigma_{ij}(\mathbf{u})\,\varepsilon_{ij}(\delta\mathbf{u})\,d\mathbf{x} \\ &= \int_\Omega a_{ijkh}\,\varepsilon_{kh}(\mathbf{u})\,\varepsilon_{ij}(\delta\mathbf{u})\,d\mathbf{x} \quad . \end{aligned} \tag{29}$$

From the symmetry property $a_{ijkh} = a_{khij}$ of the tensor of the elastic coefficients of the material, we deduce that bilinear form k is symmetric:

$$k(\mathbf{u}\,,\delta\mathbf{u}) = k(\delta\mathbf{u}\,,\mathbf{u}) \quad . \tag{30}$$

From the positive definiteness of tensor a_{ijkh}, we deduce that bilinear form k is positive semidefinite. The fixed and free cases must be distinguished.

Structure fixed on Γ_0. Bilinear form $k(\mathbf{u}\,,\delta\mathbf{u})$ defined on $\mathcal{C}_0 \times \mathcal{C}_0$ is positive definite:

$$k(\mathbf{u}\,,\mathbf{u}) > 0 \quad , \quad \text{for all} \quad \mathbf{u} \neq \mathbf{0} \quad . \tag{31}$$

We then introduce the linear operator $\mathbf{K}$, called the *stiffness operator*, such that [4]

$$<\mathbf{K}\,\mathbf{u}\,,\delta\mathbf{u}>= k(\mathbf{u}\,,\delta\mathbf{u}) \quad . \tag{32}$$

---

[3] Bilinear form $m(\mathbf{u}\,,\delta\mathbf{u})$ is continuous on $H \times H$ with $H=(L^2(\Omega))^3$ and is also continuous on $\mathcal{C}\times\mathcal{C}$. Then operator $\mathbf{M}$ is continuous from $\mathcal{C}$ into its dual space $\mathcal{C}'$ and the angle brackets in Eq. (28) denote the duality product between $\mathcal{C}'$ and $\mathcal{C}$.

[4] Bilinear form $k(\mathbf{u}\,,\delta\mathbf{u})$ is continuous on $\mathcal{C}_0\times\mathcal{C}_0$. Then operator $\mathbf{K}$ is continuous from $\mathcal{C}_0$ into its dual space $\mathcal{C}_0'$ and the angle brackets in Eq. (32) denote the duality product between $\mathcal{C}_0'$ and $\mathcal{C}_0$.

Free structure. Bilinear form $k(\mathbf{u}, \delta\mathbf{u})$ defined on $\mathcal{C} \times \mathcal{C}$ is positive semidefinite

$$k(\mathbf{u}, \mathbf{u}) \geq 0 \quad , \quad \text{for all} \quad \mathbf{u} \neq \mathbf{0} \quad . \tag{33}$$

In effect, from Eq. (29), we deduce that $k(\mathbf{u}, \mathbf{u}) = 0$ implies tensor $\varepsilon_{ij}(\mathbf{u}) = 0$ due to the positive definiteness of tensor a_{ijkh}. Since Ω is a connected domain, the general solution of the system of equations $\varepsilon_{ij}(\mathbf{u}) = 0$ is written as

$$\mathbf{u}_{\text{rig}}(\mathbf{x}) = \mathbf{t} + \boldsymbol{\theta} \times \mathbf{x} \quad , \quad \forall \mathbf{x} \in \Omega \quad , \tag{34}$$

in which $\mathbf{t}$ and $\boldsymbol{\theta}$ are two arbitrary constant vectors in $\mathbb{R}^3$. We denote as $\mathcal{C}_{\text{rig}}$ the subset of $\mathcal{C}$ spanned by $\mathbf{t}$ and $\boldsymbol{\theta}$. This space is of dimension 6 and $\mathbf{u}_{\text{rig}}$ belongs to $\mathcal{C}_{\text{rig}}$. From a mechanical point of view, $\mathbf{u}_{\text{rig}}$ represents the *rigid body displacement field* (in the linearized theory). We then introduce the linear operator $\mathbf{K}$, called as above the *stiffness operator*, such that [5]

$$<\mathbf{K}\mathbf{u}, \delta\mathbf{u}> = k(\mathbf{u}, \delta\mathbf{u}) \quad , \tag{35}$$

which is symmetric positive semidefinite.

5.3. Rewriting the variational formulation

Structure fixed on Γ_0. Using the above notations, the variational formulation defined by Eq. (24) can be rewritten as follows. Find real $\omega > 0$ and $\mathbf{u}$ in $\mathcal{C}_0$ such that

$$k(\mathbf{u}, \delta\mathbf{u}) = \omega^2\, m(\mathbf{u}, \delta\mathbf{u}) \quad , \quad \forall\, \delta\mathbf{u} \in \mathcal{C}_0 \quad . \tag{36}$$

Free structure. Similarly, the variational formulation defined by Eq. (25) can be rewritten as follows. Find real $\omega \geq 0$ and $\mathbf{u}$ in $\mathcal{C}$ such that

$$k(\mathbf{u}, \delta\mathbf{u}) = \omega^2\, m(\mathbf{u}, \delta\mathbf{u}) \quad , \quad \forall\, \delta\mathbf{u} \in \mathcal{C} \quad . \tag{37}$$

5.4. Comments about slender structures

In the case of slender structures such as beams, plates and shells, the structural mass and stiffness bilinear forms defined by Eqs. (26) and (29) for the three-dimensional case are simply replaced by appropriate expressions, see Leissa, 1993a and 1993b; Novozhilov, 1964; Soedel, 1993. For beam, plate and shell finite elements see Bathe and Wilson, 1976; Dautray and Lions, 1992; Hughes, 1987; Ohayon and Nicolas-Vullierme, 1981; Zienkiewicz and Taylor, 1989.

---

5 Bilinear form $k(\mathbf{u}, \delta\mathbf{u})$ is continuous on $\mathcal{C} \times \mathcal{C}$. Then operator $\mathbf{K}$ is continuous from $\mathcal{C}$ into its dual space $\mathcal{C}'$ and the angle brackets in Eq. (35) denote the duality product between $\mathcal{C}'$ and $\mathcal{C}$.

6. Basic Properties of the Eigenfrequencies and Structural Modes for a Structure Fixed on Γ_0

The eigenvalue problem. Setting $\lambda = \omega^2$, the spectral problem defined by Eq. (36) is rewritten as the following eigenvalue problem. Find λ and $\mathbf{u} \neq \mathbf{0}$ in $\mathcal{C}_0$ such that

$$k(\mathbf{u}\,,\delta\mathbf{u}) = \lambda\, m(\mathbf{u}\,,\delta\mathbf{u}) \quad , \quad \forall\, \delta\mathbf{u} \in \mathcal{C}_0 \quad . \tag{38}$$

Countable number of positive eigenvalues. Since Ω is a bounded domain and from the properties indicated in Section 5 for a fixed structure (symmetry and positive definiteness of the mass and stiffness bilinear forms), it can be shown that there exists an increasing sequence of positive eigenvalues

$$0 < \lambda_1 \leq \lambda_2 \leq \ldots \leq \lambda_\alpha \leq \ldots \quad . \tag{39}$$

In addition, any multiple eigenvalue has a finite multiplicity (which means that a multiple eigenvalue is repeated a finite number of times). It should be noted that the positivity of λ_α can be directly deduced from Eq. (38) by taking $\delta\mathbf{u} = \mathbf{u}$. In addition, λ_α is strictly positive because $\mathbf{u} = \mathbf{0}$ on Γ_0 (in the case of a fixed structure, there are no rigid body modes).

Completeness of the eigenfunctions. Let $\mathbf{u}_\alpha$ be the eigenfunction associated with eigenvalue λ_α. It can be shown that the eigenfunctions $\{\mathbf{u}_\alpha\}_{\alpha\geq 1}$ form a complete set in $\mathcal{C}_0$ which means that an arbitrary function $\mathbf{u}$ belonging to $\mathcal{C}_0$ can be expanded as

$$\mathbf{u} = \sum_{\alpha=1}^{+\infty} q_\alpha\, \mathbf{u}_\alpha \quad , \tag{40}$$

in which $\{q_\alpha\}_\alpha$ is a sequence of real numbers.

Orthogonality of the eigenfunctions. It can be shown that the sequence of eigenfunctions $\{\mathbf{u}_\alpha\}_\alpha$ satisfies the orthogonality conditions with respect to mass and stiffness,

$$m(\mathbf{u}_\alpha\,,\mathbf{u}_\beta) = \mu_\alpha\, \delta_{\alpha\beta} \quad , \tag{41}$$

$$k(\mathbf{u}_\alpha\,,\mathbf{u}_\beta) = \mu_\alpha\, \omega_\alpha^2\, \delta_{\alpha\beta} \quad , \tag{42}$$

in which

$$\omega_\alpha = \sqrt{\lambda_\alpha} \quad , \tag{43}$$

and where μ_α is a positive real number depending on the normalization of eigenfunction $\mathbf{u}_\alpha$, because the eigenfunctions are defined up to a multiplicative constant.

Linear operator equation. The linear operator equation corresponding to the variational formulation defined by Eq. (38) is written as

$$\mathbf{K}\,\mathbf{u} = \lambda\,\mathbf{M}\,\mathbf{u} \quad , \quad \mathbf{u} \in \mathcal{C}_0 \quad . \tag{44}$$

This problem is called a generalized eigenvalue problem because operator $\mathbf{M}$ is not the identity operator. The algebraic properties defined by Eqs. (41) and (42) can be rewritten as

$$<\mathbf{M}\,\mathbf{u}_\alpha\,,\mathbf{u}_\beta>= \mu_\alpha\,\delta_{\alpha\beta} \quad , \tag{45}$$

$$<\mathbf{K}\,\mathbf{u}_\alpha\,,\mathbf{u}_\beta>= \mu_\alpha\,\omega_\alpha^2\,\delta_{\alpha\beta} \quad . \tag{46}$$

Bibliographical comments. Concerning the mathematical aspects, we refer the reader to Dautray and Lions, 1992; Kato, 1966; Dieudonné, 1969; Hörmander, 1985; Raviart and Thomas, 1983; Sanchez-Hubert and Sanchez-Palencia, 1989; Yosida, 1966.

Terminology. In structural vibrations, ω_α defined by Eq. (43) is called the *eigenfrequency* (or the natural frequency) of *elastic structural mode* $\mathbf{u}_\alpha$ (or the eigenmode or mode shape of vibration) whose normalization is defined by *generalized mass* μ_α. On the right-hand side of Eq. (40), real numbers q_α are called the *generalized coordinates*. An *elastic structural mode* α is defined by the three quantities $\{\omega_\alpha, \mathbf{u}_\alpha, \mu_\alpha\}$.

7. Basic Properties of the Eigenfrequencies and Structural Modes for a Free Structure

7.1. The eigenvalue problem

Setting $\lambda = \omega^2$, the spectral problem defined by Eq. (37) is rewritten as the following eigenvalue problem. Find λ and $\mathbf{u} \neq \mathbf{0}$ in $\mathcal{C}$ such that

$$k(\mathbf{u}\,,\delta\mathbf{u}) = \lambda\; m(\mathbf{u}\,,\delta\mathbf{u}) \quad , \quad \forall\;\delta\mathbf{u} \in \mathcal{C} \quad . \tag{47}$$

7.2. Rigid body modes (solutions for $\lambda = 0$)

Let us show that there exist solutions of the type $\{\lambda = 0, \mathbf{u} \neq \mathbf{0}\}$. Taking $\lambda = 0$ in Eq. (47) yields $k(\mathbf{u}\,,\delta\mathbf{u}) = 0$ for all $\delta\mathbf{u} \in \mathcal{C}$. The general solution of this problem corresponds to the rigid body displacement field $\mathbf{u}_{\rm rig} \in \mathcal{C}_{\rm rig}$ (see Section 5.2 and Eq. (34)). Since the dimension of $\mathcal{C}_{\rm rig}$ is 6, then $\lambda = 0$ can be considered as a "zero eigenvalue" of multiplicity 6, denoted as $\lambda_{-5}, \ldots, \lambda_0$. Let $\mathbf{u}_{-5}, \ldots, \mathbf{u}_0$ be the corresponding eigenfunctions which are constructed such that the following orthogonality conditions with respect to the mass and stiffness bilinear forms are satisfied, for α and β in $\{-5, \ldots, 0\}$,

$$m(\mathbf{u}_\alpha \,, \mathbf{u}_\beta) = \mu_\alpha \, \delta_{\alpha\beta} \quad , \tag{48}$$

$$k(\mathbf{u}_\alpha \,, \mathbf{u}_\beta) = 0 \quad . \tag{49}$$

These eigenfunctions, called the *rigid body modes*, form a basis of $\mathcal{C}_{\rm rig} \subset \mathcal{C}$ and any rigid body displacement $\mathbf{u}_{\rm rig}$ in $\mathcal{C}_{\rm rig}$ can be expanded as

$$\mathbf{u}_{\rm rig} = \sum_{\alpha=-5}^{0} q_\alpha \, \mathbf{u}_\alpha \quad . \tag{50}$$

7.3. Elastic structural modes (solutions for $\lambda \neq 0$)

Definition of the vector space $\mathcal{C}_{\rm elas}$. We introduce the subset $\mathcal{C}_{\rm elas}$ of $\mathcal{C}$ corresponding to all the displacement fields $\mathbf{u}_{\rm elas}$ belonging to $\mathcal{C}$ and which do not belong to $\mathcal{C}_{\rm rig}$ ($\mathcal{C}_{\rm elas} = \mathcal{C} \setminus \mathcal{C}_{\rm rig}$). Consequently, $k(\mathbf{u}_{\rm elas} \,, \delta\mathbf{u}_{\rm elas})$ defined on $\mathcal{C}_{\rm elas} \times \mathcal{C}_{\rm elas}$ is positive definite and we then have

$$k(\mathbf{u}_{\rm elas} \,, \mathbf{u}_{\rm elas}) > 0 \quad , \quad \forall \; \mathbf{u}_{\rm elas} \neq \mathbf{0} \in \mathcal{C}_{\rm elas} \quad . \tag{51}$$

Eigenvalue problem restricted to $\mathcal{C}_{\rm elas}$. The eigenvalue problem defined by Eq. (47) and restricted to space $\mathcal{C}_{\rm elas}$ is written as follows. Find $\lambda \neq 0$ and $\mathbf{u}_{\rm elas} \neq \mathbf{0}$ in $\mathcal{C}_{\rm elas}$ such that

$$k(\mathbf{u}_{\rm elas} \,, \delta\mathbf{u}_{\rm elas}) = \lambda \; m(\mathbf{u}_{\rm elas} \,, \delta\mathbf{u}_{\rm elas}) \quad , \quad \forall \; \delta\mathbf{u}_{\rm elas} \in \mathcal{C}_{\rm elas} \quad . \tag{52}$$

Countable number of positive eigenvalues. We consider the solutions of Eq. (52). Since Ω is a bounded domain and due to the symmetry and positive definiteness of bilinear forms m and k on $\mathcal{C}_{\rm elas} \times \mathcal{C}_{\rm elas}$, it can be proved that there exists an increasing sequence of positive eigenvalues

$$0 < \lambda_1 \leq \lambda_2 \leq \ldots \leq \lambda_\alpha \leq \ldots \quad . \tag{53}$$

In addition, any multiple positive eigenvalue has a finite multiplicity (which means that a multiple positive eigenvalue is repeated a finite number of times). The positivity of the eigenvalues can easily be proved by taking $\delta\mathbf{u}_{\text{elas}} = \mathbf{u}_{\text{elas}}$ in Eq. (52).

Completeness of the eigenfunctions corresponding to the positive eigenvalues. Let $\mathbf{u}_\alpha$ be the eigenfunction associated with eigenvalue $\lambda_\alpha > 0$. It can be shown that eigenfunctions $\{\mathbf{u}_\alpha\}_{\alpha\geq 1}$ form a complete set in $\mathcal{C}_{\text{elas}}$ and that an arbitrary function $\mathbf{u}_{\text{elas}}$ belonging to $\mathcal{C}_{\text{elas}}$ can be expanded as

$$\mathbf{u}_{\text{elas}} = \sum_{\alpha=1}^{+\infty} q_\alpha \, \mathbf{u}_\alpha \quad , \tag{54}$$

in which $\{q_\alpha\}_\alpha$ is a sequence of real numbers. These eigenfunctions are called the *elastic structural modes.*

Orthogonality of the eigenfunctions corresponding to the positive eigenvalues. It can easily be shown that the sequence of eigenfunctions $\{\mathbf{u}_\alpha\}_\alpha$ in $\mathcal{C}_{\text{elas}}$ corresponding to the positive eigenvalues satisfies the following orthogonality conditions with respect to the mass and stiffness bilinear forms,

$$m(\mathbf{u}_\alpha \, , \mathbf{u}_\beta) = \mu_\alpha \, \delta_{\alpha\beta} \quad , \tag{55}$$

$$k(\mathbf{u}_\alpha \, , \mathbf{u}_\beta) = \mu_\alpha \, \omega_\alpha^2 \, \delta_{\alpha\beta} \quad , \tag{56}$$

in which

$$\omega_\alpha = \sqrt{\lambda_\alpha} \quad , \tag{57}$$

and where μ_α is a positive real number depending on the normalization of eigenfunction $\mathbf{u}_\alpha$, because the eigenfunctions are defined up to a multiplicative constant.

7.4. Orthogonality between the elastic structural modes and the rigid body modes

Substituting any elastic structural mode $\mathbf{u} = \mathbf{u}_\alpha \in \mathcal{C}_{\text{elas}} \subset \mathcal{C}$ into Eq. (47) yields

$$k(\mathbf{u}_\alpha \, , \delta\mathbf{u}) = \lambda_\alpha \; m(\mathbf{u}_\alpha \, , \delta\mathbf{u}) \quad , \quad \forall \; \delta\mathbf{u} \in \mathcal{C} \quad , \tag{58}$$

in which $\lambda_\alpha \neq 0$. Taking $\delta\mathbf{u} = \mathbf{u}_{\text{rig}} \in \mathcal{C}_{\text{rig}} \subset \mathcal{C}$ in Eq. (58) yields

$$k(\mathbf{u}_\alpha \, , \mathbf{u}_{\text{rig}}) = \lambda_\alpha \; m(\mathbf{u}_\alpha \, , \mathbf{u}_{\text{rig}}) \quad . \tag{59}$$

From Section 5.2, we deduce that

$$k(\mathbf{u}_{\text{rig}} \, , \delta\mathbf{u}) = 0 \quad , \quad \text{for all} \quad \delta\mathbf{u} \in \mathcal{C} \quad . \tag{60}$$

Taking $\delta\mathbf{u} = \mathbf{u}_\alpha$ in Eq. (60) and using the symmetry of k yields

$$k(\mathbf{u}_\alpha\,,\mathbf{u}_{\text{rig}}) = 0 \quad . \tag{61}$$

From Eqs. (59) and (61), we deduce that

$$m(\mathbf{u}_\alpha\,,\mathbf{u}_{\text{rig}}) = 0 \quad . \tag{62}$$

Equations (61) and (62) express the orthogonality between the elastic structural modes and the rigid body modes. Substituting Eq. (34) into Eq. (62) yields

$$\int_\Omega \mathbf{u}_\alpha(\mathbf{x})\,\rho(\mathbf{x})\,d\mathbf{x} = \mathbf{0} \quad , \tag{63}$$

$$\int_\Omega \mathbf{x} \times \mathbf{u}_\alpha(\mathbf{x})\,\rho(\mathbf{x})\,d\mathbf{x} = \mathbf{0} \quad . \tag{64}$$

Let Ω_α be the domain deduced from Ω such that $\Omega_\alpha = \{\mathbf{x}_\alpha = \mathbf{x}+\mathbf{u}_\alpha(\mathbf{x}), \mathbf{x} \in \Omega\}$ in which $\mathbf{u}_\alpha$ is any elastic structural mode. For a free structure, Eq. (63) implies that the inertial center of Ω coincides with the inertial center of Ω_α.

7.5. Expansion of the displacement field using the rigid body modes and the elastic structural modes

From the above results, we deduce that any displacement field $\mathbf{u}$ in $\mathcal{C}$ has the following unique decomposition

$$\mathbf{u} = \mathbf{u}_{\text{rig}} + \mathbf{u}_{\text{elas}} \quad \text{with} \quad \mathbf{u}_{\text{rig}} \in \mathcal{C}_{\text{rig}} \quad \text{and} \quad \mathbf{u}_{\text{elas}} \in \mathcal{C}_{\text{elas}} \quad . \tag{65}$$

Substituting Eqs. (50) and (54) into Eq. (65) yields

$$\mathbf{u} = \sum_{\alpha=-5}^{+\infty} q_\alpha\,\mathbf{u}_\alpha \quad , \tag{66}$$

in which $\{q_\alpha\}_{\alpha\geq -5}$ is a sequence of real numbers. From a mathematical point of view, Eq. (65) can be written as the direct sum

$$\mathcal{C} = \mathcal{C}_{\text{rig}} \oplus \mathcal{C}_{\text{elas}} \quad , \tag{67}$$

with the usual properties $\mathcal{C}_{\text{rig}} \subset \mathcal{C}$, $\mathcal{C}_{\text{elas}} \subset \mathcal{C}$ and $\mathcal{C}_{\text{rig}} \cap \mathcal{C}_{\text{elas}} = \{0\}$.

7.6. Linear operator equation

The linear operator equation corresponding to the variational formulation defined by Eq. (47) is written as

$$\mathbf{K}\,\mathbf{u} = \lambda\,\mathbf{M}\,\mathbf{u} \quad , \quad \mathbf{u} \in \mathcal{C} \quad , \tag{68}$$

and corresponds to a generalized eigenvalue problem. The orthogonality conditions of the rigid body modes and elastic structural modes (see Eqs. (48), (49), (55), (56), (61) and (62)) can be rewritten as

$$<\mathbf{M}\,\mathbf{u}_\alpha\,,\mathbf{u}_\beta>= \mu_\alpha\,\delta_{\alpha\beta} \quad , \tag{69}$$

$$<\mathbf{K}\,\mathbf{u}_\alpha\,,\mathbf{u}_\beta>== \mu_\alpha\,\omega_\alpha^2\,\delta_{\alpha\beta} \quad , \tag{70}$$

in which $\alpha \geq -5$ and $\beta \geq -5$.

7.7. Terminology

In structural vibrations, the terminology used for the free structure case is the same as the terminology defined in Section 6 for the fixed structure case. Nevertheless, a *structural mode* α, defined by the three quantities $\{\omega_\alpha, \mathbf{u}_\alpha, \mu_\alpha\}$, is either a *rigid body mode* associated with a zero eigenfrequency ($\omega_\alpha = 0$) or an *elastic structural mode* associated with a positive eigenfrequency ($\omega_\alpha > 0$).

8. Cases of Structures with Symmetry Properties

In this section, we briefly discuss the results concerning the structural modes of structures with a plane of symmetry, of axisymmetric structures and of structures with cyclic symmetries.

8.1. Case of a structure with a plane of symmetry

The structure has a plane of symmetry (π) if its domain Ω is symmetric with respect to (π) and if for any two symmetric points $\mathbf{x}$ and $\mathbf{x}'$ with respect to (π), the mass density and the tensor of the elastic coefficients of the material are such that $\rho(\mathbf{x}) = \rho(\mathbf{x}')$ and $a_{ijkh}(\mathbf{x}) = a_{ijkh}(\mathbf{x}')$ respectively. In addition, if the structure is fixed on Γ_0, Γ_0 must be symmetric with respect to (π). The set of structural modes is decomposed into two subsets, the subset of symmetric modes $\mathcal{C}^+$ and the subset of antisymmetric modes $\mathcal{C}^-$ (see Fig. 3).

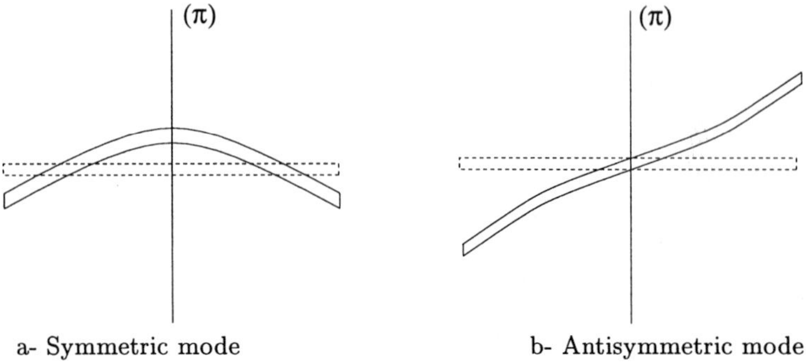

Fig. 3. Case of a structure with a plane of symmetry

The initial spectral problem can then be split into two independent spectral problems posed in a half domain. Their variational formulations are obtained by restriction to $\mathcal{C}^+$ and $\mathcal{C}^-$.

8.2. Case of an axisymmetric structure

Let $(\mathbf{i}, \mathbf{j}, \mathbf{k})$ be the Cartesian frame of reference and (x_1, x_2, x_3) the coordinates of $\mathbf{x}$ in $(\mathbf{i}, \mathbf{j}, \mathbf{k})$. We introduce the cylindrical coordinates (r, θ, z) such that $x_1 = r\cos\theta$, $x_2 = r\sin\theta$, $x_3 = z$ and the local cylindrical frame $(\mathbf{e}_r, \mathbf{e}_\theta, \mathbf{e}_z)$ attached at each point $\mathbf{x}$ such that for $\theta = 0$, we have $\mathbf{e}_r = \mathbf{i}$, $\mathbf{e}_\theta = \mathbf{j}$ and $\mathbf{e}_z = \mathbf{k}$ (see Fig. 4).

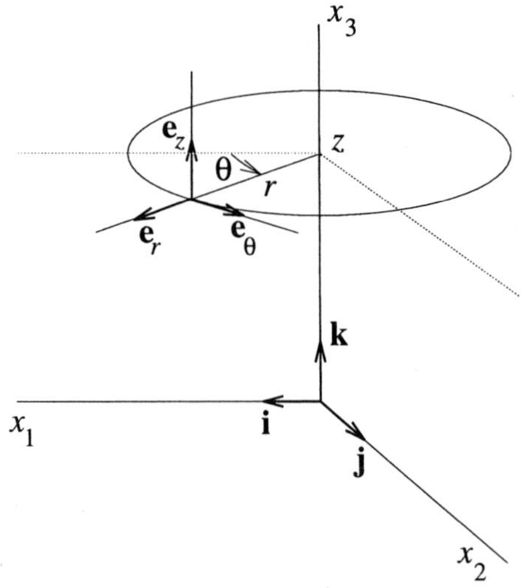

Fig. 4. Coordinate systems for an axisymmetric structure

The structure is axisymmetric with respect to the z axis if its domain Ω has the z axis as a rotational symmetry, if the mass density and the tensor of the elastic coefficients of the material are independent of θ. In addition, if the structure is fixed on Γ_0, boundary Γ_0 must have rotational symmetry with respect to the z axis. Let $\mathbf{u}(r,\theta,z)$ be the displacement field expressed in the local cylindrical frame $(\mathbf{e}_r, \mathbf{e}_\theta, \mathbf{e}_z)$. The Fourier series expansion of $\mathbf{u}$ with respect to θ yields

$$\mathbf{u}(r,\theta,z) = \sum_{n=0}^{+\infty} \{\mathbf{u}_n^+(r,\theta,z) + \mathbf{u}_n^-(r,\theta,z)\} \quad , \tag{71}$$

in which for each circumferential wave number $n \geq 0$,

$$\mathbf{u}_n^+(r,\theta,z) = \big(\mathbb{u}_n^+(r,z)\, \cos n\theta \,, \mathbb{v}_n^+(r,z)\, \sin n\theta \,, \mathbb{w}_n^+(r,z)\, \cos n\theta\big) \ , \tag{72}$$

$$\mathbf{u}_n^-(r,\theta,z) = \big(\mathbb{u}_n^-(r,z)\, \sin n\theta \,, -\mathbb{v}_n^-(r,z)\, \cos n\theta \,, \mathbb{w}_n^-(r,z)\, \sin n\theta\big) \ , \tag{73}$$

where for $n \geq 0$, $(\mathbb{u}_n^\pm, \mathbb{v}_n^\pm, \mathbb{w}_n^\pm)$ are the harmonic components of order n. The reference plane $(\mathbf{i}, \mathbf{k})$, denoted as (π), is the generating plane of three-dimensional domain Ω and is a plane of symmetry for the structure. Consequently, the set of structural modes is decomposed into two subsets, the subset of symmetric modes $\mathbf{u}_n^+, n \geq 0$, with respect to (π) and the subset of antisymmetric modes $\mathbf{u}_n^-, n \geq 0$, with respect to (π) (see Fig. 5).

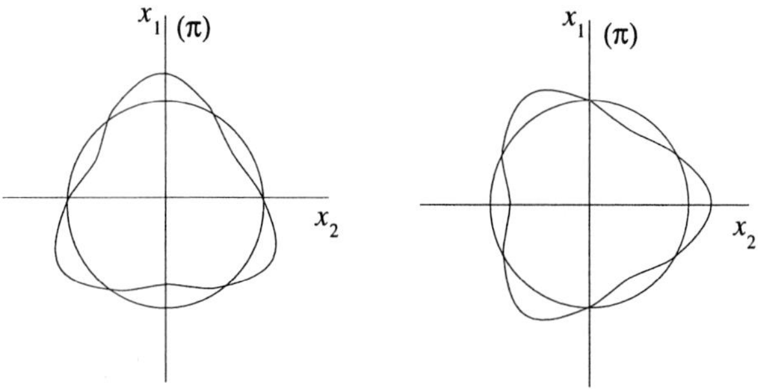

a- Symmetric mode b- Antisymmetric mode

Fig. 5. Deformation of a cross section in the plane x_1,x_2 for n=3

The subset of symmetric modes is decomposed into subsets $C_n^+, n \geq 0$. The subset of antisymmetric modes is decomposed into subsets $C_n^-, n \geq 0$. The initial three-dimensional spectral problem can then be split into a countable number of independent two-dimensional spectral problems $\mathcal{P}_n^\pm, n \geq 0$ posed in plane (π). Their variational formulations, expressed in terms of the harmonic components, are obtained by restriction to $\mathcal{C}_n^{\pm}$. For each $n \geq 1$, problems $\mathcal{P}_n^+$ and $\mathcal{P}_n^-$ are identical (same eigenvalues). Problem $\mathcal{P}_0^+$ related to harmonic components $(\mathbb{u}_0^+, 0, \mathbb{w}_0^+)$ is different from problem $\mathcal{P}_0^-$ related to harmonic components $(0, -\mathbb{v}_0^-, 0)$. In conclusion, the eigenvalue problems to be solved are related to $\mathcal{P}_0^+$, $\mathcal{P}_0^-$ and $\mathcal{P}_n^+, n \geq 1$, and the structural modes of the three-dimensional structure are reconstituted using Eqs. (72) and (73).

8.3. Case of a structure with cyclic symmetries

Let (Δ) be the rotational axis of symmetry and $N \geq 1$ be a positive integer. Domain Ω has cyclic symmetry with axis (Δ) and angle $2\pi/N$ if

$$\Omega = \cup_{m=0}^{N-1} \Omega_m \quad , \tag{74}$$

in which Ω_0 is the generating sector of angle $2\pi/N$ and if, for all $m \geq 1$, sector Ω_m is deduced from sector Ω_{m-1} by the rotation $((\Delta), 2\pi/N)$ (see Fig. 6). Let $\text{Rot}_{2\pi m/N}$ be the rotation $((\Delta), 2\pi m/N)$ which maps any point $\mathbf{x}_0$ in Ω_0 into its corresponding point $\mathbf{x}_m$ in Ω_m:

$$\mathbf{x}_m = \text{Rot}_{2\pi m/N}(\mathbf{x}_0) \quad , \quad \forall \mathbf{x}_0 \in \Omega_0 \quad . \tag{75}$$

Each sector Ω_m has its own Cartesian reference system $(\mathbf{i}, \mathbf{j}, \mathbf{k})_m$ which rotates with the sector.

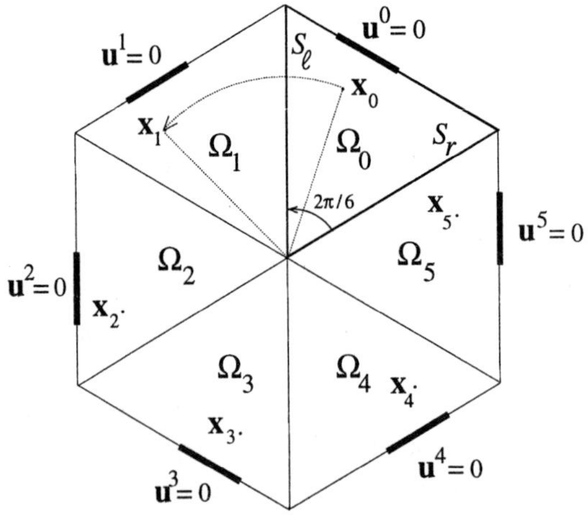

Fig. 6. Structure with cyclic symmetry

The structure has cyclic symmetry with axis (Δ) and angle $2\pi/N$ firstly, if its domain Ω has cyclic symmetry with axis (Δ) and angle $2\pi/N$ and secondly, if for all $m \geq 1$, the components of the tensor of the elastic coefficients of the material in $(\mathbf{i},\mathbf{j},\mathbf{k})_m$ at any point $\mathbf{x}_m$ in Ω_m are equal to the components of this tensor in $(\mathbf{i},\mathbf{j},\mathbf{k})_0$ at the corresponding point $\mathbf{x}_0$ in Ω_0 and thirdly, if for all $m \geq 1$, mass density $\rho(\mathbf{x}_m)$ at any point $\mathbf{x}_m$ in Ω_m is equal to $\rho(\mathbf{x}_0)$ at the corresponding point $\mathbf{x}_0$ in Ω_0. In addition, if the structure is fixed on Γ_0, boundary Γ_0 must have cyclic symmetry with axis (Δ) and angle $2\pi/N$. Let $\mathbf{u}$ be the displacement field defined on Ω. Let $\mathbf{u}^m$ be the restriction of $\mathbf{u}$ to sector Ω_m whose components are expressed in $(\mathbf{i},\mathbf{j},\mathbf{k})_m$. For two points $\mathbf{x}_0 \in \Omega_0$ and $\mathbf{x}_m \in \Omega_m$ satisfying Eq. (75), it can be shown, see Arnold (Appendix 10), 1978; Bluman and Kumei, 1989; Ludwig and Falter, 1988; Ohayon, 1985, that

$$\mathbf{u}^m(\mathbf{x}_m) = \Re e \left\{ \sum_{n=0}^{N-1} \mathbf{u}_n(\mathbf{x}_0) \exp(-\frac{2i\pi mn}{N}) \right\} \quad , \tag{76}$$

where $\mathbf{u}_n$ represents the *complex harmonic component* of order n defined on generating sector Ω_0. Eq. (76) is a Discrete Fourier Transform (DFT). Two cases must be distinguished.

N **even**. The structure has an even number N of sectors. Instead of the complex harmonic components, it is convenient to introduce $\frac{N}{2}+1$ new real variables $\{\mathbf{u}_n^{\pm}\}$ defined in Ω_0 and called the *cyclic components*, such that

$$\mathbf{u}_0^+ = \Re e \{\mathbf{u}_0\} \quad , \quad \mathbf{u}_{N/2}^+ = \Re e \{\mathbf{u}_{N/2}\} \quad , \tag{77}$$

and for $n = 1, \ldots, \frac{N}{2}-1$,

$$\mathbf{u}_n^+ = \Re e \{\mathbf{u}_n + \mathbf{u}_{N-n}\} \quad , \quad \mathbf{u}_n^- = \Re e \{-i(\mathbf{u}_n - \mathbf{u}_{N-n})\} \quad . \tag{78}$$

Using Eqs. (77) and (78), we deduce that Eq. (76) can be rewritten, for $m = 0, \ldots, N-1$, as

$$\mathbf{u}^m(\mathbf{x}_m) = \mathbf{u}_0^{m,+}(\mathbf{x}_m) + \mathbf{u}_{N/2}^{m,+}(\mathbf{x}_m) + \sum_{n=1}^{\frac{N}{2}-1} \{\mathbf{u}_n^{m,+}(\mathbf{x}_m) + \mathbf{u}_n^{m,-}(\mathbf{x}_m)\} \quad , \tag{79}$$

in which

$$\mathbf{u}_n^{m,+}(\mathbf{x}_m) = \mathbf{u}_n^+(\mathbf{x}_0) \cos(\frac{2\pi mn}{N}) \quad , \tag{80}$$

$$\mathbf{u}_n^{m,-}(\mathbf{x}_m) = \mathbf{u}_n^-(\mathbf{x}_0) \sin(\frac{2\pi mn}{N}) \quad . \tag{81}$$

It can be shown that the initial three-dimensional spectral problem can then be split into $\frac{N}{2}+1$ independent spectral problems $\{\mathcal{P}_n\,, n = 0, \ldots, N/2\}$ in terms of the cyclic components, posed in the generating sector Ω_0, with appropriate boundary conditions on the two sides of Ω_0. These sides are denoted as S_r and S_ℓ and are such that $S_\ell = \text{Rot}_{2\pi/N}(S_r)$. Let $\mathbf{K}$ and $\mathbf{M}$ be the operators defined by Eq. (32) or Eq. (35) and Eq. (28) respectively for the three-dimensional domain Ω (for a fixed or free structure). Let $\mathbf{K}_0$ and $\mathbf{M}_0$ be the corresponding operators related to domain Ω_0.

Problem $\mathcal{P}_0$. The eigenvalue problem $\mathcal{P}_0$ is defined by

$$\mathbf{K}_0\, \mathbf{u}_0^+ = \lambda\, \mathbf{M}_0\, \mathbf{u}_0^+ \quad , \quad \mathbf{u}_0^+ \in \mathcal{U}_0 \quad , \tag{82}$$

in which $\mathcal{U}_0$ is the appropriate admissible function space (including the boundary condition if the structure is fixed) containing the following constraint equation on sides S_r and S_ℓ

$$(\mathbf{u}_0^+)_{S_\ell} = (\mathbf{u}_0^+)_{S_r} \quad . \tag{83}$$

Problem $\mathcal{P}_0$ has a countable number of eigenvalues and eigenvectors denoted as $\{\lambda, \mathbf{u}_0^+\}_{\alpha \geq 1}$.

Problem $\mathcal{P}_{N/2}$. The eigenvalue problem $\mathcal{P}_{N/2}$ is defined by

$$\mathbf{K}_0\, \mathbf{u}_{N/2}^+ = \lambda\, \mathbf{M}_0\, \mathbf{u}_{N/2}^+ \quad , \quad \mathbf{u}_{N/2}^+ \in \mathcal{U}_{N/2} \quad , \tag{84}$$

in which $\mathcal{U}_{N/2}$ is the appropriate admissible function space (including the boundary condition if the structure is fixed) containing the following constraint equation on sides S_r and S_ℓ

$$(\mathbf{u}_{N/2}^+)_{S_\ell} = -(\mathbf{u}_{N/2}^+)_{S_r} \quad . \tag{85}$$

Problem $\mathcal{P}_{N/2}$ has a countable number of eigenvalues and eigenvectors denoted as $\{\lambda, \mathbf{u}_{N/2}^+\}_{\alpha \geq 1}$.

Problem $\mathcal{P}_n$. For n fixed in $\{1, \ldots, \frac{N}{2} - 1\}$, the eigenvalue problem $\mathcal{P}_n$ is defined by

$$\begin{bmatrix} \mathbf{K}_0 & \mathbf{0} \\ \mathbf{0} & \mathbf{K}_0 \end{bmatrix} \begin{bmatrix} \mathbf{u}_n^+ \\ \mathbf{u}_n^- \end{bmatrix} = \lambda \begin{bmatrix} \mathbf{M}_0 & \mathbf{0} \\ \mathbf{0} & \mathbf{M}_0 \end{bmatrix} \begin{bmatrix} \mathbf{u}_n^+ \\ \mathbf{u}_n^- \end{bmatrix} \quad , \quad (\mathbf{u}_n^+, \mathbf{u}_n^-) \in \mathcal{U}_n \quad , \tag{86}$$

in which $\mathcal{U}_n$ is the appropriate admissible function space (including the boundary condition if the structure is fixed) containing the following constraint equation on sides S_r and S_ℓ

$$\begin{bmatrix} (\mathbf{u}_n^+)_{S_\ell} \\ (\mathbf{u}_n^-)_{S_\ell} \end{bmatrix} = \begin{bmatrix} \cos(2\pi n/N) & \sin(2\pi n/N) \\ -\sin(2\pi n/N) & \cos(2\pi n/N) \end{bmatrix} \begin{bmatrix} (\mathbf{u}_n^+)_{S_r} \\ (\mathbf{u}_n^-)_{S_r} \end{bmatrix} \quad . \tag{87}$$

Problem $\mathcal{P}_n$ has a countable number of eigenvalues and eigenvectors denoted as $\{\lambda, (\mathbf{u}_n^+, \mathbf{u}_n^-)\}_{\alpha \geq 1}$.

Structural modes of the three-dimensional structure. The structural modes of the three-dimensional structure occupying domain Ω are constituted

(1)- by $\{\lambda, \mathbf{u}_0^{m,+}\}_{\alpha \geq 1}$ in which the structural modes are reconstituted by Eq. (80), that is to say by $\mathbf{u}_0^{m,+}(\mathbf{x}_m) = \mathbf{u}_0^+(\mathbf{x}_0)$, where $\{\lambda, \mathbf{u}_0^+\}_{\alpha \geq 1}$ are the solutions of problem $\mathcal{P}_0$.

(2)- by $\{\lambda, \mathbf{u}_{N/2}^{m,+}\}_{\alpha \geq 1}$ in which the structural modes are reconstituted by Eq. (80), that is to say by $\mathbf{u}_{N/2}^{m,+}(\mathbf{x}_m) = \mathbf{u}_{N/2}^+(\mathbf{x}_0)$, where $\{\lambda, \mathbf{u}_{N/2}^+\}_{\alpha \geq 1}$ are the solutions of problem $\mathcal{P}_{N/2}$.

(3)- for all n fixed in $\{1, \ldots, \frac{N}{2} - 1\}$, by $\{\lambda, \mathbf{u}_n^{m,+}\}_{\alpha \geq 1}$ and $\{\lambda, \mathbf{u}_n^{m,-}\}_{\alpha \geq 1}$ in which the structural modes are reconstituted by Eqs. (80) and (81), where $\{\lambda, (\mathbf{u}_n^+, \mathbf{u}_n^-)\}_{\alpha \geq 1}$ are the solutions of problem $\mathcal{P}_n$. Consequently, for each λ, there is a pair $\{\mathbf{u}_n^{m,+}, \mathbf{u}_n^{m,-}\}$ of structural modes associated with the same eigenvalue λ.

N **odd.** The structure has an odd number N of sectors. As was done for the case with N even, instead of the complex harmonic components, we introduce $(N+1)/2$ cyclic components $\{\mathbf{u}_n^\pm\}$ defined in Ω_0 and such that

$$\mathbf{u}_0^+ = \Re e\,\{\mathbf{u}_0\} \quad , \tag{88}$$

and for $n = 1, \ldots, (N-1)/2$,

$$\mathbf{u}_n^+ = \Re e\,\{\mathbf{u}_n + \mathbf{u}_{N-n}\} \quad , \quad \mathbf{u}_n^- = \Re e\,\{-i(\mathbf{u}_n - \mathbf{u}_{N-n})\} \quad . \tag{89}$$

Using Eqs. (88) and (89), we deduce that Eq. (76) can be rewritten, for $m = 0, \ldots, N-1$, as

$$\mathbf{u}^m(\mathbf{x}_m) = \mathbf{u}_0^{m,+}(\mathbf{x}_m) + \sum_{n=1}^{(N-1)/2} \{\mathbf{u}_n^{m,+}(\mathbf{x}_m) + \mathbf{u}_n^{m,-}(\mathbf{x}_m)\} \quad , \tag{90}$$

in which

$$\mathbf{u}_n^{m,+}(\mathbf{x}_m) = \mathbf{u}_n^+(\mathbf{x}_0)\,\cos(\frac{2\pi mn}{N}) \quad , \tag{91}$$

$$\mathbf{u}_n^{m,-}(\mathbf{x}_m) = \mathbf{u}_n^-(\mathbf{x}_0)\,\sin(\frac{2\pi mn}{N}) \quad . \tag{92}$$

It can be shown that the initial three-dimensional spectral problem can then be split into $(N+1)/2$ independent spectral problems $\{\mathcal{P}_n\,, n =$

$0,\ldots,(N-1)/2\}$ in terms of the cyclic components, posed in the generating sector Ω_0, with appropriate boundary conditions on the two sides S_r and S_ℓ of Ω_0. Let $\mathbf{K}$ and $\mathbf{M}$ be the operators defined by Eq. (32) or (35) and Eq. (28) respectively, for the three-dimensional domain Ω (for a fixed or free structure). Let $\mathbf{K}_0$ and $\mathbf{M}_0$ be the corresponding operators related to domain Ω_0.

Problem $\mathcal{P}_0$. The eigenvalue problem $\mathcal{P}_0$ is defined by Eqs. (82) and (83).

Problem $\mathcal{P}_n$. For n fixed in $\{1,\ldots,(N-1)/2\}$, the eigenvalue problem $\mathcal{P}_n$ is defined by Eqs. (86) and (87).

Structural modes of the three-dimensional structure. The structural modes of the three-dimensional structure occupying domain Ω are constituted

(1)- by $\{\lambda, \mathbf{u}_0^{m,+}\}_{\alpha\geq 1}$ in which the structural modes are reconstituted by Eq. (80), that is to say by $\mathbf{u}_0^{m,+}(\mathbf{x}_m) = \mathbf{u}_0^{+}(\mathbf{x}_0)$, where $\{\lambda, \mathbf{u}_0^{+}\}_{\alpha\geq 1}$ are the solutions of problem $\mathcal{P}_0$.

(2)- for all n fixed in $\{1,\ldots,(N-1)/2\}$, by $\{\lambda, \mathbf{u}_n^{m,+}\}_{\alpha\geq 1}$ and $\{\lambda, \mathbf{u}_n^{m,-}\}_{\alpha\geq 1}$ in which the structural modes are reconstituted by Eqs. (80) and (81), where $\{\lambda, (\mathbf{u}_n^{+}, \mathbf{u}_n^{-})\}_{\alpha\geq 1}$ are the solutions of problem $\mathcal{P}_n$. Consequently, for each λ, there is a pair $\{\mathbf{u}_n^{m,+}, \mathbf{u}_n^{m,-}\}$ of structural modes associated with the same eigenvalue λ.

9. Finite Element Discretization and Generalized Symmetric Matrix Eigenvalue Problem

The matrix equation of the generalized symmetric eigenvalue problem corresponding to the finite element discretization of the three-dimensional continuous problem is directly deduced from the eigenvalue problem defined by Eqs. (38) or (47) (see Chapter II)

$$[\,K\,]\,\mathbf{U} = \lambda\,[\,M\,]\,\mathbf{U} \quad , \tag{93}$$

in which $\mathbf{U} = (U_1,\ldots,U_n)$ is the vector of the DOFs which are the values of the displacement field at the nodes of the finite element mesh of domain Ω. The $(n\times n)$ mass matrix $[\,M\,]$ is symmetric positive definite. For a fixed structure on Γ_0, the $(n\times n)$ stiffness matrix $[\,K\,]$ is symmetric positive definite and then invertible. For a free structure, stiffness matrix $[\,K\,]$ is symmetric positive semidefinite and singular of rank $n-6$. The finite element procedures in structural vibrations can be found in Argyris and Mlejnek, 1991; Bathe and Wilson, 1976; Géradin and Rixen, 1994; Petyt,

1990; Zienkiewicz and Taylor, 1989. The numerical solution of the generalized symmetric eigenvalue problem defined by Eq. (93) requires specific eigenvalue solvers, see Parlett, 1980; Bathe and Wilson, 1976; Golub and Van Loan, 1989; Chatelin, 1993.

10. Dynamic Substructuring Procedures for Calculation of the Structural Modes

10.1. Objective of the substructuring procedures

The strategy of dynamic substructuring consists in replacing the global model of the structure by an assemblage of substructures. Dynamic substructuring techniques based on the use of the fixed-interface modes or free-interface modes of each substructure have been widely developed in the literature: see for example Hurty, 1965; Craig and Bampton, 1968; MacNeal, 1971; Rubin, 1975; Craig, 1985; Min, Igusa and Achenbach, 1992; Leung, 1993; Farhat and Geradin, 1994; Morand and Ohayon, 1992 and 1995; Ohayon, Sampaio and Soize, 1997. Below we present the main method based on the use of the fixed-interface modes and completed by static boundary functions (Hurty, 1965; Craig and Bampton, 1968; Morand and Ohayon, 1992 and 1995). To clarify the main ideas, we consider a structure composed of only two substructures that interact through a common boundary (the extension to the case of more than two substructures is straightforward). In the first subsection, we present the variational formulation for the case of a free structure decomposed into two substructures. The second subsection is devoted to a dynamic substructuring method using the fixed-interface modes and boundary static functions of each substructure, presented in a general framework allowing various other decomposition procedures to be obtained. After constructing the reduced matrix model of each substructure, we construct the eigenvalue problem for the structure as the assemblage of the substructures.

10.2. Variational formulation for a structure made of two substructures

Interface conditions. Domain Ω of the free structure is decomposed into two subdomains, Ω_1 with boundary $\partial\Omega_1 = \Gamma_1 \cup \Sigma$ and Ω_2 with boundary $\partial\Omega_2 = \Gamma_2 \cup \Sigma$, that interact through the common boundary Σ (see Fig. 7). For $r = 1$ and $r = 2$, let $\mathbf{u}^r$ be the restriction of the displacement field $\mathbf{u}$ to Ω_r. Let $\sigma^r = \{\sigma^r_{ij}\}_{ij}$ be the stress tensor related to Ω_r such that

$$\sigma^r_{ij}(\mathbf{u}^r) = a^r_{ijkh}\, \varepsilon_{kh}(\mathbf{u}^r) \quad , \tag{94}$$

in which a^r_{ijkh} is the restriction of a_{ijkh} to Ω_r. Finally, we introduce ρ^r as the restriction of mass density ρ to Ω_r. The coupling conditions on Σ are written as

$$\mathbf{u}^1 = \mathbf{u}^2 \quad \text{on} \quad \Sigma \ , \tag{95}$$

$$\sigma^1_{ij}(\mathbf{u}^1)\, n^1_j = -\,\sigma^2_{ij}(\mathbf{u}^2) n^2_j \quad \text{on} \quad \Sigma \ , \tag{96}$$

where $\mathbf{n}^r = (n^r_1, n^r_2, n^r_3)$ is the unit normal to Σ, external to Ω_r. Equation (95) expresses the continuity of the displacement on Σ and Eq. (96) expresses the surface-forces interaction through Σ.

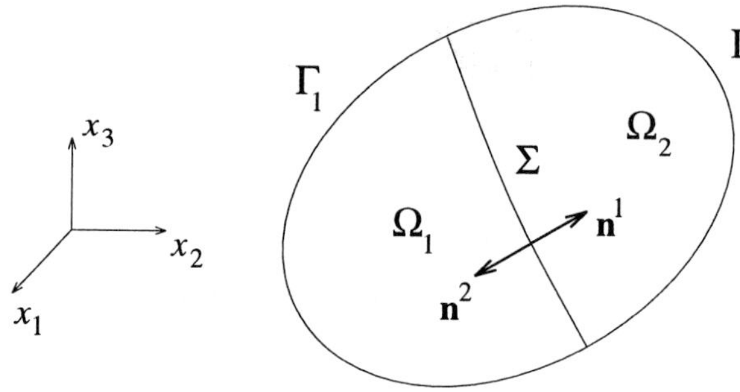

Fig. 7. Structure made of two substructures

Variational formulation in terms of $(\mathbf{u}^1, \mathbf{u}^2)$. The variational formulation of the eigenvalue problem is defined by Eq. (47) for the free structure case. Let $\mathcal{C}_{\Omega_r}$ be the admissible function space [6] defined on Ω_r (which does not contain any constraint on $\partial\Omega_r = \Gamma_r \cup \Sigma$). For $r = 1$ and $r = 2$, let

$$m^r(\mathbf{u}^r, \delta\mathbf{u}^r) = \int_{\Omega_r} \rho^r\ \mathbf{u}^r \cdot \delta\mathbf{u}^r\ d\mathbf{x} \tag{97}$$

be the structural mass bilinear form and

$$k^r(\mathbf{u}^r, \delta\mathbf{u}^r) = \int_{\Omega_r} \sigma^r_{ij}(\mathbf{u}^r)\, \varepsilon_{ij}(\delta\mathbf{u}^r)\, d\mathbf{x} \tag{98}$$

be the structural stiffness bilinear form, defined on $\mathcal{C}_{\Omega_r} \times \mathcal{C}_{\Omega_r}$. The variational formulation defined by Eq. (47) can be rewritten as follows. Find

6 Space $\mathcal{C}_{\Omega_r} = (H^1(\Omega_r))^3$.

λ, $\mathbf{u}^1 \in \mathcal{C}_{\Omega_1}$ and $\mathbf{u}^2 \in \mathcal{C}_{\Omega_2}$ satisfying $\mathbf{u}^1 = \mathbf{u}^2$ (Eq. (95)) [7] such that for all $\delta\mathbf{u}^1 \in \mathcal{C}_{\Omega_1}$ and $\delta\mathbf{u}^2 \in \mathcal{C}_{\Omega_2}$ satisfying $\delta\mathbf{u}^1 = \delta\mathbf{u}^2$ on Σ,

$$k^1(\mathbf{u}^1, \delta\mathbf{u}^1) + k^2(\mathbf{u}^2, \delta\mathbf{u}^2) = \lambda \left\{ m^1(\mathbf{u}^1, \delta\mathbf{u}^1) + m^2(\mathbf{u}^2, \delta\mathbf{u}^2) \right\} \quad . \tag{99}$$

Using Green's formula, it can easily be seen that the above variational formulation yields all the local equations of the problem.

10.3. Dynamic substructuring using the fixed-interface modes of each substructure

The substructuring technique consists in using a reduced matrix model for each substructure Ω_r ($r = 1$ or $r = 2$) based on the decomposition of displacement field $\mathbf{u}^r$ first, on elastic structural modes $\mathbf{u}^r_\alpha$ of substructure Ω_r fixed on Σ, and secondly on static functions $\mathbf{u}^r_{\text{stat}}$.

Fixed-interface modes of substructure Ω_r. A fixed-interface mode of substructure Ω_r (for $r = 1$ or $r = 2$) is defined as an elastic structural mode of substructure Ω_r fixed on Σ (see Fig. 8). This type of problem was studied in Section 6.

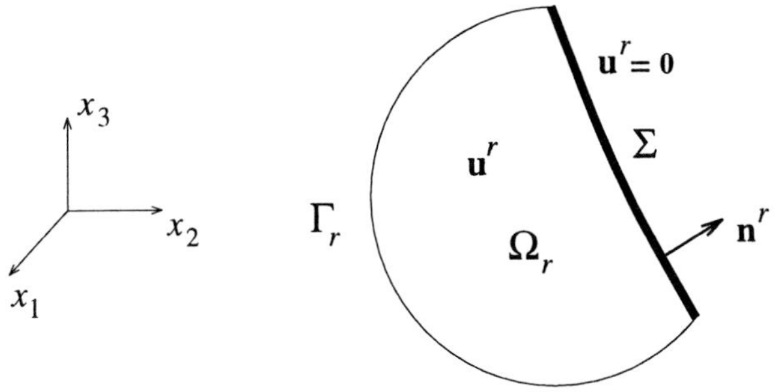

Fig. 8. Eigenvalue problem for each substructure with zero displacement on the interface

Introducing admissible function space $\mathcal{C}^0_{\Omega_r}$ defined by

$$\mathcal{C}^0_{\Omega_r} = \{ \mathbf{u}^r \in \mathcal{C}_{\Omega_r} \quad ; \quad \mathbf{u}^r = \mathbf{0} \quad \text{on} \quad \Sigma \} \quad , \tag{100}$$

---

[7] If $\partial\Omega_r$ is the boundary of Ω_r, then the trace operator from $(H^1(\Omega_r))^3$ onto $(H^{1/2}(\partial\Omega_r))^3$ is a continuous linear surjection. Equality $\mathbf{u}^1 = \mathbf{u}^2$ must be read in the sense of the traces on $\Sigma = \partial\Omega_1 \cap \partial\Omega_2$.

each eigenvalue λ^r_α and its corresponding eigenvector $\mathbf{u}^r_\alpha \in \mathcal{C}^0_{\Omega_r}$ satisfy the following variational formulation

$$k^r(\mathbf{u}^r_\alpha, \delta\mathbf{u}^r) - \lambda^r_\alpha \; m^r(\mathbf{u}^r_\alpha, \delta\mathbf{u}^r) = 0 \quad , \quad \forall\, \delta\mathbf{u}^r \in \mathcal{C}^0_{\Omega_r} \quad , \tag{101}$$

in which m^r and k^r are defined by Eqs. (97) and (98) respectively. In Eq. (101), if test function $\delta\mathbf{u}^r$ belongs to $\mathcal{C}_{\Omega_r}$ and not to subspace $\mathcal{C}^0_{\Omega_r}$, then the left-hand side of Eq. (101) is not equal to zero and we have

$$k^r(\mathbf{u}^r_\alpha, \delta\mathbf{u}^r) - \lambda^r_\alpha \; m^r(\mathbf{u}^r_\alpha, \delta\mathbf{u}^r) = \ell^r_\Sigma(\delta\mathbf{u}^r) \quad , \quad \forall\, \delta\mathbf{u}^r \in \mathcal{C}_{\Omega_r} \quad , \tag{102 – 1}$$

where $\ell^r_\Sigma(\delta\mathbf{u}^r)$ is a linear form on $\mathcal{C}_{\Omega_r}$ which represents the work of the force field induced by constraint $\mathbf{u}^r_\alpha = \mathbf{0}$ on Σ. From a mechanical point of view, if $\mathbf{F}^r_\alpha = \sigma^r(\mathbf{u}^r_\alpha)\,\mathbf{n}^r$ denotes this reaction force field applied to Σ, we have

$$\ell^r_\Sigma(\delta\mathbf{u}^r) = \int_\Sigma \mathbf{F}^r_\alpha \cdot \delta\mathbf{u}^r \, ds \quad . \tag{102 – 2}$$

The right-hand side of Eq. (102-2) is a notation which is used below.

Static functions of substructure Ω_r, introduction of the linear operator $\mathbf{S}^r$. We consider the solution $\mathbf{u}^r_{\mathrm{stat}}$ of the elastostatic problem of substructure Ω_r subjected to a prescribed displacement field $\mathbf{u}^r_{\mathrm{presc}}$ on Σ (see Fig. 9).

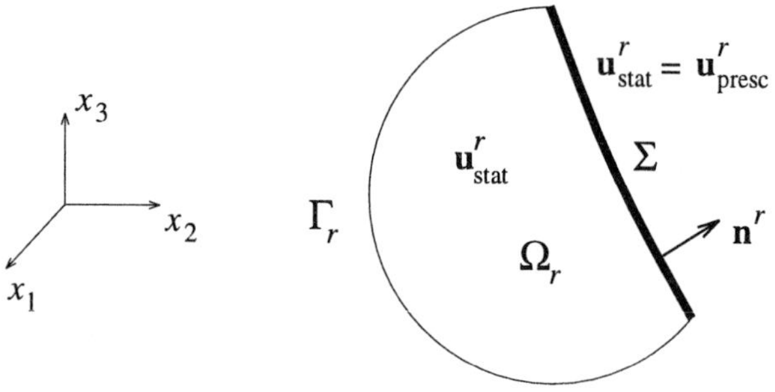

Fig. 9. Elastostatic problem for each substructure with prescribed displacement on the interface

The mapping $\mathbf{u}^r_{\mathrm{presc}} \mapsto \mathbf{u}^r_{\mathrm{stat}}$ is a linear operator [8] denoted as $\mathbf{S}^r$. We then have

$$\mathbf{u}^r_{\mathrm{stat}} = \mathbf{S}^r(\mathbf{u}^r_{\mathrm{presc}}) \quad . \tag{103}$$

[8] Linear operator $\mathbf{S}^r$ is called a lifting operator and is continuous from the trace space $V_\Sigma \subset (H^{1/2}(\partial\Omega_r))^3$ on Σ into $(H^1(\Omega_r))^3$.

By construction, the value of $\mathbf{S}^r(\mathbf{u}^r_{\text{presc}})$ on the boundary Σ is equal to $\mathbf{u}^r_{\text{presc}}$. To construct operator $\mathbf{S}^r$, we introduce the set $\mathcal{C}^{\text{presc}}_{\Omega_r}$ of all the functions $\mathbf{u}^r_{\text{stat}}$ defined on Ω_r whose values on Σ are equal to $\mathbf{u}^r_{\text{presc}}$

$$\mathcal{C}^{\text{presc}}_{\Omega_r} = \{\, \mathbf{u}^r_{\text{stat}} \in \mathcal{C}_{\Omega_r} \quad ; \quad \mathbf{u}^r_{\text{stat}} = \mathbf{u}^r_{\text{presc}} \quad \text{on} \quad \Sigma \,\} \quad . \tag{104}$$

Linear operator $\mathbf{S}^r$ defined by Eq. (103) is constructed by solving the following problem. Find $\mathbf{u}^r_{\text{stat}}$ in $\mathcal{C}^{\text{presc}}_{\Omega_r}$ such that

$$k^r(\mathbf{u}^r_{\text{stat}}, \delta\mathbf{u}^r) = 0 \quad , \quad \forall\, \delta\mathbf{u}^r \in \mathcal{C}^0_{\Omega_r} \quad , \tag{105}$$

where $\mathcal{C}^0_{\Omega_r}$ is the space defined by Eq. (100). It should be noted that (1) the space of rigid body displacement fields (introduced in Section 7.2) is a subset of the range space of operator $\mathbf{S}^r$; (2) the discretization of $\mathbf{S}^r$ by the finite element method is obtained by a classical static condensation procedure of the stiffness matrix of substructure Ω_r with respect to degrees of freedom on Σ (see Section 10.4).

Conjugate relationships between $\mathbf{u}^r_\alpha$ and $\mathbf{u}^r_{\text{stat}}$. Taking $\delta\mathbf{u}^r = \mathbf{u}^r_\alpha$ in Eq. (105) yields

$$k^r(\mathbf{u}^r_{\text{stat}}, \mathbf{u}^r_\alpha) = 0 \quad . \tag{106}$$

Since $\mathcal{C}^{\text{presc}}_{\Omega_r}$ is included in $\mathcal{C}_{\Omega_r}$, we can choose $\mathbf{u}^r_{\text{stat}}$ as $\delta\mathbf{u}^r$ in Eq. (102) where $\mathbf{u}^r_{\text{stat}}$ is the solution of Eq. (105) for an arbitrary prescribed displacement field $\mathbf{u}^r_{\text{presc}}$ on Σ. From Eqs. (102) and (106) and the symmetry property of bilinear forms k^r and m^r, we deduce that

$$m^r(\mathbf{u}^r_{\text{stat}}, \mathbf{u}^r_\alpha) = -\frac{1}{\lambda^r_\alpha}\int_\Sigma \mathbf{F}^r_\alpha \cdot \mathbf{u}^r_{\text{presc}}\, ds \quad , \tag{107}$$

because the trace of $\mathbf{u}^r_{\text{stat}}$ on Σ is $\mathbf{u}^r_{\text{presc}}$. Finally, the properties defined by Eqs. (106) and (107) can be rewritten as follows. For all prescribed displacement fields $\mathbf{u}^r_{\text{presc}}$ on Σ and eigenvector $\mathbf{u}^r_\alpha$ solution of Eq. (101), we have

$$k^r(\mathbf{S}^r(\mathbf{u}^r_{\text{presc}}), \mathbf{u}^r_\alpha) = 0 \quad , \tag{108}$$

$$m^r(\mathbf{S}^r(\mathbf{u}^r_{\text{presc}}), \mathbf{u}^r_\alpha) = -\frac{1}{\lambda^r_\alpha}\int_\Sigma \mathbf{F}^r_\alpha \cdot \mathbf{u}^r_{\text{presc}}\, ds \quad . \tag{109}$$

Decomposition of $\mathcal{C}_{\Omega_r}$. Consider an arbitrary displacement field $\mathbf{u}^r$ in $\mathcal{C}_{\Omega_r}$ whose value on Σ is $\mathbf{u}^r_\Sigma$. Taking $\mathbf{u}^r_{\text{presc}} = \mathbf{u}^r_\Sigma$ and from Eq. (103), we deduce that the displacement field $\mathbf{u}^r - \mathbf{u}^r_{\text{stat}}$ is equal to zero on Σ. Then this difference belongs to $\mathcal{C}^0_{\Omega_r}$ and can then be spanned by fixed-interface modes $\mathbf{u}^r_\alpha$ according to $\mathbf{u}^r - \mathbf{u}^r_{\text{stat}} = \sum_{\alpha=1}^{N_r} q^r_\alpha\, \mathbf{u}^r_\alpha$. Therefore, introducing the space $\mathcal{C}^{\text{stat}}_{\Omega_r}$ of all the solutions of Eq. (105) for all displacement fields $\mathbf{u}^r_\Sigma$ on Σ, we have the following decomposition into a direct sum

$$\mathcal{C}_{\Omega_r} = \mathcal{C}^{\text{stat}}_{\Omega_r} \oplus \mathcal{C}^0_{\Omega_r} \quad , \tag{110}$$

$$\mathbf{u}^r = \mathbf{S}^r(\mathbf{u}^r_\Sigma) + \sum_{\alpha=1}^{\infty} q^r_\alpha\, \mathbf{u}^r_\alpha \quad . \tag{111}$$

Construction of the reduced matrix model for substructure Ω_r. We use the Ritz-Galerkin method. Let us introduce the finite dimension subspace of $\mathcal{C}^0_{\Omega_r}$ spanned by eigenvectors $\{\mathbf{u}^r_1, \ldots, \mathbf{u}^r_{N_r}\}$ with $N_r \geq 1$. From Eq. (111) and using the same notation $\mathbf{u}^r$ to simplify the writing, we deduce that

$$\mathbf{u}^r = \mathbf{S}^r(\mathbf{u}^r_\Sigma) + \sum_{\alpha=1}^{N_r} q^r_\alpha\, \mathbf{u}^r_\alpha \quad , \tag{112}$$

and the associated test function is then denoted as

$$\delta\mathbf{u}^r = S^r(\delta\mathbf{u}^r_\Sigma) + \sum_{\alpha=1}^{N_r} \delta q^r_\alpha\, \mathbf{u}^r_\alpha \quad . \tag{113}$$

Let $\mathbf{q}^r = (q^r_1, \ldots, q^r_{N_r})$ be the vector of the generalized coordinates. Substituting Eqs. (112) and (113) into Eqs. (97) and (98), we obtain the reduced matrix model of substructure Ω_r with respect to $\{\mathbf{u}^r_\Sigma, \mathbf{q}^r\}$

$$\mathbf{M}^r_{\text{red}} = \begin{bmatrix} \mathbf{M}^r_\Sigma & {}^t\!\mathcal{A}^r \\ \mathcal{A}^r & [\mathcal{M}^r] \end{bmatrix} \quad , \quad \mathbf{K}^r_{\text{red}} = \begin{bmatrix} \mathbf{K}^r_\Sigma & \mathbf{0} \\ \mathbf{0} & [\mathcal{K}^r] \end{bmatrix} \quad , \tag{114}$$

in which $[\mathcal{M}^r]$ and $[\mathcal{K}^r]$ are diagonal matrices defined by Eq. (41) and (42) (for the fixed-interface modes of substructure Ω_r) and the linear operators $\mathbf{M}^r_\Sigma$ and $\mathbf{K}^r_\Sigma$ are defined by [9]

$$<\mathbf{M}^r_\Sigma\, \mathbf{u}^r_\Sigma \,, \delta\mathbf{u}^r_\Sigma> = m^r(\mathbf{S}^r(\mathbf{u}^r_\Sigma)\,, \mathbf{S}^r(\delta\mathbf{u}^r_\Sigma)) \quad , \tag{115}$$

[9] The trace space related to Σ is denoted as $V_\Sigma \subset (H^{1/2}(\partial\Omega_r))^3$. Linear operators $\mathbf{M}^r_\Sigma$ and $\mathbf{K}^r_\Sigma$ are continuous from V_Σ into V'_Σ in which V'_Σ is the dual space of V_Σ. In Eqs. (115) and (116), the angle brackets $<.\,,.>$ denote the duality product between V'_Σ and V_Σ.

$$<\mathbf{K}_\Sigma^r\,\mathbf{u}_\Sigma^r\,,\delta\mathbf{u}_\Sigma^r>= k^r(\mathbf{S}^r(\mathbf{u}_\Sigma^r)\,,\mathbf{S}^r(\delta\mathbf{u}_\Sigma^r)) \quad , \tag{116}$$

where m^r and k^r are defined by Eqs. (97) and (98). It should be noted that these operators are related to surface Γ and correspond to the static condensation on Σ of the mass and stiffness operators (Guyan, 1965). Finally, $\mathcal{A}^r$ is the linear operator defined by

$$(\mathcal{A}^r\,\mathbf{u}_\Sigma^r)\cdot\delta\mathbf{q}^r = \sum_{\alpha=1}^{N_r} m^r(\mathbf{S}^r(\mathbf{u}_\Sigma^r)\,,\mathbf{u}_\alpha^r)\,\delta q_\alpha^r \quad . \tag{117}$$

From Eq. (109), we deduce that

$$(\mathcal{A}^r\,\mathbf{u}_\Sigma^r)\cdot\delta\mathbf{q}^r = -\sum_{\alpha=1}^{N_r} \delta q_\alpha^r \left\{\frac{1}{\lambda_\alpha^r}\int_\Sigma \mathbf{F}_\alpha^r\cdot\mathbf{u}_\Sigma^r\,ds\right\} \quad . \tag{118}$$

Operator ${}^t\!\mathcal{A}^r$ is the adjoint [10] of operator $\mathcal{A}^r$. In conclusion, $\mathbf{M}_{\text{red}}^r$ and $\mathbf{K}_{\text{red}}^r$ defined by Eq. (114) are called the "reduced matrix model" of substructure Ω_r relative to the displacement field $\mathbf{u}_\Sigma^r$ on Σ and to the generalized coordinate vector $\mathbf{q}^r$ (which can be viewed as "internal generalized degrees of freedom").

Construction of the reduced matrix model for structure Ω. Let $\mathbf{u}_\Sigma = \mathbf{u}_{|\Sigma}^1 = \mathbf{u}_{|\Sigma}^2$ be the displacement field on Σ. Substituting Eqs. (112) and (113) into Eq. (99) and using Eq. (114) for $r = 1$ and 2, we obtain the reduced matrix model of the generalized eigenvalue problem for structure Ω:

$$\begin{bmatrix} \mathbf{K}_\Sigma^1+\mathbf{K}_\Sigma^2 & \mathbf{0} & \mathbf{0} \\ \mathbf{0} & [\mathcal{K}^1] & 0 \\ \mathbf{0} & 0 & [\mathcal{K}^2] \end{bmatrix} \begin{bmatrix} \mathbf{u}_\Sigma \\ \mathbf{q}^1 \\ \mathbf{q}^2 \end{bmatrix} = \lambda \begin{bmatrix} \mathbf{M}_\Sigma^1+\mathbf{M}_\Sigma^2 & {}^t\!\mathcal{A}^1 & {}^t\!\mathcal{A}^2 \\ \mathcal{A}^1 & [\mathcal{M}^1] & 0 \\ \mathcal{A}^2 & 0 & [\mathcal{M}^2] \end{bmatrix} \begin{bmatrix} \mathbf{u}_\Sigma \\ \mathbf{q}^1 \\ \mathbf{q}^2 \end{bmatrix} \quad . \tag{119}$$

10.4. Finite element discretization

The numerical implementation of the substructuring technique presented above needs the discretization of two types of variational formulation which are related to the calculation of fixed-interface modes of each substructure (Eq. (101)) and to linear operator $\mathbf{S}^r$ (Eq. (105)), in order to calculate the finite element discretization of the reduced matrix model of each substructure (see Eq. (114)). Below we use a compatible finite element mesh on

---

[10] Linear operator $\mathcal{A}^r$ is continuous from V_Σ into $\mathbb{R}^{N_r}$, in which V_Σ is defined in footnote 9. The adjoint operator ${}^t\!\mathcal{A}^r$ is continuous from $\mathbb{R}^{N_r}$ into the dual space V'_Σ of V_Σ.

interface Σ (for the case of an incompatible mesh, see Farhat and Geradin, 1994 and Ohayon, Sampaio and Soize, 1997).

Calculation of fixed-interface modes. The generalized eigenvalue problem defined by Eq. (101) has to be solved for each substructure. As presented in Section 9, the corresponding finite element discretization can be written as

$$[K_0^r]\,\mathbf{U}^r = \lambda^r\,[M_0^r]\,\mathbf{U}^r \quad , \tag{120}$$

in which $\mathbf{U}^r = (U_1^r, \ldots, U_{n_r}^r)$ is the vector of the DOFs which are the values of the displacement field at the nodes of the finite element mesh of domain Ω_r. Matrices $[M_0^r]$ and $[K_0^r]$ of dimension $(n_r \times n_r)$ are symmetric positive definite. In a first step, the numerical solution of this generalized symmetric matrix eigenvalue problem (see Section 9) gives $\{\lambda_\alpha^r, \mathbf{u}_\alpha^r, \mu_\alpha^r\}$ for $\alpha = 1, \ldots, N_r$, in which coefficients μ_α^r are the generalized masses defined by Eq. (41). We then deduce diagonal matrices $[\mathcal{M}^r]$ and $[\mathcal{K}^r]$. In a second step, for each α, we calculate the finite element discretization of the linear form associated with reaction force field $\mathbf{F}_\alpha^r$ on Σ using Eq. (102) which allows us to calculate matrix $[\mathcal{A}^r]$ from Eq. (118).

Calculation of linear operator $\mathbf{S}^r$. For each substructure, we construct the finite element discretisation $[S^r]$ of operator $\mathbf{S}^r$ defined by Eq. (103). Let $\mathbf{U}_{\text{stat}}^r$ be the vector of the DOFs which are the values of the elastostatic displacement field at the nodes of the finite element mesh of domain Ω_r. We then introduce vectors $\mathbf{U}_1^r$ of the DOFs related to boundary Σ and $\mathbf{U}_2^r$ of the other DOFs. Therefore, we have $\mathbf{U}_{\text{stat}}^r = (\mathbf{U}_1^r, \mathbf{U}_2^r)$. We introduce the following block splitting of the stiffness matrix $[K^r]$ corresponding to Eq. (98)

$$[K^r] = \begin{bmatrix} [K_{11}^r] & [K_{12}^r] \\ [K_{12}^r]^T & [K_{22}^r] \end{bmatrix} \quad . \tag{121}$$

From Eqs. (103) to (105), we deduce that $\mathbf{U}_2^r = -[K_{22}^r]^{-1}\,[K_{12}^r]^T\mathbf{U}_1^r$ and consequently,

$$[S^r] = \begin{bmatrix} [I] \\ -[K_{22}^r]^{-1}[K_{12}^r]^T \end{bmatrix} \quad . \tag{122}$$

From Eqs. (116) and (121), we deduce that

$$[K_\Sigma^r] = [K_{11}^r] - [K_{12}^r]\,[K_{22}^r]^{-1}\,[K_{12}^r]^T \quad , \tag{123}$$

which corresponds to the classical static condensation. Matrix $[M_\Sigma^r]$ is deduced from Eq. (115) and (121) and corresponds to the Guyan condensation.

Calculation of the structural modes of the global structure. By assemblage of the reduced matrix of each substructure, we obtain the finite element discretization of the generalized eigenvalue problem defined by Eq. (119). This generalized symmetric matrix eigenvalue problem is solved as indicated in Section 9 for a free structure (the case now considered).

CHAPTER IV

Dissipative Constitutive Equation for the Master Structure

1. Introduction

In dynamics, the master structure must always be modeled as a dissipative continuum. In Chapter III, for the conservative part of the master structure, we used the linear elasticity theory which allowed us to introduce the structural modes. This was justified by the fact that, in the low-frequency range (see Chapter I), the conservative part of the master structure can be modeled as an elastic continuum. In this chapter, we introduce damping models for the master structure.

In Section 2, we introduce some general notation for the Fourier transform with respect to time and in particular for the displacement field, and the strain and stress tensors.

In Section 3, we introduce the damping model with frequency-independent coefficients based on the linear theory of viscoelasticity without memory.

In Section 4, we present a constitutive equation with frequency-dependent coefficients based on general linear viscoelasticity theory.

Finally, in Section 5, we summarize the viscoelastic constitutive equation which will be used for low- and medium-frequency ranges. This constitutive equation is the sum of an "elastic" part and a "damping" part which are generally frequency-dependent. In the low-frequency range, the "elastic" part is frequency-independent and the "damping part" is frequency-dependent. The model presented in Section 3 for which the damping part is also frequency-independent (viscoelasticity without memory) is a particular case. On the other hand, in the medium-frequency range we use the general model based on linear viscoelasticity theory (with memory) presented in Section 4 for which both the "elastic" and the "damping" parts are frequency-dependent.

2. Notation for the Fourier Transform

Let $\mathbf{u}(\mathbf{x}, t)$ be the displacement field at time t and $\varepsilon_{ij}(\mathbf{u}(\mathbf{x}, t))$ be the strain tensor (see Eq. (III.2)). At each time t, we denote the stress tensor as $\sigma_{ij}(t)$ and we introduce the following notation for the strain tensor

$$\varepsilon_{ij}(t) = \varepsilon_{ij}(\mathbf{u}(\mathbf{x}, t)) \quad , \quad \dot{\varepsilon}_{ij}(t) = \varepsilon_{ij}(\partial_t \mathbf{u}(\mathbf{x}, t)) \quad . \tag{1}$$

As explained in Chapter I, we are interested in the formulations of structural-acoustics and vibration problems in the frequency domain. Therefore, we have to introduce the Fourier transform for various quantities (see Mathematical Notations in the appendix). For the displacement field, and the stress and strain tensors, we use the following simplified notation consisting in using the same symbol for a quantity and its Fourier transform

$$\mathbf{u}(\mathbf{x}, \omega) = \int_{\mathbb{R}} e^{-i\omega t}\, \mathbf{u}(\mathbf{x}, t)\, dt \quad , \tag{2}$$

$$\sigma_{ij}(\omega) = \int_{\mathbb{R}} e^{-i\omega t}\, \sigma_{ij}(t)\, dt \quad , \quad \varepsilon_{ij}(\omega) = \int_{\mathbb{R}} e^{-i\omega t}\, \varepsilon_{ij}(t)\, dt \quad . \tag{3}$$

From Eqs. (1) and (2), we deduce that

$$\dot{\varepsilon}_{ij}(\omega) = i\omega\, \varepsilon_{ij}(\omega) \quad , \tag{4}$$

in which $\dot{\varepsilon}_{ij}(\omega)$ denotes the Fourier transform of $\dot{\varepsilon}_{ij}(t)$.

3. Damping Model with Frequency-Independent Coefficients

The damping model with frequency-independent coefficients is based on the linear theory of viscoelasticity without memory.

Constitutive equation in the time domain. At each time t, symmetric stress tensor $\sigma_{ij}(t)$ is decomposed into an elastic part $\sigma_{ij}^{\text{elas}}(t)$ and a dissipative (damped) part $\sigma_{ij}^{\text{damp}}(t)$ such that

$$\sigma_{ij}(t) = \sigma_{ij}^{\text{elas}}(t) + \sigma_{ij}^{\text{damp}}(t) \quad , \tag{5}$$

in which, using Eq. (1),

$$\sigma_{ij}^{\text{elas}}(t) = a_{ijkh}\, \varepsilon_{kh}(t) \quad , \tag{6}$$

$$\sigma_{ij}^{\text{damp}}(t) = b_{ijkh}\, \dot{\varepsilon}_{kh}(t) \quad . \tag{7}$$

Tensor a_{ijkh} of the elastic coefficients and tensor b_{ijkh} of the damping coefficients of the material depend on **x**, are independent of t and have the usual properties of symmetry and positive definiteness. Denoting as η_{ijkh} either a_{ijkh} or b_{ijkh}, these properties can be written as

$$\eta_{ijkh} = \eta_{jikh} = \eta_{ijhk} = \eta_{khij} \quad , \tag{8}$$

and for all second-order real symmetric tensors X_{ij}

$$\eta_{ijkh}\, X_{kh}\, X_{ij} \geq c\, X_{ij}\, X_{ij} \quad , \tag{9}$$

with $c > 0$. The model defined by Eqs. (5) to (7) is said to be without memory because the stress tensor depends on the strain tensor locally in time (the past history of the strain tensor does not affect the present state of the stress tensor).

Constitutive equation in the frequency domain. Taking the Fourier transform with respect to time t of the two sides of Eqs. (5) to (7) and using Eqs. (2) to (4) yields

$$\sigma_{ij}(\omega) = \sigma_{ij}^{\text{elas}}(\omega) + i\omega\, s_{ij}^{\text{damp}}(\omega) \quad , \tag{10}$$

in which

$$\sigma_{ij}^{\text{elas}}(\omega) = a_{ijkh}\, \varepsilon_{kh}(\omega) \quad , \tag{11}$$

$$s_{ij}^{\text{damp}}(\omega) = b_{ijkh}\, \varepsilon_{kh}(\omega) \quad . \tag{12}$$

This damping model is said to be frequency-independent because coefficients a_{ijkh} and b_{ijkh} depend only on **x** and are independent of ω.

4. Model with Frequency-Dependent Coefficients Based on the Linear Theory of Viscoelasticity

The damping model with frequency-dependent coefficients is based on the linear theory of viscoelasticity with memory in the general case of a material whose constitutive equation is invariant for shifts in the time domain (see Truesdell, 1984; Mandel, 1966; Fung, 1968; Bland, 1960).

Constitutive equation in the time domain. At each **x** fixed in Ω and for all time t in $\mathbb{R}$, symmetric stress tensor $\sigma_{ij}(t)$ is written as

$$\sigma_{ij}(t) = G_{ijkh}(0)\, \varepsilon_{kh}(t) + \int_0^{+\infty} \dot{G}_{ijkh}(\tau)\, \varepsilon_{kh}(t-\tau)\, d\tau \quad , \tag{13}$$

in which $\varepsilon_{ij}(t)$ is the symmetric strain tensor defined by Eq. (1) and real-valued functions $t \mapsto G_{ijkh}(\mathbf{x}, t)$, denoted simply as $G_{ijkh}(t)$, are called the *relaxation functions* at $\mathbf{x}$ in domain Ω. At each $\mathbf{x}$ fixed in Ω and each t fixed in $\mathbb{R}$, tensor G_{ijkh} (and thus $\dot{G}_{ijkh}$) has the properties of symmetry

$$G_{ijkh} = G_{jikh} = G_{ijhk} = G_{khij} \quad . \tag{14}$$

In addition, at $\mathbf{x}$ in Ω, initial value $G_{ijkh}(0)$ of the elasticity tensor must have the property of positive definiteness, i.e., for all second-order real symmetric tensors X_{ij}, we have

$$G_{ijkh}(0)\, X_{kh}\, X_{ij} \geq c\, X_{ij}\, X_{ij} \quad , \tag{15}$$

with $c > 0$. Below (in the subsection relative to the constitutive equation in the frequency domain), we add two assumptions on the relaxation functions which are necessary to obtain a stable mechanical system. Finally, replacing $t-\tau$ by τ in the integrand of Eq. (13), we can rewrite Eq. (13) as

$$\sigma_{ij}(t) = G_{ijkh}(0)\, \varepsilon_{kh}(t) + \int_{-\infty}^{t} \dot{G}_{ijkh}(t-\tau)\, \varepsilon_{kh}(\tau)\, d\tau \quad . \tag{16}$$

The model defined by Eq. (13) or (16) is said to be with memory because the stress tensor at time t depends on the past history $\tau < t$ of the strain tensor.

Integrability properties of the relaxation functions and Fourier transform of their derivative. The relaxation functions are defined on $[0\,,+\infty[$ and differentiable with respect to t on $]0\,,+\infty[$. Their derivatives are denoted as $\dot{G}_{ijkh}(t)$. In addition, we assume that functions $t \mapsto \dot{G}_{ijkh}(t)$ are integrable on $[0\,,+\infty[$. Consequently, for all $\mathbf{x}$ fixed in Ω, $G_{ijkh}(t)$ can be written as

$$G_{ijkh}(t) = G_{ijkh}(0) + \int_{0}^{t} \dot{G}_{ijkh}(\tau)\, d\tau \quad . \tag{17}$$

The limit of $G_{ijkh}(t)$ is finite as t tends to $+\infty$ and is denoted as $G_{ijkh}(\infty)$,

$$G_{ijkh}(\infty) = G_{ijkh}(0) + \int_{0}^{+\infty} \dot{G}_{ijkh}(\tau)\, d\tau \quad . \tag{18}$$

For all $\mathbf{x}$ fixed in Ω, we introduce the functions $t \mapsto g_{ijkh}(\mathbf{x}, t)$ defined on $\mathbb{R}$ and simply denoted as $g_{ijkh}(t)$, such that

$$g_{ijkh}(t) = 0 \quad \text{if} \quad t < 0 \quad , \quad g_{ijkh}(t) = \dot{G}_{ijkh}(t) \quad \text{if} \quad t \geq 0 \quad . \tag{19}$$

Using Eq. (19), Eq. (13) can be rewritten as

$$\sigma_{ij}(t) = G_{ijkh}(0)\,\varepsilon_{kh}(t) + \int_{\mathbb{R}} g_{ijkh}(\tau)\,\varepsilon_{kh}(t-\tau)\,d\tau \quad . \tag{20}$$

Since $t \mapsto g_{ijkh}(t)$ is an integrable function from $\mathbb{R}$ into $\mathbb{R}$, its Fourier transform $\omega \mapsto g_{ijkh}(\omega)$ defined by

$$\begin{aligned} g_{ijkh}(\omega) &= \int_{\mathbb{R}} e^{-i\omega t}\, g_{ijkh}(t)\,dt \\ &= \int_0^{+\infty} e^{-i\omega t}\, \dot{G}_{ijkh}(t)\,dt \quad , \end{aligned} \tag{21}$$

is a continuous function from $\mathbb{R}$ into $\mathbb{C}$ (i.e. it belongs to $C^0(\mathbb{R},\mathbb{C})$), which converges to 0 as $|\omega| \to +\infty$ (see Dautray and Lions, 1992; Soize, 1993a),

$$\{\omega \mapsto g_{ijkh}(\omega)\} \in C^0(\mathbb{R},\mathbb{C}) \quad \text{and} \quad \lim_{|\omega|\to+\infty} |g_{ijkh}(\omega)| = 0 \quad . \tag{22}$$

Constitutive equation in the frequency domain. For all $\mathbf{x}$ fixed in Ω, taking the Fourier transform with respect to time t of the two sides of Eq. (13), rewritten as Eq. (20), and using Eqs. (3) and (21) yields

$$\sigma_{ij}(\omega) = (G_{ijkh}(0) + g_{ijkh}(\omega))\,\varepsilon_{kh}(\omega) \quad . \tag{23}$$

From Eq. (23), introducing the real part $g^R_{ijkh}(\omega)$ and the imaginary part $g^I_{ijkh}(\omega)$ of $g_{ijkh}(\omega)$

$$g_{ijkh}(\omega) = g^R_{ijkh}(\omega) + i\; g^I_{ijkh}(\omega) \quad , \tag{24}$$

we deduce that, for all $\mathbf{x}$ fixed in Ω, stress tensor $\sigma_{ij}(\omega)$ can be written in the frequency domain as

$$\sigma_{ij}(\omega) = \sigma^{\text{elas}}_{ij}(\omega) + i\omega\, s^{\text{damp}}_{ij}(\omega) \quad , \tag{25}$$

in which

$$\sigma^{\text{elas}}_{ij}(\omega) = a_{ijkh}(\omega)\;\varepsilon_{kh}(\omega) \quad , \tag{26}$$

$$s^{\text{damp}}_{ij}(\omega) = b_{ijkh}(\omega)\;\varepsilon_{kh}(\omega) \quad . \tag{27}$$

Elastic coefficients $a_{ijkh}(\omega)$ and damping coefficients $b_{ijkh}(\omega)$ depend on $\mathbf{x}$ and ω (as above, the dependence on $\mathbf{x}$ has been omitted for brevity) and are such that

$$a_{ijkh}(\omega) = G_{ijkh}(0) + g^R_{ijkh}(\omega) \quad , \tag{28}$$

$$\omega\ b_{ijkh}(\omega) = g^I_{ijkh}(\omega) \quad . \tag{29}$$

From Eq. (22), we deduce that

$$\lim_{|\omega|\to+\infty} a_{ijkh}(\omega) = G_{ijkh}(0) \quad , \tag{30}$$

$$\lim_{|\omega|\to+\infty} \omega\ b_{ijkh}(\omega) = 0 \quad . \tag{31}$$

Denoting as $\eta_{ijkh}(\omega)$ either $a_{ijkh}(\omega)$ or $b_{ijkh}(\omega)$, from Eq. (14), we deduce that the elastic and damping coefficients have the following symmetry properties

$$\eta_{ijkh}(\omega) = \eta_{jikh}(\omega) = \eta_{ijhk}(\omega) = \eta_{khij}(\omega) \quad . \tag{32}$$

Finally, for all fixed real ω, relaxation functions G_{ijkh} must be such that the elastic and damping tensors $a_{ijkh}(\omega)$ and $b_{ijkh}(\omega)$ are positive definite, i.e., for all second-order real symmetric tensors X_{ij},

$$\eta_{ijkh}(\omega)\, X_{kh}\, X_{ij} \geq c(\omega)\, X_{ij}\, X_{ij} \quad , \tag{33}$$

in which, for all real ω, positive constant $c(\omega)$ is such that $c(\omega) \geq c_0 > 0$ where c_0 is a positive real constant independent of ω. This damping model is said to be frequency-dependent because coefficients $a_{ijkh}(\omega)$ and $b_{ijkh}(\omega)$ depend on $\mathbf{x}$ and ω.

Simple example of a constitutive equation in the frequency domain. Let us consider a linear differential equation on σ_{ij} and ε_{ij} in the time domain. We limit the presentation to the following first-order linear differential operator in time

$$C_{ijkh}\,\sigma_{kh}(t) + \dot{\sigma}_{ij}(t) = A_{ijkh}\,\varepsilon_{kh}(t) + B_{ijkh}\,\dot{\varepsilon}_{kh}(t) \quad , \tag{34}$$

which can be denoted using global tensor notation as

$$\mathbf{C}\,\sigma(t) + \mathbf{I}\,\dot{\sigma}(t) = \mathbf{A}\,\varepsilon(t) + \mathbf{B}\,\dot{\varepsilon}(t) \quad , \tag{35}$$

in which $\mathbf{I}$ is the fourth-order identity tensor and where $\mathbf{A}$, $\mathbf{B}$ and $\mathbf{C}$ are time-independent fourth-order tensors. We have to find the appropriate initial condition at time $t = 0$ such that the solution of Eq. (35) for $t > 0$ yields a constitutive equation of the type defined by Eq. (13) (in the time domain), or equivalently, by Eqs. (25) to (27) (in the frequency domain). For real numbers ξ and ω, introducing the complex number $p = \xi + i\omega$ and taking the Laplace transform of the two sides of Eq. (35) yields,

$$\mathbf{C}\,\widetilde{\sigma}(p) + \mathbf{I}\,(p\,\widetilde{\sigma}(p) - \sigma(0)) = \mathbf{A}\,\widetilde{\varepsilon}(p) + \mathbf{B}\,(p\,\widetilde{\varepsilon}(p) - \varepsilon(0)) \quad , \tag{36}$$

in which

$$\widetilde{\sigma}(p) = \int_0^{+\infty} e^{-pt}\,\sigma(t)\,dt \quad , \quad \widetilde{\varepsilon}(p) = \int_0^{+\infty} e^{-pt}\,\varepsilon(t)\,dt \quad . \tag{37}$$

In this Laplace transform calculus, we assume that p belongs to the right-half complex plane ($\xi > \xi_0$) in which $\widetilde{\sigma}(p)$ and $\widetilde{\varepsilon}(p)$ are defined. To obtain a linear relation between $\widetilde{\sigma}(p)$ and $\widetilde{\varepsilon}(p)$ (transfer function), Eq. (36) shows that the initial condition must satisfy the following equation

$$\sigma(0) = \mathbf{B}\,\varepsilon(0) \quad , \tag{38}$$

and therefore Eq. (36) yields

$$(\mathbf{C} + p\,\mathbf{I})\,\widetilde{\sigma}(p) = (\mathbf{A} + p\,\mathbf{B})\,\widetilde{\varepsilon}(p) \quad . \tag{39}$$

We assume that fourth-order tensor $\mathbf{C}$ is such that, for all fixed real ω, tensor $(\mathbf{C} + i\omega\,\mathbf{I})$ is invertible. Consequently, we obtain in the frequency domain

$$\sigma(\omega) = (\mathbf{C} + i\omega\,\mathbf{I})^{-1}\,(\mathbf{A} + i\omega\,\mathbf{B})\,\varepsilon(\omega) \quad . \tag{40}$$

Identifying Eq. (40) with Eqs. (25) to (27) yields

$$\mathbf{a}(\omega) = \Re e\left\{(\mathbf{C} + i\omega\,\mathbf{I})^{-1}\,(\mathbf{A} + i\omega\,\mathbf{B})\right\} \quad , \tag{41}$$

$$\omega\,\mathbf{b}(\omega) = \Im m\left\{(\mathbf{C} + i\omega\,\mathbf{I})^{-1}\,(\mathbf{A} + i\omega\,\mathbf{B})\right\} \quad , \tag{42}$$

in which $\mathbf{a}(\omega)$ and $\mathbf{b}(\omega)$ are the fourth-order tensors $a_{ijkh}(\omega)$ and $b_{ijkh}(\omega)$ introduced in Eqs. (26) and (27). It should be noted that tensors $\mathbf{A}$, $\mathbf{B}$ and $\mathbf{C}$ must be such that $a_{ijkh}(\omega)$ and $b_{ijkh}(\omega)$ satisfy the symmetry and positivity properties defined by Eqs. (32) and (33).

5. Summary

Sections 3 and 4 can be summarized as follows. In the frequency domain, the general constitutive equation is written as

$$\sigma_{ij}(\omega) = \sigma_{ij}^{\text{elas}}(\omega) + i\omega\, s_{ij}^{\text{damp}}(\omega) \quad , \tag{43}$$

where $\sigma_{ij}(\omega)$ has complex values and in which

$$\sigma_{ij}^{\text{elas}}(\omega) = a_{ijkh}(\omega)\ \varepsilon_{kh}(\omega) \quad , \tag{44}$$

$$s_{ij}^{\text{damp}}(\omega) = b_{ijkh}(\omega)\ \varepsilon_{kh}(\omega) \quad . \tag{45}$$

Equations (43) to (45) can be rewritten as

$$\sigma_{ij}(\omega) = (a_{ijkh}(\omega) + i\omega\, b_{ijkh}(\omega))\, \varepsilon_{kh}(\omega) \quad . \tag{46}$$

For each real ω and at each $\mathbf{x}$ in domain Ω, tensors $a_{ijkh}(\omega)$ and $b_{ijkh}(\omega)$ must satisfy the symmetry properties

$$a_{ijkh}(\omega) = a_{jikh}(\omega) = a_{ijhk}(\omega) = a_{khij}(\omega) \quad , \tag{47}$$

$$b_{ijkh}(\omega) = b_{jikh}(\omega) = b_{ijhk}(\omega) = b_{khij}(\omega) \quad , \tag{48}$$

and the positive-definiteness properties, i.e., for all fixed real ω, for all second-order real symmetric tensors X_{ij},

$$a_{ijkh}(\omega)\, X_{kh}\, X_{ij} \geq c_a(\omega)\, X_{ij}\, X_{ij} \quad , \tag{49}$$

$$b_{ijkh}(\omega)\, X_{kh}\, X_{ij} \geq c_b(\omega)\, X_{ij}\, X_{ij} \quad . \tag{50}$$

in which, for all real ω, positive constants $c_a(\omega)$ and $c_b(\omega)$ are such that $c_a(\omega) \geq c_0 > 0$ and $c_b(\omega) \geq c_0 > 0$ where c_0 is a positive real constant independent of ω.

5.1. LF range: constitutive equation with frequency-independent elastic coefficients and frequency-dependent damping coefficients

This model, described in Section 3, is given by Eqs. (46) to (50) in which tensor a_{ijkh}, related to the elastic stress tensor $\sigma_{ij}^{\text{elas}}$, is independent of ω (but depends on $\mathbf{x}$) and tensor $b_{ijkh}(\omega)$, related to the damping stress tensor $s_{ij}^{\text{damp}}(\omega)$, depends on ω (and on $\mathbf{x}$). In addition, at each $\mathbf{x}$ in domain Ω, function $\omega \mapsto b_{ijkh}(\omega)$ must be continuous on $\mathbb{R}$ and, for $|\omega| \to +\infty$, must satisfy the asymptotic properties defined by Eq. (31). It should be noted that the damping model with frequency-independent coefficients presented in Section 3 is a particular case for which Eq. (31) does not need to be satisfied (as proved in Section 3).

5.2. MF range: constitutive equation with frequency-dependent coefficients

This model, described in Section 4, is given by Eqs. (46) to (50) in which tensors $a_{ijkh}(\omega)$ and $b_{ijkh}(\omega)$ depend on ω (and on $\mathbf{x}$). In addition, at each $\mathbf{x}$ in domain Ω, functions $\omega \mapsto a_{ijkh}(\omega)$ and $\omega \mapsto b_{ijkh}(\omega)$ must be continuous on $\mathbb{R}$ and, for $|\omega| \rightarrow +\infty$, must satisfy the asymptotic properties defined by Eqs. (30) and (31).

CHAPTER V

Master Structure Frequency Response Function

1. Introduction

In Chapter III, we introduced the equations of the vibrations of a conservative structure in terms of the structural displacement field. Using the test function method, we presented the variational formulation and introduced the structural mass and stiffness operators and studied their properties. In this chapter we present the variational formulation of the vibrations of the master structure submitted to given forces, for which the damping effects are modeled using the dissipative constitutive equation introduced in Chapter IV for the low- and medium-frequency ranges. Finally, we introduce the operator-valued Frequency Response Function (FRF) which is intrinsic to the structure. This operator allows calculation of the dynamical response of the structure for arbitrary deterministic or random excitations.

In Section 2, we introduce the equations of the master structure in the frequency domain.

In Sections 3 and 4, we present the variational formulation of these equations and the corresponding linear operator equation. We introduce the structural mass, damping and stiffness operators.

In Section 5, we introduce the frequency response function which is explicitly constructed in Chapters VI and VII for the LF and MF ranges respectively. In Section 6, we introduce the corresponding finite element approximation.

In Section 7, we consider the coupling between two substructures on an interface. The problem consists in eliminating the displacement field in one substructure as a function of the displacement field at the interface. For this purpose, we construct the boundary impedance operator which relates the value of the velocity field on the interface with the force field on this interface for this substructure.

2. Equations in the Frequency Domain

2.1. Master structure configuration

We consider linear vibrations of the master structure around a position of static equilibrium taken as reference configuration Ω (for the sake of brevity, we do not consider a reference configuration with prestresses: we use the natural state at rest). The physical space $\mathbb{R}^3$ is referred to a Cartesian reference system $(\mathbf{i}, \mathbf{j}, \mathbf{k})$ and we denote the generic point of $\mathbb{R}^3$ as $\mathbf{x} = (x_1, x_2, x_3)$. Let Ω be a three-dimensional bounded connected domain of $\mathbb{R}^3$ occupied by the master structure at equilibrium with a smooth boundary $\partial\Omega$. The external unit normal to $\partial\Omega$ is denoted as $\mathbf{n}$ (see Fig. 1).

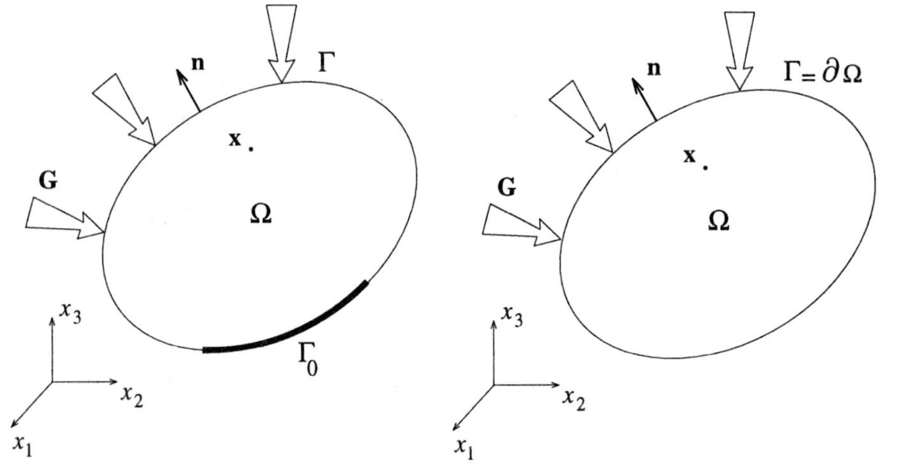

a- Master structure fixed on Γ_0 b- Free master structure

Fig. 1. Master structure configuration

Let $\mathbf{u}(\mathbf{x}, t) = (u_1(\mathbf{x}, t), u_2(\mathbf{x}, t), u_3(\mathbf{x}, t))$ be the displacement of a particle located at point $\mathbf{x}$ in Ω and at a time t. As in Chapter III, two cases are considered.

- The master structure is fixed on a part Γ_0 of boundary $\partial\Omega$ ($\mathbf{u} = \mathbf{0}$ on Γ_0), a given surface force field $\mathbf{G}(\mathbf{x}, t) = (G_1(\mathbf{x}, t), G_2(\mathbf{x}, t), G_3(\mathbf{x}, t))$ is applied to the part $\Gamma = \partial\Omega\backslash\Gamma_0$ and a given body force field $\mathbf{g}(\mathbf{x}, t) = (g_1(\mathbf{x}, t), g_2(\mathbf{x}, t), g_3(\mathbf{x}, t))$ is applied in Ω (see Fig. 1-a).

- The master structure is free ($\Gamma_0 = \emptyset$), a given surface force field $\mathbf{G}(\mathbf{x}, t)$ is applied to the total boundary $\Gamma = \partial\Omega$ and a given body force field $\mathbf{g}(\mathbf{x}, t)$ is applied in Ω (see Fig. 1-b).

2.2. Equations in the time domain

For any function $a(\mathbf{x}, t)$, we use the notation $a_{,j}(\mathbf{x}, t) = \partial a(\mathbf{x}, t)/\partial x_j$, $\partial_t a(\mathbf{x}, t) = \partial a(\mathbf{x}, t)/\partial t$ and $\partial_t^2 a(\mathbf{x}, t) = \partial^2 a(\mathbf{x}, t)/\partial t^2$. We also use the classical convention for summations over repeated Latin indices. The elastodynamic equation is written as

$$\rho\, \partial_t^2 u_i(\mathbf{x}, t) - \sigma_{ij,j}(\mathbf{x}, t) = g_i(\mathbf{x}, t) \quad \text{in} \quad \Omega \quad , \quad t > 0 \quad , \tag{1}$$

in which $\rho(\mathbf{x})$ is the mass density field on Ω at equilibrium and $\sigma_{ij}(\mathbf{x}, t)$ is the stress tensor. The constitutive equation in the frequency domain is defined in the next subsection. Concerning the boundary conditions, since there is a given surface force field $\mathbf{G}(\mathbf{x}, t)$ applied to Γ,

$$\sigma_{ij}(\mathbf{x}, t)\, n_j(\mathbf{x}) = G_i(\mathbf{x}, t) \quad \text{on} \quad \Gamma \quad . \tag{2}$$

In addition, if the master structure is fixed on Γ_0, we have

$$\mathbf{u}(\mathbf{x}, t) = \mathbf{0} \quad \text{on} \quad \Gamma_0 \quad . \tag{3}$$

For a free master structure ($\Gamma_0 = \emptyset$), the boundary condition defined by Eq. (3) does not exist and Eq. (2) holds on the total boundary $\Gamma = \partial\Omega$.

2.3. Fourier transform notation

As explained in Chapter IV, in order to obtain the equations in the frequency domain, we introduce the Fourier transform for various quantities using the same symbol for a quantity and its Fourier transform (see Mathematical Notations in the appendix). For all fixed $\mathbf{x}$, body force $\mathbf{g}(\mathbf{x}, t)$, surface force $\mathbf{G}(\mathbf{x}, t)$, displacement $\mathbf{u}(\mathbf{x}, t)$ and stress tensor $\sigma_{ij}(\mathbf{x}, t)$ have the following Fourier transforms with respect to t

$$\mathbf{g}(\mathbf{x}, \omega) = \int_{\mathbb{R}} e^{-i\omega t}\, \mathbf{g}(\mathbf{x}, t)\, dt \quad , \quad \mathbf{G}(\mathbf{x}, \omega) = \int_{\mathbb{R}} e^{-i\omega t}\, \mathbf{G}(\mathbf{x}, t)\, dt \quad , \tag{4}$$

$$\mathbf{u}(\mathbf{x}, \omega) = \int_{\mathbb{R}} e^{-i\omega t}\, \mathbf{u}(\mathbf{x}, t)\, dt \quad , \quad \sigma_{ij}(\mathbf{x}, \omega) = \int_{\mathbb{R}} e^{-i\omega t}\, \sigma_{ij}(\mathbf{x}, t)\, dt \quad . \tag{5}$$

It should be noted that these Fourier transforms are complex-valued vectors and tensors.

2.4. Equations in the frequency domain

Taking the Fourier transform of Eqs. (1) to (3) and using Eqs. (4) and (5), we obtain, for all real ω,

$$-\omega^2 \rho u_i(\mathbf{x}, \omega) - \sigma_{ij,j}(\mathbf{x}, \omega) = g_i(\mathbf{x}, \omega) \quad \text{in} \quad \Omega \quad , \tag{6}$$

$$\sigma_{ij}(\mathbf{x}, \omega) n_j(\mathbf{x}) = G_i(\mathbf{x}, \omega) \quad \text{on} \quad \Gamma \quad , \tag{7}$$

and, if the master structure is fixed on Γ_0,

$$\mathbf{u}(\mathbf{x}, \omega) = \mathbf{0} \quad \text{on} \quad \Gamma_0 \quad . \tag{8}$$

For a free master structure ($\Gamma_0 = \emptyset$ and $\Gamma = \partial\Omega$), the boundary condition defined by Eq. (8) does not exist. In the frequency domain, the constitutive equation of the master structure (which is dissipative) are given by Eqs. (IV.43) to (IV.50).

2.5. Boundary value problem in the frequency domain

The Fourier transform of the strain tensor defined by Eq. (III.2) is written as

$$\varepsilon_{kh}(\mathbf{u}) = \frac{1}{2}(u_{k,h}(\mathbf{x}, \omega) + u_{h,k}(\mathbf{x}, \omega)) \quad . \tag{9}$$

From Eqs. (IV.43) to (IV.45) defining the constitutive equation of the master structure and using Eq. (9), we deduce that

$$\sigma_{ij}^{\text{elas}}(\mathbf{u}) = a_{ijkh}(\omega)\, \varepsilon_{kh}(\mathbf{u}) \quad , \tag{10}$$

$$s_{ij}^{\text{damp}}(\mathbf{u}) = b_{ijkh}(\omega)\, \varepsilon_{kh}(\mathbf{u}) \quad , \tag{11}$$

$$\sigma_{ij}(\mathbf{u}) = \sigma_{ij}^{\text{elas}}(\mathbf{u}) + i\omega\, s_{ij}^{\text{damp}}(\mathbf{u}) \quad , \tag{12}$$

in which $a_{ijkh}(\omega)$ and $b_{ijkh}(\omega)$ depend on $\mathbf{x}$ and ω and satisfy the properties introduced in Section IV.5 (the dependence on $\mathbf{x}$ has been omitted for brevity). Substituting Eq. (12) into Eqs. (6) and (7), we obtain

$$-\omega^2 \rho u_i - \sigma_{ij,j}(\mathbf{u}) = g_i \quad \text{in} \quad \Omega \quad , \tag{13}$$

$$\sigma_{ij}(\mathbf{u}) n_j = G_i \quad \text{on} \quad \Gamma \quad , \tag{14}$$

and if the master structure is fixed on Γ_0,

$$\mathbf{u} = \mathbf{0} \quad \text{on} \quad \Gamma_0 \quad . \tag{15}$$

For a free master structure ($\Gamma_0 = \emptyset$ and $\Gamma = \partial\Omega$), the boundary condition defined by Eq. (15) does not exist. The problem can be summarized as follows. For each real ω and at each $\mathbf{x}$ in Ω, find complex-valued vector $\mathbf{u}(\mathbf{x}, \omega)$ as a solution of Eqs. (13) and (14) with the additional boundary condition defined by Eq. (15) if the master structure is fixed on Γ_0.

3. Variational Formulation

The variational formulation of the damped master structure equations in the frequency domain is constructed using the test-function method presented in Sections 4 and 5 of Chapter III for a conservative elastic structure whose principles are explained in Chapter II. As the solution $\mathbf{u}$ is a complex-valued vector, it is necessary to introduce the admissible function space of complex functions $\mathcal{C}^c$ and $\mathcal{C}_0^c$ instead of the admissible function space of real functions $\mathcal{C}$ and $\mathcal{C}_0$. Consequently, all the linear and bilinear forms related to real vector space $\mathcal{C}$ or $\mathcal{C}_0$ are replaced by the corresponding antilinear and sesquilinear forms related to complex vector space $\mathcal{C}^c$ or $\mathcal{C}_0^c$ (see Mathematical Notations in the appendix).

3.1. Construction of the variational formulation

We introduce the complex vector space $\mathcal{C}^c$ of sufficiently differentiable [1] functions defined on Ω with values in $\mathbb{C}^3$. Multiplying Eq. (13) by $\overline{\delta\mathbf{u}} \in \mathcal{C}^c$ and integrating over domain Ω yields

$$-\omega^2 \int_\Omega \rho\, u_i\, \overline{\delta u_i}\, d\mathbf{x} - \int_\Omega \sigma_{ij,j}(\mathbf{u})\, \overline{\delta u_i}\, d\mathbf{x} = \int_\Omega g_i\, \overline{\delta u_i}\, d\mathbf{x} \quad . \tag{16}$$

The second integral on the left-hand side of Eq. (16) is transformed using the same calculations performed in Section III.4.1. We then obtain

$$-\omega^2\int_\Omega \rho\, u_i\, \overline{\delta u_i}\, d\mathbf{x} + \int_\Omega \sigma_{ij}(\mathbf{u})\, \varepsilon_{ij}(\overline{\delta\mathbf{u}})\, d\mathbf{x} - \int_{\partial\Omega} \sigma_{ij}(\mathbf{u})\, n_j\, \overline{\delta u_i}\, ds = \int_\Omega g_i\, \overline{\delta u_i}\, d\mathbf{x} \,. \tag{17}$$

Finally, we introduce the *admissible function space.* Two cases must be distinguished.

Structure fixed on Γ_0. In this case, $\Gamma_0 \neq \emptyset$ and the admissible function space must satisfy the constraint on $\mathbf{u}$ defined by Eq. (15). Consequently, this admissible function space is the subspace $\mathcal{C}_0^c$ of $\mathcal{C}^c$ defined by [2]

$$\mathcal{C}_0^c = \{\, \mathbf{u} \in \mathcal{C}^c \quad ; \quad \mathbf{u} = \mathbf{0} \quad \text{on} \quad \Gamma_0 \,\} \quad . \tag{18}$$

Let $\delta\mathbf{u}$ be any test function in $\mathcal{C}_0^c \subset \mathcal{C}^c$. Taking into account Eq. (14) and since $\delta\mathbf{u} = \mathbf{0}$ on Γ_0, for all $\delta\mathbf{u}$ in $\mathcal{C}_0^c$, Eq. (17) yields

$$-\omega^2 \int_\Omega \rho\, u_i\, \overline{\delta u_i}\, d\mathbf{x} + \int_\Omega \sigma_{ij}(\mathbf{u})\, \varepsilon_{ij}(\overline{\delta\mathbf{u}})\, d\mathbf{x} = \int_\Omega g_i\, \overline{\delta u_i}\, d\mathbf{x} + \int_\Gamma G_i\, \overline{\delta u_i}\, ds \,. \tag{19}$$

---

[1] Space $\mathcal{C}^c$ is the Sobolev space $(H^1(\Omega))^3$.

[2] Space $\mathcal{C}_0^c$ is the Sobolev space $\{\mathbf{u} \in (H^1(\Omega))^3 \,, \mathbf{u}=0 \text{ on } \Gamma_0\}$.

Free structure. In this case, $\Gamma_0 = \emptyset$ and Eq. (15) does not exist. Therefore, the admissible function space of the problem is $\mathcal{C}^c$ (there is no constraint on $\mathbf{u}$). Let $\delta\mathbf{u}$ be any test function in $\mathcal{C}^c$. Taking into account Eq. (14) and for all $\delta\mathbf{u}$ in $\mathcal{C}^c$, Eq. (17) yields

$$-\omega^2 \int_\Omega \rho\, u_i\, \overline{\delta u_i}\, d\mathbf{x} + \int_\Omega \sigma_{ij}(\mathbf{u})\, \varepsilon_{ij}(\overline{\delta \mathbf{u}})\, d\mathbf{x} = \int_\Omega g_i\, \overline{\delta u_i}\, d\mathbf{x} + \int_\Gamma G_i\, \overline{\delta u_i}\, ds\,. \quad (20)$$

3.2. Variational formulation

Structure fixed on Γ_0. Substituting Eq. (12) into Eq. (19), the variational formulation of the boundary value problem defined by Eqs. (13), (14) and (15) is stated as follows. For all fixed real ω, find $\mathbf{u}$ in $\mathcal{C}_0^c$ such that, for all $\delta\mathbf{u}$ in $\mathcal{C}_0^c$,

$$-\omega^2 \int_\Omega \rho\; \mathbf{u} \cdot \overline{\delta \mathbf{u}}\; d\mathbf{x} + i\omega \int_\Omega s_{ij}^{\text{damp}}(\mathbf{u})\, \varepsilon_{ij}(\overline{\delta \mathbf{u}})\, d\mathbf{x} + \int_\Omega \sigma_{ij}^{\text{elas}}(\mathbf{u})\, \varepsilon_{ij}(\overline{\delta \mathbf{u}})\, d\mathbf{x}$$
$$= \int_\Omega \mathbf{g} \cdot \overline{\delta \mathbf{u}}\; d\mathbf{x} + \int_\Gamma \mathbf{G} \cdot \overline{\delta \mathbf{u}}\; ds \quad , \qquad (21)$$

in which $\sigma_{ij}^{\text{elas}}(\mathbf{u})$ and $s_{ij}^{\text{damp}}(\mathbf{u})$ are defined by Eqs. (10) and (11) respectively.

Free structure. Substituting Eq. (12) into Eq. (20), the variational formulation of the boundary value problem defined by Eqs. (13) and (14) is stated as follows. For all fixed real ω, find $\mathbf{u}$ in $\mathcal{C}^c$ such that, for all $\delta\mathbf{u}$ in $\mathcal{C}^c$,

$$-\omega^2 \int_\Omega \rho\; \mathbf{u} \cdot \overline{\delta \mathbf{u}}\; d\mathbf{x} + i\omega \int_\Omega s_{ij}^{\text{damp}}(\mathbf{u})\, \varepsilon_{ij}(\overline{\delta \mathbf{u}})\, d\mathbf{x} + \int_\Omega \sigma_{ij}^{\text{elas}}(\mathbf{u})\, \varepsilon_{ij}(\overline{\delta \mathbf{u}})\, d\mathbf{x}$$
$$= \int_\Omega \mathbf{g} \cdot \overline{\delta \mathbf{u}}\; d\mathbf{x} + \int_\Gamma \mathbf{G} \cdot \overline{\delta \mathbf{u}}\; ds \quad , \qquad (22)$$

in which $\sigma_{ij}^{\text{elas}}(\mathbf{u})$ and $s_{ij}^{\text{damp}}(\mathbf{u})$ are defined by Eqs. (10) and (11) respectively.

4. Linear Operators and Algebraic Properties

We introduce the linear operator equation corresponding to the variational formulation defined by Eq. (21) or Eq. (22).

4.1. Mass operator

We introduce the sesquilinear form $m(\mathbf{u}, \delta\mathbf{u})$ on $\mathcal{C}^c \times \mathcal{C}^c$, called the structural mass sesquilinear form, defined by

$$m(\mathbf{u}, \delta\mathbf{u}) = \int_\Omega \rho\, \mathbf{u} \cdot \overline{\delta\mathbf{u}}\, d\mathbf{x} \quad , \tag{23}$$

which is Hermitian,

$$m(\mathbf{u}, \delta\mathbf{u}) = \overline{m(\delta\mathbf{u}, \mathbf{u})} \quad , \tag{24}$$

and which is positive definite because $\rho(\mathbf{x}) > 0$ (for all $\mathbf{u} \neq \mathbf{0}$, we have $m(\mathbf{u}, \mathbf{u}) > 0$). This result holds for the fixed and free cases. The linear operator $\mathbf{M}$, called the *mass operator*, is defined by [3]

$$<\mathbf{M}\mathbf{u}, \delta\mathbf{u}> = m(\mathbf{u}, \delta\mathbf{u}) \quad . \tag{25}$$

4.2. Stiffness operator

Using Eq. (10), for all fixed real ω, we introduce the sesquilinear form $k(\omega; \mathbf{u}, \delta\mathbf{u})$ defined on $\mathcal{C}^c \times \mathcal{C}^c$ and called the structural stiffness sesquilinear form, defined by

$$\begin{aligned} k(\omega; \mathbf{u}, \delta\mathbf{u}) &= \int_\Omega \sigma_{ij}^{\text{elas}}(\mathbf{u})\, \varepsilon_{ij}(\overline{\delta\mathbf{u}})\, d\mathbf{x} \\ &= \int_\Omega a_{ijkh}(\omega)\, \varepsilon_{kh}(\mathbf{u})\, \varepsilon_{ij}(\overline{\delta\mathbf{u}}) d\mathbf{x} \quad . \end{aligned} \tag{26}$$

From the properties defined by Eqs. (IV.47) and (IV.49), we deduce that sesquilinear form $k(\omega; \mathbf{u}, \delta\mathbf{u})$ is Hermitian:

$$k(\omega; \mathbf{u}, \delta\mathbf{u}) = \overline{k(\omega; \delta\mathbf{u}, \mathbf{u})} \quad . \tag{27}$$

The fixed and free cases must be distinguished.

Structure fixed on Γ_0. For all fixed real ω, the restriction of sequilinear form $k(\omega; \mathbf{u}, \delta\mathbf{u})$ to $\mathcal{C}_0^c \times \mathcal{C}_0^c$ is positive definite,

$$k(\omega; \mathbf{u}, \mathbf{u}) > 0 \quad , \quad \forall \mathbf{u} \neq \mathbf{0} \quad . \tag{28}$$

Sesquilinear form $m(\mathbf{u},\delta\mathbf{u})$ is continuous on $H^c \times H^c$ with $H^c = (L^2(\Omega))^3$ and is also continuous on $\mathcal{C}^c \times \mathcal{C}^c$. This sesquilinear form defines a continuous operator $\mathbf{M}$ from $\mathcal{C}^c$ into its antidual space $\mathcal{C}^{c\prime}$ and in Eq. (25), the angle brackets denote the antiduality product between $\mathcal{C}^{c\prime}$ and $\mathcal{C}^c$.

Stiffness operator $\mathbf{K}(\omega)$ is defined by [4]

$$<\mathbf{K}(\omega)\,\mathbf{u}\,,\delta\mathbf{u}> = k(\omega\,;\mathbf{u}\,,\delta\mathbf{u}) \quad . \tag{29}$$

Free structure. For all fixed real ω, sesquilinear form $k(\omega\,;\mathbf{u}\,,\delta\mathbf{u})$ defined on $\mathcal{C}^c \times \mathcal{C}^c$ is positive semidefinite

$$k(\omega\,;\mathbf{u}\,,\mathbf{u}) \geq 0 \quad , \quad \forall \mathbf{u} \neq \mathbf{0} \quad . \tag{30}$$

Stiffness operator $\mathbf{K}(\omega)$ is then defined by [5]

$$<\mathbf{K}(\omega)\,\mathbf{u}\,,\delta\mathbf{u}> = k(\omega\,;\mathbf{u}\,,\delta\mathbf{u}) \quad . \tag{31}$$

4.3. Damping operator

For the "damping" part of the model, we proceed as for the "elastic" part (stiffness operator). Using Eq. (11), for all fixed real ω, we introduce the sesquilinear form $d(\omega\,;\mathbf{u}\,,\delta\mathbf{u})$ defined on $\mathcal{C}^c \times \mathcal{C}^c$ and called the structural damping sesquilinear form, defined by

$$\begin{aligned} d(\omega\,;\mathbf{u}\,,\delta\mathbf{u}) &= \int_\Omega s_{ij}^{\mathrm{damp}}(\mathbf{u})\,\varepsilon_{ij}(\overline{\delta\mathbf{u}})\,d\mathbf{x} \\ &= \int_\Omega b_{ijkh}(\omega)\,\varepsilon_{kh}(\mathbf{u})\,\varepsilon_{ij}(\overline{\delta\mathbf{u}})\,d\mathbf{x} \quad . \end{aligned} \tag{32}$$

From the properties defined by Eqs. (IV.48) and (IV.50), we deduce that sesquilinear form $d(\omega\,;\mathbf{u}\,,\delta\mathbf{u})$ is Hermitian,

$$d(\omega\,;\mathbf{u}\,,\delta\mathbf{u}) = \overline{d(\omega\,;\delta\mathbf{u}\,,\mathbf{u})} \quad . \tag{33}$$

The fixed and free cases must be distinguished.

Structure fixed on Γ_0. For all fixed real ω, the restriction of sesquilinear form $d(\omega\,;\mathbf{u}\,,\delta\mathbf{u})$ to $\mathcal{C}_0^c \times \mathcal{C}_0^c$ is positive definite,

$$d(\omega\,;\mathbf{u}\,,\mathbf{u}) > 0 \quad , \quad \forall \mathbf{u} \neq \mathbf{0} \quad . \tag{34}$$

---

[4] Sesquilinear form $k(\omega\,;\mathbf{u},\delta\mathbf{u})$ is continuous on $\mathcal{C}_0^c \times \mathcal{C}_0^c$, coercive on $\mathcal{C}_0^c$ due to Eq. (IV.49). This sesquilinear form defines a continuous operator $\mathbf{K}(\omega)$ from $\mathcal{C}_0^c$ into its antidual space $\mathcal{C}_0^{c\,\prime}$ and in Eq. (29), the angle brackets denote the antiduality product between $\mathcal{C}_0^{c\,\prime}$ and $\mathcal{C}_0^c$.

[5] See footnote 4, replacing $\mathcal{C}_0^c$ by $\mathcal{C}^c$. It should be noted that $k(\omega\,;\mathbf{u},\mathbf{u})$ is coercive on $\mathcal{C}^c \backslash \mathcal{C}_{\mathrm{rig}}$ in which $\mathcal{C}_{\mathrm{rig}} \subset \mathcal{C}$ is the set of the rigid body displacement fields (see Section III.5.2).

Damping operator $\mathbf{D}(\omega)$ is defined by [6]

$$<\mathbf{D}(\omega)\,\mathbf{u}\,,\delta\mathbf{u}> = d(\omega\,;\mathbf{u}\,,\delta\mathbf{u}) \quad . \tag{35}$$

Free structure. For all fixed real ω, sesquilinear bilinear form $d(\omega\,;\mathbf{u}\,,\delta\mathbf{u})$ defined on $\mathcal{C}^c \times \mathcal{C}^c$ is positive semidefinite

$$d(\omega\,;\mathbf{u}\,,\mathbf{u}) \geq 0 \quad , \quad \forall \mathbf{u} \neq \mathbf{0} \quad . \tag{36}$$

Damping operator $\mathbf{D}(\omega)$ is then defined by [7]

$$<\mathbf{D}(\omega)\,\mathbf{u}\,,\delta\mathbf{u}> = d(\omega\,;\mathbf{u}\,,\delta\mathbf{u}) \quad . \tag{37}$$

4.4. Given force vector

For all fixed real ω, we introduce the antilinear form $f(\omega\,;\delta\mathbf{u})$ defined on $\mathcal{C}^c$ by

$$f(\omega\,;\delta\mathbf{u}) = \int_\Omega \mathbf{g}(\mathbf{x},\omega)\cdot\overline{\delta\mathbf{u}(\mathbf{x})}\;d\mathbf{x} + \int_\Gamma \mathbf{G}(\mathbf{x},\omega)\cdot\overline{\delta\mathbf{u}(\mathbf{x})}\;ds(\mathbf{x}) \quad , \tag{38}$$

which defines the element $\mathbf{f}(\omega)$, called the given *force vector*, such that [8]

$$<\mathbf{f}(\omega)\,,\delta\mathbf{u}> = f(\omega\,;\delta\mathbf{u}) \quad . \tag{39}$$

4.5. Rewriting of the variational formulation

Structure fixed on Γ_0. With the above notation, the variational formulation defined by Eq. (21) can be rewritten as follows. For all fixed real ω, find $\mathbf{u}$ in $\mathcal{C}_0^c$ such that, for all $\delta\mathbf{u} \in \mathcal{C}_0^c$,

$$-\omega^2\, m(\mathbf{u}\,,\delta\mathbf{u}) + i\omega\, d(\omega\,;\mathbf{u}\,,\delta\mathbf{u}) + k(\omega\,;\mathbf{u}\,,\delta\mathbf{u}) = f(\omega\,;\delta\mathbf{u}) \quad . \tag{40}$$

---

[6] Sesquilinear form $d(\omega;\mathbf{u},\delta\mathbf{u})$ is continuous on $\mathcal{C}_0^c \times \mathcal{C}_0^c$, coercive on $\mathcal{C}_0^c$ due to Eq. (IV.50). This sesquilinear form defines a continuous operator $\mathbf{D}(\omega)$ from $\mathcal{C}_0^c$ into its antidual space $\mathcal{C}_0^{c\,\prime}$ and in Eq. (35), the angle brackets denote the antiduality product between $\mathcal{C}_0^{c\,\prime}$ and $\mathcal{C}_0^c$.

[7] See footnote 6, replacing $\mathcal{C}_0^c$ by $\mathcal{C}^c$. It should be noted that $d(\omega;\mathbf{u},\mathbf{u})$ is coercive on $\mathcal{C}^c \backslash \mathcal{C}_{\mathrm{rig}}$ in which $\mathcal{C}_{\mathrm{rig}} \subset \mathcal{C}$ is the set of the rigid body displacement fields (see Section III.5.2).

[8] For all fixed real ω, mapping $\mathbf{x} \mapsto \mathbf{g}(\mathbf{x},\omega)$ defined in Ω and mapping $\mathbf{x} \mapsto \mathbf{G}(\mathbf{x},\omega)$ defined in Γ are assumed to be such that $\delta\mathbf{u} \mapsto f(\omega\,;\delta\mathbf{u})$ is continuous on $\mathcal{C}^c$ (or $\mathcal{C}_0^c$). Element $\mathbf{f}(\omega) \in \mathcal{C}^{c\,\prime}$ (or $\mathcal{C}_0^{c\,\prime}$). In Eq. (39), the angle brackets denote the antiduality product between $\mathcal{C}^{c\,\prime}$ and $\mathcal{C}^c$ (or between $\mathcal{C}_0^{c\,\prime}$ and $\mathcal{C}_0^c$).

The associated linear operator equation is written as [9]

$$\left(-\omega^2 \mathbf{M} + i\omega\, \mathbf{D}(\omega) + \mathbf{K}(\omega)\right) \mathbf{u} = \mathbf{f}(\omega) \quad . \tag{41}$$

Free structure. The variational formulation defined by Eq. (22) can be rewritten as follows. For all fixed real ω, find $\mathbf{u}$ in $\mathcal{C}^c$ such that, for all $\delta\mathbf{u} \in \mathcal{C}^c$,

$$-\omega^2\, m(\mathbf{u}\,,\delta\mathbf{u}) + i\omega\, d(\omega\,;\mathbf{u}\,,\delta\mathbf{u}) + k(\omega\,;\mathbf{u}\,,\delta\mathbf{u}) = f(\omega\,;\delta\mathbf{u}) \quad . \tag{42}$$

The associated linear operator equation is then written as [10]

$$\left(-\omega^2 \mathbf{M} + i\omega\, \mathbf{D}(\omega) + \mathbf{K}(\omega)\right) \mathbf{u} = \mathbf{f}(\omega) \quad . \tag{43}$$

5. Frequency Response Function

In this section, we introduce the operator-valued Frequency Response Function (FRF). This operator allows the calculation of the dynamical response of the structure for arbitrary deterministic or random excitations.

5.1. Dynamic stiffness and impedance operators

For the two cases defined in Section 4.5, we introduce the so-called dynamic stiffness sesquilinear form $a(\omega\,;\mathbf{u}\,,\delta\mathbf{u})$ and the impedance sesquilinear form $z(\omega\,;\mathbf{u}\,,\delta\mathbf{u})$ defined by

$$a(\omega\,;\mathbf{u}\,,\delta\mathbf{u}) = -\omega^2\, m(\mathbf{u}\,,\delta\mathbf{u}) + i\omega\, d(\omega\,;\mathbf{u}\,,\delta\mathbf{u}) + k(\omega\,;\mathbf{u}\,,\delta\mathbf{u}) \quad , \tag{44}$$

$$i\omega\, z(\omega\,;\mathbf{u}\,,\delta\mathbf{u}) = a(\omega\,;\mathbf{u}\,,\delta\mathbf{u}) \quad . \tag{45}$$

We then define the associated dynamic stiffness operator $\mathbf{A}(\omega)$ and impedance operator $\mathbf{Z}(\omega)$ such that [11]

$$<\mathbf{A}(\omega)\,\mathbf{u}\,,\delta\mathbf{u}> = a(\omega\,;\mathbf{u}\,,\delta\mathbf{u}) \quad , \tag{46}$$

$$\mathbf{A}(\omega) = -\omega^2 \mathbf{M} + i\omega\, \mathbf{D}(\omega) + \mathbf{K}(\omega) \quad , \quad i\omega\, \mathbf{Z}(\omega) = \mathbf{A}(\omega) \quad . \tag{47}$$

---

[9] Eq. (41) is an equality in the antidual space $\mathcal{C}_0^{c\,\prime}$ of $\mathcal{C}_0^c$.

[10] Eq. (43) is an equality in the antidual space $\mathcal{C}^{c\,\prime}$ of $\mathcal{C}^c$.

[11] Sesquilinear form $a(\omega;\mathbf{u},\delta\mathbf{u})$ is continuous on $\mathcal{C}_0^c \times \mathcal{C}_0^c$ or $\mathcal{C}^c \times \mathcal{C}^c$. Linear operator $\mathbf{A}(\omega) \in \mathcal{L}(\mathcal{C}_0^c, \mathcal{C}_0^{c\,\prime})$ or $\mathbf{A}(\omega) \in \mathcal{L}(\mathcal{C}^c, \mathcal{C}^{c\,\prime})$ (see Mathematical Notations in the appendix).

5.2. Operator-valued FRF for a structure fixed on Γ_0

The operator-valued frequency response function is defined by the linear operator which associates each given force vector with the displacement field as the solution of the variational formulation defined by Eq. (40). From Section 4, it can be proved[12] that for all fixed real ω and given force vector $\mathbf{f}(\omega)$, the variational formulation defined by Eq. (40) has a unique solution $\mathbf{u}(\omega)$ in $\mathcal{C}_0^c$ which depends linearly on $\mathbf{f}$. From Eq. (39), we deduce that there exists a linear operator $\mathbf{f}(\omega) \mapsto \mathbf{u}(\omega)$ such that[13]

$$\mathbf{u}(\omega) = \mathbf{T}(\omega)\,\mathbf{f}(\omega) \quad , \tag{48}$$

in which $\omega \mapsto \mathbf{T}(\omega) = \mathbf{A}(\omega)^{-1}$ is called the operator-valued frequency response function (FRF).

5.3. Operator-valued FRF for a free structure

In this case, we use the variational formulation defined by Eq. (42). For all real ω, we assume that given body force field $\mathbf{g}(\omega)$ and given surface force field $\mathbf{G}(\omega)$ are in equilibrium in the following sense. Let $\mathcal{C}_{\mathrm{rig}} \subset \mathcal{C}^c$ be the set of the rigid body displacement fields $\mathbf{u}_{\mathrm{rig}}$ (see Section III.5.2). Taking $\delta\mathbf{u} = \mathbf{u}_{\mathrm{rig}}$ in Eq. (38), for all fixed real ω, the equilibrium hypothesis is written as

$$\begin{aligned} f(\omega;\mathbf{u}_{\mathrm{rig}}) &= \int_\Omega \mathbf{g}(\mathbf{x},\omega)\cdot\mathbf{u}_{\mathrm{rig}}(\mathbf{x})\,d\mathbf{x} + \int_\Gamma \mathbf{G}(\mathbf{x},\omega)\cdot\mathbf{u}_{\mathrm{rig}}(\mathbf{x})\,ds(\mathbf{x}) \\ &= 0 \ . \end{aligned} \tag{49}$$

We deduce that the condition defined by Eq. (49) is equivalent to the set of linear equations obtained for $\alpha = \{-5,\ldots,0\}$

$$\int_\Omega \mathbf{g}(\mathbf{x},\omega)\cdot\mathbf{u}_\alpha(\mathbf{x})\,d\mathbf{x} + \int_\Gamma \mathbf{G}(\mathbf{x},\omega)\cdot\mathbf{u}_\alpha(\mathbf{x})\,ds(\mathbf{x}) = 0 \quad , \tag{50}$$

in which vectors $\mathbf{u}_{-5},\ldots,\mathbf{u}_0$ span $\mathcal{C}_{\mathrm{rig}}$ (see Section III.7.2). Two cases must be considered.

---

[12] For $\omega>0$, the proof of the existence and uniqueness of the solution in $\mathcal{C}_0^c$ is obtained by applying the Lax-Milgram theorem to the variational formulation $a_c(\omega;\mathbf{u},\delta\mathbf{u})=f_c(\omega;\delta\mathbf{u})$ in which $a_c(\omega;\mathbf{u},\delta\mathbf{u})=-i\,a(\omega;\mathbf{u},\delta\mathbf{u})$ and $f_c(\omega;\delta\mathbf{u})=-i\,f(\omega;\delta\mathbf{u})$, because $\Re e\, a_c(\omega;\mathbf{u},\mathbf{u})=\omega\, d(\omega;\mathbf{u},\mathbf{u})$ is coercive due to Eq. (IV.50). For $\omega<0$, the proof is similar considering $a_c(\omega;\mathbf{u},\delta\mathbf{u})=i\,a(\omega\,;\mathbf{u},\delta\mathbf{u})$.

[13] Footnote 12 proves that, for all real ω, operator $\mathbf{A}(\omega)\in\mathcal{L}(\mathcal{C}_0^c,\mathcal{C}_0^{c\,\prime})$ is invertible. Consequently, $\mathbf{T}(\omega)=\mathbf{A}(\omega)^{-1}\in\mathcal{L}(\mathcal{C}_0^{c\,\prime},\mathcal{C}_0^c)$.

First case $\omega = 0$. Taking $\omega = 0$ in Eq. (42), the variational formulation consists in finding $\mathbf{u}$ in $\mathcal{C}^c$ such that,

$$k(0\,;\mathbf{u}\,,\delta\mathbf{u}) = f(0\,;\delta\mathbf{u}) \quad , \quad \forall\delta\mathbf{u} \in \mathcal{C}^c \quad . \tag{51}$$

The variational formulation defined by Eq. (51) has a unique solution [14] $\mathbf{u}_{\text{elas}}$ in space $\mathcal{C}^c_{\text{elas}}$ defined by

$$\mathcal{C}^c_{\text{elas}} = \mathcal{C}^c \setminus \mathcal{C}_{\text{rig}} \quad , \quad \mathcal{C}^c = \mathcal{C}_{\text{rig}} \oplus \mathcal{C}^c_{\text{elas}} \quad . \tag{52}$$

Second case $\omega \neq 0$. It can easily be proved [15] that for all fixed real $\omega \neq 0$ and given $f(\omega\,;\delta\mathbf{u})$, the variational formulation defined by Eq. (42) has a unique solution $\mathbf{u}(\omega)$ in $\mathcal{C}^c$.

Summary. Considering the two above cases, for all fixed real ω, unique solution $\mathbf{u}_{\text{elas}}(\omega)$ depends linearly on $\mathbf{f}(\omega)$ and defines a linear operator $\mathbf{f}(\omega) \mapsto \mathbf{u}_{\text{elas}}(\omega)$ such that [16]

$$\mathbf{u}_{\text{elas}}(\omega) = \mathbf{T}(\omega)\,\mathbf{f}(\omega) \quad , \tag{53}$$

in which $\omega \mapsto \mathbf{T}(\omega) = \mathbf{A}_{\text{elas}}(\omega)^{-1}$ where operator $\mathbf{A}_{\text{elas}}(\omega)$ is the restriction to $\mathcal{C}^c_{\text{elas}}$ of operator $\mathbf{A}(\omega)$. Operator $\mathbf{T}(\omega)$ is called the operator-valued frequency response function (FRF).

Remark. If for all real ω, the given force vector was not in equilibrium, we should replace force vector $\mathbf{f}(\omega)$ by an associated force vector $\widetilde{\mathbf{f}}(\omega)$ in equilibrium in the sense defined by Eq. (50). This new force vector would have to be constructed such that, for all real ω,

$$\widetilde{\mathbf{f}}(\omega) = \mathbf{f}(\omega) - \sum_{\alpha=-5}^{0} \frac{1}{\mu_\alpha}\, f(\omega; \mathbf{u}_\alpha)\, \rho(\mathbf{x})\, \mathbf{u}_\alpha(\mathbf{x}) \quad , \tag{54}$$

---

14 $k(0\,;\mathbf{u}\,,\mathbf{u})$ is coercive in $\mathcal{C}^c_{\text{elas}}$.

15 For $\omega>0$, the proof of the existence and uniqueness of the solution in $\mathcal{C}^c_{\text{elas}}$, is obtained by applying the Lax-Milgram theorem to the variational formulation $a_c(\omega;\mathbf{u},\delta\mathbf{u})= f_c(\omega;\delta\mathbf{u})$ in which $a_c(\omega;\mathbf{u},\delta\mathbf{u})=-i\,a(\omega;\mathbf{u},\delta\mathbf{u})$ and $f_c(\omega;\delta\mathbf{u})=-i\,f(\omega;\delta\mathbf{u})$, because $\Re e\, a_c(\omega;\mathbf{u},\mathbf{u})=\omega\, d(\omega;\mathbf{u},\mathbf{u})$ is coercive in $\mathcal{C}^c_{\text{elas}}$ due to Eq. (IV.50). For $\omega<0$, the proof is similar considering $a_c(\omega;\mathbf{u},\delta\mathbf{u})= i\,a(\omega;\mathbf{u},\delta\mathbf{u})$. It should be noted that, for $\omega\neq 0$, there is a unique solution in $\mathcal{C}^c$ because, if $\mathbf{u}\in\mathcal{C}^c_{\text{rig}}$, then $a(\omega;\mathbf{u},\mathbf{u})=-\omega^2 m(\mathbf{u},\mathbf{u})\neq 0$.

16 The restriction $\mathbf{A}_{\text{elas}}(\omega)$ to $\mathcal{C}^c_{\text{elas}}$ of operator $\mathbf{A}(\omega)\in\mathcal{L}(\mathcal{C}^c, \mathcal{C}^{c\prime})$ is continuous and invertible from $\mathcal{C}^c_{\text{elas}}$ into $\mathcal{C}^{c\prime}_{\text{elas}}$. Consequently, $\mathbf{T}(\omega)=\mathbf{A}_{\text{elas}}(\omega)^{-1}\in\mathcal{L}(\mathcal{C}^{c\,\prime}_{\text{elas}}, \mathcal{C}^c_{\text{elas}})$.

in which $\mu_\alpha = m(\mathbf{u}_\alpha\,, \mathbf{u}_\alpha)$ is the generalized mass associated with rigid body mode $\mathbf{u}_\alpha$ (see Section III.7.2). Consequently, antilinear form $f(\omega\,;\delta\mathbf{u})$ must be replaced by antilinear form $\widetilde{f}(\omega\,;\delta\mathbf{u})$ in the variational formulation, such that

$$\widetilde{f}(\omega\,;\delta\mathbf{u}) = <\widetilde{\mathbf{f}}(\omega)\,,\delta\mathbf{u}> \quad . \tag{55}$$

From Eq. (54), we deduce that $\widetilde{f}(\omega;\mathbf{u}_{\text{rig}}) = 0$, which proves that force vector $\widetilde{\mathbf{f}}(\omega)$ is associated with a system of external forces in equilibrium.

5.4. Introduction of a unified notation for the fixed and free structure cases

We introduce the admissible function space $\mathcal{C}_S^c$ such that

$$\mathcal{C}_S^c = \begin{cases} \mathcal{C}_0^c & \text{for a structure fixed on } \Gamma_0 \\ \mathcal{C}_{\text{elas}}^c & \text{for a free structure} \end{cases} \quad , \tag{56}$$

in which $\mathcal{C}_0^c$ and $\mathcal{C}_{\text{elas}}^c$ are defined by Eqs. (18) and (52) respectively. From Sections 5.2 and 5.3, response $\mathbf{u}(\omega)$ in $\mathcal{C}_S^c$ defined by

$$\mathbf{u}(\omega) = \mathbf{T}(\omega)\,\mathbf{f}(\omega) \quad , \tag{57}$$

is the unique solution of the following variational formulation written in the unified notation. For all fixed real ω, find $\mathbf{u} \in \mathcal{C}_S^c$ such that,

$$a(\omega\,;\mathbf{u}\,,\delta\mathbf{u}) = f(\omega\,;\,\delta\mathbf{u}) \quad , \quad \forall\delta\mathbf{u} \in \mathcal{C}_S^c \quad , \tag{58}$$

in which $a(\omega\,;\mathbf{u}\,,\delta\mathbf{u})$ is defined by Eq. (44) and it is assumed that $f(\omega\,;\,\delta\mathbf{u})$ satisfies Eq. (49).

5.5. Associated linear filter and operator-valued impulse response function

In this section, we define the linear filter whose operator-valued frequency response function is $\mathbf{T}(\omega)$. Let $\mathbf{f}(t)$ be such that its Fourier transform is $\mathbf{f}(\omega)$,

$$\mathbf{f}(\omega) = \int_{\mathbb{R}} e^{-i\omega t}\,\mathbf{f}(t)\,dt \quad . \tag{59}$$

For all fixed real ω, let $\mathbf{u}(\omega)$ be the unique solution in $\mathcal{C}_S^c$ of Eq. (58). Let $\mathbf{u}(t)$ be such that its Fourier transform is $\mathbf{u}(\omega)$,

$$\mathbf{u}(\omega) = \int_{\mathbb{R}} e^{-i\omega t}\,\mathbf{u}(t)\,dt \quad . \tag{60}$$

It should be noted that Eq. (60) corresponds to Eq. (5) in which all the values of $\mathbf{x}$ are considered. Let us consider the linear filter whose input is $\mathbf{f}(t)$, whose output is $\mathbf{u}(t)$ and whose operator-valued impulse response function is $\mathbf{h}(t)$. Consequently, its operator-valued frequency response function is $\mathbf{T}(\omega)$ such that,

$$\mathbf{T}(\omega) = \int_{\mathbb{R}} e^{-i\omega t}\, \mathbf{h}(t)\, dt \quad . \tag{61}$$

Since $\mathbf{h}(t)$ is a function with values in a real vector space, we then have

$$\mathbf{T}(-\omega) = \overline{\mathbf{T}(\omega)} \quad . \tag{62}$$

This linear filter denoted as $\mathbf{h} *_t$ is such that

$$\mathbf{u}(t) = (\mathbf{h} *_t \mathbf{f})(t) \quad , \tag{63}$$

which means that

$$\mathbf{u}(t) = \int_{\mathbb{R}} \mathbf{h}(\tau)\, \mathbf{f}(t-\tau)\, d\tau = \int_{\mathbb{R}} \mathbf{h}(t-\tau)\, \mathbf{f}(\tau)\, d\tau \quad . \tag{64}$$

Eq. (63) or (64) appears as a convolution product with respect to the time variable. Due to the properties of the viscoelastic constitutive equation introduced in Chapter IV and used in this chapter, it can be verified that linear filter $\mathbf{h} *_t$ is causal (or physically realizable), which means that

$$\mathbf{h}(t) = \mathbf{0} \quad , \quad \forall t < 0 \quad . \tag{65}$$

Consequently, Eq. (64) can be rewritten as

$$\mathbf{u}(t) = \int_0^{+\infty} \mathbf{h}(\tau)\, \mathbf{f}(t-\tau)\, d\tau = \int_{-\infty}^{t} \mathbf{h}(t-\tau)\, \mathbf{f}(\tau)\, d\tau \quad . \tag{66}$$

For more details concerning linear filters, linear filtering, impulse response function, frequency response function, causal filter, etc., within the context of mathematical methods in signal analysis, see for instance Bendat and Piersol, 1971; Bucy and Joseph, 1968; Guikhman and Skorokhod, 1979; Papoulis, 1977; Priestley, 1981; Soize, 1994 and 1993a.

6. Finite Element Discretization

Since the finite element method uses a real basis, the discretization of sesquilinear forms $m(\mathbf{u}, \delta\mathbf{u})$, $k(\omega; \mathbf{u}, \delta\mathbf{u})$ and $d(\omega; \mathbf{u}, \delta\mathbf{u})$ defined by Eqs. (23), (26) and (32) respectively leads to mass, stiffness and damping real matrices $[M]$, $[K(\omega)]$ and $[D(\omega)]$ respectively. Consequently, the finite element discretization of Eq. (40) or (42) is written as

$$\left(-\omega^2 [M] + i\omega [D(\omega)] + [K(\omega)]\right) \mathbf{U} = \mathbf{F}(\omega) \quad . \tag{67}$$

in which $\mathbf{U} = (U_1, \ldots, U_n)$ is the complex-valued vector of the DOFs which are the values of the displacement field at the nodes of the finite element mesh of domain Ω and $\mathbf{F} = (F_1, \ldots, F_n)$ is the complex-valued vector corresponding to the finite element discretization of antilinear form $f(\omega; \delta\mathbf{u})$ defined by Eq. (38). For a finite element discretization in $\mathcal{C}_S^c$ (see Eq. (56)), $(n \times n)$ real mass, damping and stiffness matrices $[M]$, $[D(\omega)]$ and $[K(\omega)]$ are symmetric positive definite. Eq. (67) can be rewritten as

$$[A(\omega)]\, \mathbf{U} = \mathbf{F}(\omega) \quad , \tag{68}$$

in which the dynamic stiffness and impedance $(n \times n)$ symmetric complex matrices are such that

$$[A(\omega)] = -\omega^2 [M] + i\omega [D(\omega)] + [K(\omega)] \quad , \quad i\omega [Z(\omega)] = [A(\omega)] \quad . \tag{69}$$

For all fixed real ω, Eq. (68) corresponds to the finite element discretization of Eq. (58) and matrix $[A(\omega)]$ is invertible. We then deduce that

$$\mathbf{U}(\omega) = [T(\omega)]\, \mathbf{F}(\omega) \quad , \tag{70}$$

in which $[T(\omega)]$ is an $(n \times n)$ symmetric complex matrix such that

$$\begin{aligned} [T(\omega)] &= [A(\omega)]^{-1} \\ &= \left[-\omega^2 M + i\omega\, D(\omega) + K(\omega)\right]^{-1} \quad . \end{aligned} \tag{71}$$

It should be noted that Eq. (70) corresponds to the finite element discretization of Eq. (57). From Eq. (61), we deduce that the $(n \times n)$ real matrix-valued impulse response function $t \mapsto [\,h(t)\,]$ is such that

$$[T(\omega)] = \int_{\mathbb{R}} e^{-i\omega t}\, [h(t)]\, dt \quad . \tag{72}$$

For more details concerning the finite element procedures in structural vibrations, see for instance Argyris and Mlejnek, 1991; Bathe and Wilson, 1976; Zienkiewicz and Taylor, 1989.

7. Boundary Impedance Operator

7.1. Boundary value problem for a structure made of two substructures

Consider a free structure with domain Ω which is decomposed into two subdomains, Ω_1 with boundary $\partial\Omega_1 = \Gamma_1 \cup \Sigma$ and Ω_2 with boundary $\partial\Omega_2 = \Gamma_2 \cup \Sigma$, that interact through the common boundary Σ (see Fig. 2). Let $\mathbf{n}^r = (n_1^r, n_2^r, n_3^r)$ be the unit normal to $\partial\Omega_r$, external to Ω_r. For the sake of brevity, we assume that only substructure Ω_1 is submitted to external applied forces satisfying the equilibrium hypothesis defined by Eq. (49) (the extension to the general case is straightforward).

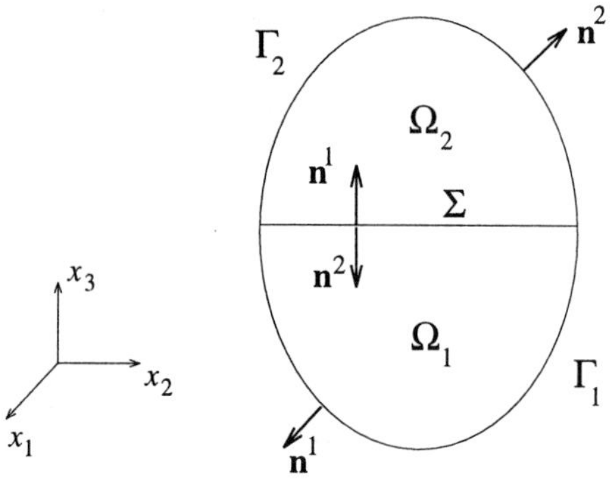

Fig. 2. Structure decomposed into two substructures

The boundary value problem defined by Eqs. (13) and (14) is rewritten as

$$-\omega^2 \rho_1 u_i^1 - \sigma_{ij,j}^1(\mathbf{u}^1) = g_i^1 \quad \text{in} \quad \Omega_1 \quad , \tag{73}$$

$$\sigma_{ij}^1(\mathbf{u}^1)\, n_j^1 = G_i^1 \quad \text{on} \quad \Gamma_1 \quad , \tag{74}$$

$$\sigma_{ij}^1(\mathbf{u}^1)\, n_j^1 = -\,\sigma_{ij}^2(\mathbf{u}^2) n_j^2 \quad \text{on} \quad \Sigma \quad , \tag{75}$$

$$\mathbf{u}^2 = \mathbf{u}^1 \quad \text{on} \quad \Sigma \quad , \tag{76}$$

$$-\omega^2 \rho_2 u_i^2 - \sigma_{ij,j}^2(\mathbf{u}^2) = 0 \quad \text{in} \quad \Omega_2 \quad , \tag{77}$$

$$\sigma_{ij}^2(\mathbf{u}^2)\, n_j^2 = 0 \quad \text{on} \quad \Gamma_2 \quad , \tag{78}$$

in which, for $r = 1$ and $r = 2$, $\mathbf{u}^r$ is the restriction of the displacement field $\mathbf{u}$ to Ω_r, σ^r is the stress tensor defined by Eq. (12) related to Ω_r and ρ_r is the restriction of the mass density ρ to Ω_r.

7.2. Variational formulation in terms of $(\mathbf{u}^1, \mathbf{u}^2)$

Let $\mathcal{C}^c_{\Omega_r}$ be the admissible function space [17] defined on Ω_r (which does not contain any constraint on $\partial\Omega_r = \Gamma_r \cup \Sigma$). For $r = 1$ and $r = 2$, we introduce the structural mass sesquilinear form,

$$m^r(\mathbf{u}^r, \delta\mathbf{u}^r) = \int_{\Omega_r} \rho_r \, \mathbf{u}^r \cdot \overline{\delta\mathbf{u}^r} \, d\mathbf{x} \quad , \tag{79}$$

the structural stiffness sesquilinear form,

$$k^r(\omega\,;\mathbf{u}^r, \delta\mathbf{u}^r) = \int_{\Omega_r} a^r_{ijkh}(\omega)\, \varepsilon_{kh}(\mathbf{u}^r)\, \varepsilon_{ij}(\overline{\delta\mathbf{u}^r})\, d\mathbf{x} \quad , \tag{80}$$

and the structural damping sesquilinear form,

$$d^r(\omega\,;\mathbf{u}^r, \delta\mathbf{u}^r) = \int_{\Omega_r} b^r_{ijkh}(\omega)\, \varepsilon_{kh}(\mathbf{u}^r)\, \varepsilon_{ij}(\overline{\delta\mathbf{u}^r})\, d\mathbf{x} \quad , \tag{81}$$

the three sesquilinear forms being defined on $\mathcal{C}^c_{\Omega_r} \times \mathcal{C}^c_{\Omega_r}$. We introduce the antilinear form $f^1(\omega\,;\delta\mathbf{u}^1)$ on $\mathcal{C}^c_{\Omega_1}$,

$$f^1(\omega\,;\delta\mathbf{u}^1) = \int_{\Omega_1} \mathbf{g}^1(\omega)\cdot\overline{\delta\mathbf{u}^1}\, d\mathbf{x} + \int_{\Gamma_1} \mathbf{G}^1(\omega)\cdot\overline{\delta\mathbf{u}^1}\, ds \quad . \tag{82}$$

Finally, we introduce the dynamic stiffness sesquilinear form and the corresponding impedance sesquilinear form on $\mathcal{C}^c_{\Omega_r} \times \mathcal{C}^c_{\Omega_r}$,

$$a^r(\omega\,;\mathbf{u}^r, \delta\mathbf{u}^r) = -\omega^2 m^r(\mathbf{u}^r, \delta\mathbf{u}^r) + i\omega\, d^r(\omega\,;\mathbf{u}^r, \delta\mathbf{u}^r) + k^r(\omega\,;\mathbf{u}^r, \delta\mathbf{u}^r), \tag{83}$$

$$i\omega\, z^r(\omega\,;\mathbf{u}^r, \delta\mathbf{u}^r) = a^r(\omega\,;\mathbf{u}^r, \delta\mathbf{u}^r) \quad . \tag{84}$$

The variational formulation of the boundary value problem defined by Eqs. (73) to (78) is the following. Let $\mathcal{C}^c_\Omega$ be the admissible function space [18] defined by

$$\mathcal{C}^c_\Omega = \{\mathbf{u}^1 \in \mathcal{C}^c_{\Omega_1} \ , \ \mathbf{u}^2 \in \mathcal{C}^c_{\Omega_2} \ ; \ \mathbf{u}^1 = \mathbf{u}^2 \text{ on } \Sigma\} \quad . \tag{85}$$

---

[17] Space $\mathcal{C}^c_{\Omega_r} = (H^1(\Omega_r))^3$.

[18] Introducing the trace space $\mathcal{C}^c_\Sigma \subset (H^{1/2}(\partial\Omega_r))^3$ of the traces on Σ of all the functions in $(H^1(\Omega_r))^3$, $\mathbf{u}^1 = \mathbf{u}^2$ is an equality in $\mathcal{C}^c_\Sigma$.

For all fixed real $\omega \neq 0$, find $(\mathbf{u}^1, \mathbf{u}^2)$ in $\mathcal{C}^c_\Omega$ such that

$$a^1(\omega\,;\mathbf{u}^1, \delta\mathbf{u}^1) + a^2(\omega\,;\mathbf{u}^2, \delta\mathbf{u}^2) = f^1(\omega\,;\delta\mathbf{u}^1) \quad , \quad \forall(\delta\mathbf{u}^1, \delta\mathbf{u}^2) \in \mathcal{C}^c_\Omega \quad . \tag{86}$$

Let $\mathbf{A}^r(\omega)$ be the operator [19] defined by sesquilinear form $a^r(\omega\,;\mathbf{u}^r, \delta\mathbf{u}^r)$, such that

$$<\mathbf{A}^r(\omega)\,\mathbf{u}^r, \delta\mathbf{u}^r> = a^r(\omega\,;\mathbf{u}^r, \delta\mathbf{u}^r) \quad . \tag{87}$$

Let $\mathbf{f}^1(\omega)$ be the element [20] defined by antilinear form $<\mathbf{f}^1(\omega)\,, \delta\mathbf{u}^1>$, such that

$$<\mathbf{f}^1(\omega)\,, \delta\mathbf{u}^1> = f^1(\omega\,;\delta\mathbf{u}^1) \quad . \tag{88}$$

Eq. (86) can then be rewritten as

$$<\mathbf{A}^1(\omega)\,\mathbf{u}^1, \delta\mathbf{u}^1> + <\mathbf{A}^2(\omega)\,\mathbf{u}^2, \delta\mathbf{u}^2> = <\mathbf{f}^1(\omega)\,, \delta\mathbf{u}^1> \quad . \tag{89}$$

For all fixed real $\omega \neq 0$, the variational formulation defined by Eq. (86) of the boundary value problem defined by Eqs. (73) to (78) has a unique solution $(\mathbf{u}^1, \mathbf{u}^2)$ in $\mathcal{C}^c_\Omega$.

7.3. Boundary impedance operator related to Σ of substructure Ω_2

Statement of the problem. The problem consists in replacing the boundary value problem in $(\mathbf{u}^1, \mathbf{u}^2)$ defined by Eqs. (73) to (78) by a boundary value problem expressed only in terms of $\mathbf{u}^1$. Field $\mathbf{u}^2$ must therefore be eliminated by constructing the boundary impedance operator $\widetilde{\mathbf{Z}}_\Sigma(\omega)$ related to interface Σ such that

$$\mathbf{f}_\Sigma = i\omega\,\widetilde{\mathbf{Z}}_\Sigma(\omega)\,\mathbf{u}_\Sigma \quad , \tag{90}$$

which relates the trace $\mathbf{u}_\Sigma$ of displacement field $\mathbf{u}^2$ on boundary Σ with force field $\mathbf{f}_\Sigma$ on Σ.

---

[19] Sesquilinear form $a^r(\omega;\mathbf{u}^r,\delta\mathbf{u}^r)$ is continuous on $\mathcal{C}^c_{\Omega_r}\times\mathcal{C}^c_{\Omega_r}$ and defines a continuous operator $\mathbf{A}^r(\omega)$ from $\mathcal{C}^c_{\Omega_r}$ into its antidual space $\mathcal{C}^{c\,\prime}_{\Omega_r}$ and in Eq. (87), the angle brackets denote the antiduality product between $\mathcal{C}^{c\,\prime}_{\Omega_r}$ and $\mathcal{C}^c_{\Omega_r}$.

[20] For all fixed real ω, $\mathbf{g}^1(\omega)$ and $\mathbf{G}^1(\omega)$ are assumed to be such that $\delta\mathbf{u}^1 \mapsto f^1(\omega;\delta\mathbf{u}^1)$ is continuous on $\mathcal{C}^c_{\Omega_1}$. Then element $\mathbf{f}^1(\omega) \in \mathcal{C}^{c\,\prime}_{\Omega_1}$ and the brackets in Eq. (88) denote the antiduality product between $\mathcal{C}^{c\,\prime}_{\Omega_1}$ and $\mathcal{C}^c_{\Omega_1}$.

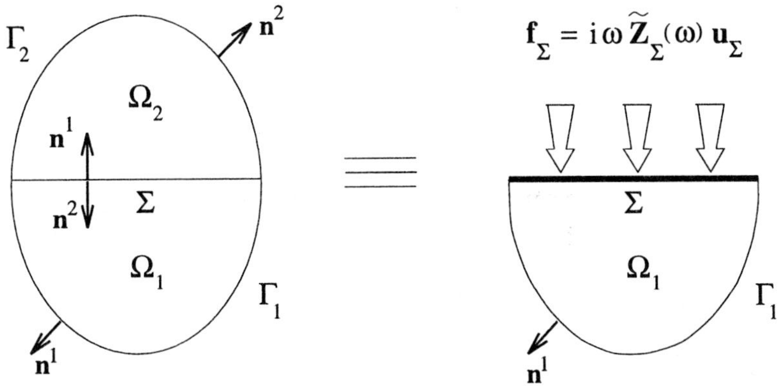

Fig. 3. Structure Ω, equivalent to substructure Ω_1 with a boundary impedance operator on interface Σ

Boundary value problem for substructure Ω_2 and operator $\mathbf{S}(\omega)$. To construct operator $\mathbf{Z}_\Sigma(\omega)$, we consider the boundary value problem defined by Eqs. (76) to (78) as a boundary value problem with prescribed displacement on boundary Σ,

$$-\omega^2\,\rho_2\,u_i^2 - \sigma_{ij,j}^2(\mathbf{u}^2) = 0 \quad \text{in} \quad \Omega_2 \quad , \tag{91}$$

$$\sigma_{ij}^2(\mathbf{u}^2)\,n_j^2 = 0 \quad \text{on} \quad \Gamma_2 \quad , \tag{92}$$

$$\mathbf{u}^2 = \mathbf{u}_\Sigma \quad \text{on} \quad \Sigma \quad , \tag{93}$$

in which $\mathbf{u}_\Sigma$ belongs to the admissible set [21] $\mathcal{C}_\Sigma^c$ of the $\mathbb{C}^3$-valued displacement fields defined on Σ. For the construction of the variational formulation of the boundary value problem defined by Eqs. (91) to (93), we introduce the following admissible sets

$$\mathcal{C}_{\Omega_2,0}^c = \left\{ \mathbf{u}^2 \in \mathcal{C}_{\Omega_2}^c \;;\; \mathbf{u}^2 = \mathbf{0} \text{ on } \Sigma \right\} \quad , \tag{94}$$

$$\mathcal{C}_{\Omega_2,\mathbf{u}_\Sigma}^c = \left\{ \mathbf{u}^2 \in \mathcal{C}_{\Omega_2}^c \;;\; \mathbf{u}^2 = \mathbf{u}_\Sigma \text{ on } \Sigma,\ \text{for } \mathbf{u}_\Sigma \in \mathcal{C}_\Sigma^c \right\} \quad . \tag{95}$$

The variational formulation is then the following. For all fixed real $\omega \neq 0$, find $\mathbf{u}^2$ in $\mathcal{C}_{\Omega_2,\mathbf{u}_\Sigma}^c$ such that

$$a^2(\omega\,;\mathbf{u}^2,\delta\mathbf{u}^2) = 0 \quad , \quad \forall\,\delta\mathbf{u}^2 \in \mathcal{C}_{\Omega_2,0}^c \quad . \tag{96}$$

[21] Space $\mathcal{C}_\Sigma^c \subset (H^{1/2}(\partial\Omega_2))^3$ is the set of the traces on Σ of all the functions in $(H^1(\Omega_2))^3$.

Eq. (96) has a unique solution $\mathbf{u}^2$ in $\mathcal{C}^c_{\Omega_2, \mathbf{u}_\Sigma}$ whose trace on boundary Σ is $\mathbf{u}_\Sigma$. Therefore, this solution defines a unique operator $\mathbf{u}_\Sigma \mapsto \mathbf{u}^2$ denoted as $\mathbf{S}(\omega)$ which is linear [22] from $\mathcal{C}^c_\Sigma$ into $\mathcal{C}^c_{\Omega_2}$:

$$\mathbf{u}^2 = \mathbf{S}(\omega)\, \mathbf{u}_\Sigma \quad . \tag{97}$$

Definition of a boundary impedance operator. We introduce the sesquilinear form $\widetilde{a}_\Sigma(\omega\,; \mathbf{u}_\Sigma\,, \delta\mathbf{u}_\Sigma)$ defined on $\mathcal{C}^c_\Sigma \times \mathcal{C}^c_\Sigma$ by

$$\widetilde{a}_\Sigma(\omega\,; \mathbf{u}_\Sigma, \delta\mathbf{u}_\Sigma) = a^2(\omega\,; \mathbf{S}(\omega)\, \mathbf{u}_\Sigma\,, \mathbf{S}(\omega)\, \delta\mathbf{u}_\Sigma) \quad . \tag{98}$$

This sesquilinear form defines a linear operator [23] $\widetilde{\mathbf{A}}_\Sigma(\omega)$ such that

$$< \widetilde{\mathbf{A}}_\Sigma(\omega)\, \mathbf{u}_\Sigma\,, \delta\mathbf{u}_\Sigma) >= \widetilde{a}_\Sigma(\omega\,; \mathbf{u}_\Sigma\,, \delta\mathbf{u}_\Sigma) \quad . \tag{99}$$

For $\mathbf{u}_\Sigma$ fixed in $\mathcal{C}^c_\Sigma$, the antilinear form $\delta\mathbf{u}_\Sigma \mapsto < \widetilde{\mathbf{A}}_\Sigma(\omega)\, \mathbf{u}_\Sigma\,, \delta\mathbf{u}_\Sigma) >$ on $\mathcal{C}^c_\Sigma$ defines a unique element [24] $\mathbf{f}_\Sigma$ such that, for all $\delta\mathbf{u}_\Sigma$ in $\mathcal{C}^c_\Sigma$,

$$< \mathbf{f}_\Sigma, \delta\mathbf{u}_\Sigma >=< \widetilde{\mathbf{A}}_\Sigma(\omega)\, \mathbf{u}_\Sigma\,, \delta\mathbf{u}_\Sigma) > \quad , \tag{100}$$

which yields

$$\mathbf{f}_\Sigma = \widetilde{\mathbf{A}}_\Sigma(\omega)\, \mathbf{u}_\Sigma \quad . \tag{101}$$

The boundary impedance operator [25] $\widetilde{\mathbf{Z}}_\Sigma(\omega)$ is then defined by

$$i\omega\, \widetilde{\mathbf{Z}}_\Sigma(\omega) = \widetilde{\mathbf{A}}_\Sigma(\omega) \quad . \tag{102}$$

From Eqs. (101) and (102), we deduce that

$$\mathbf{f}_\Sigma = i\omega\, \widetilde{\mathbf{Z}}_\Sigma(\omega)\, \mathbf{u}_\Sigma \quad . \tag{103}$$

It should be noted that boundary impedance operator $\widetilde{\mathbf{Z}}_\Sigma(\omega)$, related to Σ, does not depend on substructure Ω_1 but only on substructure Ω_2.

---

22 Operator $\mathbf{S}_\Sigma(\omega)$ is continuous from $\mathcal{C}^c_\Sigma$ into $\mathcal{C}^c_{\Omega_2} = (H^1(\Omega_2))^3$.

23 Sesquilinear form $\widetilde{a}_\Sigma(\omega\,; \mathbf{u}_\Sigma, \delta\mathbf{u}_\Sigma)$ is continuous on $\mathcal{C}^c_\Sigma \times \mathcal{C}^c_\Sigma$ and defines a continuous linear operator $\widetilde{\mathbf{A}}_\Sigma(\omega)$ from $\mathcal{C}^c_\Sigma$ into its antidual space $\mathcal{C}^{c\,\prime}_\Sigma$ and in Eq. (99), the angle brackets denote the antiduality product between $\mathcal{C}^{c\,\prime}_\Sigma$ and $\mathcal{C}^c_\Sigma$.

24 Since for all $\mathbf{u}_\Sigma$ fixed in $\mathcal{C}^c_\Sigma$, antilinear form $\delta\mathbf{u}_\Sigma \mapsto < \widetilde{\mathbf{A}}_\Sigma(\omega)\, \mathbf{u}_\Sigma, \delta\mathbf{u}_\Sigma >$ is continuous on $\mathcal{C}^c_\Sigma$, element $\mathbf{f}_\Sigma \in \mathcal{C}^{c\,\prime}_\Sigma$. In Eq. (100), the angle brackets denote the antiduality product between $\mathcal{C}^{c\,\prime}_\Sigma$ and $\mathcal{C}^c_\Sigma$.

25 Operator $\widetilde{\mathbf{Z}}_\Sigma(\omega)$ is continuous from $\mathcal{C}^c_\Sigma$ into its antidual space $\mathcal{C}^{c\,\prime}_\Sigma$.

Variational formulation in terms of $\mathbf{u}^1$. Substituting Eq. (97) into Eq. (86), using Eq. (98) and since $\mathbf{u}_\Sigma = \mathbf{u}^1|_\Sigma$ and $\delta\mathbf{u}_\Sigma = \delta\mathbf{u}^1|_\Sigma$, we obtain

$$a^1(\omega\,;\mathbf{u}^1,\delta\mathbf{u}^1) + \widetilde{a}_\Sigma(\omega\,;\mathbf{u}^1|_\Sigma,\delta\mathbf{u}^1|_\Sigma) = f^1(\omega\,;\delta\mathbf{u}^1)\ ,\ \ \forall\delta\mathbf{u}^1 \in \mathcal{C}^c_{\Omega_1}\ . \tag{104}$$

Using Eqs. (87) and (99), Eq. (104) can be rewritten as

$$<\mathbf{A}^1(\omega)\,\mathbf{u}^1,\delta\mathbf{u}^1> + <\widetilde{\mathbf{A}}_\Sigma(\omega)\,\mathbf{u}^1|_\Sigma,\delta\mathbf{u}^1|_\Sigma> = f^1(\omega\,;\delta\mathbf{u}^1)\ ,\ \forall\delta\mathbf{u}^1 \in \mathcal{C}^c_{\Omega_1}\,. \tag{105}$$

Introducing the *boundary dynamic stiffness operator* $\mathbf{A}_\Sigma$ such that [26]

$$<\mathbf{A}_\Sigma(\omega)\,\mathbf{u}^1\,,\delta\mathbf{u}^1> = <\widetilde{\mathbf{A}}_\Sigma(\omega)\,\mathbf{u}^1|_\Sigma\,,\delta\mathbf{u}^1|_\Sigma> \quad , \tag{106}$$

and using Eq. (88), Eq. (105) yields the following operator equation [27]

$$(\mathbf{A}^1(\omega) + \mathbf{A}_\Sigma(\omega))\,\mathbf{u}^1 = \mathbf{f}^1 \quad . \tag{107}$$

We introduce impedance operators $\mathbf{Z}^1(\omega)$ and $\mathbf{Z}_\Sigma(\omega)$ such that

$$i\omega\,\mathbf{Z}^1(\omega) = \mathbf{A}^1(\omega) \quad , \quad i\omega\,\mathbf{Z}_\Sigma(\omega) = \mathbf{A}_\Sigma(\omega) \quad . \tag{108}$$

It should be noted that operators $\mathbf{Z}_\Sigma(\omega)$ and $\widetilde{\mathbf{Z}}_\Sigma(\omega)$ are related by

$$<\mathbf{Z}_\Sigma(\omega)\,\mathbf{u}^1\,,\delta\mathbf{u}^1> = <\widetilde{\mathbf{Z}}_\Sigma(\omega)\,\mathbf{u}^1|_\Sigma\,,\delta\mathbf{u}^1|_\Sigma> \quad . \tag{109}$$

Eq. (107) can then be rewritten as

$$i\omega\,(\mathbf{Z}^1(\omega) + \mathbf{Z}_\Sigma(\omega))\,\mathbf{u}^1 = \mathbf{f}^1 \quad . \tag{110}$$

Existence and uniqueness of the boundary impedance operator. The above construction shows that boundary impedance operator $\mathbf{Z}_\Sigma(\omega)$ exists and is unique.

Symmetry property of the boundary impedance operator. It can easily be verified that boundary impedance operator $\mathbf{Z}_\Sigma(\omega)$ satisfies the following symmetry property,

$$^t\mathbf{Z}_\Sigma(\omega) = \overline{\mathbf{Z}}_\Sigma(\omega) \quad . \tag{110 – 1}$$

---

[26] Operator $\mathbf{A}_\Sigma(\omega)$ is continuous from $\mathcal{C}^c_{\Omega_1}$ into its antidual space $\mathcal{C}^{c\,\prime}_{\Omega_1}$.

[27] Eq. (107) is an equality in space $\mathcal{C}^{c\,\prime}_{\Omega_1}$.

Reduced matrix model for substructure Ω_2. The dynamic substructuring method presented in Section III.10.3 for conservative structures can be used to construct a reduced matrix model for dissipative substructure Ω_2. The reduced matrix model $\mathbf{A}^2_{\rm red}(\omega)$ of operator $\mathbf{A}^2(\omega)$ defined by Eq. (87) is then written as

$$\mathbf{A}^2_{\rm red}(\omega) = \begin{bmatrix} [\mathcal{A}_{\rm int}(\omega)] & \boldsymbol{\mathcal{A}}_{{\rm int},\Sigma}(\omega) \\ {}^t\overline{\boldsymbol{\mathcal{A}}}_{{\rm int},\Sigma}(\omega) & \boldsymbol{\mathcal{A}}_{\Sigma,\Sigma}(\omega) \end{bmatrix} \quad , \tag{110 - 2}$$

in which the detailed expressions of the operators which appear in Eq. (110-2) are given in Ohayon, Sampaio and Soize, 1997. From Eq. (110-2), we deduce the reduced boundary-dynamic-stiffness-matrix model $\widetilde{\mathbf{A}}_{\Sigma,\rm red}(\omega)$ of $\widetilde{\mathbf{A}}_{\Sigma}(\omega)$ defined by Eq. (99),

$$\widetilde{\mathbf{A}}_{\Sigma,\rm red}(\omega) = \boldsymbol{\mathcal{A}}_{\Sigma,\Sigma}(\omega) - {}^t\overline{\boldsymbol{\mathcal{A}}}_{{\rm int},\Sigma}(\omega)\,[\mathcal{A}_{\rm int}(\omega)]^{-1}\,\boldsymbol{\mathcal{A}}_{{\rm int},\Sigma}(\omega) \quad . \tag{110 - 3}$$

7.4. Finite element discretization of the boundary impedance operator

We consider a finite element mesh of substructure Ω_2. The complex-valued vector $\mathbf{U}^2$ of the DOFs which are the values of the displacement field $\mathbf{u}^2$ at the nodes of the finite element mesh of domain Ω_2 is written as $\mathbf{U}^2 = (\mathbf{U}^2_{\rm int}, \mathbf{U}_\Sigma)$ in which $\mathbf{U}^2_{\rm int}$ is related to the nodes in Ω_2 and on Γ_2 and $\mathbf{U}_\Sigma$ is related to the nodes on Σ. The finite element discretization of the variational formulation defined by Eq. (96) yields

$$[\delta\mathbf{U}^2]^T\,[A^2(\omega)]\,\mathbf{U}^2 = 0 \quad , \quad \forall\,\delta\mathbf{U}^2_{\rm int} \quad , \quad \delta\mathbf{U}_\Sigma = \mathbf{0} \quad , \tag{111}$$

in which $[A^2(\omega)]$ is the matrix corresponding to the finite element discretization of sesquilinear form $a^2(\omega\,;\mathbf{u}^2,\delta\mathbf{u}^2)$ defined by Eq. (83). We introduce the block splitting of $[A^2(\omega)]$ relative to $\mathbf{U}^2 = (\mathbf{U}^2_{\rm int}, \mathbf{U}_\Sigma)$,

$$[A^2(\omega)] = \begin{bmatrix} [A_{\rm int}(\omega)] & [A_{{\rm int},\Sigma}(\omega)] \\ [A_{{\rm int},\Sigma}(\omega)]^T & [A_{\Sigma,\Sigma}(\omega)] \end{bmatrix} \quad . \tag{112}$$

From Eqs. (111) and (112), we deduce that

$$\mathbf{U}^2_{\rm int} = [S(\omega)]\,\mathbf{U}_\Sigma \quad , \tag{113}$$

in which $[S(\omega)]$ is the matrix defined by

$$[S(\omega)] = -[A_{\rm int}(\omega)]^{-1}\,[A_{{\rm int},\Sigma}(\omega)] \quad . \tag{114}$$

It should be noted that Eqs. (113) and (114) correspond to the finite element discretization of Eq. (97). The finite element discretization of Eq. (98) yields $\delta\mathbf{U}_\Sigma^T\,[\widetilde{A}_\Sigma(\omega)]\,\mathbf{U}_\Sigma = [\delta\mathbf{U}^2]^T\,[A^2(\omega)]\,\mathbf{U}^2$ with $\mathbf{U}^2_{\text{int}} = [S(\omega)]\,\mathbf{U}_\Sigma$ and $\delta\mathbf{U}^2_{\text{int}} = [S(\omega)]\,\delta\mathbf{U}_\Sigma$. Using Eq. (114), we obtain the symmetric complex matrix $[\widetilde{A}_\Sigma(\omega)]$ of the finite element discretization of operator $\widetilde{\mathbf{A}}_\Sigma(\omega)$,

$$[\widetilde{A}_\Sigma(\omega)] = [A_{\Sigma,\Sigma}(\omega)] - [A_{\text{int},\Sigma}(\omega)]^T\,[A_{\text{int}}(\omega)]^{-1}\,[A_{\text{int},\Sigma}(\omega)] \quad . \tag{115}$$

The construction of Eq. (115) corresponds to the Schur complement (see Bathe and Wilson, 1976; Golub and Van Loan, 1989; Guyan, 1965). From Eq. (102), we deduce the symmetric complex matrix $[\widetilde{Z}_\Sigma(\omega)]$ corresponding to the finite element discretization of boundary impedance operator $\widetilde{\mathbf{Z}}_\Sigma(\omega)$,

$$i\omega\,[\widetilde{Z}_\Sigma(\omega)] = [\widetilde{A}_\Sigma(\omega)] \quad . \tag{116}$$

Finally, the finite element discretization of Eq. (103) is

$$\mathbf{F}_\Sigma = i\omega\,[\widetilde{Z}_\Sigma(\omega)]\,\mathbf{U}_\Sigma \quad . \tag{117}$$

7.5. Finite element discretization of structure Ω

We consider a finite element mesh of substructure Ω_1 compatible on interface Σ with the finite element mesh of substructure Ω_2 introduced in Section 7.4. Let $\mathbf{U}^1$ be the complex-valued vector of the DOFs which are the values of the displacement field $\mathbf{u}^1$ at the nodes of the finite element mesh of domain Ω_1. Vector $\mathbf{U}^1$ is written as $\mathbf{U}^1 = (\mathbf{U}^1_{\text{int}}, \mathbf{U}_\Sigma)$ in which $\mathbf{U}^1_{\text{int}}$ is related to the nodes in Ω_1 and on Γ_1 and $\mathbf{U}_\Sigma$ is related to the nodes on Σ. Let $[A^1(\omega)]$ be the matrix corresponding to the finite element discretization of sesquilinear form $a^1(\omega\,;\mathbf{u}^1,\delta\mathbf{u}^1)$ defined by Eq. (83). Then the finite element discretization of Eq. (104) yields

$$([A^1(\omega)] + [A_\Sigma(\omega)])\,\mathbf{U}^1 = \mathbf{F}^1 \quad , \tag{118}$$

in which $\mathbf{F}^1$ corresponds to the finite element discretization of the antilinear form $f^1(\omega\,;\delta\mathbf{u}^1)$ and $[A_\Sigma(\omega)]$ is the symmetric complex matrix corresponding to the finite element discretization of operator $\mathbf{A}_\Sigma(\omega)$ defined by Eq. (106),

$$[A_\Sigma(\omega)] = \begin{bmatrix} [0] & [0] \\ [0]^T & [\widetilde{A}_\Sigma(\omega)] \end{bmatrix} \quad . \tag{119}$$

In terms of symmetric complex impedance matrices, Eq. (118) yields

$$i\omega\,([Z^1(\omega)] + [Z_\Sigma(\omega)])\,\mathbf{U}^1 = \mathbf{F}^1 \quad , \tag{120}$$

in which

$$i\omega\,[Z^1(\omega)] = [A^1(\omega)] \quad , \quad i\omega\,[Z_\Sigma(\omega)] = [A_\Sigma(\omega)] \quad . \tag{121}$$

CHAPTER VI

Calculation of the Master Structure Frequency Response Function in the LF Range

1. Introduction

In Chapter V, we introduced the operator-valued frequency response function (FRF) for the master structure, which completely characterizes the response of the master structure and which allows calculation of the dynamical response of the structure submitted to arbitrary deterministic or random forces. An important problem is the effective calculation of this FRF. The present chapter is devoted to this calculation in the low-frequency (LF) range. Since this range corresponds to a low modal density case (see Chapter I), the methodology is mainly based on the use of the Ritz-Galerkin method (see Chapter II) in which the basis of the admissible function space is a finite set of elastic structural modes (see Chapter III). This method leads to a reduced model of the FRF for which a frequency-by-frequency construction is appropriate.

In Section 2, we deduce the expression of the FRF in the LF range from Chapters IV and V using the appropriate constitutive equation.

In Section 3, we construct a projection of the FRF on a subspace of finite dimension spanned by a set of elastic structural modes of the associated conservative system.

Section 4 deals with the particular cases for which the damping operator is diagonalized by the elastic structural modes.

In Section 5, we present the quasi-static correction terms due to the error induced by the modal truncation procedure.

Finally, in Section 6 we introduce the numerical procedure for constructing the FRF in the LF range.

2. FRF Model in the LF Range

2.1. Dynamic stiffness and impedance operators in the LF range

The constitutive equation of the master structure in the LF range is defined in Section IV.5.1. This model is given by Eqs. (IV.46) to (IV.50) in which tensor a_{ijkh} is independent of ω and tensor $b_{ijkh}(\omega)$ depends on ω. We then deduce that the stiffness sesquilinear form and its associated operator is independent of ω. The dynamic stiffness and impedance sesquilinear forms defined by Eqs. (V.44) and (V.45) are rewritten as

$$a(\omega\,;\mathbf{u}\,,\delta\mathbf{u}) = -\omega^2\, m(\mathbf{u}\,,\delta\mathbf{u}) + i\omega\, d(\omega\,;\mathbf{u}\,,\delta\mathbf{u}) + k(\mathbf{u}\,,\delta\mathbf{u}) \quad , \tag{1}$$

$$i\omega\, z(\omega\,;\mathbf{u}\,,\delta\mathbf{u}) = a(\omega\,;\mathbf{u}\,,\delta\mathbf{u}) \quad , \tag{2}$$

and from Eq. (V.47), the dynamic stiffness operator $\mathbf{A}(\omega)$ and impedance operator $\mathbf{Z}(\omega)$ are rewritten as

$$\mathbf{A}(\omega) = -\omega^2\,\mathbf{M} + i\omega\,\mathbf{D}(\omega) + \mathbf{K} \quad , \tag{3}$$

$$i\omega\,\mathbf{Z}(\omega) = \mathbf{A}(\omega) \quad . \tag{4}$$

2.2. Operator-valued FRF in the LF range

The operator-valued frequency response function $\mathbf{T}(\omega)$ is defined by Eq. (V.57) and is such that

$$\mathbf{u}(\omega) = \mathbf{T}(\omega)\,\mathbf{f}(\omega) \quad , \tag{5}$$

in which $\mathbf{f}(\omega)$ is defined by Eq. (V.39) and $\mathbf{u}(\omega)$ is the unique solution, in admissible function space $\mathcal{C}_S^c$ defined by Eq. (V.56), of the variational formulation

$$a(\omega\,;\mathbf{u}\,,\delta\mathbf{u}) = f(\omega\,;\delta\mathbf{u}) \quad , \quad \forall\delta\mathbf{u}\in\mathcal{C}_S^c \quad , \tag{6}$$

in which $a(\omega\,;\mathbf{u}\,,\delta\mathbf{u})$ is defined by Eq. (1) and $f(\omega\,;\delta\mathbf{u})$ by

$$f(\omega\,;\delta\mathbf{u}) = \int_\Omega \mathbf{g}(\mathbf{x},\omega)\cdot\overline{\delta\mathbf{u}(\mathbf{x})}\;d\mathbf{x} + \int_\Gamma \mathbf{G}(\mathbf{x},\omega)\cdot\overline{\delta\mathbf{u}(\mathbf{x})}\;ds(\mathbf{x}) \quad . \tag{7}$$

3. Projection of the FRF on the Elastic Structural Modes of the Associated Conservative System

3.1. Response to given force vector

We use the Ritz-Galerkin approximation presented in Chapter II. We introduce the subspace $\mathcal{C}^c_{S,N}$ of $\mathcal{C}^c_S$, of dimension $N \geq 1$, spanned by the finite family $\{\mathbf{u}_1, \ldots, \mathbf{u}_N\}$ of elastic structural modes $\mathbf{u}_\alpha$ of the associated conservative system defined in Sections 6 and 7 of Chapter III. The projection $\mathbf{u}^N$ on $\mathcal{C}^c_{S,N}$ of $\mathbf{u}$ in $\mathcal{C}^c_S$ can be written as

$$\mathbf{u}^N(\mathbf{x}, \omega) = \sum_{\alpha=1}^{N} q_\alpha(\omega)\, \mathbf{u}_\alpha(\mathbf{x}) \quad , \tag{8}$$

in which

$$\mathbf{q} = (q_1, \ldots, q_N) \tag{9}$$

is a complex-valued vector of the generalized coordinates. Any test function $\delta\mathbf{u}$ in $\mathcal{C}^c_{S,N}$ can be written as

$$\delta\mathbf{u}(\mathbf{x}) = \sum_{\beta=1}^{N} \delta q_\beta \; \mathbf{u}_\beta(\mathbf{x}) \quad , \tag{10}$$

in which $\delta\mathbf{q} = (\delta q_1, \ldots, \delta q_N)$ is a complex-valued vector. Substituting Eqs. (8) and (10) into Eq. (6) yields

$$\sum_{\alpha=1}^{N} \sum_{\beta=1}^{N} a(\omega\,; \mathbf{u}_\alpha\,, \mathbf{u}_\beta)\, q_\alpha\, \overline{\delta q_\beta} = \sum_{\beta=1}^{N} f(\omega\,; \mathbf{u}_\beta)\, \overline{\delta q_\beta} \quad , \quad \forall\, \delta\mathbf{q} \in \mathbb{C}^N \quad . \tag{11}$$

Using the orthogonality properties of the elastic structural modes established in Sections 6 and 7 of Chapter III, we deduce the following matrix equation of dimension N

$$[\mathcal{A}(\omega)]\, \mathbf{q}(\omega) = \boldsymbol{\mathcal{F}}(\omega) \quad , \tag{12}$$

in which $[\mathcal{A}(\omega)]$ is an $(N \times N)$ complex symmetric matrix

$$[\mathcal{A}(\omega)]^T = [\mathcal{A}(\omega)] \quad , \tag{13}$$

invertible for all real ω, such that

$$[\mathcal{A}(\omega)] = -\omega^2\, [\mathcal{M}\,] + i\omega\, [\mathcal{D}(\omega)] + [\,\mathcal{K}\,] \quad , \tag{14}$$

where $[\mathcal{M}]$, $[\mathcal{D}(\omega)]$ and $[\mathcal{K}]$ are $(N \times N)$ real symmetric positive-definite matrices. Matrices $[\mathcal{M}]$ and $[\mathcal{K}]$ are diagonal and such that

$$[\mathcal{M}]_{\alpha\beta} = m(\mathbf{u}_\beta, \mathbf{u}_\alpha) = \mu_\alpha\, \delta_{\alpha\beta} \quad , \tag{15}$$

$$[\mathcal{K}]_{\alpha\beta} = k(\mathbf{u}_\beta, \mathbf{u}_\alpha) = \mu_\alpha\, \omega_\alpha^2 \delta_{\alpha\beta} \quad . \tag{16}$$

For all fixed real ω, matrix $[\mathcal{D}(\omega)]$ is dense and such that

$$[\mathcal{D}(\omega)]_{\alpha\beta} = d(\omega\,; \mathbf{u}_\beta, \mathbf{u}_\alpha) \quad . \tag{17}$$

The component $\mathcal{F}_\alpha$ of the complex-valued vector of the generalized forces

$$\boldsymbol{\mathcal{F}} = (\mathcal{F}_1, \ldots, \mathcal{F}_N) \quad , \tag{18}$$

is written as

$$\begin{aligned} \mathcal{F}_\alpha(\omega) &= f(\omega\,; \mathbf{u}_\alpha) \\ &= \int_\Omega \mathbf{g}(\mathbf{x}, \omega) \cdot \mathbf{u}_\alpha(\mathbf{x})\, d\mathbf{x} + \int_\Gamma \mathbf{G}(\mathbf{x}, \omega) \cdot \mathbf{u}_\alpha(\mathbf{x})\, ds(\mathbf{x}) \quad . \end{aligned} \tag{19}$$

From Eq. (12) and the invertibility of matrix $[\mathcal{A}(\omega)]$, we deduce that

$$\mathbf{q}(\omega) = [\mathcal{T}(\omega)]\, \boldsymbol{\mathcal{F}}(\omega) \quad , \tag{20}$$

in which $[\mathcal{T}(\omega)]$ is an $(N \times N)$ complex symmetric matrix such that

$$[\mathcal{T}(\omega)] = [\mathcal{A}(\omega)]^{-1} \quad , \quad [\mathcal{T}(\omega)]^T = [\mathcal{T}(\omega)] \quad . \tag{21}$$

From Eqs. (8), (20) and (21), we deduce that

$$\mathbf{u}^N(\mathbf{x}, \omega) = \sum_{\alpha=1}^{N} \sum_{\beta=1}^{N} [\mathcal{T}(\omega)]_{\alpha\beta}\, \mathcal{F}_\beta(\omega)\, \mathbf{u}_\alpha(\mathbf{x}) \quad . \tag{22}$$

3.2. Projection of the FRF on the elastic structural modes

As explained in Section V.5, operator-valued frequency response function $\mathbf{T}(\omega)$ is an intrinsic operator characterizing the dynamics of the structure independently of the values of the applied forces. It should be noted that in the context of three-dimensional viscoelasticity, the input of the linear filter whose $\mathbf{T}(\omega)$ is the FRF is a body force field in the three-dimensional domain

Ω. For this reason, we can take $\mathbf{G} = \mathbf{0}$ to construct $\mathbf{T}(\omega)$. Consequently, force vector $\mathbf{f}(\omega)$ can be written as [1]

$$f(\omega\,;\delta\mathbf{u}) = <\mathbf{f}(\omega)\,,\delta\mathbf{u}> = \int_\Omega \mathbf{g}(\mathbf{x},\omega)\cdot\overline{\delta\mathbf{u}(\mathbf{x})}\;d\mathbf{x} \quad , \tag{23}$$

and Eq. (19) becomes

$$\mathcal{F}_\beta(\omega) = \int_\Omega \mathbf{g}(\mathbf{x},\omega)\cdot\mathbf{u}_\beta(\mathbf{x})\;d\mathbf{x} \quad . \tag{24}$$

Let $\mathbf{T}_N(\omega)$ be the projection on $\mathcal{C}^c_{S,N}$ of operator-valued frequency response function $\mathbf{T}(\omega)$ defined by Eq. (5). From Eq. (22), we deduce that

$$\mathbf{T}_N(\omega)\,\mathbf{f}(\omega) = \sum_{\alpha=1}^N\sum_{\beta=1}^N\,[\mathcal{T}(\omega)]_{\alpha\beta}\,\mathcal{F}_\beta(\omega)\,\mathbf{u}_\alpha \quad . \tag{25}$$

Substituting Eq. (24) into Eq. (25) yields, for all $\mathbf{x}$ in Ω,

$$\big(\mathbf{T}_N(\omega)\,\mathbf{f}(\omega)\big)(\mathbf{x}) = \int_\Omega [\,\tau_N(\omega;\mathbf{x},\mathbf{x}')]\;\mathbf{g}(\mathbf{x}')\,d\mathbf{x}' \quad , \tag{26}$$

in which $[\,\tau_N(\omega;\mathbf{x},\mathbf{x}')]$ is the (3×3) complex matrix defined by

$$[\,\tau_N(\omega;\mathbf{x},\mathbf{x}')] = \sum_{\alpha=1}^N\sum_{\beta=1}^N\,[\mathcal{T}(\omega)]_{\alpha\beta}\,\mathbf{u}_\alpha(\mathbf{x})\,\mathbf{u}_\beta(\mathbf{x}')^T \quad . \tag{27}$$

Equation (26) shows that linear operator $\mathbf{T}_N(\omega)$ is an integral operator [2] with a kernel $[\,\tau_N(\omega;\mathbf{x},\mathbf{x}')]$ satisfying the following property

$$[\,\tau_N(\omega;\mathbf{x},\mathbf{x}')] = [\,\tau_N(\omega;\mathbf{x}',\mathbf{x})]^T \quad , \tag{28}$$

and consequently, for all fixed real ω, we have [3]

$$\mathbf{T}_N(\omega)^* = \overline{\mathbf{T}_N(\omega)} \quad . \tag{29}$$

---

[1] If $\mathbf{x}\mapsto\mathbf{g}(\mathbf{x},\omega)$ belongs to $H^c=(L^2(\Omega))^3$, then the antiduality product $<\mathbf{g}\,,\delta\mathbf{u}>$ between $\mathcal{C}^{c\,\prime}_S$ and $\mathcal{C}^c_S$ of $\mathbf{g}$ with $\delta\mathbf{u}$ is equal to the usual inner product $(\mathbf{g}\,,\delta\mathbf{u})_{H^c}=\int_\Omega \mathbf{g}(\mathbf{x})\cdot\overline{\delta\mathbf{u}(\mathbf{x})}\,d\mathbf{x}$ of H^c. In this case, we have $\mathbf{f}(\omega)=\mathbf{g}(\omega)$.

[2] Integral operator $\mathbf{T}_N(\omega)$ considered as an operator in $H^c=(L^2(\Omega))^3$ can be written as $\mathbf{T}_N(\omega)=\sum_{\alpha=1}^N\sum_{\beta=1}^N\,[\mathcal{T}(\omega)]_{\alpha\beta}\;\mathbf{u}_\alpha\otimes_{H^c}\mathbf{u}_\beta$.

[3] From a mathematical point of view, $\mathbf{T}_N(\omega)^*$ is the adjoint operator of bounded operator $\mathbf{T}_N(\omega)$ (see Mathematical Notations in the appendix).

3.3. Finite element discretization of the projection of the FRF on elastic structural modes

Consider a finite element discretization of domain Ω. Let n be the number of degrees of freedom. Let $\{\mathbf{U}_\alpha\}_{\alpha=1,\ldots,N}$ be the set of N eigenvectors defined by Eq. (III.93) which corresponds to the finite element discretization of the continuous eigenvalue problem. Each $\mathbf{U}_\alpha$ is an $\mathbb{R}^n$ vector representing the finite element discretization of elastic structural mode $\mathbf{u}_\alpha(\mathbf{x})$. The matrix $[T_N(\omega)]$ of operator $\mathbf{T}_N(\omega)$ obtained from the finite element discretization, can then be written as

$$[T_N(\omega)] = \sum_{\alpha=1}^{N}\sum_{\beta=1}^{N} [\mathcal{T}(\omega)]_{\alpha\beta}\, \mathbf{U}_\alpha\, \mathbf{U}_\beta^T \quad , \tag{30}$$

in which $[\mathcal{T}(\omega)]$ is the $(N \times N)$ complex symmetric matrix defined by Eq. (21). Matrix $[T_N(\omega)]$ is an $(n \times n)$ complex symmetric matrix,

$$[T_N(\omega)]^T = [T_N(\omega)] \quad . \tag{31}$$

This property corresponds to Eq. (29).

4. Remark on a Nonviscoelastic Model Diagonalized by the Elastic Structural Modes

4.1. Statement of the problem

We saw above that with the theory of linear viscoelasticity, damping operator $\mathbf{D}(\omega)$ was not diagonalized by the elastic structural modes (matrix $[\mathcal{D}(\omega)]$ is dense with $[\mathcal{D}(\omega)]_{\alpha\beta} = d(\omega\,; \mathbf{u}_\beta, \mathbf{u}_\alpha)$ and there is no reason for matrix $[\mathcal{D}(\omega)]$ to be diagonal). However, this does not raise any problems because the inversion of $(N \times N)$ dense matrix $[\mathcal{A}(\omega)]$ to construct $[\mathcal{T}(\omega)] = [\mathcal{A}(\omega)]^{-1}$ is straightforward because $N \ll n$. Nevertheless, if tensor $\{b_{ijkh}(\omega, \mathbf{x})\}_{ijkh}$ introduced in the constitutive equation for the viscoelastic material (see Section IV.5.1) is not available for some reason, a mathematical model which is not based on the linear theory of viscoelasticity can sometimes be used for small damped structures in the low-frequency range. In this case, a convenient mathematical model can be constructed so that damping operator $\mathbf{D}(\omega)$ is diagonalized by the elastic structural modes, i.e., $[\mathcal{D}(\omega)]$ is an $(N \times N)$ diagonal matrix.

4.2. Diagonalization of the damping operator

When damping operator $\mathbf{D}(\omega)$ is diagonalized by the elastic structural modes, matrix $[\mathcal{D}(\omega)]$ defined by Eq. (17) is an $(N \times N)$ diagonal matrix and we write

$$<\mathbf{D}(\omega)\,\mathbf{u}_\beta\,,\mathbf{u}_\alpha> = d(\omega\,;\mathbf{u}_\beta,\mathbf{u}_\alpha) \tag{32}$$
$$= [\mathcal{D}(\omega)]_{\alpha\beta} \tag{33}$$
$$= 2\,\mu_\alpha\,\omega_\alpha\,\xi_\alpha(\omega)\,\delta_{\alpha\beta} \tag{34}$$

in which μ_α and ω_α are defined by Eqs. (15) and (16) respectively. The critical damping rate $\xi_\alpha(\omega)$ of elastic structural mode $\mathbf{u}_\alpha$ is a positive real number depending on Ω,

$$\xi_\alpha(\omega) > 0 \quad , \quad \forall\omega \in \mathbb{R} \quad , \quad \forall\alpha \in \{1, 2, \ldots\} \quad . \tag{35}$$

A small damped structure is a structure such that

$$0 < \xi_\alpha(\omega) \ll 1 \quad , \quad \forall\omega \in \mathbb{R} \quad , \quad \forall\alpha \in \{1, 2, \ldots\} \quad . \tag{36}$$

The problem is then to establish general algebraic expressions of operator $\mathbf{D}(\omega)$ such that $\mathbf{D}(\omega)$ is diagonalized by the elastic structural modes. Below, we present two algebraic models. The first corresponds to the situation where the elastic structural modes and the mass operator are used. The second corresponds to the case where both the mass and stiffness operators are used but not the elastic structural modes.

4.3. General algebraic expression for a diagonalizable damping operator based on the use of the elastic structural modes and the mass operator

Let $d(\omega\,;\mathbf{u},\delta\mathbf{u})$ be the structural damping sesquilinear form related to the damping operator by (see Section 4.3 of Chapter V)

$$<\mathbf{D}(\omega)\,\mathbf{u}\,,\delta\mathbf{u}> = d(\omega\,;\mathbf{u},\delta\mathbf{u}) \quad . \tag{37}$$

We propose a general algebraic expression for $d(\omega\,;\mathbf{u},\delta\mathbf{u})$ based on the use of the finite family $\{\mathbf{u}_1,\ldots,\mathbf{u}_N\}$ of the elastic structural modes $(N \geq 1)$ and the mass operator, such that [4]

$$d(\omega\,;\mathbf{u},\delta\mathbf{u}) = \sum_{\gamma=1}^{N} 2\,\xi_\gamma(\omega)\,\frac{\omega_\gamma}{\mu_\gamma}\,m(\mathbf{u}_\gamma,\mathbf{u})\,m(\mathbf{u}_\gamma,\delta\mathbf{u}) \quad . \tag{38}$$

---

[4] Operator $\mathbf{D}(\omega)$ considered as an operator in $H^c = (L^2(\Omega))^3$ is written as $\mathbf{D}(\omega) = \sum_{\gamma=1}^{N} 2\xi_\gamma(\omega)\,\frac{\omega_\gamma}{\mu_\gamma}\,(\rho\,\mathbf{u}_\gamma) \otimes_{H^c} (\rho\,\mathbf{u}_\gamma)$. This is an equivalent expression of Eq. (38).

Using the orthogonality condition defined by Eq. (15), it can then be verified that Eq. (34) holds. If critical damping rate $\xi_\alpha(\omega)$ is known for each elastic structural mode $\mathbf{u}_\alpha$, then Eq. (38) allows $d(\omega\,;\mathbf{u},\delta\mathbf{u})$, i.e. damping operator $\mathbf{D}(\omega)$, to be constructed. It should be noted that a more general expression can also be considered [5].

4.4. General algebraic expression for a diagonalizable damping operator based on the use of the mass and stiffness operators

We propose a similar algebraic representation of $d(\omega\,;\mathbf{u},\delta\mathbf{u})$ for the continuous case based on the matrix expression given by Caughey, 1960. This expression uses the mass and stiffness operators $\mathbf{M}$ and $\mathbf{K}$, such that [6]

$$\mathbf{D}(\omega) = \mathbf{M}\ \ P(\mathbf{M}^{-1}\mathbf{K}) \quad , \tag{39}$$

in which $P(z)$ is a real polynomial of degree $\nu \geq 0$:

$$P(z) = a_0(\omega) + a_1(\omega)\, z + \ldots + a_\nu(\omega)\, z^\nu \quad , \tag{40}$$

where $a_0(\omega), a_1(\omega), \ldots, a_\nu(\omega)$ are real constants, not equal to zero simultaneously, depending on ω and such that operator $\mathbf{M}\,P(\mathbf{M}^{-1}\mathbf{K})$ is positive. Using the orthogonality conditions defined by Eqs. (15) and (16), it can then be verified that Eq. (34) holds and that, for all $\alpha \geq 1$, we have

$$2\,\omega_\alpha\,\xi_\alpha(\omega) = a_0(\omega) + a_1(\omega)\,\omega_\alpha^2 + \ldots + a_\nu(\omega)\,\omega_\alpha^{2^\nu} \quad . \tag{41}$$

Equations (39) to (41) allow the following *inverse problem* to be solved for all fixed real ω. This problem is to construct damping operator $\mathbf{D}(\omega)$ for $\nu = N-1$ and for given critical damping rates $\xi_1(\omega), \ldots, \xi_N(\omega)$ of elastic structural modes $\mathbf{u}_1, \ldots, \mathbf{u}_N$. The procedure consists in solving the linear system of equations defined by Eq. (41) for $\alpha = \{1, \ldots, N\}$, in which the unknowns are $a_0(\omega), a_1(\omega), \ldots, a_{N-1}(\omega)$. It should be noted that the

---

5 Let $f(z_1,z_2)$ be a real-valued function on $\mathbb{R}\times\mathbb{R}$ such that $f(z_1,z_2)=\sum_{\nu_1,\nu_2} f_{\nu_1,\nu_2}(\omega)\, z_1^{\nu_1}\, z_2^{\nu_2}$ in which $\nu_1\geq 0$, $\nu_2\geq 0$ and $\nu_1+\nu_2\geq 1$. It is assumed that the real constants $f_{\nu_1,\nu_2}(\omega)$ which depend on ω are such that, $\forall\alpha=\{1,2,\ldots\}$, the series $\sum_{\nu_1,\nu_2} f_{\nu_1,\nu_2}(\omega)\,\mu_\alpha^{2\nu_1}(\mu_\alpha\omega_\alpha^2)^{2\nu_2}$ is convergent. The limit of this series is then denoted as $\xi_\alpha(\omega)$. A general expression allowing $d(\omega;\mathbf{u},\delta\mathbf{u})$ to be diagonalized by the basis $\{\mathbf{u}_\alpha\}_{\alpha\geq 1}$ can be written as $d(\omega;\mathbf{u},\delta\mathbf{u})= \sum_{\gamma=1}^{N} 2\mu_\gamma\omega_\gamma f(m(\mathbf{u}_\gamma,\mathbf{u})\, m(\mathbf{u}_\gamma,\delta\mathbf{u})\,,\, k(\mathbf{u}_\gamma,\mathbf{u})\, k(\mathbf{u}_\gamma,\delta\mathbf{u}))$. It can then be verified that Eq. (34) holds.

6 Operators $\mathbf{D}(\omega)$ and $\mathbf{K}$, which are bounded operators from $\mathcal{C}_S^c$ into its antidual space $\mathcal{C}_S^{c\,\prime}$, are considered in Eq. (39) as unbounded operators in $H^c=(L^2(\Omega))^3$ for $\nu\geq 1$. We recall that $\mathbf{M}$ is a bounded operator in H^c which is invertible.

above procedure can be extended to an overdetermined case for which the degree ν of $P(z)$ is less than the number N of given critical damping rates. A minimization procedure can then be applied.

Particular cases of a proportional damping operator. This is the particular case of the above model for which $\nu = 1$ and therefore, Eqs. (39) and (40) yield

$$\mathbf{D}(\omega) = a_0(\omega)\,\mathbf{M} + a_1(\omega)\,\mathbf{K} \quad , \tag{42}$$

which is sometimes called *Rayleigh damping.* If, in addition, $a_0(\omega) = 0$, we obtain

$$\mathbf{D}(\omega) = a_1(\omega)\,\mathbf{K} \quad , \tag{43}$$

which is sometimes called *hysteretic damping.* In this last case, tensor $b_{ijkh}(\omega)$ is such that $b_{ijkh}(\omega) = a_1(\omega)\,a_{ijkh}$ where tensor a_{ijkh} does not depend on ω. In the same way, if $a_1(\omega) = 0$, we obtain

$$\mathbf{D}(\omega) = a_0(\omega)\,\mathbf{M} \quad . \tag{44}$$

4.5. Finite element discretization of the representations of the damping operator

We consider the finite element discretization of domain Ω introduced in Section 3.3 in which n is the number of degrees of freedom. The finite element discretization of Eqs. (37) and (38) is

$$[D(\omega)] = \sum_{\gamma=1}^{N} 2\,\xi_\gamma(\omega)\,\frac{\omega_\gamma}{\mu_\gamma}\,[M]\,\mathbf{U}_\gamma\,\mathbf{U}_\gamma^T\,[M] \quad . \tag{45}$$

The finite element discretization of Eq. (39) is

$$[D(\omega)] = [M]\;\; P([M]^{-1}\,[K]) \quad . \tag{46}$$

If $N = n$ in Eq. (45), then the two above discretized models coincide [7], i.e.

$$[M]\;\; P([M]^{-1}\,[K]) = \sum_{\gamma=1}^{n} 2\,\xi_\gamma(\omega)\,\frac{\omega_\gamma}{\mu_\gamma}\,[M]\,\mathbf{U}_\gamma\,\mathbf{U}_\gamma^T\,[M] \quad . \tag{47}$$

In order to prove Eq. (47), we use the representation $[K]=\sum_{\gamma=1}^{n} \omega_\gamma^2\,\frac{1}{\mu_\gamma}\,[M]\,\mathbf{U}_\gamma\,\mathbf{U}_\gamma^T\,[M]$ which, for all integers $\ell\geq 0$, yields $\left([M]^{-1}[K]\right)^\ell=\sum_{\gamma=1}^{n}(\omega_\gamma^2)^\ell\,\frac{1}{\mu_\gamma}\,[M]\,\mathbf{U}_\gamma\,\mathbf{U}_\gamma^T\,[M]$.

5. Introduction of Quasi-Static Correction Terms

5.1. Construction of the quasi-static correction terms

Let $N \geq 1$ be an integer such that

$$0 < \omega_1 \leq \ldots \leq \omega_N < \omega_{\mathrm{LF,\,final}} < \omega_{N+1} \leq \ldots \quad , \tag{48}$$

in which $\mathbb{B}_{\mathrm{LF}} = [\,0\,, \omega_{\mathrm{LF,\,final}}\,]$ is the low-frequency band of analysis. The projection $\mathbf{T}_N(\omega)$ of the operator-valued frequency response function $\mathbf{T}(\omega)$ defined by Eqs. (26)-(27), represents the approximation corresponding to the first N elastic structural modes $\{\mathbf{u}_\alpha\}_{\alpha=1,\ldots,N}$ whose eigenfrequencies $\{\omega_\alpha\}_{\alpha=1,\ldots,N}$ lie inside $\mathbb{B}_{\mathrm{LF}}$. We introduce the operator $\mathbf{R}_N(\omega)$ defined by

$$\mathbf{R}_N(\omega) = \mathbf{T}(\omega) - \mathbf{T}_N(\omega) \quad , \tag{49}$$

which represents the projection of $\mathbf{T}(\omega)$ on the remaining elastic structural modes $\{\mathbf{u}_\alpha\}_{\alpha=N+1,\ldots,+\infty}$. We then have the identity

$$\mathbf{T}(\omega) = \mathbf{T}_N(\omega) + \mathbf{R}_N(\omega) \quad . \tag{50}$$

For $\omega \in \mathbb{B}_{\mathrm{LF}}$, operator $\mathbf{R}_N(\omega)$ can be approximated by $\mathbf{R}_N(0)$. Operator $\mathbf{R}_N(0)$ represents the quasi-static contribution of the elastic structural modes whose eigenfrequencies are greater than $\omega_{\mathrm{LF,\,final}}$. Consequently, in order to accelerate the convergence, we introduce the sequence of operators including quasi-static correction terms such that

$$\mathbf{T}_{N,\,\mathrm{acc}}(\omega) = \mathbf{T}_N(\omega) + \mathbf{R}_N(0) \quad , \tag{51}$$

with $\mathbf{R}_N(0)$ given by

$$\mathbf{R}_N(0) = \mathbf{T}(0) - \mathbf{T}_N(0) \quad . \tag{52}$$

From Sections 2.1 and 2.2, we deduce that

$$\mathbf{T}(0) = \mathbf{K}^{-1} \quad , \tag{53}$$

in which $\mathbf{K}^{-1}$ denotes the inverse [8] of $\mathbf{K}$, considered as an operator defined on $\mathcal{C}_S^c$. From Eqs. (26) and (27), we deduce that

$$\big(\mathbf{T}_N(0)\,\mathbf{f}(\omega)\big)(\mathbf{x}) = \int_\Omega [\,\tau_N(0;\mathbf{x},\mathbf{x}')]\ \mathbf{g}(\mathbf{x}',\omega)\,d\mathbf{x}' \quad , \tag{54}$$

[8] Operator $\mathbf{K}$ is continuous and invertible from $\mathcal{C}_S^c$ into $\mathcal{C}_S^{c\,\prime}$. Consequently, operator $\mathbf{K}^{-1}$ is continuous from $\mathcal{C}_S^{c\,\prime}$ into $\mathcal{C}_S^c$.

$$[\tau_N(0;\mathbf{x},\mathbf{x}')] = \sum_{\alpha=1}^{N} \frac{1}{\mu_\alpha\,\omega_\alpha^2}\,\mathbf{u}_\alpha(\mathbf{x})\,\mathbf{u}_\alpha(\mathbf{x}')^T \quad . \tag{55}$$

Finally, operator $\mathbf{T}_{N,\,\mathrm{acc}}(\omega)$ can be written as

$$\mathbf{T}_{N,\,\mathrm{acc}}(\omega) = \mathbf{T}_N(\omega) + \mathbf{T}_{N,\,\mathrm{stat}} \quad , \tag{56}$$

in which $\mathbf{T}_N(\omega)$ is defined by Eqs. (26)-(27) and $\mathbf{T}_{N,\,\mathrm{stat}} = \mathbf{R}_N(0)$ is written as

$$\mathbf{T}_{N,\,\mathrm{stat}} = \mathbf{K}^{-1} - \mathbf{T}_N(0) \quad , \tag{57}$$

where $\mathbf{T}_N(0)$ is defined by Eqs. (54) and (55). For N fixed and for all ω fixed in $\mathbb{B}_{\mathrm{LF}}$, projection $\mathbf{T}_{N,\,\mathrm{acc}}(\omega)$ is a better approximation of $\mathbf{T}(\omega)$ than $\mathbf{T}_N(\omega)$.

5.2. Finite element discretization of the operator-valued FRF

We consider the finite element discretization of domain Ω introduced in Section 3.3 in which n is the number of degrees of freedom. The finite element discretization of operator $\mathbf{K}^{-1}$ is the $(n \times n)$ real symmetric matrix denoted as $[K]^{-1}$ (matrix $[K]^{-1}$ denotes the inverse of $[K]$ in subspace $\mathcal{C}^c_{S,n}$ of $\mathcal{C}^c_S$). The finite element discretization of Eq. (56) is written as

$$[T_{N,\,\mathrm{acc}}(\omega)] = [T_N(\omega)] + [T_{N,\,\mathrm{stat}}] \quad , \tag{58}$$

in which $[T_N(\omega)]$ is defined by Eq. (30) and $[T_{N,\,\mathrm{stat}}]$ is such that

$$[T_{N,\,\mathrm{stat}}] = [K]^{-1} - \sum_{\alpha=1}^{N} \frac{1}{\mu_\alpha\,\omega_\alpha^2}\,\mathbf{U}_\alpha\,\mathbf{U}_\alpha^T \quad . \tag{59}$$

6. Frequency-by-Frequency Construction of the FRF

The numerical procedure for constructing an approximation of operator-valued frequency response function $\omega \mapsto \mathbf{T}(\omega)$ on a low-frequency band

$$\mathbb{B}_{\mathrm{LF}} = [\,0\,,\omega_{\mathrm{LF,final}}\,] \quad , \tag{60}$$

is based on the use of projection $\mathbf{T}_N(\omega)$ of $\mathbf{T}(\omega)$ on the first N elastic structural modes, defined by Eqs. (26) and (27) (or by taking into account quasi-static correction terms, using $\mathbf{T}_{N,\,\mathrm{acc}}(\omega)$ defined by Eqs. (56) and (57)). We then have to calculate the elastic structural modes (see Chapter

III) whose eigenfrequencies lie inside low-frequency band $\mathbb{B}_{LF}$. In the low-frequency range, N is small. This means that in the context of the finite element discretization with n DOFs, we have $N \ll n$. Consequently, matrix $[\mathcal{A}(\omega)]$ can be inverted numerically frequency by frequency in $\mathbb{B}_{LF}$ in order to construct $[\mathcal{T}(\omega)] = [\mathcal{A}(\omega)]^{-1}$ (see Eq. (21)). If $[\mathcal{A}(\omega)]$ is a diagonal matrix, i.e. if generalized damping matrix $[\mathcal{D}(\omega)]$ is diagonal (see Section 4), then this inversion is explicit. In addition, if $\mathbf{T}_{N,\,\mathrm{acc}}(\omega)$ is used, we have to calculate $\mathbf{T}_{N,\,\mathrm{stat}}$ defined by Eq. (57), i.e. basically, an approximation of $\mathbf{K}^{-1}$. Eq. (59) is used for finite element discretization (but matrix $[\,K\,]$ is generally not inverted explicitely, see Chapter IX).

CHAPTER VII

Calculation of the Master Structure Frequency Response Function in the MF Range

1. Introduction

In Chapter V, we introduced the operator-valued frequency response function (FRF) for the master structure, which allows calculation of the dynamical response of the structure submitted to arbitrary deterministic or random forces. The present chapter concerns the important problem of effective calculation of the FRF in the medium-frequency (MF) range. Since this frequency range does not correspond to a low modal density case (see Chapter I), the methodology presented in Chapter VI, which is based on the use of the elastic structural modes, is not applicable for practical reasons. In addition, it should be noted that a frequency-by-frequency construction of the solution of the initial unreduced system is not appropriate. Consequently, below we present an appropriate MF method for constructing the FRF in the MF range.

The general concepts of this MF method were initially introduced by Soize, 1982a and 1982b, Chabas and Soize, 1987, for structural vibrations, then extended to structural acoustics by Soize et al., 1986b and 1992. This method was also analyzed by Liu, Zhang and Ramirez, 1991. Applications of this method were made by Vasudevan, 1991; Vasudevan and Liu, 1991.

In Section 2, we present the model of the FRF in the MF range, deduced from Chapter V.

In Section 3, we give the definition of an MF narrow band. The MF range is considered as a finite union of MF narrow bands.

Section 4 deals with the definition of a class of excitation force fields for a given MF narrow band. This class of particular excitations allows the FRF to be constructed in an MF narrow band.

In Section 5, we introduce an approximation of the FRF on an MF narrow band due to the frequency-dependent damping and stiffness operators.

In Section 6, we introduce the MF method for constructing the FRF on an MF narrow band. This method is based on a frequency transform technique which uses two time scales, a short time scale associated with the center frequency and a long time scale associated with the width of the MF narrow band.

Finally, in Section 7 we present the construction of the FRF on an MF broad band.

2. FRF Model in the MF Range

2.1. Dynamic stiffness and impedance operators in the MF range

The constitutive equation for the master structure in the MF range is defined in Section IV.5.2. This model is given by Eqs. (IV.46) to (IV.50) in which tensor $a_{ijkh}(\omega)$ and tensor $b_{ijkh}(\omega)$ both depend on ω. The dynamic stiffness and impedance sesquilinear forms are then defined by Eqs. (V.44) and (V.45),

$$a(\omega\,;\mathbf{u}\,,\delta\mathbf{u}) = -\omega^2\, m(\mathbf{u}\,,\delta\mathbf{u}) + i\omega\, d(\omega\,;\mathbf{u}\,,\delta\mathbf{u}) + k(\omega\,;\mathbf{u}\,,\delta\mathbf{u}) \quad , \tag{1}$$

$$i\omega\, z(\omega\,;\mathbf{u}\,,\delta\mathbf{u}) = a(\omega\,;\mathbf{u}\,,\delta\mathbf{u}) \quad , \tag{2}$$

and dynamic stiffness operator $\mathbf{A}(\omega)$ and impedance operator $\mathbf{Z}(\omega)$ defined by Eq. (V.47) are such that

$$\mathbf{A}(\omega) = -\omega^2\,\mathbf{M} + i\omega\,\mathbf{D}(\omega) + \mathbf{K}(\omega) \quad , \tag{3}$$

$$i\omega\,\mathbf{Z}(\omega) = \mathbf{A}(\omega) \quad . \tag{4}$$

2.2. Operator-valued FRF in the MF range

The operator-valued frequency response function $\mathbf{T}(\omega)$ is defined by Eq. (V.57) and is such that

$$\mathbf{u}(\omega) = \mathbf{T}(\omega)\,\mathbf{f}(\omega) \quad , \tag{5}$$

in which $\mathbf{f}(\omega)$ is defined by Eq. (V.39) and $\mathbf{u}(\omega)$ is the unique solution, in admissible function space $\mathcal{C}_S^c$ defined by Eq. (V.56), of the variational formulation,

$$a(\omega\,;\mathbf{u}\,,\delta\mathbf{u}) = f(\omega\,;\delta\mathbf{u}) \quad , \quad \forall\delta\mathbf{u} \in \mathcal{C}_S^c \quad , \tag{6}$$

in which $a(\omega\,;\mathbf{u}\,,\delta\mathbf{u})$ is defined by Eq. (1) and $f(\omega\,;\delta\mathbf{u})$ by

$$f(\omega\,;\delta\mathbf{u}) = \int_\Omega \mathbf{g}(\mathbf{x},\omega)\cdot\overline{\delta\mathbf{u}(\mathbf{x})}\;d\mathbf{x} + \int_\Gamma \mathbf{G}(\mathbf{x},\omega)\cdot\overline{\delta\mathbf{u}(\mathbf{x})}\;ds(\mathbf{x}) \quad . \tag{7}$$

We recall that the relation between $\mathbf{f}(\omega)$ and $f(\omega\,;\delta\mathbf{u})$ is written as [1]

$$<\mathbf{f}(\omega)\,,\delta\mathbf{u}> = f(\omega\,;\delta\mathbf{u}) \quad , \quad \forall\delta\mathbf{u}\in\mathcal{C}_S^c \quad . \tag{8}$$

3. Definition of an MF Narrow Band

Let $\mathbb{B}_\nu$ be the limited frequency band of $\mathbb{R}^+$ defined by

$$\mathbb{B}_\nu = [\,\Omega_\nu - \Delta\omega/2\,,\,\Omega_\nu + \Delta\omega/2\,] \quad , \tag{9}$$

in which $\Omega_\nu > 0$ is the center frequency of band $\mathbb{B}_\nu$ and $\Delta\omega$ is its bandwidth such that $0 < \Delta\omega < 2\,\Omega_\nu$. With band $\mathbb{B}_\nu$, we associate band $\mathbb{B}_0$ having the same bandwidth and defined by

$$\mathbb{B}_0 = [\,-\Delta\omega/2\,,\Delta\omega/2\,] \quad . \tag{10}$$

By definition, limited frequency band $\mathbb{B}_\nu$ is an MF narrow band if

$$\frac{\Delta\omega}{\Omega_\nu} \ll 1 \quad . \tag{11}$$

We introduce two time scales τ_{long} and τ_{short} such that

$$\tau_{\text{long}} = \frac{2\pi}{\Delta\omega} \quad , \quad \tau_{\text{short}} = \frac{2\pi}{\Omega_\nu} \quad . \tag{12}$$

From Eqs. (11) and (12), we deduce that

$$\frac{\tau_{\text{short}}}{\tau_{\text{long}}} = \frac{\Delta\omega}{\Omega_\nu} \ll 1 \quad . \tag{13}$$

Consequently, τ_{short} appears as a short time scale associated with the MF center frequency Ω_ν and τ_{long} as a long time scale associated with the bandwith $\Delta\omega$ of MF narrow band $\mathbb{B}_\nu$. Band $\mathbb{B}_0$ defined by Eq. (10) is called the LF band associated with MF narrow band $\mathbb{B}_\nu$ because it involves only one time scale which is τ_{long}.

---

1 Element $\mathbf{f}(\omega)$ belongs to antidual space $\mathcal{C}_S^{c\,\prime}$ of $\mathcal{C}_S^c$ (see Chapter V), and in Eq. (8), the angle brackets denote the antiduality product between $\mathcal{C}_S^{c\,\prime}$ and $\mathcal{C}_S^c$.

4. Class of MF Narrow Band Excitation Force Fields

4.1. MF narrow band excitation force field in the frequency domain

In Eq. (5), element $\mathbf{f}(\omega)$ is the excitation force field defined by Eqs. (7) and (8). We define a particular class of excitation force field $\mathbf{f}(\omega)$ related to a given MF narrow band $\mathbb{B}_\nu$, by

$$\mathbf{f}(\omega) = \theta_\nu(\omega)\,\mathbf{b} \quad , \tag{14}$$

in which spatial part[2] $\mathbf{b}$ is independent of Ω and frequency part $\theta_\nu(\omega)$ is a given square integrable function from $\mathbb{R}$ into $\mathbb{C}$ such that[3]

$$\theta_\nu(\omega) = 0 \quad , \quad \forall\omega \notin \mathbb{B}_\nu \quad . \tag{15}$$

Consequently, the following integral is finite

$$\int_{\mathbb{B}_\nu} |\theta_\nu(\omega)|^2\, d\omega < +\infty \quad . \tag{16}$$

4.2. MF narrow band excitation force field in the time domain

We define the square integrable time function $\theta_\nu(t)$ from $\mathbb{R}$ into $\mathbb{C}$ as the inverse Fourier transform[4] of $\theta_\nu(\omega)$ which is written as

$$\theta_\nu(t) = \frac{1}{2\pi}\int_{\mathbb{B}_\nu} e^{i\omega t}\,\theta_\nu(\omega)\, d\omega \quad . \tag{16}$$

The Fourier transform of $\theta_\nu(t)$ is

$$\theta_\nu(\omega) = \int_{\mathbb{R}} e^{-i\omega t}\,\theta_\nu(t)\, dt \quad . \tag{17}$$

The square integrable function $\theta_\nu(t)$ such that its Fourier transform $\theta_\nu(\omega)$ satisfies Eq. (15) is called an *MF narrow band signal* related to band $\mathbb{B}_\nu$. For an MF narrow band signal, we then have the Plancherel equality (see Mathematical Notations in the appendix)

$$\int_{\mathbb{R}} |\theta_\nu(t)|^2\, dt = \frac{1}{2\pi}\int_{\mathbb{B}_\nu} |\theta_\nu(\omega)|^2\, d\omega \quad . \tag{18}$$

---

[2] Element $\mathbf{b}$ belongs to $\mathcal{C}_S^{c\,\prime}$ (see Footnote 1).

[3] Function θ_ν has a compact support $\mathbb{B}_\nu$.

[4] We use the same symbol for a quantity and its Fourier transform (see Mathematical Notations in the appendix).

4.3. LF signal associated with the MF narrow band signal

Let $\theta_\nu(t)$ be an MF narrow band signal related to band $\mathbb{B}_\nu$. The LF frequency signal $\theta_0(t)$ associated with $\theta_\nu(t)$ is defined by

$$\theta_0(t) = \theta_\nu(t)\, e^{-i\,\Omega_\nu t} \quad . \tag{19}$$

Taking the Fourier transform of the two sides of Eq. (19) yields

$$\theta_0(\omega) = \theta_\nu(\omega + \Omega_\nu) \quad , \quad \forall \omega \in \mathbb{R} \quad . \tag{20}$$

From Eqs. (15) and (20), we deduce that

$$\theta_0(\omega) = 0 \quad , \quad \forall \omega \notin \mathbb{B}_0 \quad , \tag{21}$$

in which $\mathbb{B}_0$ is the band defined by Eq. (10). Consequently, signal $\theta_0(t)$ is an LF signal on LF band $\mathbb{B}_0$, which has only one time scale, the long time scale τ_{long}.

5. Approximation of the FRF on an MF Narrow Band Due to the Frequency-Dependent Damping and Stiffness Operators

For a given MF narrow band $\mathbb{B}_\nu$, the construction of $\mathbf{T}(\omega)$ on $\mathbb{B}_\nu$ requires solving Eq. (6) on $\mathbb{B}_\nu$ for the class of MF narrow band excitation force fields defined in Section 4, that is to say, for all ω in $\mathbb{B}_\nu$,

$$(-\omega^2\, \mathbf{M} + i\omega\, \mathbf{D}(\omega) + \mathbf{K}(\omega))\, \mathbf{u}(\omega) = \theta_\nu(\omega)\, \mathbf{b} \quad , \tag{22}$$

in which $\theta_\nu(\omega)$ and $\mathbf{b}$ are defined in Section 4. Since it is not efficient to solve Eq. (22) frequency by frequency (see Section 1), we propose the use of an appropriate MF method (see Section 6). This method can only be applied if the mass, damping and stiffness operators are frequency-independent. Consequently, the following approximation of Eq. (22) is introduced

$$(-\omega^2\, \mathbf{M} + i\omega\, \mathbf{D}_\nu + \mathbf{K}_\nu)\, \mathbf{u}_\nu(\omega) = \theta_\nu(\omega)\, \mathbf{b} \quad , \tag{23}$$

in which the operators $\mathbf{D}_\nu$ and $\mathbf{K}_\nu$ are independent of ω and defined by

$$\mathbf{D}_\nu = \mathbf{D}(\Omega_\nu) \quad , \quad \mathbf{K}_\nu = \mathbf{K}(\Omega_\nu) \quad . \tag{24}$$

Since functions $\omega \mapsto \mathbf{D}_\nu(\omega)$ and $\omega \mapsto \mathbf{K}_\nu(\omega)$ are continuous on $\mathbb{R}$ (see Section IV.5.2), there is always a bandwith $\Delta\omega$ of MF narrow band $\mathbb{B}_\nu$ such that, for all ω in $\mathbb{B}_\nu$, solution $\mathbf{u}_\nu(\omega)$ of Eq. (23) approximates solution $\mathbf{u}(\omega)$ of Eq. (22). This means that for any fixed $\varepsilon > 0$, there exists a $\Delta\omega$ such that, for all ω in $\mathbb{B}_\nu$, we have [5]

$$\|\mathbf{u}(\omega) - \mathbf{u}_\nu(\omega)\|_{\mathcal{C}_S^c} \leq \varepsilon \quad . \tag{25}$$

Norm $\|\,.\,\|_{\mathcal{C}_S^c}$ used in Eq. (25) is the norm of $\mathcal{C}_S^c \subset (H^1(\Omega))^3$.

6. Analytical Processing of the Short Time Scale in the Frequency Domain and Numerical Processing of the Long Time Scale in the Time Domain

The proposed MF method is based on analytical processing of the short time scale in the frequency domain and numerical processing of the long time scale in the time domain. In this section, we construct the solution of Eq. (23) for all ω in $\mathbb{B}_\nu$, rewritten as

$$\mathbf{A}_\nu(\omega)\,\mathbf{u}_\nu(\omega) = \theta_\nu(\omega)\,\mathbf{b} \quad , \quad \forall\omega \in \mathbb{B}_\nu \quad , \tag{26}$$

in which the operator $\mathbf{A}_\nu(\omega)$ is defined by

$$\mathbf{A}_\nu(\omega) = -\omega^2\,\mathbf{M} + i\omega\,\mathbf{D}_\nu + \mathbf{K}_\nu \quad . \tag{27}$$

6.1. Frequency transform technique and associated LF equation in the frequency domain

Analytical processing of the short time scale is performed in the frequency domain using a frequency transform technique which allows the so-called associated LF equation to be constructed in the frequency domain. Equation (26) related to the MF narrow band $\mathbb{B}_\nu$ is transformed into an LF equation on the LF band $\mathbb{B}_0$ using the following transform technique

$$\mathbf{u}_\nu(t) = \mathbf{u}_0(t)\,e^{i\,\Omega_\nu t} \quad . \tag{28}$$

Taking the Fourier transform [6] of the two sides of Eq. (28) yields

$$\mathbf{u}_0(\omega) = \mathbf{u}_\nu(\omega + \Omega_\nu) \quad , \quad \forall\omega \in \mathbb{R} \quad . \tag{29}$$

From Eqs. (15) and (26), we deduce that $\mathbf{u}_\nu(\omega) = \mathbf{0}$ for all $\omega \notin \mathbb{B}_\nu$ and using Eq. (29) yields

$$\mathbf{u}_0(\omega) = 0 \quad \text{for all} \quad \omega \notin \mathbb{B}_0 \quad , \tag{30}$$

in which $\mathbb{B}_0$ is the band defined by Eq. (10). Consequently, $\mathbf{u}_0(t)$ is an LF signal on LF band $\mathbb{B}_0$, which has only one time scale, the long time scale τ_{long}. Substituting Eqs. (20) and (29) into Eq. (26), we deduce that Eq. (26), related to the MF narrow band $\mathbb{B}_\nu$, is transformed into an

[6] We use the same symbol for a quantity and its Fourier transform (see Mathematical Notations in the appendix).

equation related to the LF band $\mathbb{B}_0$, called the *associated LF equation in the frequency domain,*

$$\mathbf{A}_\nu(\omega + \Omega_\nu)\, \mathbf{u}_0(\omega) = \theta_0(\omega)\, \mathbf{b} \quad , \quad \forall\, \omega \in \mathbb{B}_0 \quad . \tag{31}$$

Using Eq. (27), Eq. (31) can be written as

$$(-\omega^2\, \mathbf{M} + i\omega\, \widetilde{\mathbf{D}}_\nu + \widetilde{\mathbf{K}}_\nu)\, \mathbf{u}_0(\omega) = \theta_0(\omega)\, \mathbf{b} \quad , \quad \forall\, \omega \in \mathbb{B}_0 \quad , \tag{32}$$

in which frequency-independent operators $\widetilde{\mathbf{D}}_\nu$ and $\widetilde{\mathbf{K}}_\nu$ are defined by

$$\widetilde{\mathbf{D}}_\nu = \mathbf{D}_\nu + 2\, i\, \Omega_\nu\, \mathbf{M} \quad , \tag{33}$$

$$\widetilde{\mathbf{K}}_\nu = -\Omega_\nu^2\, \mathbf{M} + i\, \Omega_\nu\, \mathbf{D}_\nu + \mathbf{K}_\nu \quad . \tag{34}$$

This frequency transform technique corresponds to explicit analytical processing of short time scale τ_{short} in the frequency domain.

6.2. Associated LF equation in the time domain

The long time scale occurring in the problem is processed numerically by solving a so-called associated LF equation in the time domain. The inverse Fourier transform $\mathbf{u}_0(t)$ of solution $\mathbf{u}_0(\omega)$ of Eq. (32) satisfies the following second-order differential equation with respect to time t (with time-independent operators)

$$\mathbf{M}\, \partial_t^2 \mathbf{u}_0(t) + \widetilde{\mathbf{D}}_\nu\, \partial_t \mathbf{u}_0(t) + \widetilde{\mathbf{K}}_\nu\, \mathbf{u}_0(t) = \theta_0(t)\, \mathbf{b} \quad , \quad \forall\, t \in \mathbb{R} \quad . \tag{35}$$

Since $\theta_0(t)$ and $\mathbf{u}_0(t)$ are LF signals related to LF band $\mathbb{B}_0$, Eq. (35) is called the *associated LF equation in the time domain.* This LF equation is related only to long time scale τ_{long} in the time domain. In the proposed MF method, the associated LF Eq. (35) in the time domain is solved numerically using an appropriate time integration scheme for a second-order differential equation with respect to time with time-independent operators. The associated LF equation in the time domain corresponds to numerical processing of the long time scale.

6.3. Expression of $\mathbf{u}_\nu(\omega)$ in an MF narrow band in terms of the time sampling of solution $\mathbf{u}_0(t)$ of the associated LF equation in the time domain

Using an appropriate time integration scheme, the numerical calculation of Eq. (35) leads to constructing $\mathbf{u}_0(t)$ at discrete points in time corresponding to an integration time step δt. This calculation gives $\mathbf{u}_0(t)$ at the sampling points $t_m = m\,\Delta t$, in which $\Delta t = p\,\delta t$ (with $p \geq 1$ a given positive integer) is the sampling time step given by Shannon's theorem. The problem consists in constructing the expression of $\mathbf{u}_\nu(\omega)$, for all ω in MF narrow band $\mathbb{B}_\nu$, in terms of $\{\mathbf{u}_0(m\,\Delta t)\,,\, m \in \mathbb{Z}\}$ in which $\mathbb{Z}$ denotes the set of all positive and negative integers. Since $\mathbf{u}_0(t)$ is a square integrable time signal whose Fourier transform $\mathbf{u}_0(\omega)$ is such that $\mathbf{u}_0(\omega) = 0$ for $\omega \notin \mathbb{B}_0 = [\,-\Delta\omega/2\,,\Delta\omega/2\,]$, Shannon's theorem can be used and yields (see Papoulis, 1977; Soize, 1993a),

$$\mathbf{u}_0(t) = \sum_{m\in\mathbb{Z}} \mathbf{u}_0(m\,\Delta t)\,\frac{\sin\{\omega_L(t - m\,\Delta t)\}}{\omega_L(t - m\,\Delta t)} \quad , \quad \forall t \in \mathbb{R} \quad , \tag{36}$$

in which ω_L and Δt are such that

$$\omega_L = \frac{\Delta\omega}{2} \quad , \quad \Delta t = \frac{\pi}{\omega_L} \quad . \tag{37}$$

From Eqs. (28) and (36), we deduce that

$$\mathbf{u}_\nu(t) = e^{i\,\Omega_\nu t} \sum_{m\in\mathbb{Z}} \mathbf{u}_0(m\,\Delta t)\,\frac{\sin\{\omega_L(t - m\,\Delta t)\}}{\omega_L(t - m\,\Delta t)} \quad , \quad \forall t \in \mathbb{R} \quad . \tag{38}$$

Taking the Fourier transform of the two sides of Eq. (38), we obtain the expression of $\mathbf{u}_\nu(\omega)$ in MF narrow band $\mathbb{B}_\nu$ in terms of the time sampling points of solution $\mathbf{u}_0(t)$ of the associated LF equation, i.e.

$$\mathbf{u}_\nu(\omega) = \mathbf{1}_{\mathbb{B}_\nu}(\omega)\;\Delta t \sum_{m\in\mathbb{Z}} \mathbf{u}_0(m\,\Delta t)\; e^{-im\,\Delta t\,(\omega-\Omega_\nu)} \quad , \quad \forall\omega \in \mathbb{R} \quad , \tag{39}$$

in which $\mathbf{1}_{\mathbb{B}_\nu}(\omega) = 1$ if $\omega \in \mathbb{B}_\nu$ and $= 0$ if $\omega \notin \mathbb{B}_\nu$. It should be noted that Eq. (39) is an exact representation of $\mathbf{u}_\nu(\omega)$, not an approximation.

6.4. Time integration of the associated LF equation

The first problem is to transform the problem defined by Eq. (35) into an equivalent Cauchy problem

$$\mathbf{M}\,\partial_t^2\mathbf{u}_0(t) + \widetilde{\mathbf{D}}_\nu\,\partial_t\mathbf{u}_0(t) + \widetilde{\mathbf{K}}_\nu\,\mathbf{u}_0(t) = \theta_0(t)\,\mathbf{b} \quad , \quad \forall t > t_{\mathrm{i}} \quad , \tag{40}$$

$$\{\mathbf{u}_0(t)\}_{t_{\mathrm{i}}} = \mathbf{u}_{\mathrm{i}} \quad , \quad \{\partial_t\mathbf{u}_0(t)\}_{t_{\mathrm{i}}} = \mathbf{v}_{\mathrm{i}} \quad , \tag{41}$$

in which $\mathbf{u}_{\mathrm{i}}$ and $\mathbf{v}_{\mathrm{i}}$ are appropriate initial Cauchy conditions at the finite initial time t_{i} and operators $\widetilde{\mathbf{D}}_\nu$ and $\widetilde{\mathbf{K}}_\nu$ are given by Eqs. (33) and (34) respectively.

Choice of the initial time and Cauchy conditions. In order to construct the equivalent problem defined by Eqs. (40) and (41), two conditions must be satisfied. First, the energy of the input signal $\theta_0(t)$ over time interval $]-\infty\,, t_{\mathrm{i}}\,[$ must be negligible with respect to the total energy of signal $\theta_0(t)$. Secondly, it is necessary to avoid the introduction of spurious transient responses induced by the initial Cauchy conditions. The second condition is satisfied by taking $\mathbf{u}_{\mathrm{i}} = \mathbf{0}$ and $\mathbf{v}_{\mathrm{i}} = \mathbf{0}$. Concerning the first one, we introduce a negative integer m_{i} such that $t_{\mathrm{i}} = m_{\mathrm{i}}\,\Delta t$. Since $\theta_0(t)$ is a square integrable function on $\mathbb{R}$, for any fixed $\varepsilon > 0$, there exists m_{i} such that

$$\int_{-\infty}^{m_{\mathrm{i}}\,\Delta t} |\theta_0(t)|^2\, dt < \varepsilon \quad . \tag{42}$$

Choice of the final time. The final integration time t_{f} is written as $t_{\mathrm{f}} = m_{\mathrm{f}}\,\Delta t$ in which m_{f} is a positive integer. In order to determine m_{f}, the energy of signal $\mathbf{u}_0(t)$ over $]t_{\mathrm{f}}\,, +\infty[$ is neglected with respect to the total energy of signal $\mathbf{u}_0(t)$. Since $\mathbf{u}_0(t)$ is a square integrable function on $\mathbb{R}$ with respect to t, for any fixed $\varepsilon > 0$, there exists m_{f} such that

$$\int_{m_{\mathrm{f}}\Delta t}^{+\infty} \int_{\Omega} \|\mathbf{u}_0(\mathbf{x},t)\|^2\, d\mathbf{x}\, dt < \varepsilon \quad . \tag{43}$$

From the Cauchy problem defined by Eqs. (40) and (41) (dissipative second-order dynamical system), it is known that the value of m_{f} satisfying Eq. (43) is controlled by the damping operator $\widetilde{\mathbf{D}}_\nu$.

Step-by-step integration method for solving the time-variable equation. The solution of Eqs. (40) and (41) can be constructed by using an unconditionally stable implicit time integration scheme (in the context of the MF method presented, see Soize, 1982a and, Soize et al., 1986b; for general considerations on time integration schemes, see Argyris and Mlejnek, 1991; Bathe and Wilson, 1976; Belytschko and Hughes, 1983; Dautray and Lions, 1992; Zienkiewicz and Taylor, 1989). Integration time step δt must be such that $\delta t < \Delta t$. Consequently, for a fixed integer $p > 1$, δt must be chosen such that $\delta t = \Delta t/p$. For example, $p = 3$ or $p = 5$ is an appropriate value.

Expression of $\mathbf{u}_\nu(\omega)$ in terms of $\mathbf{u}_0(m\,\Delta t)$ for m in $\{m_{\mathrm{i}}, \ldots, m_{\mathrm{f}}\}$. From Eq. (39) and the above considerations, we deduce that

$$\mathbf{u}_\nu(\omega) \simeq \mathbf{1}_{\mathbb{B}_\nu}(\omega)\,\Delta t \sum_{m=m_{\mathrm{i}}}^{m_{\mathrm{f}}} \mathbf{u}_0(m\,\Delta t)\; e^{-im\,\Delta t\,(\omega-\Omega_\nu)} \quad , \quad \forall\omega \in \mathbb{R} \quad . \tag{44}$$

Example of signal $\theta_0(t)$. If $\theta_0(\omega)$ is defined by

$$\theta_0(\omega) = \mathbf{1}_{\mathbb{B}_0}(\omega) \quad , \tag{45}$$

in which $\mathbf{1}_{\mathbb{B}_0}(\omega)$ is the indicator function of band $\mathbb{B}_0$ defined by Eq. (10), we then have

$$\theta_0(t) = \frac{1}{\pi t} \sin(\frac{\Delta\omega}{2} t) \quad . \tag{46}$$

Using the Plancherel formula (see Mathematical Notations in the appendix), the total energy of signal $\theta_0(t)$ is then given by

$$\|\theta_0\|_{L^2}^2 = \int_{\mathbb{R}} |\theta_0(t)|^2 \, dt = \frac{1}{2\pi} \int_{\mathbb{R}} |\theta_0(\omega)|^2 \, d\omega = \frac{\Delta\omega}{2\pi} \quad . \tag{47}$$

For this particular signal and since $m_{\mathrm{i}} < 0$, Eq. (42) yields

$$\int_{-\infty}^{m_{\mathrm{i}}\,\Delta t} |\theta_0(t)|^2 \, dt \leq \frac{1}{\pi^2} \int_{-\infty}^{m_{\mathrm{i}}\,\Delta t} \frac{dt}{t^2} = \frac{\Delta\omega}{2\pi} \times \frac{1}{|m_{\mathrm{i}}|\,\pi^2} \quad . \tag{48}$$

From Eqs. (47) and (48), we deduce that the relative error introduced by the use of m_{i} in place of $-\infty$, is equal to $1/(|m_{\mathrm{i}}|\,\pi^2)$. For example, $m_{\mathrm{i}} = -3$ is an appropriate value.

7. Construction of the FRF on an MF broad band

A limited MF broad band

$$\mathbb{B}_{\mathrm{MF}} = [\omega_{\mathrm{MF,init}} \, , \omega_{\mathrm{MF,final}}] \subset \mathbb{R}^+ \quad , \tag{49}$$

is written as a finite union of MF narrow bands $\mathbb{B}_\nu$

$$\mathbb{B}_{\mathrm{MF}} = \cup_\nu \, \mathbb{B}_\nu \quad . \tag{50}$$

To choose the bandwidth $(\Delta\omega)_\nu$ of each MF narrow band $\mathbb{B}_\nu$ appearing in Eq. (50), we generally use the following rule

$$\frac{(\Delta\omega)_\nu}{\Omega_\nu} = \text{constant} \quad \text{for all} \quad \nu \quad . \tag{51}$$

Eq. (51) shows that bandwidth $(\Delta\omega)_\nu$ increases with the value of center frequency Ω_ν. For each MF narrow band $\mathbb{B}_\nu$, we use the method presented in Sections 3 to 6 to construct the operator-valued FRF $\mathbf{T}(\omega)$. The function $\theta_0(t)$ defined by Eq. (46) can be used. For $\mathbf{b}$, we can choose the function bases spanning the vector space of $\mathbf{f}(\omega)$ [7].

---

[7] This vector space is $\mathcal{C}_S^{c\,\prime}$ (see footnote 1) or a subspace which is generally $H^c = (L^2(\Omega))^3 \subset \mathcal{C}_S^{c\,\prime}$.

8. Finite Element Discretization

The finite element discretization of Eq. (22) is written as (see Section V.6)

$$\left(-\omega^2\,[M] + i\omega\,[D(\omega)] + [K(\omega)]\right)\mathbf{U} = \theta_\nu(\omega)\,\mathbf{B} \quad , \tag{52}$$

in which $\mathbf{U} = (U_1, \ldots, U_n)$ is the complex-valued vector of the DOFs which are the values of the displacement field at the nodes of the finite element mesh of domain Ω and $\mathbf{B} = (B_1, \ldots, B_n)$ is the complex-valued vector independent of ω corresponding to the finite element discretization of the antilinear form $\delta\mathbf{u} \mapsto <\mathbf{b}\,,\delta\mathbf{u}>$. For the finite element discretization in $\mathcal{C}_S^c$ (see Eq. (V.56)), $(n \times n)$ real mass, damping and stiffness matrices $[M]$, $[D(\omega)]$ and $[K(\omega)]$ are symmetric and positive definite. We then deduce the finite element discretization of Eq. (23),

$$\left(-\omega^2\,[M] + i\omega\,[D_\nu] + [K_\nu]\right)\mathbf{U}_\nu = \theta_\nu(\omega)\,\mathbf{B} \quad , \tag{53}$$

in which $(n \times n)$ real matrices $[D_\nu]$ and $[K_\nu]$ are such that (see Eq. (24))

$$[D_\nu] = [D(\Omega_\nu)] \quad , \quad [K_\nu] = [K(\Omega_\nu)] \quad . \tag{54}$$

The finite element discretization of Eqs. (40) and (41) yields

$$[M]\,\ddot{\mathbf{U}}_0(t) + [\widetilde{D}_\nu]\,\dot{\mathbf{U}}_0(t) + [\widetilde{K}_\nu]\mathbf{U}_0(t) = \theta_0(t)\,\mathbf{B} \quad , \quad \forall t > t_{\mathrm{i}} \ , \tag{55}$$

$$\mathbf{U}_0(t_{\mathrm{i}}) = \mathbf{0} \quad , \quad \dot{\mathbf{U}}_0(t_{\mathrm{i}}) = \mathbf{0} \quad , \tag{56}$$

in which $(n \times n)$ complex symmetric matrices $[\widetilde{D}_\nu]$ and $[\widetilde{K}_\nu]$ are such that (see Eqs. (33) and (34))

$$[\widetilde{D}_\nu] = [D_\nu] + 2\,i\,\Omega_\nu\,[M] \quad , \tag{57}$$

$$[\widetilde{K}_\nu] = -\Omega_\nu^2\,[M] + i\,\Omega_\nu\,[D_\nu] + [K_\nu] \quad . \tag{58}$$

The finite element discretisation of Eq. (44) yields the expression of $\mathbf{U}_\nu(\omega)$ in MF narrow band $\mathbb{B}_\nu$ in terms of the time sampling points of solution $\mathbf{U}_0(t)$ of Eqs. (55) and (56),

$$\mathbf{U}_\nu(\omega) \simeq \mathbf{1}_{\mathbb{B}_\nu}(\omega)\ \Delta t \sum_{m=m_{\mathrm{i}}}^{m_{\mathrm{f}}} \mathbf{U}_0(m\,\Delta t)\ e^{-im\,\Delta t\,(\omega-\Omega_\nu)} \quad , \quad \forall\omega \in \mathbb{R} \quad , \tag{59}$$

in which $\mathbf{1}_{\mathbb{B}_\nu}(\omega) = 1$ if $\omega \in \mathbb{B}_\nu$ and $= 0$ if $\omega \notin \mathbb{B}_\nu$.

CHAPTER VIII

Reduced Model in the MF Range

1. Introduction

It is well known (see for instance Argyris and Mlejnek, 1991; Bathe and Wilson, 1976; Clough and Penzien, 1975; Leung, 1993; Meirovitch, 1980; Morand and Ohayon, 1995; Roseau, 1980) that, for low-frequency dynamic analysis in structural dynamics, reduced models are a very efficient tool for constructing the solution. These techniques correspond to a Ritz-Galerkin reduction of the structural-dynamics model using, for instance, the normal modes corresponding to the lowest eigenfrequencies of the associated conservative structure. The efficiency of this kind of reduced model is due to the small number of generalized degrees of freedom used in the representation and, in addition, the model is obtained by solving a well-stated generalized symmetric eigenvalue problem for which only the first eigenvalues and the corresponding eigenfunctions have to be calculated. Furthermore, when such a reduced model is obtained, responses to deterministic or random excitations can be calculated at no significant additional numerical cost, and the reduced model can be used directly for solving various structural-acoustic problems in the low-frequency range. The fundamental problem related to construction of a reduced model in the medium-frequency range for general dissipative structural-dynamic systems has not yet been solved. An efficient solution for constructing such a reduced model for general structural-dynamic problems was proposed recently (Soize, 1997a). In this chapter, we present a summary of the ideas and the main results.

In Section 2, we recall the definition of a narrow MF band and in Section 3, we illustrate the main ideas using a simple linear oscillator, introducing different quantities related to the kinetic, elastic and mechanical energy.

Section 4 is devoted to the three-dimensional continuous case for which we recall the variational formulation of the master structure in the MF range developed in Chapters V and VII.

In Section 5, we construct an appropriate functional basis allowing the reduced model to be derived in the MF range. We define an energy operator. Its dominant eigenspace is defined as the finite dimension subspace spanned by the eigenfunctions associated with its highest eigenvalues.

In Section 6, we define the reduce model in the MF range and in Section 7, we introduce the finite element discretization.

Finally, Section 8 deals with the construction of the dominant eigensubspace using the subspace iteration method.

2. Definition of a Narrow MF Band

Let $\mathbb{B}_\nu$ be the limited frequency band of $\mathbb{R}^+$ defined by

$$\mathbb{B}_\nu = [\,\Omega_\nu - \Delta\omega/2\,,\ \Omega_\nu + \Delta\omega/2\,] \quad , \tag{1}$$

in which $\Omega_\nu > 0$ is the center frequency of band $\mathbb{B}_\nu$ and $\Delta\omega$ its bandwidth such that $0 < \Delta\omega < 2\,\Omega_\nu$. With band $\mathbb{B}_\nu$ we associate band $\widetilde{\mathbb{B}}_\nu$ having the same bandwidth and defined by

$$\widetilde{\mathbb{B}}_\nu = [\,-\Omega_\nu - \Delta\omega/2\,,\ -\Omega_\nu + \Delta\omega/2\,] \quad , \tag{2}$$

and band $\mathbb{B}_0$ having the same bandwidth and defined by

$$\mathbb{B}_0 = [\,-\Delta\omega/2\,, \Delta\omega/2\,] \quad . \tag{3}$$

By definition, limited frequency band $\mathbb{B}_\nu$ is a narrow MF band if

$$\frac{\Delta\omega}{\Omega_\nu} \ll 1 \quad . \tag{4}$$

3. Energy Properties of a Simple-Linear-Oscillator Response

In this section, we present results concerning the simple linear oscillator in order to introduce the definition of the energy of its forced response. In Section 5, these results will be used to introduce the corresponding concept of energy operator for the case of a continuous three-dimensional master structure.

3.1. Definition of the excitation force

The external force applied to the simple linear oscillator is denoted as $f(t)$ in the time domain. In the frequency domain, we have

$$f(\omega) = \int_{\mathbb{R}} e^{-i\omega t}\, f(t)\, dt \quad . \tag{5}$$

We assume that $f(\omega)$ is equal to a real constant g over frequency band $\mathbb{B}_\nu \cup \widetilde{\mathbb{B}}_\nu$ and to zero outside,

$$f(\omega) = \eta(\omega)\, g \quad , \tag{6 – 1}$$

in which $\eta(\omega)$ is the function defined by

$$\eta(\omega) = \mathbf{1}_{\mathbb{B}_\nu \cup \widetilde{\mathbb{B}}_\nu}(\omega) \quad , \tag{6 – 2}$$

where $\mathbf{1}_B(\omega) = 1$ for $\omega \in B$ and $= 0$ for $\omega \notin B$.

3.2. Frequency response function of the simple linear oscillator

The mass, stiffness and damping of the simple linear oscillator are denoted as M, K and D respectively. The mass displacement is denoted as $u(t)$. In the frequency domain, we have

$$u(\omega) = \int_{\mathbb{R}} e^{-i\omega t}\, u(t)\, dt \quad . \tag{7}$$

The frequency response function $T(\omega)$ is then written as

$$T(\omega) = (-\omega^2\, M + i\omega\, D + K)^{-1} \quad . \tag{8}$$

Denoting the critical damping rate as ξ, the eigenfrequency of the associated conservative oscillator as ω_α and the resonant frequency as ω_r, we have

$$D = 2\xi\, M\, \omega_\alpha \quad , \quad \omega_\alpha = (K/M)^{1/2} \quad , \quad \omega_r = \omega_\alpha(1 - 2\xi^2)^{1/2} \quad . \tag{9}$$

Below, we assume that

$$0 < \xi \ll 1 \quad . \tag{10}$$

The equivalent passband of the linear filter whose frequency response function is $T(\omega)$ is denoted as b_w. It is defined by (see Fig. 1)

$$2\, b_w\, |T(\omega_r)|^2 = \int_{\mathbb{R}} |T(\omega)|^2\, d\omega \quad , \tag{11}$$

and is written as

$$b_w = \pi\,\xi\,\omega_\alpha(1-\xi^2) \quad . \tag{12}$$

Finally, we assume that

$$[\,\omega_r - b_w/2\,,\omega_r + b_w/2\,] \subset \mathbb{B}_\nu \quad . \tag{13}$$

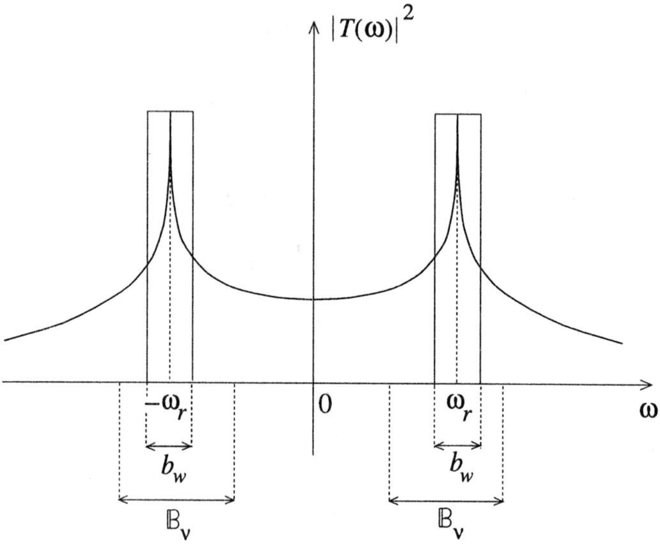

Fig. 1. Equivalent passband

3.3. Forced response of the simple linear oscillator

We consider the forced response $u^g(t)$ of the oscillator submitted to excitation $f(t)$ defined in Section 3.1 and such that

$$u^g(t) = (h * f)(t) = \int_{\mathbb{R}} h(t-\tau)\, f(\tau)\, d\tau \quad , \quad \forall t \in \mathbb{R} \quad , \tag{14}$$

in which $h(t)$ is the impulse response function of the linear filter whose frequency response function

$$T(\omega) = \int_{\mathbb{R}} e^{-i\omega t}\, h(t)\, dt \quad , \tag{15}$$

is defined by Eq. (8). It can be seen that the linear filter is causal (or physically realizable), which means that

$$h(t) = 0 \quad , \quad \forall t < 0 \quad , \tag{16}$$

and is stable [1] because h is an integrable function [2] on $\mathbb{R}$,

$$\int_{\mathbb{R}} |h(t)|\, dt < +\infty \quad . \tag{17}$$

The Fourier transform of $u^g(t)$ is then written as

$$u^g(\omega) = \eta(\omega)\, T(\omega)\, f(\omega) \quad , \quad \forall\, \omega \in \mathbb{R} \quad . \tag{18}$$

From Eq. (6), we deduce that functions $-\omega^2 u^g(\omega)$, $i\omega\, u^g(\omega)$ and $u^g(\omega)$ have [3] a limited frequency band $\mathbb{B}_\nu \cup \widetilde{\mathbb{B}}_\nu$,

$$-\omega^2 u^g(\omega) = i\omega\, u^g(\omega) = u^g(\omega) = 0 \quad \text{as} \quad \omega \notin \mathbb{B}_\nu \cup \widetilde{\mathbb{B}}_\nu \quad , \tag{19}$$

and are square integrable on $\mathbb{R}$,

$$\int_{\mathbb{R}} |u^g(\omega)|^2\, d\omega <+\infty, \int_{\mathbb{R}} \omega^2 |u^g(\omega)|^2\, d\omega <+\infty, \int_{\mathbb{R}} \omega^4 |u^g(\omega)|^2\, d\omega <+\infty . \tag{20}$$

From Eqs. (19) and (20), we deduce [4] that $u^g(t)$ is twice differentiable on $\mathbb{R}$ and that u^g, $\dot{u}^g$ and $\ddot{u}^g$ are continuous and square integrable functions on $\mathbb{R}$ which approach 0 as $t \to \pm\infty$,

$$u^g, \dot{u}^g, \ddot{u}^g \in C^0(\mathbb{R}) \cap L^2(\mathbb{R}) \ ; \ u^g(t), \dot{u}^g(t), \ddot{u}^g(t) \to 0 \text{ as } t \to \pm\infty \ . \tag{21}$$

We then deduce that for all t in $\mathbb{R}$, $u^g(t)$ defined by Eq. (14) satisfies the second-order differential equation

$$M\ddot{u}^g(t) + D\dot{u}^g(t) + Ku^g(t) = f(t) \quad . \tag{22}$$

---

[1] If the fact that f is a bounded function on $\mathbb{R}$ implies that u^g is a bounded function on $\mathbb{R}$, then the linear filter is stable. Since $h \in L^1(\mathbb{R})$, then the linear filter is stable (see for instance, Soize, 1994 and 1993a; Priestley, 1981; Papoulis, 1977; Halmos, 1976).

[2] The explicit calculation of function $h(t)=(1/2\pi)\int_{\mathbb{R}} e^{i\omega t} T(\omega)\, d\omega$ shows that $h \in L^1(\mathbb{R})$. The following lemma can also be used: if T and its derivatives $T', T'' \in L^1(\mathbb{R})$, then $h \in L^1(\mathbb{R})$ (for the proof of this lemma, see Soize, 1993a, pp. 15).

[3] From Eqs. (6) and (18), we deduce that function $\omega \mapsto \omega^2 u^g(\omega)$ has a compact support $\mathbb{B}_\nu \cup \widetilde{\mathbb{B}}_\nu$. Since function $\omega \mapsto \omega^2 T(\omega)$ is bounded on $\mathbb{R}$ and $\omega \mapsto f(\omega)$ belongs to $L^2(\mathbb{R})$, then $\omega \mapsto \omega^2 u^g(\omega)$ belongs to $L^2(\mathbb{R})$. The same results hold for $\omega \mapsto i\omega u^g(\omega)$ and $\omega \mapsto u^g(\omega)$.

[4] From Eqs. (19) and (20), we deduce that $\omega \mapsto \omega^2 u^g(\omega) \in L^1(\mathbb{R})$ because $L^2(K) \subset L^1(K)$ for any compact $K \subset \mathbb{R}$. Consequently, $\omega \mapsto \omega^2 u^g(\omega) \in L^1(\mathbb{R}) \cap L^2(\mathbb{R})$. Therefore, the inverse Fourier transform of $\omega \mapsto \omega^2 u^g(\omega)$ is in $L^2(\mathbb{R})$ and is the second derivative $\ddot{u}^g(t)$ of $u^g(t)$. Since $\ddot{u}^g(t)=(1/2\pi)\int_{\mathbb{R}} e^{i\omega t}\{-\omega^2 u^g(\omega)\}\, d\omega$ and $\omega \mapsto -\omega^2 u^g(\omega) \in L^1(\mathbb{R})$, then $\ddot{u}^g \in C^0(\mathbb{R})$ and $\ddot{u}^g \to 0$ as $t \to \pm\infty$. This result holds for $u^g(t)$ and $\dot{u}^g(t)$.

3.4. Kinetic, elastic and mechanical energy

Let $u^g(t)$ be the response of the simple linear oscillator, defined by Eq. (14), whose Fourier transform is given by Eq. (18). The kinetic energy $\varepsilon_{\text{kin}}(t)$, the elastic energy $\varepsilon_{\text{elas}}(t)$ and the mechanical energy $\varepsilon_{\text{mech}}(t)$ of the simple oscillator at time t for vibration u^g are defined by

$$\varepsilon_{\text{kin}}(t) = \frac{1}{2} M \, \dot{u}^g(t)^2 \quad , \tag{23}$$

$$\varepsilon_{\text{elas}}(t) = \frac{1}{2} K \, u^g(t)^2 \quad , \tag{24}$$

$$\varepsilon_{\text{mech}}(t) = \varepsilon_{\text{kin}}(t) + \varepsilon_{\text{elas}}(t) \quad . \tag{25}$$

The total kinetic energy $\mathcal{E}_{\text{kin}}$, the total elastic energy $\mathcal{E}_{\text{elas}}$ and the total mechanical energy $\mathcal{E}_{\text{mech}}$ of the simple linear oscillator for vibration u^g are defined by

$$\mathcal{E}_{\text{kin}} = \int_{\mathbb{R}} \varepsilon_{\text{kin}}(t) \, dt = \int_{\mathbb{R}} \frac{1}{2} M \, \dot{u}^g(t)^2 \, dt \quad , \tag{26}$$

$$\mathcal{E}_{\text{elas}} = \int_{\mathbb{R}} \varepsilon_{\text{elas}}(t) \, dt = \int_{\mathbb{R}} \frac{1}{2} K \, u^g(t)^2 \, dt \quad , \tag{27}$$

$$\mathcal{E}_{\text{mech}} = \int_{\mathbb{R}} \varepsilon_{\text{mech}}(t) \, dt = \mathcal{E}_{\text{kin}} + \mathcal{E}_{\text{elas}} \quad . \tag{28}$$

From Eqs. (19) and (20) and using the Plancherel formula (see Mathematical Notations in the appendix), Eqs. (26) and (27) can be rewritten as

$$\mathcal{E}_{\text{kin}} = \frac{1}{2\pi} \int_{\mathbb{B}_\nu \cup \widetilde{\mathbb{B}}_\nu} \frac{1}{2} M \, \omega^2 |u^g(\omega)|^2 \, d\omega \quad , \tag{29}$$

$$\mathcal{E}_{\text{elas}} = \frac{1}{2\pi} \int_{\mathbb{B}_\nu \cup \widetilde{\mathbb{B}}_\nu} \frac{1}{2} K \, |u^g(\omega)|^2 \, d\omega \quad . \tag{30}$$

3.5. Relationship between kinetic and elastic energy

Substituting Eq. (18) into the right-hand sides of Eqs. (29) and (30) yields

$$\mathcal{E}_{\text{kin}} = \frac{g^2}{2\pi}\frac{M}{2}\int_{\mathbb{B}_\nu \cup \widetilde{\mathbb{B}}_\nu} \omega^2\,|\eta(\omega)|^2\,|T(\omega)|^2\,d\omega \quad , \tag{31}$$

$$\mathcal{E}_{\text{elas}} = \frac{g^2}{2\pi}\frac{K}{2}\int_{\mathbb{B}_\nu \cup \widetilde{\mathbb{B}}_\nu} |\eta(\omega)|^2\,|T(\omega)|^2\,d\omega \quad . \tag{32}$$

We use the following formulas to calculate an approximation of the integrals in the right-hand side of Eqs. (31) and (32),

$$\int_{\mathbb{R}} \omega^2 |T(\omega|^2\,d\omega = \frac{\pi}{DM} \quad , \tag{33}$$

$$\int_{\mathbb{R}} |T(\omega)|^2\,d\omega = \frac{\pi}{DK} \quad . \tag{34}$$

Taking Eqs. (11) and (13) into account and since $|\eta(\omega)|^2 = 1$ over $\mathbb{B}_\nu \cup \widetilde{\mathbb{B}}_\nu$, it can be proved that

$$\int_{\mathbb{B}_\nu \cup \widetilde{\mathbb{B}}_\nu} \omega^2\,|\eta(\omega)|^2\,|T(\omega)|^2\,d\omega \simeq \int_{\mathbb{R}} \omega^2|T(\omega|^2\,d\omega = \frac{\pi}{DM} \quad , \tag{35}$$

$$\int_{\mathbb{B}_\nu \cup \widetilde{\mathbb{B}}_\nu} |\eta(\omega)|^2\,|T(\omega)|^2\,d\omega \simeq \int_{\mathbb{R}} |T(\omega)|^2\,d\omega = \frac{\pi}{DK} \quad . \tag{36}$$

Substituting Eq. (35) in Eq. (31) and substituting Eq. (36) in Eq. (32) yields

$$\mathcal{E}_{\text{kin}} \simeq \mathcal{E}_{\text{elas}} \simeq \frac{g^2}{4D} \quad . \tag{37}$$

From Eqs. (28) and (37), we deduce that

$$\mathcal{E}_{\text{mech}} \simeq 2\,\mathcal{E}_{\text{kin}} \quad . \tag{38}$$

3.6. Definition of the energy of vibration u^g

The energy of vibration u^g is denoted as $\mathcal{E}_{\mathbb{B}_\nu}(u^g)$ and is defined as twice the value of the kinetic energy of the simple linear oscillator for vibration u^g. From Eq. (31), we deduce that

$$\mathcal{E}_{\mathbb{B}_\nu}(u^g) = 2\,\mathcal{E}_{\text{kin}} = E_{\mathbb{B}_\nu}\,g^2 \quad , \tag{39}$$

in which $E_{\mathbb{B}_\nu}$ can be written as

$$E_{\mathbb{B}_\nu} = \frac{1}{2\pi} \int_{\mathbb{B}_\nu \cup \widetilde{\mathbb{B}}_\nu} \omega^2 \, |\eta(\omega)|^2 \, \overline{T(\omega)} \, M \, T(\omega) \, d\omega \quad . \tag{40}$$

Since $T(-\omega) = \overline{T(\omega)}$, Eq. (40) can be rewritten as

$$E_{\mathbb{B}_\nu} = \frac{1}{\pi} \int_{\mathbb{B}_\nu} \omega^2 \, |\eta(\omega)|^2 \, \overline{T(\omega)} \, M \, T(\omega) \, d\omega \quad . \tag{41}$$

Comparing Eq. (38) with Eq. (39) yields

$$\mathcal{E}_{\text{mech}} \simeq \mathcal{E}_{\mathbb{B}_\nu}(u^g) = E_{\mathbb{B}_\nu} \, g^2 \quad . \tag{42}$$

3.7. Additional comments concerning the power equation

Multiplying Eq. (22) by $\dot{u}^g(t)$, we obtain

$$\frac{d}{dt} \left\{ \frac{1}{2} M \, \dot{u}^g(t)^2 + \frac{1}{2} K \, u^g(t)^2 \right\} + D \, \dot{u}^g(t)^2 = f(t) \, \dot{u}^g(t) \quad . \tag{43}$$

The power dissipated in the oscillator is defined by

$$\pi_{\text{diss}}(t) = D \, \dot{u}^g(t)^2 \quad , \tag{44}$$

and the input power is defined by

$$\pi_{\text{in}}(t) = f(t) \, \dot{u}^g(t) \quad . \tag{45}$$

Using Eqs. (23) to (25), Eq. (43) can be rewritten as

$$\dot{\varepsilon}_{\text{mech}}(t) + \pi_{\text{diss}}(t) = \pi_{\text{in}}(t) \quad . \tag{46}$$

Let $\mathcal{P}_{\text{diss}}$ be the total power dissipated in the oscillator and let $\mathcal{P}_{\text{in}}$ be the total input power defined by

$$\mathcal{P}_{\text{diss}} = \int_{\mathbb{R}} \pi_{\text{diss}}(t) \, dt \quad , \quad \mathcal{P}_{\text{in}} = \int_{\mathbb{R}} \pi_{\text{in}}(t) \, dt \quad . \tag{47}$$

The integration of Eq. (46) yields

$$\int_{\mathbb{R}} \dot{\varepsilon}_{\text{mech}}(t) \, dt + \mathcal{P}_{\text{diss}} = \mathcal{P}_{\text{in}} \quad , \tag{48}$$

which can be written as

$$\varepsilon_{\text{mech}}(+\infty) - \varepsilon_{\text{mech}}(-\infty) + \mathcal{P}_{\text{diss}} = \mathcal{P}_{\text{in}} \quad . \tag{49}$$

From Eqs. (21) and (23) to (25), we deduce that $\varepsilon_{\text{mech}}(+\infty) = 0$ and $\varepsilon_{\text{mech}}(-\infty) = 0$ and consequently,

$$\mathcal{P}_{\text{diss}} = \mathcal{P}_{\text{in}} \quad . \tag{50}$$

Using the Plancherel formula, the total input power is written as

$$\mathcal{P}_{\text{in}} = \int_{\mathbb{R}} f(t)\, \dot{u}^g(t)\, dt = \frac{1}{2\pi} \Re e \int_{\mathbb{R}} f(\omega)\, \overline{i\omega\, u^g(\omega)}\, d\omega \quad . \tag{51}$$

Substituting Eq. (18) on the right-hand side of Eq. (51) yields

$$\mathcal{P}_{\text{in}} = -\frac{1}{2\pi}\, g^2\, \Re e \int_{\mathbb{B}_\nu \cup \widetilde{\mathbb{B}}_\nu} i\omega\, |\eta(\omega)|^2\, \overline{T(\omega)}\, d\omega \quad . \tag{52}$$

Applying the method of approximation introduced in Section 3.5 yields

$$\mathcal{P}_{\text{in}} \simeq \frac{g^2}{2M} \quad . \tag{53}$$

4. Variational Formulation of the Master Structure in the MF Range

We consider the master structure whose geometry and mechanical model are defined in Chapter V. For the MF range, we recall that the constitutive equation of the master structure corresponds to the viscoelastic model described in Section IV.5.2. We consider the case of a master structure fixed on Γ_0 (see Fig. 2)

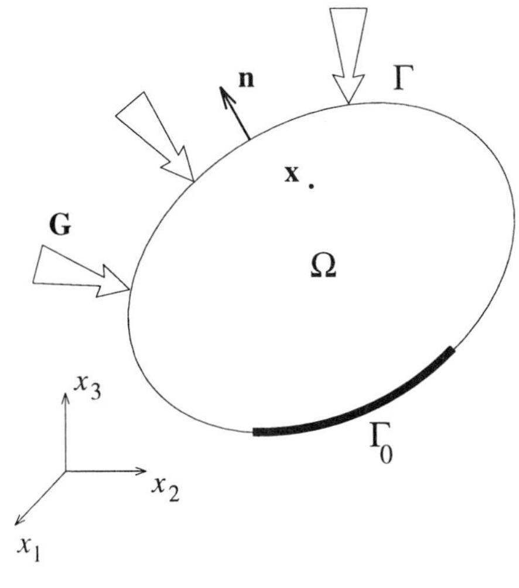

Fig. 2. Master structure configuration

The variational formulation concerning the master structure fixed on Γ_0 is given in Chapters V and VII and is summarized below.

4.1. Admissible function space

Let H be the space of all the square integrable functions defined on Ω with values in $\mathbb{C}^3$, equipped with the inner product

$$(\mathbf{u}\,,\mathbf{v})_H = \int_\Omega \mathbf{u}(\mathbf{x})\cdot\overline{\mathbf{v}(\mathbf{x})}\,d\mathbf{x} \quad , \tag{54}$$

and the associated norm

$$\|\mathbf{u}\|_H = (\mathbf{u}\,,\mathbf{u})_H^{1/2} \quad . \tag{55}$$

We introduce the complex vector space [5] $\mathcal{C}^c \subset H$ of functions defined on Ω with values in $\mathbb{C}^3$ and its subspace [6] $\mathcal{C}_0^c \subset \mathcal{C}^c \subset H$ of admissible displacement fields $\mathbf{u}$ defined on Ω with values in $\mathbb{C}^3$ such that $\mathbf{u} = \mathbf{0}$ on Γ_0,

$$\mathcal{C}_0^c = \{\mathbf{u} \in \mathcal{C}^c \quad ; \quad \mathbf{u} = \mathbf{0} \quad \text{on} \quad \Gamma_0\} \quad . \tag{56}$$

4.2. Excitation forces in the MF range

The master structure is submitted to a body force field $\eta(\omega)\,\mathbf{g}(\mathbf{x},\omega)$ defined on Ω with values in $\mathbb{C}^3$ and to a surface force field $\eta(\omega)\,\mathbf{G}(\mathbf{x},\omega)$ defined on Γ with values in $\mathbb{C}^3$. Function $\eta(\omega)$ is defined on $\mathbb{R}$ with values in $\mathbb{C}$, and has a limited support $\mathbb{B}_\nu \cup \widetilde{\mathbb{B}}_\nu$, which means that

$$\eta(\omega) = 0 \quad \text{as} \quad \omega \notin \mathbb{B}_\nu \cup \widetilde{\mathbb{B}}_\nu \quad . \tag{57}$$

In addition, we assume that function $\eta(\omega)$ is continuous on $\mathbb{B}_\nu$ and is such that

$$|\eta(-\omega)| = |\eta(\omega)| \quad , \tag{58}$$

$$|\eta(\omega)| \neq 0 \quad , \quad \forall\omega \in \mathbb{B}_\nu \quad . \tag{59}$$

Finally, we assume that for all fixed ω, function $\mathbf{x} \mapsto \mathbf{g}(\mathbf{x},\omega)$ belongs to H and function $\mathbf{x} \mapsto \mathbf{G}(\mathbf{x},\omega)$ is a square integrable function on Γ.

---

[5] Space $\mathcal{C}^c$ is the Sobolev space $(H^1(\Omega))^3$.

[6] Space $\mathcal{C}_0^c$ is the Hilbert space $\{\mathbf{u}\in(H^1(\Omega))^3\,,\mathbf{u}=0 \text{ on } \Gamma_0\}$.

4.3. Variational formulation

The variational formulation is written as follows. For all ω in $\mathbb{B}_\nu \cup \widetilde{\mathbb{B}}_\nu$, find $\mathbf{u}(\omega)$ in $\mathcal{C}_0^c$ such that

$$a(\omega\,;\mathbf{u}\,,\delta\mathbf{u}) = f(\omega\,;\delta\mathbf{u}) \quad , \quad \forall\,\delta\mathbf{u} \in \mathcal{C}_0^c \quad , \tag{60}$$

in which sesquilinear form $a(\omega\,;\mathbf{u}\,,\delta\mathbf{u})$ defined on $\mathcal{C}_0^c \times \mathcal{C}_0^c$ is such that [7]

$$a(\omega\,;\mathbf{u}\,,\delta\mathbf{u}) = -\omega^2\, m(\mathbf{u}\,,\delta\mathbf{u}) + i\omega\, d(\omega\,;\mathbf{u}\,,\delta\mathbf{u}) + k(\omega\,;\mathbf{u}\,,\delta\mathbf{u}) \quad , \tag{61}$$

where mass, damping and stiffness structural sesquilinear forms m, d and k are defined on $\mathcal{C}_0^c \times \mathcal{C}_0^c$ and are such that

$$m(\mathbf{u}\,,\delta\mathbf{u}) = \int_\Omega \rho\; \mathbf{u}\cdot \overline{\delta\mathbf{u}}\; d\mathbf{x} \quad , \tag{62}$$

$$d(\omega\,;\mathbf{u}\,,\delta\mathbf{u}) = \int_\Omega b_{ijkh}(\omega)\, \varepsilon_{kh}(\mathbf{u})\, \varepsilon_{ij}(\overline{\delta\mathbf{u}})\, d\mathbf{x} \quad , \tag{63}$$

$$k(\omega\,;\mathbf{u}\,,\delta\mathbf{u}) = \int_\Omega a_{ijkh}(\omega)\, \varepsilon_{kh}(\mathbf{u})\, \varepsilon_{ij}(\overline{\delta\mathbf{u}})\, d\mathbf{x} \quad . \tag{64}$$

It is assumed that coefficients $a_{ijkh}(\mathbf{x},\omega)$ and $b_{ijkh}(\mathbf{x},\omega)$ are such that $d(\mathbf{u},\mathbf{v}\,;\omega)$ and $k(\mathbf{u},\mathbf{v}\,;\omega)$ are continuous functions on band $\mathbb{B}_\nu$ with respect to ω. Antilinear form $f(\omega\,;\delta\mathbf{u})$ is defined on $\mathcal{C}_0^c$ and is such that [8]

$$f(\omega\,;\delta\mathbf{u}) = \eta(\omega)\Big\{\int_\Omega \mathbf{g}(\mathbf{x},\omega)\cdot \overline{\delta\mathbf{u}(\mathbf{x})}\; d\mathbf{x} + \int_\Gamma \mathbf{G}(\mathbf{x},\omega)\cdot \overline{\delta\mathbf{u}(\mathbf{x})}\; ds(\mathbf{x})\Big\} \quad . \tag{65}$$

For all ω in $\mathbb{B}_\nu \cup \widetilde{\mathbb{B}}_\nu$, Eq. (60) has a unique solution $\mathbf{u}(\omega)$ in $\mathcal{C}_0^c$.

5. Construction of an Appropriate Functional Basis for the Reduced Model in the MF Range

The reduced model is constructed by applying the Ritz-Galerkin method to the variational formulation defined by Eq. (60). As explained in Section 1, the structural modes cannot be used in the MF range and consequently a

---

[7] Sesquilinear form $a(\omega;\mathbf{u},\delta\mathbf{u})$ is continuous on $\mathcal{C}_0^c \times \mathcal{C}_0^c$ and is coercive ($\mathcal{C}_0^c$-elliptic), $a(\omega;\mathbf{u},\mathbf{u}) \geq c\|\mathbf{u}\|^2_{\mathcal{C}_0^c}$.

[8] Since $\mathbf{g}(\omega)\in(L^2(\Omega))^3$ and $\mathbf{G}(\omega)\in(L^2(\Gamma))^3$, we then deduce that $\delta\mathbf{u}\mapsto f(\omega\,;\delta\mathbf{u})$ is a continuous antilinear form on $\mathcal{C}_0^c$.

basis adapted to the MF range must be constructed. Below, we present a new approach for constructing such a basis (Soize, 1997a). We begin by introducing the operator-valued frequency response function defined for a special class of mechanical excitations. We then define the energy operator and finally we construct the basis as the eigenfunctions of this energy operator.

5.1. Operator-valued frequency response function

Let us assume that $\mathbf{G} = \mathbf{0}$ and that $\mathbf{g}$ is independent of ω. Then $f(\omega\,;\delta\mathbf{u})$ defined by Eq. (65) can be rewritten as

$$f(\omega\,;\delta\mathbf{u}) = \eta(\omega)\,(\mathbf{g}\,,\delta\mathbf{u})_H \quad . \tag{66}$$

For all ω fixed in $\mathbb{B}_\nu \cup \widetilde{\mathbb{B}}_\nu$ and $\mathbf{g}$ given, it can be proved (see Soize 1982a, 1982b and 1997a) that Eq. (60) has a unique solution denoted as $\mathbf{u}^{\mathbf{g}}(\omega)$ belonging to $\mathcal{C}_0^c \subset H$ which can be written as

$$\mathbf{u}^{\mathbf{g}}(\omega) = \eta(\omega)\,\mathbf{T}_H(\omega)\,\mathbf{g} \quad , \tag{67}$$

in which $\mathbf{T}_H(\omega)$ is an operator[9] in H having a countable number of complex eigenvalues $\lambda_\alpha(\omega)$ such that

$$\sum_{\alpha=1}^{+\infty} |\lambda_\alpha(\omega)|^2 < +\infty \quad . \tag{68}$$

Solution $\mathbf{u}^{\mathbf{g}}(\omega)$ corresponds to the vibration induced by excitation $\eta(\omega)\,\mathbf{g}$ and $\mathbf{T}_H(\omega)$ is the operator-valued frequency response function.

5.2. Definition of an energy operator

For the present continuous three-dimensional case, based on the definition given in Section 3.6 for a simple linear oscillator, the energy $\mathcal{E}_{\mathbb{B}_\nu}(\mathbf{u}^{\mathbf{g}})$ of vibration $\mathbf{u}^{\mathbf{g}}$ given by Eq. (67) is defined as twice the value of the total kinetic energy of the master structure for vibration $\mathbf{u}^{\mathbf{g}}$,

$$\mathcal{E}_{\mathbb{B}_\nu}(\mathbf{u}^{\mathbf{g}}) = \frac{1}{2\pi}\int_{\mathbb{B}_\nu\cup\widetilde{\mathbb{B}}_\nu} \omega^2\,(\mathbf{M}\,\mathbf{u}^{\mathbf{g}}(\omega), \mathbf{u}^{\mathbf{g}}(\omega))_H\, d\omega \quad , \tag{69}$$

---

[9] It is proved that $\mathbf{T}_H(\omega)$ is a Hilbert-Schmidt operator in H, i.e. a compact operator in H whose spectrum is countable and whose eigenvalues satisfy Eq. (68). For general notions on Hilbert-Schmidt operators, see for instance Kato, 1966; Reed and Simon, 1980.

in which $\mathbf{M}$ is the mass operator in H such that, for all $\mathbf{u}$ and $\delta\mathbf{u}$ in H,

$$(\mathbf{M}\,\mathbf{u}\,,\delta\mathbf{u})_H = m(\mathbf{u},\delta\mathbf{u}) \quad . \tag{70}$$

We introduce the solutions

$$\mathbf{u}^{\mathbf{g}_1}(\omega) = \eta(\omega)\,\mathbf{T}_H(\omega)\,\mathbf{g}_1 \tag{71}$$

$$\mathbf{u}^{\mathbf{g}_2}(\omega) = \eta(\omega)\,\mathbf{T}_H(\omega)\,\mathbf{g}_2 \tag{72}$$

corresponding to the vibrations due to excitations $\eta(\omega)\,\mathbf{g}_1$ and $\eta(\omega)\,\mathbf{g}_2$ respectively, where $\mathbf{g}_1$ and $\mathbf{g}_2$ are in H and are assumed to be independent of ω. The energy operator $\mathbf{E}_{\mathbb{B}_\nu}$ related to band $\mathbb{B}_\nu$ is defined by

$$(\mathbf{E}_{\mathbb{B}_\nu}\mathbf{g}_1\,,\mathbf{g}_2)_H = \frac{1}{2\pi}\int_{\mathbb{B}_\nu\cup\widetilde{\mathbb{B}}_\nu} \omega^2\,(\mathbf{M}\,\mathbf{u}^{\mathbf{g}_1}(\omega),\mathbf{u}^{\mathbf{g}_2}(\omega))_H\,d\omega \quad . \tag{73}$$

From Eqs. (69) and (71), we deduce that

$$\mathcal{E}_{\mathbb{B}_\nu}(\mathbf{u}^{\mathbf{g}}) = (\mathbf{E}_{\mathbb{B}_\nu}\mathbf{g}\,,\mathbf{g})_H \quad . \tag{74}$$

It should be noted that operator $\mathbf{E}_{\mathbb{B}_\nu}$ depends on $\mathbb{B}_\nu$ and η, but does not depend on the spatial parts $\mathbf{g}_1$ and $\mathbf{g}_2$ of the excitations.

5.3. Mathematical properties of the energy operator

In order to characterize the energy operator, we need to introduce the mathematical notion of trace operators in space H (see, for instance, Reed and Simon, 1980). Roughly speaking, a trace operator in H is an operator in H which has a countable spectrum whose eigenvalues λ_α are such that $\sum_{\alpha=1}^{+\infty}|\lambda_\alpha| < +\infty$. It is proved that energy operator $\mathbf{E}_{\mathbb{B}_\nu}$ defined by Eq. (73) is a positive-definite symmetric trace operator [10] in H whose range space [11] is a subspace of $\mathcal{C}_0^c$, which can be written as

$$\mathbf{E}_{\mathbb{B}_\nu} = \frac{1}{2\pi}\int_{\mathbb{B}_\nu\cup\widetilde{\mathbb{B}}_\nu} \omega^2\,|\eta(\omega)|^2\,\mathbf{T}_H(\omega)^*\,\mathbf{M}\,\mathbf{T}_H(\omega)\,d\omega \quad . \tag{75}$$

In Eq. (75), $\mathbf{T}_H(\omega)^*$ is the adjoint [12] of $\mathbf{T}_H(\omega)$ and is a trace operator in H. Operator $\mathbf{E}_{\mathbb{B}_\nu}$ can also be written as

$$\mathbf{E}_{\mathbb{B}_\nu} = \frac{1}{\pi}\int_B \omega^2\,|\eta(\omega)|^2\,\Re e\,\{\mathbf{T}_H(\omega)^*\,\mathbf{M}\,\mathbf{T}_H(\omega)\}\,d\omega \quad , \tag{76}$$

---

[10] Operator $\mathbf{E}_{\mathbb{B}_\nu}$ is compact, symmetric ($\mathbf{E}_{\mathbb{B}_\nu}^*=\mathbf{E}_{\mathbb{B}_\nu}$), positive-definite ($(\mathbf{E}_{\mathbb{B}_\nu}\mathbf{g}\,,\mathbf{g})_H>0$ for all $\mathbf{g}\neq 0$ in H) and finally is a trace operator, i.e. the series of its eigenvalues is convergent.

[11] The range space of operator $\mathbf{E}_{\mathbb{B}_\nu}$ is such that $R\{\mathbf{E}_{\mathbb{B}_\nu}\}=\{\mathbf{E}_{\mathbb{B}_\nu}\mathbf{e}\,,\forall\,\mathbf{e}\in H\}=(H^2(\Omega))^3\cap\mathcal{C}_0^c$.

[12] The adjoint of a bounded operator in a Hilbert space is defined in Section 6.2 of Mathematical Notations in the appendix.

in which $\Re e$ denotes the real part. For the proofs of these results, we refer the reader to Soize, 1997a.

5.4. Spectral properties of the energy operator

From the mathematical properties of energy operator $\mathbf{E}_{\mathcal{B}}$ established in Section 5.3, we deduce that operator $\mathbf{E}_{\mathbb{B}_\nu}$ has a countable number of decreasing positive eigenvalues with finite multiplicity, possibly excepting zero,

$$\lambda_1 \geq \lambda_2 \geq \ldots \to 0 \quad , \tag{77}$$

in which the λ_α terms are the repeated eigenvalues of $\mathbf{E}_{\mathbb{B}_\nu}$. The corresponding eigenfunctions $\{\mathbf{e}_\alpha\}_{\alpha \geq 1}$, such that

$$\mathbf{E}_{\mathbb{B}_\nu} \mathbf{e}_\alpha = \lambda_\alpha \, \mathbf{e}_\alpha \quad , \tag{78}$$

are functions $\mathbf{e}_\alpha(\mathbf{x})$ from Ω into $\mathbb{R}^3$ and form a complete orthonormal family in H,

$$(\mathbf{e}_\alpha \, , \mathbf{e}_\beta)_H = \delta_{\alpha\beta} \quad . \tag{79}$$

Since $\mathbf{E}_{\mathbb{B}_\nu}$ is a positive-definite symmetric trace operator, we have

$$\sum_{\alpha=1}^{+\infty} |\lambda_\alpha| = \sum_{\alpha=1}^{+\infty} \lambda_\alpha < +\infty \quad . \tag{80}$$

The trace norm of $\mathbf{E}_{\mathbb{B}_\nu}$ denoted as tr $\mathbf{E}_{\mathbb{B}_\nu}$ is such that

$$\text{tr } \mathbf{E}_{\mathbb{B}_\nu} = \sum_{\alpha=1}^{+\infty} \lambda_\alpha < +\infty \quad , \tag{81}$$

and $\mathbf{E}_{\mathbb{B}_\nu}$ can be written as

$$\mathbf{E}_{\mathbb{B}_\nu} = \sum_{\alpha=1}^{+\infty} \lambda_\alpha \, (\, . \, , \mathbf{e}_\alpha)_H \, \mathbf{e}_\alpha \quad , \tag{82}$$

which means that, for all $\mathbf{g}$ in H,

$$\mathbf{E}_{\mathbb{B}_\nu} \mathbf{g} = \sum_{\alpha=1}^{+\infty} \lambda_\alpha \, (\mathbf{g} \, , \mathbf{e}_\alpha)_H \, \mathbf{e}_\alpha \quad . \tag{83}$$

5.5. Fundamental property of the eigenfunctions of the energy operator

The set $\{\mathbf{e}_\alpha\}_{\alpha=1,\ldots,+\infty}$ of eigenfunctions of operator $\mathbf{E}_{\mathbb{B}_\nu}$ is a complete family in admissible space $\mathcal{C}_0^c \subset H$, orthonormal for the inner product of H. In addition, each eigenfunction $\mathbf{e}_\alpha$ is a continuous function from Ω into $\mathbb{R}^3$. For the proof of these properties, we refer the reader to Soize, 1997a. This result allows a Ritz-Galerkin projection of the variational formulation defined in Section 4 to be constructed using a truncation strategy based on the use of the property defined by Eq. (81).

6. Construction of a Reduced Model in the MF Range

Taking into account the result of Section 5.5, the reduced model adapted to medium-frequency band $\mathbb{B}_\nu$ is obtained using the Ritz-Galerkin projection of the variational formulation defined by Eq. (60) on the subspace $\mathcal{C}_{0,N}^c$ of $\mathcal{C}_0^c$ spanned by the eigenfunctions $\{\mathbf{e}_1,\ldots,\mathbf{e}_N\}$ which correspond to the N highest eigenvalues $\{\lambda_1,\ldots,\lambda_N\}$ of energy operator $\mathbf{E}_{\mathbb{B}_\nu}$. Let $\mathbf{u}(\omega) \in \mathcal{C}_0^c$ be the unique solution of Eq. (60) and let $\mathbf{u}^N(\omega)$ be the projection of $\mathbf{u}(\omega)$ on $\mathcal{C}_{0,N}^c$ such that

$$\mathbf{u}^N(\mathbf{x},\omega) = \sum_{\alpha=1}^{N} \theta_\alpha(\omega)\,\mathbf{e}_\alpha(\mathbf{x}) \quad , \tag{84}$$

in which $\theta_\alpha(\omega) \in \mathbb{C}$. From Eq. (60), we deduce that for all ω in $\mathbb{B}_\nu \cup \widetilde{\mathbb{B}}_\nu$, $\boldsymbol{\theta}(\omega) = (\theta_1(\omega),\ldots,\theta_N(\omega)) \in \mathbb{C}^N$ is the solution of the linear equation

$$[\mathcal{A}_N(\omega)]\,\boldsymbol{\theta}(\omega) = \eta(\omega)\,\boldsymbol{\mathcal{F}}(\omega) \quad , \tag{85}$$

in which $[\mathcal{A}_N(\omega)]$ is the symmetric $(N\times N)$ complex matrix defined by

$$[\mathcal{A}_N(\omega)]_{\beta\alpha} = a(\omega\,;\mathbf{e}_\alpha,\mathbf{e}_\beta) \quad , \tag{86}$$

and where $\boldsymbol{\mathcal{F}}(\omega) = (\mathcal{F}_1(\omega),\ldots,\mathcal{F}_N(\omega)) \in \mathbb{C}^N$ is such that

$$\eta(\omega)\,\mathcal{F}_\alpha(\omega) = f(\omega\,;\mathbf{e}_\alpha) \quad . \tag{87}$$

For all ω in $\mathbb{B}_\nu \cup \widetilde{\mathbb{B}}_\nu$, matrix $[\mathcal{A}_N(\omega)]$ is invertible and the solution of Eq. (85) is written as

$$\boldsymbol{\theta}(\omega) = \eta(\omega)\,[\mathcal{T}_N(\omega)]\,\boldsymbol{\mathcal{F}}(\omega) \quad , \tag{88}$$

in which $[\mathcal{T}_N(\omega)]$ is the symmetric $(N\times N)$ complex matrix such that

$$[\mathcal{T}_N(\omega)] = [\mathcal{A}_N(\omega)]^{-1} \quad . \tag{89}$$

For all ω in $\mathbb{B}_\nu \cup \widetilde{\mathbb{B}}_\nu$, $\mathbf{u}^N(\omega) \to \mathbf{u}(\omega)$ in $\mathcal{C}_0^c$ as $N \to +\infty$. The reduced model adapted to medium-frequency band $\mathbb{B}_\nu$ of the master structure described by Eq. (60) is constituted by the set

$$\left\{\boldsymbol{\theta}(\omega)\,,[\mathcal{A}_N(\omega)]\,,\boldsymbol{\mathcal{F}}(\omega)\right\} \quad . \tag{90}$$

7. Finite Element Discretization

An explicit construction of the eigenfunctions $\{\mathbf{e}_1,\dots,\mathbf{e}_N\}$ of energy operator $\mathbf{E}_{\mathbb{B}_\nu}$ cannot be obtained in the general case. A finite dimension approximation $\mathbf{E}_{\mathbb{B}_\nu,n}$ of $\mathbf{E}_{\mathbb{B}_\nu}$ must be introduced and the eigenfunctions $\{\mathbf{e}_1^n \dots, \mathbf{e}_N^n\}$ of $\mathbf{E}_{B,n}$ (associated with the N highest eigenvalues) constitute the approximation of $\{\mathbf{e}_1 \dots, \mathbf{e}_N\}$. This finite approximation can be obtained using the finite element method. We consider a finite element mesh of master structure Ω and we introduce the subspace $\mathcal{C}_{0,n}^c \subset \mathcal{C}_0^c$ of finite dimension n. Let $\mathbf{U} = (U_1,\dots,U_n)$ be the complex vector of the DOFs which are the values of $\mathbf{u}$ at the nodes of the finite element mesh of domain Ω. Since the finite element method uses a real basis for constructing the finite element matrices, the finite element discretization of the variational formulation defined by Eq. (60) yields the complex symmetric matrix equation

$$[A(\omega)]\,\mathbf{U} = \eta(\omega)\,\mathbf{F}(\omega) \quad , \tag{91}$$

in which $[A(\omega)]$ is the dynamic stiffness matrix of the master structure. It is an invertible symmetric $(n\times n)$ complex matrix such that

$$[A(\omega)] = -\omega^2[\,M\,] + i\omega\,[D(\omega)] + [K(\omega)] \quad . \tag{92}$$

Mass, damping and stiffness matrices $[\,M\,]$, $[D(\omega)]$ and $[K(\omega)]$ are positive-definite symmetric $(n\times n)$ real matrices. The finite element discretization of antilinear form $f(\omega\,;\delta\mathbf{u})$ yields the complex vector $\eta(\omega)\,\mathbf{F}(\omega) \in \mathbb{C}^n$.

7.1. Finite element discretization of the energy operator

The finite element discretization of operator $\mathbf{E}_{\mathbb{B}_\nu}$ defined by Eq. (76) is the positive-definite symmetric $(n\times n)$ real matrix $[E_{\mathbb{B}_\nu,n}]$ such that

$$[E_{\mathbb{B}_\nu,n}] = [\,G\,]\,[E_n]\,[\,G\,] \quad , \tag{93}$$

in which the invertible symmetric $(n{\times}n)$ real matrix $[\,G\,]$ corresponds to the finite element discretization of the sesquilinear form $\int_\Omega \mathbf{u}\cdot\overline{\delta\mathbf{u}}\;d\mathbf{x}$ defined on

$\mathcal{C}_0^c \times \mathcal{C}_0^c$. The positive-definite symmetric $(n\times n)$ real matrix $[E_n]$ is written as

$$[E_n] = \int_B [e_n(\omega)]\, d\omega \quad , \tag{94}$$

in which $(n\times n)$ real matrix $[e_n(\omega)]$ is given by

$$[e_n(\omega)] = \frac{1}{\pi}\omega^2\, |\eta(\omega)|^2\, \Re e\,\{[T(\omega)]^*\,[M]\,[T(\omega)]\} \quad , \tag{95}$$

in which $[T(\omega)]$ is the symmetric $(n\times n)$ complex matrix such that

$$[T(\omega)] = [A(\omega)]^{-1} \quad , \quad [T(\omega)]^* = \overline{[T(\omega)]} \quad . \tag{96}$$

In the particular case where $f(\omega\,;\delta\mathbf{u}) = \eta(\omega)\,(\mathbf{g}\,,\delta\mathbf{u})_H$, denoting its finite element discretization as $\eta(\omega)\,\mathbf{F}(\omega)$, we have

$$(\mathbf{E}_{\mathbb{B}_\nu}\,\mathbf{g}\,,\mathbf{g})_H = \mathbf{F}(\omega)^*\,[E_n]\,\mathbf{F}(\omega) \quad . \tag{97}$$

It should be noted that the right-hand side of Eq. (97) includes $[E_n]$, not $[E_{\mathbb{B}_\nu,n}]$. For the proof of the results of Section 7.1, we refer the reader to Soize, 1997a.

7.2. Finite element discretization of the spectral problem associated with the energy operator

Let $\mathbf{P}^\alpha = (P_1^\alpha,\ldots,P_n^\alpha)$ be the real vector of the DOFs which are the values of $\mathbf{e}_\alpha(\mathbf{x})$ at the nodes of the finite element mesh of domain Ω. The finite element discretization of Eq. (78) yields

$$[E_{\mathbb{B}_\nu,n}]\,\mathbf{P}^\alpha = \lambda_\alpha^n\,[\,G\,]\,\mathbf{P}^\alpha \quad . \tag{98}$$

The eigenvectors $\{\mathbf{P}^1,\ldots,\mathbf{P}^n\}$ of Eq. (98) form a basis of $\mathbb{R}^n$ and satisfy the orthogonality conditions

$$\mathbf{P}^{\alpha T}[\,G\,]\,\mathbf{P}^\beta = \delta_{\alpha\beta} \quad , \quad \mathbf{P}^{\alpha T}[E_{\mathbb{B}_\nu,n}]\,\mathbf{P}^\beta = \lambda_\alpha^n\,\delta_{\alpha\beta} \quad . \tag{99}$$

7.3. Reduced model adapted to medium-frequency band $\mathbb{B}_\nu$

Let $N \ll n$. The finite element discretization of Eq. (84) is written as

$$\mathbf{U} = [\,P\,]\,\boldsymbol{\theta} \quad , \tag{100}$$

in which $[\,P\,]$ is the $(n\times N)$ real matrix whose columns are of the N eigenvectors $\{\mathbf{P}^1,\ldots,\mathbf{P}^N\}$ corresponding to the N highest eigenvalues $\lambda_1^n \geq$

$\ldots \geq \lambda_N^n$. In the finite element discretization, the reduced model defined by Eq. (90) becomes

$$\left\{\boldsymbol{\theta}(\omega)\,,[\mathcal{A}_N^n(\omega)]\,,\boldsymbol{\mathcal{F}}^n(\omega)\right\} \quad , \tag{101}$$

in which

$$[\mathcal{A}_N^n(\omega)] = [\,P\,]^T\,[A(\omega)]\,[\,P\,] \quad , \tag{102}$$

$$\boldsymbol{\mathcal{F}}^n(\omega) = [\,P\,]^T\,\mathbf{F}(\omega) \quad . \tag{103}$$

7.4. Dominant eigensubspace of the energy operator and order of the reduced model

Let $\mathcal{C}_{0,n}^{N} \subset \mathcal{C}_{0,n}^{c} \subset \mathcal{C}_0^{c}$ be the space spanned by $\{\mathbf{e}_1^n \ldots, \mathbf{e}_N^n\}$. Let $\mathbf{u}_n(\omega)$ be the projection of $\mathbf{u}(\omega)$ on $\mathcal{C}_{0,n}^{c}$ in which $\mathbf{u}(\omega)$ is the solution of Eq. (60). The energy of vibration $\mathbf{u}_n$ is then written as

$$\mathcal{E}_{\mathbb{B}_\nu}(\mathbf{u}_n) = \sum_{\alpha=1}^{n} \lambda_\alpha^n\,|\mathcal{F}_\alpha^n|^2 \quad , \tag{104}$$

and the energy of the projection $\mathbf{u}_n^N$ of $\mathbf{u}(\omega)$ on $\mathcal{C}_{0,n}^{N}$ is written as

$$\mathcal{E}_{\mathbb{B}_\nu}(\mathbf{u}_n^N) = \sum_{\alpha=1}^{N} \lambda_\alpha^n\,|\mathcal{F}_\alpha^n|^2 \quad . \tag{105}$$

We then have

$$\mathcal{E}_{\mathbb{B}_\nu}(\mathbf{u}_n) \leq \lambda_1^n\,\|\boldsymbol{\mathcal{F}}^n\|^2 \quad , \tag{106}$$

and since the upper bound is effectively reached,

$$\varepsilon_{\max} = \max_{\boldsymbol{\mathcal{F}}^n \in \mathbb{C}^n} \mathcal{E}_{\mathbb{B}_\nu}(\mathbf{u}_n) = \lambda_1^n\,\|\boldsymbol{\mathcal{F}}^n\|^2 \quad . \tag{107}$$

We then deduce that

$$\frac{\mathcal{E}_{\mathbb{B}_\nu}(\mathbf{u}_n) - \mathcal{E}_{\mathbb{B}_\nu}(\mathbf{u}_n^N)}{\varepsilon_{\max}} \leq \frac{\lambda_{N+1}^n}{\lambda_1^n} \quad . \tag{108}$$

Since $\{\lambda_\alpha\}_\alpha$ is a decreasing sequence of positive numbers as $\alpha \to +\infty$, if n is sufficiently large, then there exists $N < n$ such that

$$\frac{\lambda_{N+1}^n}{\lambda_1^n} \ll 1 \quad . \tag{109}$$

If $N < n$ is such that Eq. (109) holds, then subspace $\mathcal{C}_{0,n}^N$ is called the *dominant eigensubspace* of operator $\mathbf{E}_{\mathbb{B}_\nu,n}$ corresponding to the N highest eigenvalues $\lambda_1^n \geq \ldots \geq \lambda_N^n$ and N is the order of the reduced model. Fig. 3 shows a typical example of the distribution of eigenvalues λ_α^n as a function of their rank α. In this case, the order N of the reduced model would be 25 or 30.

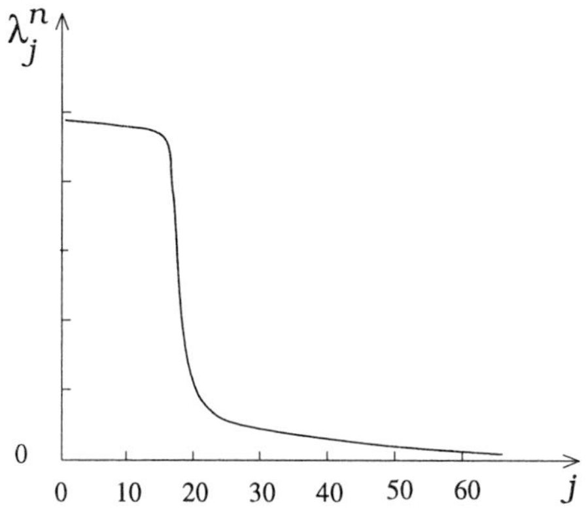

Fig. 3. Example of distribution of the energy-operator eigenvalues

8. Construction of the Dominant Eigensubspace Using the Subspace Iteration Method

The reduced model defined by Eq. (101) requires construction of the dominant eigensubspace of $\mathbf{E}_{B,n}$, i.e. calculation of the eigenvectors $\mathbf{P}^1, \ldots, \mathbf{P}^N$ in $\mathbb{R}^n$ corresponding to the highest eigenvalues $\lambda_1^n \geq \ldots \geq \lambda_N^n$ of the generalized symmetric eigenvalue problem defined by Eq. (98). Since n is large and $N \ll n$, the subspace iteration method or the Lanczos method (see for instance Bathe and Wilson, 1976; Golub and Van Loan, 1989) can *a priori* be used. The algebraic structure of matrix $[E_n]$ defined by Eqs. (94) and (95) shows that the use of the subspace iteration method allows a very efficient solver to be constructed, avoiding explicit calculation of matrix $[E_n]$. This solver is detailed extensively in Soize, 1997a. Two procedures are proposed. The first one is a procedure in the frequency domain. The second, a more efficient procedure, consists in using the MF solution method presented in Chapter VII.

CHAPTER IX

Response to Deterministic and Random Excitations

1. Introduction

In Chapter V, we introduced the operator-valued frequency response function (FRF) for the master structure. In Chapters VI and VII, we saw how to calculate this FRF in the LF and MF ranges. In the present chapter, using this FRF, we calculate the response of the master structure submitted to deterministic or random excitations.

In Section 2, we recall the main results of Chapter V useful for understanding this chapter.

In Sections 3 and 4, we calculate the response of the master structure submitted to time-periodic and time square integrable deterministic excitations.

In Section 5, we consider the case of a random excitation which is stationary with respect to time.

Finally, in Section 6, we present the calculation of the response of the master structure submitted to a nonstationary random excitation.

2. Operator-Valued FRF and Associated Linear Filter in the Time Domain

2.1. Continuous case

Frequency domain. We consider the master structure submitted to an excitation force field generated by body and surface force fields (see Chapter V). For all fixed real ω, the response $\mathbf{u}(\omega)$ is the unique solution in $\mathcal{C}_S^c$ of the variational formulation defined by Eq. (V.58),

$$a(\omega\,;\mathbf{u}\,,\delta\mathbf{u}) = f(\omega\,;\,\delta\mathbf{u}) \quad , \quad \forall\,\delta\mathbf{u} \in \mathcal{C}_S^c \quad , \tag{1}$$

in which $\mathcal{C}_S^c$ is the admissible space defined by Eq. (V.56) for the fixed and free master structure cases, $a(\omega\,;\mathbf{u}\,,\delta\mathbf{u})$ is defined by Eq. (V.44) and $f(\omega\,;\,\delta\mathbf{u})$ is defined by

$$f(\omega\,;\delta\mathbf{u}) = \int_\Omega \mathbf{g}(\mathbf{x},\omega)\cdot\overline{\delta\mathbf{u}(\mathbf{x})}\;d\mathbf{x} + \int_\Gamma \mathbf{G}(\mathbf{y},\omega)\cdot\overline{\delta\mathbf{u}(\mathbf{y})}\;ds(\mathbf{y}) \quad , \tag{2}$$

where $\mathbf{g}$ and $\mathbf{G}$ denote the body and surface force fields respectively. The linear operator equation associated with Eq. (1) is written as

$$\mathbf{A}(\omega)\,\mathbf{u}(\omega) = \mathbf{f}(\omega) \quad , \tag{3}$$

in which $\mathbf{A}(\omega)$ is the dynamic stiffness operator defined by Eq. (V.47) and $\mathbf{f}(\omega)$ is the excitation force field [1] such that

$$<\mathbf{f}(\omega)\,,\delta\mathbf{u}> = f(\omega\,;\delta\mathbf{u}) \quad , \quad \forall\,\delta\mathbf{u}\in\mathcal{C}_S^c \quad , \tag{4}$$

in which $f(\omega\,;\delta\mathbf{u})$ is defined by Eq. (2). The unique solution of Eq. (1) or Eq. (3) is written as

$$\mathbf{u}(\omega) = \mathbf{T}(\omega)\,\mathbf{f}(\omega) \quad , \tag{5}$$

in which $\mathbf{T}(\omega)$ is the operator-valued frequency response function (see Section V.5). Response $\mathbf{u}(\omega)$ of the master structure must be calculated distinguishing the low-frequency band case (Chapter VI) and the medium-frequency band case (Chapter VII) of the frequency band of analysis. We then denote the LF band as $\mathbb{B}_{\mathrm{LF}}$ and the MF broad band as $\mathbb{B}_{\mathrm{MF}}$ such that

$$\mathbb{B}_{\mathrm{LF}} = [\,0\,,\,\omega_{\mathrm{LF,final}}\,] \quad , \quad \mathbb{B}_{\mathrm{MF}} = [\,\omega_{\mathrm{MF,init}}\,,\,\omega_{\mathrm{MF,final}}\,] \quad . \tag{6}$$

In the frequency domain, the velocity field $\mathbf{v}(\omega)$ and acceleration field $\mathbf{w}(\omega)$ are then deduced from $\mathbf{u}(\omega)$ by the formulas

$$\mathbf{v}(\omega) = i\omega\,\mathbf{u}(\omega) \quad , \quad \mathbf{w}(\omega) = i\omega\,\mathbf{v}(\omega) \quad . \tag{7}$$

Time domain. In the time domain, we recall that given body force field $\mathbf{g}$ and surface force field $\mathbf{G}$ are with values in $\mathbb{R}^3$ (and not in $\mathbb{C}^3$). Using the Fourier transform,

$$\mathbf{g}(\mathbf{x},\omega) = \int_{\mathbb{R}} e^{-i\omega t}\,\mathbf{g}(\mathbf{x},t)\,dt \quad , \quad \mathbf{G}(\mathbf{y},\omega) = \int_{\mathbb{R}} e^{-i\omega t}\,\mathbf{G}(\mathbf{y},t)\,dt \quad . \tag{8}$$

[1] Element $\mathbf{f}(\omega)$ belongs to antidual space $\mathcal{C}_S^{c\,\prime}$ of $\mathcal{C}_S^c$.

The corresponding expression of $f(\omega\,;\delta\mathbf{u})$ defined by Eq. (2) is given by

$$f(t\,;\delta\mathbf{u}) = \int_\Omega \mathbf{g}(\mathbf{x},t)\cdot\overline{\delta\mathbf{u}(\mathbf{x})}\,d\mathbf{x} + \int_\Gamma \mathbf{G}(\mathbf{y},t)\cdot\overline{\delta\mathbf{u}(\mathbf{y})}\,ds(\mathbf{y}) \quad , \tag{9}$$

in which $\delta\mathbf{u}$ belongs to $\mathcal{C}_S^c$. Let $\mathbf{f}(t)$ be such that

$$\mathbf{f}(\omega) = \int_{\mathbb{R}} e^{-i\omega t}\,\mathbf{f}(t)\,dt \quad . \tag{10}$$

From Eq. (4), we deduce that excitation force field $\mathbf{f}(t)$ is such that [2]

$$<\mathbf{f}(t)\,,\delta\mathbf{u}> = f(t\,;\delta\mathbf{u}) \quad , \quad \forall\,\delta\mathbf{u}\in\mathcal{C}_S^c \quad , \tag{11}$$

in which $f(t\,;\delta\mathbf{u})$ is defined by Eq. (9). Let $\mathbf{u}(t)$ be such that

$$\mathbf{u}(\omega) = \int_{\mathbb{R}} e^{-i\omega t}\,\mathbf{u}(t)\,dt \quad . \tag{12}$$

From Section V.5.5, we deduce that response $\mathbf{u}(t)$ whose Fourier transform $\mathbf{u}(\omega)$ is given by Eq. (5), is written as

$$\mathbf{u}(t) = (\mathbf{h} *_t \mathbf{f})(t) \quad , \tag{13}$$

i.e,

$$\mathbf{u}(t) = \int_{\mathbb{R}} \mathbf{h}(t')\,\mathbf{f}(t-t')\,dt' = \int_{\mathbb{R}} \mathbf{h}(t-t')\,\mathbf{f}(t')\,dt' \quad , \tag{14}$$

in which operator-valued impulse response function $t\mapsto\mathbf{h}(t)$ satisfies

$$\mathbf{h}(t) = \mathbf{0} \quad , \quad \forall\, t<0 \quad . \tag{15}$$

The operator-valued frequency response function $\mathbf{T}(\omega)$ of the causal linear filter $\mathbf{h}\,*_t$ is then given by

$$\mathbf{T}(\omega) = \int_{\mathbb{R}} e^{-i\omega t}\,\mathbf{h}(t)\,dt \quad . \tag{16}$$

In the time domain, velocity field $\mathbf{v}(\mathbf{x},t)$ and acceleration field $\mathbf{w}(\mathbf{x},t)$, corresponding to Eq. (7), are such that

$$\mathbf{v}(\mathbf{x},t) = \partial_t\mathbf{u}(\mathbf{x},t) \quad , \quad \mathbf{w}(\mathbf{x},t) = \partial_t\mathbf{v}(\mathbf{x},t) \quad . \tag{17}$$

---

[2] From a mathematical point of view, $\mathbf{f}(t)$ belongs to antidual space $\mathcal{C}_S^{c\,\prime}$ of $\mathcal{C}_S^c$.

2.2. Ritz-Galerkin approximation

In order to construct the response of the master structure submitted to a given excitation field, we introduce a subspace $\mathcal{C}^c_{S,n}$ of $\mathcal{C}^c_S$ of finite dimension n spanned by a real basis $\{\mathbf{e}_1, \ldots, \mathbf{e}_n\}$. Consequently,

$$\delta\mathbf{u} = \sum_{j=1}^{n} \delta U_j \ \mathbf{e}_j \quad , \tag{18}$$

in which $\delta\mathbf{U} = (\delta U_1, \ldots, \delta U_n)$ belongs to $\mathbb{C}^n$. It should be noted that this basis can be constructed in the context of the finite element method or by any other method (for instance a finite set of eigenfunctions of an appropriate operator, etc.). We then use the Ritz-Galerkin method introduced in Section II.7.

Frequency domain. Let $\mathbf{F}(\omega) = (F_1(\omega), \ldots, F_n(\omega)) \in \mathbb{C}^n$ be the vector such that, $\forall j \in \{1, \ldots, n\}$,

$$F_j(\omega) = f(\omega\,; \mathbf{e}_j) = <\mathbf{f}(\omega)\,, \mathbf{e}_j> \quad . \tag{19}$$

From Eqs. (2) and (19), we deduce that

$$F_j(\omega) = \int_\Omega \mathbf{g}(\mathbf{x}, \omega) \cdot \mathbf{e}_j(\mathbf{x}) \ d\mathbf{x} + \int_\Gamma \mathbf{G}(\mathbf{y}, \omega) \cdot \mathbf{e}_j(\mathbf{y}) \ ds(\mathbf{y}) \quad . \tag{20}$$

The projection $\mathbf{u}^n(\omega)$ of $\mathbf{u}(\omega)$ in $\mathcal{C}^c_{S,n}$ can be written as

$$\mathbf{u}^n(\omega) = \sum_{j=1}^{n} U_j(\omega) \ \mathbf{e}_j \quad , \tag{21}$$

in which $\mathbf{U}(\omega) = (U_1(\omega), \ldots, U_n(\omega))$ belongs to $\mathbb{C}^n$. Substituting Eqs. (18) and (21) into Eq. (1) and using Eq. (19) yields

$$[A(\omega)]\,\mathbf{U}(\omega) = \mathbf{F}(\omega) \quad , \tag{22}$$

in which the elements of $(n \times n)$ complex symmetric matrix $[A(\omega)]$ are such that

$$[A(\omega)]_{jk} = a(\omega\,; \mathbf{e}_k\,, \mathbf{e}_j) \quad . \tag{23}$$

Consequently, the discretization of Eq. (5) is

$$\mathbf{U}(\omega) = [T(\omega)]\,\mathbf{F}(\omega) \quad , \tag{24}$$

in which for all real ω,

$$[T(\omega)] = [A(\omega)]^{-1} \quad . \tag{25}$$

The projections $\mathbf{v}^n(\omega)$ and $\mathbf{w}^n(\omega)$ in $\mathcal{C}^c_{S,n}$ of velocity field $\mathbf{v}(\omega)$ and acceleration field $\mathbf{w}(\omega)$ can be written as

$$\mathbf{v}^n(\mathbf{x},\omega) = \sum_{j=1}^{n} V_j(\omega)\ \mathbf{e}_j(\mathbf{x}) \quad , \quad \mathbf{w}^n(\mathbf{x},\omega) = \sum_{j=1}^{n} W_j(\omega)\ \mathbf{e}_j(\mathbf{x}) \quad , \tag{26}$$

in which $\mathbf{V}(\omega) = (V_1(\omega), \ldots, V_n(\omega))$ and $\mathbf{W}(\omega) = (W_1(\omega), \ldots, W_n(\omega))$ are in $\mathbb{C}^n$ and such that

$$\mathbf{V}(\omega) = i\omega\, \mathbf{U}(\omega) \quad , \quad \mathbf{W}(\omega) = i\omega\, \mathbf{V}(\omega) \quad . \tag{27}$$

Time domain. Let $\mathbf{F}(t) = (F_1(t), \ldots, F_n(t)) \in \mathbb{R}^n$ be the vector such that, $\forall j \in \{1, \ldots, n\}$,

$$F_j(t) = f(t\,; \mathbf{e}_j) = <\mathbf{f}(t)\,, \mathbf{e}_j> \quad . \tag{28}$$

From Eqs. (9) and (28), we deduce that

$$F_j(t) = \int_\Omega \mathbf{g}(\mathbf{x},t) \cdot \mathbf{e}_j(\mathbf{x})\ d\mathbf{x} + \int_\Gamma \mathbf{G}(\mathbf{y},t) \cdot \mathbf{e}_j(\mathbf{y})\ ds(\mathbf{y}) \quad . \tag{29}$$

The projection $\mathbf{u}^n(t)$ in $\mathcal{C}^c_{S,n}$ of $\mathbf{u}(t)$ (given by Eq (14)) can be written as

$$\mathbf{u}^n(\mathbf{x},t) = \sum_{j=1}^{n} U_j(t)\ \mathbf{e}_j \quad , \tag{30}$$

in which $\mathbf{U}(t) = (U_1(t), \ldots, U_n(t))$ belonging to $\mathbb{R}^n$ is written as

$$\mathbf{U}(t) = \int_{\mathbb{R}} [\,h(t')\,]\ \mathbf{F}(t-t')\, dt' \quad . \tag{31}$$

The impulse response function $[\,\mathbf{h}(t)\,]$ with values in the $(n \times n)$ real symmetric matrices satisfies

$$[\,\mathbf{h}(t)\,] = [\,0\,] \quad , \quad \forall t < 0 \quad . \tag{32}$$

The frequency response function $[T(\omega)]$ of the causal linear filter $[\,h\,]*$, with values in the $(n \times n)$ complex symmetric matrices, is then given by

$$[T(\omega)] = \int_{\mathbb{R}} e^{-i\omega t}\,[\,h(t)\,]\,dt \quad . \tag{33}$$

The projections $\mathbf{v}^n(t)$ and $\mathbf{w}^n(t)$ in $\mathcal{C}^c_{S,n}$ of velocity field $\mathbf{v}(t)$ and acceleration field $\mathbf{w}(t)$ can be written as

$$\mathbf{v}^n(t) = \sum_{j=1}^{n} V_j(t)\,\mathbf{e}_j \quad , \quad \mathbf{w}^n(t) = \sum_{j=1}^{n} W_j(t)\,\mathbf{e}_j \quad , \tag{34}$$

in which $\mathbf{V}(t) = (V_1(t), \ldots, V_n(t))$ and $\mathbf{W}(t) = (W_1(t), \ldots, W_n(t))$ belong to $\mathbb{R}^n$ and are such that

$$\mathbf{V}(t) = \dot{\mathbf{U}}(t) \quad , \quad \mathbf{W}(t) = \dot{\mathbf{V}}(t) \quad . \tag{35}$$

3. LF and MF Deterministic Cases: Time-Periodic Excitation

We directly construct the numerical procedures for calculating the response of the master structure. We then present the developments in the context of the Ritz-Galerkin approximation introduced in Section 2.

3.1. Definition of the excitation

We assume that the body and surface force fields $\mathbf{g}(\mathbf{x}, t)$ and $\mathbf{G}(\mathbf{y}, t)$ are such that excitation force vector $\mathbf{F}(t)$ defined by Eq. (29) is an $\mathbb{R}^n$-valued function on $\mathbb{R}$, periodic with respect to time, with period $T = 2\pi/\omega_0$ and square integrable[3] over T. The Fourier expansion of $\mathbf{F}(t)$ is then written as

$$\mathbf{F}(t) = \sum_{\nu=-\infty}^{+\infty} \mathbf{X}_\nu\, e^{i\,\Omega_\nu t} \quad , \tag{36}$$

where for all $\nu \in \mathbb{Z}$ (the set of all positive and negative integers)

$$\Omega_\nu = \nu\,\omega_0 \quad , \tag{37}$$

[3] This means that $\frac{1}{T}\int_{-T/2}^{T/2} \|\mathbf{F}(t)\|^2\, dt < +\infty$.

and where each Fourier coefficient $\mathbf{X}_\nu \in \mathbb{C}^n$ is such that [4]

$$\mathbf{X}_\nu = \frac{1}{T}\int_{-T/2}^{T/2} \mathbf{F}(t)\, e^{-i\Omega_\nu t}\, dt \quad , \quad \mathbf{X}_{-\nu} = \overline{\mathbf{X}}_\nu \quad . \tag{38}$$

The spectrum of $\mathbf{F}(t)$ is defined as the (generalized) Fourier transform $\mathbf{F}(\omega) = \int_{\mathbb{R}} e^{-i\omega t}\, \mathbf{F}(t)\, dt$ and can be written as

$$\mathbf{F}(\omega) = 2\pi \sum_{\nu=-\infty}^{+\infty} \mathbf{X}_\nu\, \delta_{\Omega_\nu}(\omega) \quad , \tag{39}$$

in which $\delta_{\Omega_\nu}(\omega)$ is the (generalized) delta function (or Dirac measure) at the point Ω_ν. Equation (39) shows that $\mathbf{F}(t)$ has a spectrum of lines consisting of the set $\{\Omega_\nu, \nu \in \mathbb{Z}\}$, each line being at frequency Ω_ν and having a complex vector amplitude equal to $2\pi\mathbf{X}_\nu$.

3.2. Theoretical formula for the response

We substitute Eq. (36) into Eq. (31) to calculate the response of the master structure submitted to time-periodic excitation force vector $\mathbf{F}(t)$. Using Eq. (33), we deduce that time response $\mathbf{U}(t)$ is a time-periodic function with values in $\mathbb{R}^n$, whose Fourier expansion is

$$\mathbf{U}(t) = \sum_{\nu=-\infty}^{+\infty} \mathbf{Y}_\nu\, e^{i\Omega_\nu t} \quad , \tag{40}$$

in which, for each ν, the Fourier coefficient $\mathbf{Y}_\nu$ with values in $\mathbb{C}^n$ is written as

$$\mathbf{Y}_\nu = [T(\Omega_\nu)]\, \mathbf{X}_\nu \quad , \quad \mathbf{Y}_{-\nu} = \overline{\mathbf{Y}}_\nu \quad . \tag{41}$$

Fourier coefficient $\mathbf{Y}_\nu$ is called the harmonic of rank ν of time-periodic response $\mathbf{U}(t)$ whose norm approaches zero as $|\nu|$ approaches infinity. Time-response $\mathbf{U}(t)$ has then a spectrum of lines which is written as

$$\mathbf{U}(\omega) = 2\pi \sum_{\nu=-\infty}^{+\infty} \mathbf{Y}_\nu\, \delta_{\Omega_\nu}(\omega) \quad . \tag{42}$$

The displacement, velocity and acceleration fields in the frequency and time domains are deduced from Eqs. (21), (26) and (30), (34) respectively.

---

[4] We have $\sum_{\nu=-\infty}^{+\infty} \|\mathbf{X}_\nu\|^2 < +\infty$ and the series on the right-hand side of Eq. (36) is convergent in $L^2(T, \mathbb{C}^n)$.

It should be noted that a time-harmonic excitation is a particular case of a time-periodic excitation.

3.3 Calculation of a harmonic of rank ν in low-frequency band $\mathbb{B}_{\text{LF}}$

Let ν be the rank of harmonic $\mathbf{Y}_\nu$ of the response such that $\Omega_\nu \in \mathbb{B}_{\text{LF}}$ (see Eq. (6)). It should be noted that knowing $\mathbf{Y}_\nu$ we directly deduce $\mathbf{Y}_{-\nu}$ from the second Eq. (41). To calculate $\mathbf{Y}_\nu$, we use $[T_{N,\,\text{acc}}(\omega)] = [T_N(\omega)] + [T_{N,\,\text{stat}}]$ (see Eq. (VI.58)). From Eq. (41), we then deduce that the approximation $\mathbf{Y}^\nu_{N,\,\text{acc}}$ of $\mathbf{Y}_\nu$ is such that $\mathbf{Y}^\nu_{N,\,\text{acc}} = [T_{N,\,\text{acc}}(\Omega_\nu)]\,\mathbf{X}_\nu$ which is rewritten as

$$\mathbf{Y}^\nu_{N,\,\text{acc}} = \mathbf{Y}^\nu_{N,\,\text{dyn}} + \mathbf{Y}^\nu_{N,\,\text{stat}} \quad . \tag{43}$$

The dynamic part $\mathbf{Y}^\nu_{N,\,\text{dyn}} = [T_N(\Omega_\nu)]\,\mathbf{X}_\nu$ of the response is given by (see Eq. (VI.30))

$$\mathbf{Y}^\nu_{N,\,\text{dyn}} = \sum_{\alpha=1}^{N}\sum_{\beta=1}^{N} [\mathcal{T}(\Omega_\nu)]_{\alpha\beta}\,\{\mathbf{X}_\nu^T\,\mathbf{U}_\beta\}\,\mathbf{U}_\alpha \quad , \tag{44}$$

where $(N \times N)$ complex symmetrix matrix $[\mathcal{T}(\Omega_\nu)]$ is defined by Eq. (VI.21). The quasi-static correction term $\mathbf{Y}^\nu_{N,\,\text{stat}} = [T_{N,\,\text{stat}}]\,\mathbf{X}_\nu$ of the response is such that (see Eq. (VI.59)),

$$\mathbf{Y}^\nu_{N,\,\text{stat}} = [\,K\,]^{-1}\,\mathbf{X}_\nu - \sum_{\alpha=1}^{N} \frac{1}{\mu_\alpha\,\omega_\alpha^2}\,\{\mathbf{X}_\nu^T\,\mathbf{U}_\alpha\}\,\mathbf{U}_\alpha \quad . \tag{45}$$

The first term on the right-hand side of Eq. (45) is the solution $\mathbf{Y}^\nu_{\text{stat}} \in \mathbb{C}^n$ of $[\,K\,]\,\mathbf{Y}^\nu_{\text{stat}} = \mathbf{X}_\nu$, corresponding to the static solution of the master structure submitted to given forces $\mathbf{X}_\nu$.

3.4. Calculation of a harmonic of rank ν in medium-frequency band $\mathbb{B}_{\text{MF}}$

Let ν be the rank of harmonic $\mathbf{Y}_\nu$ of the response such that $\Omega_\nu \in \mathbb{B}_{\text{MF}}$ (see Eq. (6)). As for the LF case, knowing $\mathbf{Y}_\nu$, i.e. $\mathbf{Y}_\nu = [T(\Omega_\nu)]\mathbf{X}_\nu$, we directly deduce $\mathbf{Y}_{-\nu}$ from the second Eq. (41). The first Eq. (41) cannot generally be used to calculate $\mathbf{Y}_\nu$. In effect, due to the fact that the uncertainties increase in the model as the frequency increases, $[T(\Omega_\nu)]$ cannot be interpreted in the MF range as the value of $[T(\omega)]$ for $\omega = \Omega_\nu$. Consequently, we introduce an MF narrow band

$$\mathbb{B}_\nu = [\,\Omega_\nu - \Delta\omega/2\,,\; \Omega_\nu + \Delta\omega/2\,] \quad , \tag{46}$$

in which center frequency $\Omega_\nu = \nu\,\omega_0$ (see Eq. (37)) and bandwidth $\Delta\omega$ is related to the magnitude of uncertainties. It should be noted that by definition of an MF narrow band (see Eq. (VII.11)),

$$\frac{\Delta\omega}{\Omega_\nu} \ll 1 \quad . \tag{47}$$

We then replace $\mathbf{Y}_\nu = [T(\Omega_\nu)]\,\mathbf{X}_\nu$ by first-order and second-order moments using an average over frequency band $\mathbb{B}_\nu$. This approach has a probabilistic interpretation. Instead of using a random FRF to model the uncertainties of the master structure submitted to a deterministic harmonic excitation in the MF range, we consider a deterministic FRF of the master structure submitted to a harmonic excitation with random frequency Ω_ν uniformly distributed on $\mathbb{B}_\nu$. For this we need the expression of the MF response, on narrow MF band $\mathbb{B}_\nu$ of the master structure submitted to the excitation defined by harmonic $\mathbf{X}_\nu$ of rank ν. The discretization of Eq. (VII.14) yields

$$\mathbf{F}(\omega) = \theta_\nu(\omega)\,\mathbf{X}_\nu \quad , \tag{48}$$

in which $\theta_\nu(\omega)$ is chosen such that

$$\theta_\nu(\omega) = 0 \;,\; \forall\,\omega \notin \mathbb{B}_\nu \quad \text{and} \quad \theta_\nu(\omega) = 1 \;,\; \forall\,\omega \in \mathbb{B}_\nu \quad . \tag{49}$$

Denoting this MF response as $\widetilde{\mathbf{Y}}_\nu(\omega)$, Eq. (VII.59) yields

$$\widetilde{\mathbf{Y}}_\nu(\omega) \simeq \mathbf{1}_{\mathbb{B}_\nu}(\omega)\,\Delta t \sum_{m=m_{\mathrm{i}}}^{m_{\mathrm{f}}} \widetilde{\mathbf{Y}}_0(m\,\Delta t)\; e^{-im\,\Delta t\,(\omega-\Omega_\nu)} \;,\; \forall\,\omega \in \mathbb{R} \;, \tag{50}$$

in which $\widetilde{\mathbf{Y}}_0(t)$ is the solution of the associated LF equation in the time domain (see Section VII.8). It should be noted that $\widetilde{\mathbf{Y}}_\nu(\omega)$ and $\widetilde{\mathbf{Y}}_0(m\,\Delta t)$ in Eq. (50) must be carefully distinguished from harmonic $\mathbf{Y}_\nu$ of rank ν considered in this section.

First-order moment. The first-order moment of harmonic $\mathbf{Y}_\nu$ of rank ν is defined by

$$<\mathbf{Y}_\nu>_{\Delta\omega} = \frac{1}{\Delta\omega}\int_{\mathbb{B}_\nu} \widetilde{\mathbf{Y}}_\nu(\omega)\,d\omega \quad , \tag{51}$$

in which $\widetilde{\mathbf{Y}}_\nu(\omega)$ in the right-hand side of Eq. (51) is given by Eq. (50). Substituting Eq. (50) in Eq. (51) and using the property $\Delta t \times \Delta\omega = 2\pi$ deduced from Eq. (VII-37) yields

$$<\mathbf{Y}_\nu>_{\Delta\omega} = \frac{2\pi}{\Delta\omega}\{\widetilde{\mathbf{Y}}_0(m\,\Delta t)\}_{m=0} \quad . \tag{52}$$

Second-order moment. The second-order moment of harmonic $\mathbf{Y}_\nu$ of rank ν is defined by

$$< \mathbf{Y}_\nu \, \mathbf{Y}_\nu^* >_{\Delta\omega} = \frac{1}{\Delta\omega} \int_{\mathbb{B}_\nu} \widetilde{\mathbf{Y}}_\nu(\omega) \, \widetilde{\mathbf{Y}}_\nu(\omega)^* \, d\omega \quad , \tag{53}$$

in which $\widetilde{\mathbf{Y}}_\nu(\omega)$ in the right-hand side of Eq. (53) is given by Eq. (50). Substituting Eq. (50) in Eq. (53) yields

$$< \mathbf{Y}_\nu \, \mathbf{Y}_\nu^* >_{\Delta\omega} \simeq \left(\frac{2\pi}{\Delta\omega} \right)^2 \sum_{m=m_\mathrm{i}}^{m_\mathrm{f}} \widetilde{\mathbf{Y}}_0(m\,\Delta t) \, \widetilde{\mathbf{Y}}_0(m\,\Delta t)^* \quad . \tag{54}$$

The square norm of $\mathbf{Y}_\nu$ is obtained by taking the trace of the two members of Eq. (54),

$$\begin{aligned} \|\mathbf{Y}_\nu\|^2 &= \mathrm{tr} < \mathbf{Y}_\nu \, \mathbf{Y}_\nu^* >_{\Delta\omega} \\ &\simeq \left(\frac{2\pi}{\Delta\omega} \right)^2 \sum_{m=m_\mathrm{i}}^{m_\mathrm{f}} \| \widetilde{\mathbf{Y}}_0(m\,\Delta t) \|^2 \quad . \end{aligned} \tag{55}$$

4. LF and MF Deterministic Cases: Time Square Integrable Excitation

We directly construct the numerical procedures for calculating the response of the master structure. We then present the developments within the context of the Ritz-Galerkin approximation introduced in Section 2.

4.1. Definition of the excitation

We assume that the body and surface force fields $\mathbf{g}(\mathbf{x}, t)$ and $\mathbf{G}(\mathbf{y}, t)$ are such that excitation force vector $\mathbf{F}(t)$ defined by Eq. (29), is an $\mathbb{R}^n$-valued square integrable function on $\mathbb{R}$ with respect to time:

$$\|\mathbf{F}\|_{L^2}^2 = \int_{\mathbb{R}} \|\mathbf{F}(t)\|^2 \, dt < +\infty \quad . \tag{56}$$

Using the same symbol for a quantity and its Fourier transform, the spectrum of $\mathbf{F}(t)$ is defined as the Fourier transform of $\mathbf{F}(t)$ with respect to t:

$$\mathbf{F}(\omega) = \int_{\mathbb{R}} e^{-i\,\omega t} \, \mathbf{F}(t) \, dt \quad . \tag{57}$$

Function $\mathbf{F}(\omega)$ is then square integrable on $\mathbb{R}$ with respect to ω:

$$\|\mathbf{F}\|_{L^2}^2 = \int_{\mathbb{R}} \|\mathbf{F}(\omega)\|^2 \, d\omega < +\infty \quad , \tag{58}$$

and from the Plancherel equality (see Mathematical Notations in the appendix)

$$\int_{\mathbb{R}} \|\mathbf{F}(t)\|^2 \, dt = \frac{1}{2\pi} \int_{\mathbb{R}} \|\mathbf{F}(\omega)\|^2 \, d\omega \quad . \tag{59}$$

4.2. Theoretical formula for the response

The spectrum of the response of the master structure submitted to the time square integrable excitation force vector $\mathbf{F}(t)$, is defined as the Fourier transform $\mathbf{U}(\omega)$ with values in $\mathbb{C}^n$ of the time response $\mathbf{U}(t)$ with values in $\mathbb{R}^n$, and is given by Eq. (24), i.e., for all real ω,

$$\mathbf{U}(\omega) = [T(\omega)]\,\mathbf{F}(\omega) \quad , \quad \mathbf{U}(-\omega) = \overline{\mathbf{U}(\omega)} \quad . \tag{60}$$

4.3. Calculation of the response in low-frequency band $\mathbb{B}_{\text{LF}}$

Let ω be in $\mathbb{B}_{\text{LF}}$ (see Eq. (6)). It should be noted that knowing $\mathbf{U}(\omega)$, we directly deduce $\mathbf{U}(-\omega)$ from the second Eq. (60). To calculate $\mathbf{U}(\omega)$, we use $[T_{N,\,\text{acc}}(\omega)] = [T_N(\omega)] + [T_{N,\,\text{stat}}]$ (see Eq. (VI.58)). From Eq. (60), we then deduce that the approximation $\mathbf{U}_{N,\,\text{acc}}(\omega)$ of $\mathbf{U}(\omega)$ is such that $\mathbf{U}_{N,\,\text{acc}}(\omega) = [T_{N,\,\text{acc}}(\omega)]\,\mathbf{F}(\omega)]$ which is rewritten as

$$\mathbf{U}_{N,\,\text{acc}}(\omega) = \mathbf{U}_{N,\,\text{dyn}}(\omega) + \mathbf{U}_{N,\,\text{stat}}(\omega) \quad . \tag{61}$$

The dynamic part $\mathbf{U}_{N,\,\text{dyn}}(\omega) = [T_N(\omega)]\,\mathbf{f}(\omega)$ of the response is given by (see Eq. (VI.30))

$$\mathbf{U}_N(\omega) = \sum_{\alpha=1}^{N}\sum_{\beta=1}^{N} [\mathcal{T}(\omega)]_{\alpha\beta}\,\{\mathbf{F}^T\,\mathbf{U}_\beta\}\,\mathbf{U}_\alpha \quad , \tag{62}$$

where $(N \times N)$ complex symmetric matrix $[\mathcal{T}(\omega)]$ is defined by Eq. (VI.21). The quasi-static correction term $\mathbf{U}_{N,\,\text{stat}}(\omega) = [T_{N,\,\text{stat}}]\,\mathbf{F}(\omega)$ of the response is such that (see Eq. (VI.59))

$$\mathbf{U}_{N,\,\text{stat}}(\omega) = [\,K\,]^{-1}\,\mathbf{F}(\omega) - \sum_{\alpha=1}^{N} \frac{1}{\mu_\alpha\,\omega_\alpha^2}\,\{\mathbf{F}(\omega)^T\,\mathbf{U}_\alpha\}\,\mathbf{U}_\alpha \quad . \tag{63}$$

The first term on the right-hand side of Eq. (63) is the solution $\mathbf{U}_{\text{stat}}(\omega) \in \mathbb{C}^n$ of $[K]\,\mathbf{U}_{\text{stat}}(\omega) = \mathbf{F}(\omega)$, corresponding to the static solution of the master structure submitted to given forces $\mathbf{F}(\omega)$.

4.4. Calculation of the response in medium-frequency band $\mathbb{B}_{\text{MF}}$

Let ω be in $\mathbb{B}_{\text{MF}}$ (see Eq. (6)). As for the LF case, knowing $\mathbf{U}(\omega)$, we directly deduce $\mathbf{U}(-\omega)$ from the second Eq. (60). To calculate $\mathbf{U}(\omega)$, we use the method presented in Chapter VII. Consequently, band $\mathbb{B}_{\text{MF}}$ is written as a finite union of MF narrow bands $\mathbb{B}_\nu$ (see Section VII.7)

$$\mathbb{B}_{\text{MF}} = \cup_\nu \, \mathbb{B}_\nu \quad . \tag{64}$$

For each MF narrow band $\mathbb{B}_\nu$, we assume that

$$\mathbf{F}(\omega) = \sum_{j=1}^{M} F_{j,\nu}(\omega)\, \mathbf{B}_{j,\nu} \quad , \quad \forall\, \omega \in \mathbb{B}_\nu \quad , \tag{65}$$

in which $\mathbb{C}^n$-vectors $\mathbf{B}_{j,\nu}$ are independent of ω. Substituting Eq. (65) into Eq. (60) yields

$$\mathbf{U}_\nu(\omega) = \sum_{j=1}^{M} F_{j,\nu}(\omega)\, \mathbf{U}_{j,\nu}(\omega) \quad , \tag{66}$$

in which

$$\mathbf{U}_{j,\nu}(\omega) = [T(\omega)]\, \mathbf{B}_{j,\nu} \quad . \tag{67}$$

For each MF narrow band $\mathbb{B}_\nu$, the responses $\{\mathbf{U}_{1,\nu}, \ldots, \mathbf{U}_{M,\nu}\}$ are simultaneously constructed with M second members $\{\theta_\nu(\omega)\mathbf{B}_{1,\nu}, \ldots, \theta_\nu(\omega)\mathbf{B}_{M,\nu}\}$, using the MF method presented in Section VII.8, in which $\theta_\nu(\omega) = \mathbf{1}_{\mathbb{B}_\nu}(\omega)$, i.e. $\theta_0(t) = (\pi t)^{-1} \sin(\Delta\omega\, t/2)$. For all $j = 1, \ldots, M$, response $\mathbf{U}_{j,\nu}(\omega)$ is written as (see Eq. (VII.59))

$$\mathbf{U}_{j,\nu}(\omega) \simeq \mathbf{1}_{\mathbb{B}_\nu}(\omega)\, \Delta t \sum_{m=m_{\text{i}}}^{m_{\text{f}}} \mathbf{U}_{j,0}(m\,\Delta t)\; e^{-im\,\Delta t\,(\omega - \Omega_\nu)} \quad , \tag{68}$$

in which $\mathbf{U}_{j,0}(t)$ is the solution of the associated LF equation in the time domain (see Section VII.8). In the frequency domain, velocity $\mathbf{V}_{j,\nu}(\omega)$ and acceleration $\mathbf{W}_{j,\nu}(\omega)$ can be calculated from Eq. (27) using Eq. (68). However, from a numerical point of view and in the context of the proposed

MF methodology, it is more efficient to calculate $\mathbf{V}_{j,\nu}(\omega)$ and $\mathbf{W}_{j,\nu}(\omega)$ by the following formulas (see Soize, 1982a)

$$\mathbf{V}_{j,\nu}(\omega) \simeq \mathbf{1}_{\mathbb{B}_\nu}(\omega)\,\Delta t \sum_{m=m_{\mathrm{i}}}^{m_{\mathrm{f}}} \mathbf{V}_{j,0}(m\,\Delta t)\; e^{-im\,\Delta t\,(\omega-\Omega_\nu)} \quad , \tag{69}$$

$$\mathbf{W}_{j,\nu}(\omega) \simeq \mathbf{1}_{\mathbb{B}_\nu}(\omega)\,\Delta t \sum_{m=m_{\mathrm{i}}}^{m_{\mathrm{f}}} \mathbf{W}_{j,0}(m\,\Delta t)\; e^{-im\,\Delta t\,(\omega-\Omega_\nu)} \quad , \tag{70}$$

in which $\mathbf{V}_{j,0}(m\,\Delta t)$ and $\mathbf{W}_{j,0}(m\,\Delta t)$ are defined by

$$\mathbf{V}_{j,0}(t) = \dot{\mathbf{U}}_{j,0}(t) + i\Omega_\nu\,\mathbf{U}_{j,0}(t) \quad , \tag{71}$$

$$\mathbf{W}_{j,0}(t) = \ddot{\mathbf{U}}_{j,0}(t) + 2i\Omega_\nu\,\dot{\mathbf{U}}_{j,0}(t) - \Omega_\nu^2\,\mathbf{U}_{j,0}(t) \quad . \tag{72}$$

5. LF and MF Random Cases: Time Stationary Stochastic Excitation

5.1. Definition of the excitation

Second-order description of the stochastic body and surface force fields. In the time domain, the body and surface force fields $\{\mathbf{g}(\mathbf{x},t), \mathbf{x} \in \Omega, t \in \mathbb{R}\}$ and $\{\mathbf{G}(\mathbf{y},t), \mathbf{y} \in \Gamma, t \in \mathbb{R}\}$ are given stochastic fields defined on the same probability space and are indexed by $\Omega \times \mathbb{R}$ and $\Gamma \times \mathbb{R}$ respectively, with values in $\mathbb{R}^3$. We assume that $\mathbf{g}$ and $\mathbf{G}$ are second-order stochastic fields, i.e., for all t in $\mathbb{R}$, $\mathbf{x}$ in Ω and $\mathbf{y}$ in Γ,

$$E\{\|\mathbf{g}(\mathbf{x},t)\|^2\} < +\infty \quad , \quad E\{\|\mathbf{G}(\mathbf{y},t)\|^2\} < +\infty \quad , \tag{73}$$

in which E denotes the mathematical expectation (the mathematical expectation of a random variable is equal to the integral of this random variable with respect to its probability distribution). In addition, we assume that these stochastic fields are centered,

$$E\{\mathbf{g}(\mathbf{x},t)\} = \mathbf{0} \quad , \quad E\{\mathbf{G}(\mathbf{y},t)\} = \mathbf{0} \quad , \tag{74}$$

mean-square stationary with respect to t and statistically dependent. Consequently, for all t and τ in $\mathbb{R}$, $\mathbf{x}$ and $\mathbf{x}'$ in Ω, $\mathbf{y}$ and $\mathbf{y}'$ in Γ, the cross-correlation functions with values [5] in $Mat_{\mathbb{R}}(3,3)$ are written as [6]

---

[5] We denote as $Mat_{\mathbb{R}}(3,3)$ (or $Mat_{\mathbb{C}}(3,3)$) the set of all the (3×3) real (or complex) matrices (see Mathematical Notations in the appendix).

[6] Mean-square stationary in time implies that the mean functions are independent of t (see Eq. (74)) and that the cross-correlation functions depend only on τ (but not on t). Since the stochastic fields are statistically dependent, the cross-correlation functions of $\mathbf{g}$ with $\mathbf{G}$ are not identically zero.

$$[R_{\mathbf{g}}(\mathbf{x},\mathbf{x}',\tau)] = E\{\mathbf{g}(\mathbf{x},t+\tau)\,\mathbf{g}(\mathbf{x}',t)^T\} \quad , \tag{75}$$

$$[R_{\mathbf{G}}(\mathbf{y},\mathbf{y}',\tau)] = E\{\mathbf{G}(\mathbf{y},t+\tau)\,\mathbf{G}(\mathbf{y}',t)^T\} \quad , \tag{76}$$

$$[R_{\mathbf{gG}}(\mathbf{x},\mathbf{y}',\tau)] = E\{\mathbf{g}(\mathbf{x},t+\tau)\,\mathbf{G}(\mathbf{y}',t)^T\} \quad , \tag{77}$$

$$[R_{\mathbf{Gg}}(\mathbf{y},\mathbf{x}',\tau)] = E\{\mathbf{G}(\mathbf{y},t+\tau)\,\mathbf{g}(\mathbf{x}',t)^T\} \quad . \tag{78}$$

From Eqs. (77) and (78), we deduce that

$$[R_{\mathbf{Gg}}(\mathbf{y},\mathbf{x}',\tau)] = [R_{\mathbf{gG}}(\mathbf{x}',\mathbf{y},-\tau)]^T \quad . \tag{79}$$

Finally, we assume that there exist [7] cross-spectral density functions with values in $Mat_{\mathbb{C}}(3,3)$ such that, for all τ and ω in $\mathbb{R}$, $\mathbf{x}$ and $\mathbf{x}'$ in Ω, $\mathbf{y}$ and $\mathbf{y}'$ in Γ,

$$[R_{\mathbf{g}}(\mathbf{x},\mathbf{x}',\tau)] = \int_{\mathbb{R}} e^{i\omega\tau}\,[S_{\mathbf{g}}(\mathbf{x},\mathbf{x}',\omega)]\,d\omega \quad , \tag{80}$$

$$[R_{\mathbf{G}}(\mathbf{y},\mathbf{y}',\tau)] = \int_{\mathbb{R}} e^{i\omega\tau}\,[S_{\mathbf{G}}(\mathbf{y},\mathbf{y}',\omega)]\,d\omega \quad , \tag{81}$$

$$[R_{\mathbf{gG}}(\mathbf{x},\mathbf{y}',\tau)] = \int_{\mathbb{R}} e^{i\omega\tau}\,[S_{\mathbf{gG}}(\mathbf{x},\mathbf{y}',\omega)]\,d\omega \quad , \tag{82}$$

$$[R_{\mathbf{Gg}}(\mathbf{y},\mathbf{x}',\tau)] = \int_{\mathbb{R}} e^{i\omega\tau}\,[S_{\mathbf{Gg}}(\mathbf{y},\mathbf{x}',\omega)]\,d\omega \quad . \tag{83}$$

From Eqs. (79), (82) and (83), and since the cross-correlation functions are real matrices, we deduce that [8]

$$[S_{\mathbf{Gg}}(\mathbf{y},\mathbf{x}',\omega)] = [S_{\mathbf{gG}}(\mathbf{x}',\mathbf{y},-\omega)]^T \quad , \tag{84}$$

$$[S_{\mathbf{Gg}}(\mathbf{y},\mathbf{x}',\omega)] = [S_{\mathbf{gG}}(\mathbf{x}',\mathbf{y},\omega)]^* \quad . \tag{85}$$

The data required for a second-order description of the stochastic body and surface force fields are the cross-spectral density functions $[S_{\mathbf{g}}(\mathbf{x},\mathbf{x}',\omega)]$, $[S_{\mathbf{G}}(\mathbf{y},\mathbf{y}',\omega)]$ and $[S_{\mathbf{gG}}(\mathbf{x},\mathbf{y}',\omega)]$. Using Eq. (85), cross-spectral density function $[S_{\mathbf{Gg}}(\mathbf{y},\mathbf{x}',\omega)]$ can be deduced from $[S_{\mathbf{gG}}(\mathbf{x},\mathbf{y}',\omega)]$.

---

7 This means that the stochastic fields are mean-square continuous and that each matrix-valued cross-spectral measure has a density with respect to the Lebesgue measure $d\omega$.

8 Concerning notations T and $*$, see Mathematical Notations in the appendix.

Second-order representation of the stochastic excitation in the time domain. In order to construct the second-order representation of the stochastic excitation, we use the Ritz-Galerkin approximation introduced in Section 2. Excitation force $\mathbf{F}(t) = (F_1(t), \ldots, F_n(t))$ such that

$$F_j(t) = \int_\Omega \mathbf{g}(\mathbf{x}, t) \cdot \mathbf{e}_j(\mathbf{x})\, d\mathbf{x} + \int_\Gamma \mathbf{G}(\mathbf{y}, t) \cdot \mathbf{e}_j(\mathbf{y})\, ds(\mathbf{y}) \quad , \tag{86}$$

is a second-order mean-square stationary stochastic process indexed by $\mathbb{R}$ with values in $\mathbb{R}^n$. From Eqs. (74) and (86), we deduce that for all t in $\mathbb{R}$,

$$E\{\mathbf{F}(t)\} = \mathbf{0} \quad . \tag{87}$$

From Eqs. (75) to (78) and (86), we deduce that for all t and τ in $\mathbb{R}$, the $Mat_{\mathbb{R}}(n,n)$-valued autocorrelation function $[R_{\mathbf{F}}(\tau)]$ of stochastic process $\mathbf{F}(t)$, defined by

$$[R_{\mathbf{F}}(\tau)] = E\{\mathbf{F}(t+\tau)\, \mathbf{F}(t)^T\} \quad , \tag{88}$$

is such that

$$\begin{aligned}
[R_{\mathbf{F}}(\tau)]_{jk} =& E\{F_j(t+\tau)\, F_k(t)^T\} \\
=& \int_\Omega \int_\Omega \mathbf{e}_j(\mathbf{x})^T\, [R_{\mathbf{g}}(\mathbf{x}, \mathbf{x}', \tau)]\; \mathbf{e}_k(\mathbf{x}')\, d\mathbf{x}\, d\mathbf{x}' \\
&+ \int_\Gamma \int_\Gamma \mathbf{e}_j(\mathbf{y})^T\, [R_{\mathbf{G}}(\mathbf{y}, \mathbf{y}', \tau)]\; \mathbf{e}_k(\mathbf{y}')\, ds(\mathbf{y})\, ds(\mathbf{y}') \\
&+ \int_\Omega \int_\Gamma \mathbf{e}_j(\mathbf{x})^T\, [R_{\mathbf{gG}}(\mathbf{x}, \mathbf{y}', \tau)]\; \mathbf{e}_k(\mathbf{y}')\, d\mathbf{x}\, ds(\mathbf{y}') \\
&+ \int_\Gamma \int_\Omega \mathbf{e}_j(\mathbf{y})^T\, [R_{\mathbf{Gg}}(\mathbf{y}, \mathbf{x}', \tau)]\; \mathbf{e}_k(\mathbf{x}')\, ds(\mathbf{y})\, d\mathbf{x}' \quad .
\end{aligned} \tag{89}$$

Second-order representation of the stochastic excitation in the frequency domain. The matrix-valued autocorrelation function $[R_{\mathbf{F}}(\tau)]$ can be written as

$$[R_{\mathbf{F}}(\tau)] = \int_{\mathbb{R}} e^{i\omega\tau}\, [S_{\mathbf{F}}(\omega)]\, d\omega \quad , \tag{90}$$

in which the matrix-valued spectral density function $[S_{\mathbf{F}}(\omega)]$ of mean-square stationary stochastic process $\mathbf{F}$ is an $(n \times n)$ complex positive Hermitian matrix such that

$$
\begin{aligned}
[S_{\mathbf{F}}(\omega)]_{jk} &= \int_\Omega \int_\Omega \mathbf{e}_j(\mathbf{x})^T \,[S_{\mathbf{g}}(\mathbf{x},\mathbf{x}',\omega)]\; \mathbf{e}_k(\mathbf{x}')\, d\mathbf{x}\, d\mathbf{x}' \\
&+ \int_\Gamma \int_\Gamma \mathbf{e}_j(\mathbf{y})^T \,[S_{\mathbf{G}}(\mathbf{y},\mathbf{y}',\omega)]\; \mathbf{e}_k(\mathbf{y}')\, ds(\mathbf{y})\, ds(\mathbf{y}') \\
&+ \int_\Omega \int_\Gamma \mathbf{e}_j(\mathbf{x})^T \,[S_{\mathbf{gG}}(\mathbf{x},\mathbf{y}',\omega)]\; \mathbf{e}_k(\mathbf{y}')\, d\mathbf{x}\, ds(\mathbf{y}') \\
&+ \int_\Gamma \int_\Omega \mathbf{e}_j(\mathbf{y})^T \,[S_{\mathbf{Gg}}(\mathbf{y},\mathbf{x}',\omega)]\; \mathbf{e}_k(\mathbf{x}')\, ds(\mathbf{y})\, d\mathbf{x}' \quad . \qquad (91)
\end{aligned}
$$

5.2. Theoretical formula for second-order moments of the response

General properties of the stochastic response. Eqs. (31) and (32) show that response $\mathbf{U}(t) = (U_1(t), \ldots, U_n(t))$ of the master structure appears as a linear filtering [9] with respect to t

$$\mathbf{U}(t) = \int_{\mathbb{R}} [\, h(t')\,]\; \mathbf{F}(t-t')\, dt' \qquad (92)$$

of second-order mean-square stationary stochastic input $\mathbf{F}$ by a stable causal filter with matrix-valued impulse response $[\, h(t)\,]$. Consequently, it can be shown [10] that stochastic output $\mathbf{U}$ is a second-order mean-square stationary stochastic process.

Second-order representation of the stochastic response in the time domain. From Eqs. (87) and (92), we deduce that stochastic process $\mathbf{U}$ and thus stochastic field $\mathbf{u}^n$ defined by Eq. (30), are centered, i.e., for all fixed $\mathbf{x}$ in Ω and t in $\mathbb{R}$,

$$E\{\mathbf{U}(t)\} = \mathbf{0} \quad , \quad E\{\mathbf{u}^n(\mathbf{x},t)\} = \mathbf{0} \quad . \qquad (93)$$

For all τ and t in $\mathbb{R}$, the $Mat_{\mathbb{R}}(n,n)$-valued autocorrelation function $[R_{\mathbf{U}}(\tau)]$ of stochastic process $\mathbf{U}$ which is defined by

$$[R_{\mathbf{U}}(\tau)] = E\{\mathbf{U}(t+\tau)\, \mathbf{U}(t)^T\} \quad , \qquad (94)$$

[9] The integral in the right-hand side of Eq. (92) is defined as a mean-square integral.

[10] $\mathbf{U}=[\,h\,]*\mathbf{F}$ is a second-order mean-square time-stationary stochastic process indexed by $\mathbb{R}$ with values in $\mathbb{R}^n$, because $\mathbf{F}$ is a second-order mean-square time-stationary stochastic process and $[\,h\,]$ is an integrable function on $\mathbb{R}$, i.e., for all j and k, $\int_{\mathbb{R}} |[\,h(t)\,]_{jk}|\, dt = \int_0^{+\infty} |[\,h(t)\,]_{jk}|\, dt < +\infty$ (see, for instance, Soize, 1994, Chapter V, Proposition 4, p. 90).

can be written as

$$[R_{\mathbf{U}}(\tau)] = \int_{\mathbb{R}} \int_{\mathbb{R}} [\, h(t')\,]\, [R_{\mathbf{F}}(\tau + t'' - t')]\, [\, h(t'')\,]^T \, dt' \, dt'' \quad . \tag{95}$$

From Eqs. (30) and (94), we deduce that, for all $\mathbf{x}$ and $\mathbf{x}'$ in Ω and τ and t in $\mathbb{R}$, the $Mat_{\mathbb{R}}(3,3)$-valued cross-correlation function of stochastic field $\mathbf{u}^n(\mathbf{x}, t)$ can be written as

$$\begin{aligned}[R_{\mathbf{u}^n}(\mathbf{x}, \mathbf{x}', \tau)] &= E\{\mathbf{u}^n(\mathbf{x}, t+\tau)\, \mathbf{u}^n(\mathbf{x}', t)^T\} \\ &= \sum_{j,k=1}^{n} [R_{\mathbf{U}}(\tau)]_{jk}\, \mathbf{e}_j(\mathbf{x})\, \mathbf{e}_k(\mathbf{x}')^T \quad . \end{aligned} \tag{96}$$

Let $\mathbf{v}^n$ and $\mathbf{w}^n$ be the velocity and acceleration fields such that

$$\mathbf{v}^n(\mathbf{x}, t) = \partial_t \mathbf{u}^n(\mathbf{x}, t) \quad , \quad \mathbf{w}^n(\mathbf{x}, t) = \partial_t \mathbf{v}^n(\mathbf{x}, t) \quad . \tag{97}$$

Fields $\mathbf{v}^n$ and $\mathbf{w}^n$ are second-order mean-square time-stationary centered stochastic fields indexed by $\Omega \times \mathbb{R}$ with values in $\mathbb{R}^3$. Their matrix-valued cross-correlation functions are such that

$$\begin{aligned}[R_{\mathbf{v}^n}(\mathbf{x}, \mathbf{x}', \tau)] &= E\{\mathbf{v}^n(\mathbf{x}, t+\tau)\, \mathbf{v}^n(\mathbf{x}', t)^T\} \\ &= -\frac{\partial^2}{\partial \tau^2} [R_{\mathbf{u}^n}(\mathbf{x}, \mathbf{x}', \tau)] \quad , \end{aligned} \tag{98}$$

$$\begin{aligned}[R_{\mathbf{w}^n}(\mathbf{x}, \mathbf{x}', \tau)] &= E\{\mathbf{w}^n(\mathbf{x}, t+\tau)\, \mathbf{w}^n(\mathbf{x}', t)^T\} \\ &= -\frac{\partial^2}{\partial \tau^2} [R_{\mathbf{v}^n}(\mathbf{x}, \mathbf{x}', \tau)] \quad . \end{aligned} \tag{99}$$

Second-order representation of the stochastic response in the frequency domain. Substituting Eq. (90) into Eq. (95), we deduce that, for all τ in $\mathbb{R}$, matrix-valued autocorrelation function $[R_{\mathbf{U}}(\tau)]$ is written as

$$[R_{\mathbf{U}}(\tau)] = \int_{\mathbb{R}} e^{i\omega\tau}\, [S_{\mathbf{U}}(\omega)]\, d\omega \quad , \tag{100}$$

in which the spectral density function $[S_{\mathbf{U}}(\omega)]$ of stochastic process $\mathbf{U}$ is an $(n \times n)$ complex positive Hermitian matrix, related to $[S_{\mathbf{F}}(\omega)]$ by the classical spectral analysis formula for linear filtering of stationary stochastic processes (see for instance Soize, 1994, p.90; Soize 1993a, p.382; Kree and Soize, 1986, p.86; Guikhman and Skorokhod, 1979)

$$[S_{\mathbf{U}}(\omega)] = [T(\omega)]\,[S_{\mathbf{F}}(\omega)]\,[T(\omega)]^* \quad , \tag{101}$$

where $[T(\omega)]$ is the $(n \times n)$ complex symmetric matrix defined by Eq. (25). From Eqs. (96) and (100), we deduce that, for all $\mathbf{x}$ and $\mathbf{x}'$ in Ω and τ in $\mathbb{R}$, matrix-valued cross-correlation function $[R_{\mathbf{u}^n}(\mathbf{x},\mathbf{x}',\tau)]$ of stochastic field $\mathbf{u}^n(\mathbf{x},t)$ is such that

$$[R_{\mathbf{u}^n}(\mathbf{x},\mathbf{x}',\tau)] = \int_{\mathbb{R}} e^{i\omega\tau}\,[S_{\mathbf{u}^n}(\mathbf{x},\mathbf{x}',\omega)]\,d\omega \quad , \tag{102}$$

in which the $Mat_{\mathbb{C}}(3,3)$-valued cross-spectral density function $[S_{\mathbf{u}^n}(\mathbf{x},\mathbf{x}',\omega)]$ of stochastic field $\mathbf{u}^n(\mathbf{x},t)$ is such that

$$[S_{\mathbf{u}^n}(\mathbf{x},\mathbf{x}',\omega)] = \sum_{j,k=1}^{n} [S_{\mathbf{U}}(\omega)]_{jk}\;\mathbf{e}_j(\mathbf{x})\;\mathbf{e}_k(\mathbf{x}')^T \quad , \tag{103}$$

and satisfies

$$[S_{\mathbf{u}^n}(\mathbf{x},\mathbf{x}',-\omega)] = \overline{[S_{\mathbf{u}^n}(\mathbf{x},\mathbf{x}',\omega)]} = [S_{\mathbf{u}^n}(\mathbf{x}',\mathbf{x},\omega)]^T \quad . \tag{104}$$

The matrix-valued cross-spectral density functions of velocity field $\mathbf{v}^n(\mathbf{x},t)$ and acceleration field $\mathbf{w}^n(\mathbf{x},t)$ are such that

$$[S_{\mathbf{v}^n}(\mathbf{x},\mathbf{x}',\omega)] = \omega^2\,[S_{\mathbf{u}^n}(\mathbf{x},\mathbf{x}',\omega)] \quad , \tag{105}$$

$$[S_{\mathbf{w}^n}(\mathbf{x},\mathbf{x}',\omega)] = \omega^2\,[S_{\mathbf{v}^n}(\mathbf{x},\mathbf{x}',\omega)] \quad . \tag{106}$$

5.3. Reduced representation of the stochastic excitation

In this section, we present a reduction procedure related to the stochastic excitation, which we use in Sections 5.4 and 5.5 to calculate the stationary stochastic response in bands $\mathbb{B}_{\mathrm{LF}}$ and $\mathbb{B}_{\mathrm{MF}}$ respectively. This method was introduced by Soize et al., 1986a and 1986b.

Introduction of the limited frequency bands $\mathbb{B}_\nu$ and $\underline{\mathbb{B}}_\nu$. Let $\mathbb{B}_\nu$ be the limited frequency band of $\mathbb{R}^+$ defined by

$$\mathbb{B}_\nu = [\,\Omega_\nu - \Delta\omega/2\,,\;\Omega_\nu + \Delta\omega/2\,] \quad , \tag{107}$$

in which $\Omega_\nu > 0$ is the center frequency of band $\mathbb{B}_\nu$ and $\Delta\omega$ is its bandwidth such that $0 < \Delta\omega < 2\,\Omega_\nu$. We denote as $\underline{\mathbb{B}}_\nu$

$$\underline{\mathbb{B}}_\nu = [\,-\Omega_\nu - \Delta\omega/2\,,\; -\Omega_\nu + \Delta\omega/2\,] \quad . \tag{108}$$

Definition of a reduced representation of the stochastic excitation on limited frequency band $\mathbb{B}_\nu \cup \underline{\mathbb{B}}_\nu$. We consider the second-order centered mean-square stationary stochastic process $\mathbf{F}$ indexed by $\mathbb{R}$ with values in $\mathbb{R}^n$ whose matrix-valued spectral density function is $[S_{\mathbf{F}}(\omega)]$ defined by Eq. (91). A reduced representation of stochastic excitation $\mathbf{F}$ on band $\mathbb{B}_\nu$ is, by definition, a second-order centered mean-square stationary stochastic process $\mathbf{F}_\nu$ indexed by $\mathbb{R}$ with values in $\mathbb{R}^n$, which is written as

$$\mathbf{F}_\nu(t) = \sum_{j=1}^{M} X_j(t)\,\mathbf{\Phi}_j \quad , \tag{109}$$

in which $1 \leq M \ll n$ and where the matrix-valued spectral density function $[S_{\mathbf{F}_\nu}(\omega)]$ of $\mathbf{F}_\nu$ is such that

$$[S_{\mathbf{F}_\nu}(\omega)] \begin{cases} \simeq [S_{\mathbf{F}}(\omega)] & ,\forall\,\omega \in \mathbb{B}_\nu \cup \underline{\mathbb{B}}_\nu \\ \\ = \;\; [\,0\,] & ,\forall\,\omega \notin \mathbb{B}_\nu \cup \underline{\mathbb{B}}_\nu \quad . \end{cases} \tag{110}$$

In Eq. (109), $\{\mathbf{\Phi}_1, \ldots, \mathbf{\Phi}_M\}$ is a set of M given vectors in $\mathbb{R}^n$ and $\mathbf{X}(t) = (X_1(t), \ldots, X_M(t))$ is a given second-order centered mean-square stationary stochastic process indexed by $\mathbb{R}$ with values in $\mathbb{R}^M$, whose matrix-valued autocorrelation function

$$[R_{\mathbf{X}}(\tau)] = E\{\mathbf{X}(t+\tau)\,\mathbf{X}(t)^T\} \tag{111}$$

is the $(M \times M)$ real matrix such that

$$[R_{\mathbf{X}}(\tau)] = \int_{\mathbb{R}} e^{i\omega\tau}\,[S_{\mathbf{X}}(\omega)]\,d\omega = \int_{\mathbb{B}_\nu \cup \underline{\mathbb{B}}_\nu} e^{i\omega\tau}\,[S_{\mathbf{X}}(\omega)]\,d\omega \quad , \tag{112}$$

and $[S_{\mathbf{X}}(\omega)]$ is the spectral density function with values in the $(M \times M)$ complex matrices such that

$$[S_{\mathbf{X}}(\omega)] = [\,0\,] \quad , \quad \forall\,\omega \notin \mathbb{B}_\nu \cup \underline{\mathbb{B}}_\nu \quad . \tag{113}$$

Matrix-valued autocorrelation and spectral density functions of the reduced representation. From Eqs. (109) and (111), we deduce the expression of the $Mat_{\mathbb{R}}(n,n)$-valued autocorrelation function of stochastic process $\mathbf{F}_\nu$

$$[R_{\mathbf{F}_\nu}(\tau)] = E\{\mathbf{F}_\nu(t+\tau)\,\mathbf{F}_\nu(t)^T\} = \sum_{j=1}^{M}\sum_{k=1}^{M} [R_{\mathbf{X}}(\tau)]_{jk}\,\mathbf{\Phi}_j\,\mathbf{\Phi}_k^T \quad . \tag{114}$$

Substituting Eq. (112) into Eq. (114) yields

$$[S_{\mathbf{F}_\nu}(\omega)] = \sum_{j=1}^{M}\sum_{k=1}^{M} [S_{\mathbf{X}}(\omega)]_{jk}\,\mathbf{\Phi}_j\,\mathbf{\Phi}_k^T \quad . \tag{115}$$

Construction of a reduced representation in a particular case. For $M < n$, let us consider the case for which $\mathbf{F}$ can be written as

$$\mathbf{F}(t) = (\mathbf{X}(t)\,,\,\mathbf{0}) \quad , \tag{116}$$

in which $\mathbf{X}$ is a stochastic process with values in $\mathbb{R}^M$. Comparing Eq. (109) with Eq. (116), terms $\mathbf{\Phi}_j$ appear as the canonical vector basis, i.e. $\mathbf{\Phi}_1 = (1,0,0,\ldots)$, $\mathbf{\Phi}_2 = (0,1,0,0,\ldots)$, etc. This case corresponds, for instance, to a master structure submitted to a nonzero surface force field on a part of boundary Γ.

Construction of a reduced representation in the general case. Given $[S_{\mathbf{F}_\nu}(\omega)]$ defined by Eq. (110), the problem is to calculate dimension M, matrix $[S_{\mathbf{X}}(\omega)]$ for $\omega \in \mathbb{B}_\nu \cup \underline{\mathbb{B}}_\nu$ and vector basis $\{\mathbf{\Phi}_1,\ldots,\mathbf{\Phi}_M\}$ related to the representation defined by Eq. (109). Let $[\,C_{\mathbf{F}_\nu}]$ be the $(n\times n)$ real positive symmetric matrix defined by

$$[\,C_{\mathbf{F}_\nu}] = [R_{\mathbf{F}_\nu}(0)] = E\{\mathbf{F}_\nu(t)\,\mathbf{F}_\nu(t)^T\} \quad . \tag{117}$$

Eqs. (112) and (117) yield

$$[\,C_{\mathbf{F}_\nu}] = \int_{\mathbb{B}_\nu\cup\underline{\mathbb{B}}_\nu} [S_{\mathbf{X}}(\omega)]\,d\omega \quad . \tag{118}$$

Since $\mathbf{X}$ is an $\mathbb{R}^M$-valued stochastic process,

$$[S_{\mathbf{X}}(-\omega)] = \overline{[S_{\mathbf{X}}(\omega)]} \quad . \tag{119}$$

From Eqs. (118) and (119), we deduce that $[\,C_{\mathbf{F}_\nu}]$ can be rewritten as

$$[\,C_{\mathbf{F}_\nu}] = 2\,\Re e \int_{\mathbb{B}_\nu} [S_{\mathbf{X}}(\omega)]\,d\omega \quad . \tag{120}$$

We consider the eigenvalue problem related to $[\,C_{\mathbf{F}_\nu}]$,

$$[\,C_{\mathbf{F}_\nu}]\,\boldsymbol{\Phi}_j = \lambda_j\,\boldsymbol{\Phi}_j \quad . \tag{121}$$

Since $[\,C_{\mathbf{F}_\nu}]$ is a real positive symmetric matrix, $[\,C_{\mathbf{F}_\nu}]$ is diagonalizable, the eigenvalues are real and positive

$$\lambda_1 \geq \lambda_2 \geq \ldots \lambda_n \geq 0 \quad , \tag{122}$$

and the associated eigenvectors $\{\boldsymbol{\Phi}_1, \ldots, \boldsymbol{\Phi}_n\}$ form an orthonormal basis of $\mathbb{R}^n$

$$\boldsymbol{\Phi}_j^T\,\boldsymbol{\Phi}_k = \delta_{jk} \quad . \tag{123}$$

To define the value of M, we introduce the total power $\mathcal{P}_n$ of stochastic process $\mathbf{F}_\nu$

$$\mathcal{P}_n = E\{\|\mathbf{F}_\nu(t)\|^2\} \quad . \tag{124}$$

Using Eq. (117), we deduce that

$$\mathcal{P}_n = \mathrm{tr}[\,C_{\mathbf{F}_\nu}] \quad . \tag{125}$$

For any fixed $\varepsilon > 0$, there exists an integer $M \leq n$ such that

$$0 \leq \mathcal{P}_n - \sum_{j=1}^{M} \lambda_j \leq \varepsilon \quad . \tag{126}$$

Since $\mathcal{P}_n = \sum_{j=1}^{n} \lambda_j$, the difference $\mathcal{P}_n - \sum_{j=1}^{M} \lambda_j$ appearing in Eq. (126) is equal to $\sum_{j=M+1}^{n} \lambda_j$ which can be estimated without calculating the eigenvalues $\lambda_{M+1}, \ldots, \lambda_n$. Eq. (126) allows order M of the representation defined by Eq. (109) to be calculated. It should be noted that, if for ε fixed, Eq. (126) implies that $M \ll n$, then Eq. (109) defines a reduced representation of the stochastic excitation. Consequently, we have to calculate only the first $M \ll n$ eigenvalues and eigenvectors of Eq. (121). Once $\{\lambda_1, \ldots, \lambda_M\}$ and $\{\boldsymbol{\Phi}_1, \ldots, \boldsymbol{\Phi}_M\}$ have been calculated from Eqs. (115) and (123), we deduce that, for all j and k in $\{1, \ldots, M\}$, the spectral density function $[S_{\mathbf{X}}(\omega)]$ with values in the $(M \times M)$ positive Hermitian complex matrices is such that

$$[S_{\mathbf{X}}(\omega)]_{jk} = \mathbf{\Phi}_j^T \, [S_{\mathbf{F}_\nu}(\omega] \, \mathbf{\Phi}_k \quad . \tag{127}$$

5.4. Calculation of the stationary stochastic response in low-frequency band $\mathbb{B}_{\text{LF}}$

For ω belonging to $\mathbb{B}_{\text{LF}}$ (see Eq. (6)), we calculate the matrix-valued spectral density function $[S_{\mathbf{U}}(\omega)]$ allowing $[S_{\mathbf{u}^n}(\mathbf{x}, \mathbf{x}', \omega)]$, $[S_{\mathbf{v}^n}(\mathbf{x}, \mathbf{x}', \omega)]$ and $[S_{\mathbf{w}^n}(\mathbf{x}, \mathbf{x}', \omega)]$ to be deduced from Eqs. (103), (105) and (106) respectively.

Expression of $[S_{\mathbf{U}}(\omega)]$ in low-frequency band $\mathbb{B}_{\text{LF}}$. In band $\mathbb{B}_{\text{LF}}$, matrix $[S_{\mathbf{U}}(\omega)]$ given by Eq. (101) is approximated by matrix $[S_{\mathbf{U}_{N,\,\text{acc}}}(\omega)]$ such that

$$[S_{\mathbf{U}_{N,\,\text{acc}}}(\omega)] = [T_{N,\,\text{acc}}(\omega)] \, [S_{\mathbf{F}}(\omega)] \, [T_{N,\,\text{acc}}(\omega)]^* \quad , \tag{128}$$

in which $[T_{N,\,\text{acc}}(\omega)]$ is defined by Eq. (VI.58)

$$[T_{N,\,\text{acc}}(\omega)] = [T_N(\omega)] + [T_{N,\,\text{stat}}] \quad , \tag{129}$$

where $[T_N(\omega)]$ and $[T_{N,\,\text{stat}}]$ are given by Eqs. (VI.30) and (VI.59) respectively,

$$[T_N(\omega)] = \sum_{\alpha=1}^{N} \sum_{\beta=1}^{N} [\mathcal{T}(\omega)]_{\alpha\beta} \, \mathbf{U}_\alpha \, \mathbf{U}_\beta^T \quad , \tag{130}$$

$$[T_{N,\,\text{stat}}] = [\,K\,]^{-1} - \sum_{\alpha=1}^{N} \frac{1}{\mu_\alpha \, \omega_\alpha^2} \, \mathbf{U}_\alpha \, \mathbf{U}_\alpha^T \quad . \tag{131}$$

Substituting Eq. (129) into Eq. (128) yields

$$[S_{\mathbf{U}_{N,\,\text{acc}}}(\omega)] = [S_{\,\text{dyn}}(\omega)] + [S_{\,\text{qstat}}(\omega)] \quad , \tag{132}$$

in which $[S_{\,\text{dyn}}(\omega)]$ and $[S_{\,\text{qstat}}(\omega)]$ are the dynamic and quasi-static parts calculated below.

Calculation of the dynamic part. The dynamic part $[S_{\,\text{dyn}}(\omega)]$ is such that

$$\begin{aligned}[S_{\,\text{dyn}}(\omega)] &= [T_N(\omega)] \, [S_{\mathbf{F}}(\omega)] \, [T_N(\omega)]^* \\ &= \sum_{\alpha,\beta=1}^{N} \; \sum_{\alpha',\beta'=1}^{N} [\mathcal{T}(\omega)]_{\alpha\beta} \overline{[\mathcal{T}(\omega)]}_{\alpha'\beta'} \, \{\mathbf{U}_\beta^T \, [S_{\mathbf{F}}(\omega)] \, \mathbf{U}_{\beta'}\} \, \mathbf{U}_\alpha \mathbf{U}_{\alpha'}^T \quad , \end{aligned} \tag{133}$$

and allows the explicit calculation of $[S_{\text{dyn}}(\omega)]$.

Calculation of the quasi-static part. This calculation is not required if quasi-static correction terms are not introduced. Otherwise, quasi-static part $[S_{\text{qstat}}(\omega)]$ is such that

$$[S_{\text{qstat}}(\omega)] = [S_{\text{stat, stat}}(\omega)] + [S_{\text{stat, dyn}}(\omega)] + [S_{\text{stat, dyn}}(\omega)]^* \quad , \tag{134}$$

in which

$$[S_{\text{stat, stat}}(\omega)] = [T_{N,\text{stat}}]\,[S_{\mathbf{F}}(\omega)]\,[T_{N,\text{stat}}]^* \quad , \tag{135}$$

$$[S_{\text{stat, dyn}}(\omega)] = [T_{N,\text{stat}}]\,[S_{\mathbf{F}}(\omega)]\,[T_N(\omega)]^* \quad . \tag{136}$$

If dimension n is large, then Eqs. (135) and (136) do not allow an explicit calculation to be carried out due to the presence of $[K]^{-1}$ in these equations. In this case, we can use the method based on the reduced representation of stochastic excitation that we introduced in Section 5.3. To do so, low-frequency band $\mathbb{B}_{\text{LF}}$ is written as $\mathbb{B}_{\text{LF}} = \cup_\nu \mathbb{B}_\nu$ in which each $\mathbb{B}_\nu$ is a limited frequency band defined by Eq. (107). A reduced representation $\mathbf{F}_\nu$ of stochastic excitation $\mathbf{F}$ is constructed For each band $\mathbb{B}_\nu$ with $M \ll n$ and, matrix $[S_{\mathbf{F}}(\omega)]$ is replaced by matrix $[S_{\mathbf{F}_\nu}(\omega)]$ defined by Eq. (115). We then obtain

$$[S_{\text{stat, stat}}(\omega)] = \sum_{j=1}^{M}\sum_{k=1}^{M} [S_{\mathbf{X}}(\omega)]_{jk}\,\mathbf{Y}_j\,\mathbf{Y}_k^T \quad , \tag{137}$$

$$[S_{\text{stat, dyn}}(\omega)] = \sum_{j=1}^{M}\sum_{\alpha=1}^{N} [B(\omega)]_{j\alpha}\,\mathbf{Y}_j\,\mathbf{U}_\alpha^T \quad , \tag{138}$$

in which, for all $j = 1, \ldots, M$,

$$\mathbf{Y}_j = [T_{N,\text{stat}}]\,\mathbf{\Phi}_j \quad . \tag{139}$$

1- Expression of $[B(\omega)]$. The $(M \times N)$ complex matrix $[B(\omega)]$ is such that

$$[B(\omega)]_{j\alpha} = \sum_{k=1}^{M}\sum_{\beta=1}^{N} [S_{\mathbf{X}}(\omega)]_{jk}\overline{[T(\omega)]}_{\alpha\beta}\{\mathbf{\Phi}_k^T\mathbf{U}_\beta\} \quad . \tag{140}$$

2- Calculation of $\mathbf{Y}_j$. To calculate $\mathbf{Y}_j$, we substitute Eq. (131) into Eq. (139)

$$\mathbf{Y}_j = \mathbf{Z}_j - \sum_{\alpha=1}^{N} \frac{1}{\mu_\alpha\,\omega_\alpha^2}\,\{\mathbf{U}_\alpha^T\mathbf{\Phi}_j\}\,\mathbf{U}_\alpha \quad , \tag{141}$$

in which $\mathbf{Z}_j \in \mathbb{R}^n$ is the solution of the static problem

$$[\,K\,]\,\mathbf{Z}_j = \mathbf{\Phi}_j \quad . \tag{142}$$

The numerical procedure can then be summarized as follows. First, we proceed as described in Section 5.3 for determining $M \ll n$ and $\{\mathbf{\Phi}_1, \ldots, \mathbf{\Phi}_M\}$. Secondly, $\mathbf{Z}_1, \ldots, \mathbf{Z}_M$ are calculated by solving Eq. (142) with M second members simultaneously. Then $\{\mathbf{Y}_1, \ldots, \mathbf{Y}_M\}$ are calculated using Eq. (141). Finally, $[S_{\text{stat,stat}}(\omega)]$ and $[S_{\text{stat,dyn}}(\omega)]$ are deduced from Eqs. (137) and (138) respectively, and the quasi-static correction term $[S_{\text{qstat}}(\omega)]$ is obtained from Eq. (134).

5.5. Calculation of the response in medium-frequency band $\mathbb{B}_{\text{MF}}$

Statement of the problem. For ω in $\mathbb{B}_{\text{MF}}$ defined by Eq. (6), we have to calculate the matrix-valued spectral density function $[S_{\mathbf{U}}(\omega)]$ allowing $[S_{\mathbf{u}^n}(\mathbf{x}, \mathbf{x}', \omega)]$, $[S_{\mathbf{v}^n}(\mathbf{x}, \mathbf{x}', \omega)]$ and $[S_{\mathbf{w}^n}(\mathbf{x}, \mathbf{x}', \omega)]$ to be deduced from Eqs. (103), (105) and (106) respectively. In band $\mathbb{B}_{\text{MF}}$, matrix $[S_{\mathbf{U}}(\omega)]$ is given by

$$[S_{\mathbf{U}}(\omega)] = [T(\omega)]\,[S_{\mathbf{F}}(\omega)]\,[T(\omega)]^* \quad . \tag{143}$$

For large values of dimension n, a direct numerical procedure based on the use of Eq. (143) would require the construction of $[T(\omega)]$ for a large set of discrete frequencies covering frequency band $\mathbb{B}_{\text{MF}}$ plus calculation of the product of three dense matrices for each frequency. From a numerical point of view, it is more efficient to use the MF method presented in Chapter VII in conjunction with the reduced representation of the stochastic excitation introduced in Section 5.3.

Construction of $[S_{\mathbf{U}}(\omega)]$ using auxiliary deterministic MF problems. Band $\mathbb{B}_{\text{MF}}$ is written as a finite union of MF narrow bands $\mathbb{B}_\nu$ (see Section VII.7)

$$\mathbb{B}_{\text{MF}} = \cup_\nu \, \mathbb{B}_\nu \quad . \tag{144}$$

The restriction of $[S_{\mathbf{U}}(\omega)]$ to $\mathbb{B}_\nu$ is denoted as $[S_{\mathbf{U}_\nu}(\omega)]$. For each MF narrow band $\mathbb{B}_\nu$, a reduced representation $\mathbf{F}_\nu$ of stochastic excitation $\mathbf{F}$ is constructed with $M \ll n$, and matrix $[S_{\mathbf{F}}(\omega)]$ is replaced by matrix $[S_{\mathbf{F}_\nu}(\omega)]$ defined by Eq. (115). We then obtain

$$[S_{\mathbf{U}_\nu}(\omega)] = \sum_{j=1}^{M}\sum_{k=1}^{M} [S_{\mathbf{X}}(\omega)]_{jk}\,\mathbf{Y}_{j,\nu}(\omega)\,\mathbf{Y}_{k,\nu}(\omega)^* \quad , \tag{145}$$

in which, for all ω in $\mathbb{B}_\nu$,

$$\mathbf{Y}_{j,\nu}(\omega) = [T(\omega)]\,\mathbf{\Phi}_j \quad , \quad \forall j = 1,\ldots,M \quad . \tag{146}$$

Eq. (146) appears as an MF problem related to MF narrow band $\mathbb{B}_\nu$ which is solved for M deterministic excitations simultaneously.

Solving procedure. For each MF narrow band $\mathbb{B}_\nu$, functions $\{\mathbf{Y}_{1,\nu},..,\mathbf{Y}_{M,\nu}\}$ are constructed simultaneously considering M second members defined by $\{\theta_\nu(\omega)\mathbf{\Phi}_{1,\nu},\ldots,\theta_\nu(\omega)\mathbf{\Phi}_{M,\nu}\}$, using the MF method presented in Section VII.8, in which function $\theta_\nu(\omega) = \mathbf{1}_{\mathbb{B}_\nu}(\omega)$, i.e. $\theta_0(t) = \frac{1}{\pi t}\sin(\frac{\Delta\omega}{2}\,t)$. For all $j = 1,\ldots,M$, $\mathbf{Y}_{j,\nu}(\omega)$ is written as (see Eq. (VII.59))

$$\mathbf{Y}_{j,\nu}(\omega) \simeq \mathbf{1}_{\mathbb{B}_\nu}(\omega)\,\Delta t \sum_{m=m_{\mathrm{i}}}^{m_{\mathrm{f}}} \mathbf{Y}_{j,0}(m\,\Delta t)\; e^{-im\,\Delta t\,(\omega-\Omega_\nu)} \quad , \tag{147}$$

in which $\mathbf{Y}_{j,0}(t)$ is the solution of the associated LF equation in the time domain (see Section VII.8).

5.6. Bibliographical comments

For further details concerning stochastic processes, their transformations (linear filtering, derivatives, etc.), random vibration and stochastic dynamics, we refer the reader to Bendat and Piersol, 1980; Crandall and Mark, 1973; Doob, 1953; Elishakoff, 1983; Guikhman and Skorokhod, 1979; Jenkins and Watt, 1968; Kree and Soize, 1983 and 1986; Lin, 1967; Priestley, 1981; Roberts and Spanos, 1990; Soize, 1988, 1993a, 1994; Soong, 1973; Vanmarcke, 1983.

6. Random Case: Nonstationary Stochastic Excitation

6.1. Definition of the excitation

Second-order description of the time-nonstationary stochastic body and surface force fields. Let $\Theta = [\,0\,,T\,]$ be a finite time interval of $\mathbb{R}^+$. In the time domain, the body and surface force fields $\{\mathbf{g}(\mathbf{x},t), \mathbf{x}\in\Omega,\, t\in\Theta\}$ and $\{\mathbf{G}(\mathbf{y},t), \mathbf{y}\in\Gamma,\, t\in\Theta\}$ are given time-nonstationary stochastic fields defined on the same probability space, statistically dependent and indexed by $\Omega\times\Theta$ and $\Gamma\times\Theta$ respectively, with values in $\mathbb{R}^3$. We assume that $\mathbf{g}$ and $\mathbf{G}$ are second-order stochastic fields, i.e., for all t in Θ, $\mathbf{x}$ in Ω and $\mathbf{y}$ in Γ,

$$E\{\|\mathbf{g}(\mathbf{x},t)\|^2\} < +\infty \quad , \quad E\{\|\mathbf{G}(\mathbf{y},t)\|^2\} < +\infty \quad , \tag{148}$$

in which E denotes the mathematical expectation (the mathematical expectation of a random variable is equal to the integral of this random variable with respect to its probability distribution). The mean functions $\mathbf{m}_\mathbf{g}(\mathbf{x}, t)$ of $\mathbf{g}$ and $\mathbf{m}_\mathbf{G}(\mathbf{y}, t)$ of $\mathbf{G}$, defined on $\Omega \times \Theta$ and $\Gamma \times \Theta$ respectively, with values in $\mathbb{R}^3$, are such that

$$\mathbf{m}_\mathbf{g}(\mathbf{x}, t) = E\{\mathbf{g}(\mathbf{x}, t)\} \quad , \tag{149}$$

$$\mathbf{m}_\mathbf{G}(\mathbf{y}, t) = E\{\mathbf{G}(\mathbf{y}, t)\} \quad . \tag{150}$$

For all t and t' in Θ, $\mathbf{x}$ and $\mathbf{x}'$ in Ω, $\mathbf{y}$ and $\mathbf{y}'$ in Γ, the cross-correlation functions with values in $Mat_\mathbb{R}(3, 3)$ are written as

$$[R_\mathbf{g}(\mathbf{x}, \mathbf{x}', t, t')] = E\{\mathbf{g}(\mathbf{x}, t)\, \mathbf{g}(\mathbf{x}', t')^T\} \quad , \tag{151}$$

$$[R_\mathbf{G}(\mathbf{y}, \mathbf{y}', t, t')] = E\{\mathbf{G}(\mathbf{y}, t)\, \mathbf{G}(\mathbf{y}', t')^T\} \quad , \tag{152}$$

$$[R_{\mathbf{gG}}(\mathbf{x}, \mathbf{y}', t, t')] = E\{\mathbf{g}(\mathbf{x}, t)\, \mathbf{G}(\mathbf{y}', t')^T\} \quad , \tag{153}$$

$$[R_{\mathbf{Gg}}(\mathbf{y}, \mathbf{x}', t, t')] = E\{\mathbf{G}(\mathbf{y}, t)\, \mathbf{g}(\mathbf{x}', t')^T\} \quad . \tag{154}$$

From Eqs. (153) and (154), we deduce that

$$[R_{\mathbf{Gg}}(\mathbf{y}, \mathbf{x}', t, t')] = [R_{\mathbf{gG}}(\mathbf{x}', \mathbf{y}, t', t)]^T \quad . \tag{155}$$

The data allowing the second-order description of the time-nonstationary stochastic body and surface force fields are the mean functions and the cross-correlation functions defined by Eqs. (149) to (155).

Second-order representation of the nonstationary stochastic excitation in the time domain. To construct the second-order representation of the nonstationary stochastic excitation, we use the Ritz-Galerkin approximation introduced in Section 2. Excitation force $\mathbf{F}(t) = (F_1(t), \ldots, F_n(t))$ such that

$$F_j(t) = \int_\Omega \mathbf{g}(\mathbf{x}, t) \cdot \mathbf{e}_j(\mathbf{x})\, d\mathbf{x} + \int_\Gamma \mathbf{G}(\mathbf{y}, t) \cdot \mathbf{e}_j(\mathbf{y})\, ds(\mathbf{y}) \quad , \tag{156}$$

is a second-order nonstationary stochastic process indexed by Θ with values in $\mathbb{R}^n$. Let

$$\mathbf{m}_\mathbf{F}(t) = E\{\mathbf{F}(t)\} \tag{157}$$

be the $\mathbb{R}^n$-valued mean function of components $m_{F_1}(t), \ldots, m_{F_n}(t)$ of stochastic process $\mathbf{F}$. From Eqs. (149), (150) and (156), we deduce that

$$\begin{aligned} m_{F_j}(t) &= E\{F_j(t)\} \\ &= \int_\Omega \mathbf{m_g}(\mathbf{x}, t) \cdot \mathbf{e}_j(\mathbf{x})\, d\mathbf{x} + \int_\Gamma \mathbf{m_G}(\mathbf{y}, t) \cdot \mathbf{e}_j(\mathbf{y})\, ds(\mathbf{y}) \quad . \end{aligned} \tag{158}$$

From Eqs. (151)-(154) and (156), we deduce that for all t and t' in Θ, the $Mat_{\mathbb{R}}(n,n)$-valued autocorrelation function $[R_{\mathbf{F}}(t,t')]$ of stochastic process $\mathbf{F}$, defined by

$$[R_{\mathbf{F}}(t,t')] = E\{\mathbf{F}(t)\, \mathbf{F}(t')^T\} \quad , \tag{159}$$

is such that

$$\begin{aligned} [R_{\mathbf{F}}(t,t')]_{jk} =& E\{F_j(t)\, F_k(t')\} \\ =& \int_\Omega \int_\Omega \mathbf{e}_j(\mathbf{x})^T\, [R_{\mathbf{g}}(\mathbf{x}, \mathbf{x}', t, t')]\, \mathbf{e}_k(\mathbf{x}')\, d\mathbf{x}\, d\mathbf{x}' \\ &+ \int_\Gamma \int_\Gamma \mathbf{e}_j(\mathbf{y})^T\, [R_{\mathbf{G}}(\mathbf{y}, \mathbf{y}', t, t')]\, \mathbf{e}_k(\mathbf{y}')\, ds(\mathbf{y})\, ds(\mathbf{y}') \\ &+ \int_\Omega \int_\Gamma \mathbf{e}_j(\mathbf{x})^T\, [R_{\mathbf{gG}}(\mathbf{x}, \mathbf{y}', t, t')]\, \mathbf{e}_k(\mathbf{y}')\, d\mathbf{x}\, ds(\mathbf{y}') \\ &+ \int_\Gamma \int_\Omega \mathbf{e}_j(\mathbf{y})^T\, [R_{\mathbf{Gg}}(\mathbf{y}, \mathbf{x}', t, t')]\, \mathbf{e}_k(\mathbf{x}')\, ds(\mathbf{y})\, d\mathbf{x}' \quad . \end{aligned} \tag{160}$$

Construction of an associated centered nonstationary stochastic excitation in the time domain. Stochastic process $\mathbf{F}(t)$ can be written as

$$\mathbf{F}(t) = \mathbf{m_F}(t) + \mathbf{F}_c(t) \quad , \tag{161}$$

in which $\mathbf{m_F}(t)$ is the mean function of $\mathbf{F}$ defined by Eqs. (157) and (158), and the random fluctuation $\mathbf{F}_c$ is a second-order centered stochastic process indexed by Θ with values in $\mathbb{R}^n$,

$$\mathbf{m}_{\mathbf{F}_c}(t) = E\{\mathbf{F}_c(t)\} = \mathbf{0} \quad . \tag{162}$$

The $Mat_{\mathbb{R}}(n,n)$-valued autocorrelation function $[R_{\mathbf{F}_c}(t,t')]$ of centered stochastic process $\mathbf{F}_c$, defined by

$$[R_{\mathbf{F}_c}(t,t')] = E\{\mathbf{F}_c(t)\, \mathbf{F}_c(t')^T\} \quad , \tag{163}$$

is then equal to the $Mat_{\mathbb{R}}(n,n)$-valued covariance function $[C_{\mathbf{F}}(t,t')]$ of stochastic process $\mathbf{F}$, which is defined by

$$\begin{aligned}[C_{\mathbf{F}}(t,t')] =& E\{(\mathbf{F}(t) - \mathbf{m_F}(t))\,(\mathbf{F}(t') - \mathbf{m_F}(t'))^T\} \\ =& [R_{\mathbf{F}}(t,t')] - \mathbf{m_F}(t)\,\mathbf{m_F}(t')^T \quad . \end{aligned} \tag{164}$$

Representation of the mean function of the nonstationary stochastic excitation in the frequency domain. In order to define the Fourier transform of the mean function of the noncentered second-order stochastic process $\mathbf{F}$, we extend function $\mathbf{m_F}(t)$ by zero outside its domain of definition Θ,

$$\underline{\mathbf{m}}_{\mathbf{F}}(t) = \mathbf{1}_\Theta(t)\,\mathbf{m_F}(t) \quad , \tag{165}$$

in which $\mathbf{1}_\Theta(t)$ is the indicator function of Θ,

$$\mathbf{1}_\Theta(t) = 1 \quad \text{if} \quad t \in \Theta \quad \text{and} \quad = 0 \quad \text{if} \quad t \notin \Theta \quad . \tag{166}$$

We assume that nonstationary stochastic process $\mathbf{F}$ is such that function $\mathbf{m_F}(t)$ is square integrable on Θ. Consequently, $\underline{\mathbf{m}}_{\mathbf{F}}(t)$ is square integrable on $\mathbb{R}$,

$$\int_{\mathbb{R}} \|\underline{\mathbf{m}}_{\mathbf{F}}(t)\|^2\, dt < +\infty \quad . \tag{167}$$

Using the Fourier transform theory of square integrable functions, we define the function $\widetilde{\underline{\mathbf{m}}}_{\mathbf{F}}(\omega)$ on $\mathbb{R}$ such that

$$\begin{aligned}\widetilde{\underline{\mathbf{m}}}_{\mathbf{F}}(\omega) &= \frac{1}{2\pi}\int_{\mathbb{R}} e^{-i\omega t}\,\underline{\mathbf{m}}_{\mathbf{F}}(t)\,dt \\ &= \frac{1}{2\pi}\int_{\Theta} e^{-i\omega t}\,\mathbf{m_F}(t)\,dt \quad , \end{aligned} \tag{168}$$

which is square integrable on $\mathbb{R}$. Using the inverse Fourier transform, we obtain

$$\underline{\mathbf{m}}_{\mathbf{F}}(t) = \int_{\mathbb{R}} e^{i\omega t}\,\widetilde{\underline{\mathbf{m}}}_{\mathbf{F}}(\omega)\,d\omega \quad . \tag{169}$$

It should be noted that Eqs. (168) and (169) do not correspond to the Fourier transform convention used in this book. This is why we introduced a new notation using a tilda.

Representation of the autocorrelation function of centered random fluctuation $\mathbf{F}_c$ in the frequency domain. In order to define the Fourier transform of the autocorrelation function of stochastic process $\mathbf{F}_c$, we extend function $[R_{\mathbf{F}_c}(t,t')]$ by zero outside its domain of definition $\Theta \times \Theta$

$$[\underline{R}_{\mathbf{F}_c}(t,t')] = \mathbf{1}_\Theta(t)\, \mathbf{1}_\Theta(t')\, [R_{\mathbf{F}_c}(t,t')] \quad . \tag{170}$$

We assume that centered nonstationary stochastic process $\mathbf{F}_c(t)$ is such that function $[R_{\mathbf{F}_c}(t,t')]$ is bounded and square integrable on $\Theta \times \Theta$. Consequently, $[\underline{R}_{\mathbf{F}_c}(t,t')]$ is bounded and square integrable on $\mathbb{R} \times \mathbb{R}$,

$$\sup_{t,t'} \|\underline{R}_{\mathbf{F}_c}(t,t')\|_F < +\infty \quad , \quad \int_{\mathbb{R}}\int_{\mathbb{R}} \|\underline{R}_{\mathbf{F}_c}(t,t')\|_F^2 \, dt\, dt' < +\infty \quad , \tag{171}$$

in which $\|\,.\,\|_F$ is the Frobenius norm defined in Section 3 of Mathematical Notations in the appendix. Using the Fourier transform theory of square integrable functions, we define the function $[\widetilde{\underline{R}}_{\mathbf{F}_c}(\omega,\omega')]$ on $\mathbb{R}\times\mathbb{R}$, such that

$$\begin{aligned}[\widetilde{\underline{R}}_{\mathbf{F}_c}(\omega,\omega')] &= \frac{1}{4\pi^2}\int_{\mathbb{R}}\int_{\mathbb{R}} e^{-i(\omega t-\omega' t')}\, [\underline{R}_{\mathbf{F}_c}(t,t')]\, dt\, dt' \\ &= \frac{1}{4\pi^2}\int_{\Theta}\int_{\Theta} e^{-i(\omega t-\omega' t')}\, [R_{\mathbf{F}_c}(t,t')]\, dt\, dt' \quad ,\end{aligned} \tag{172}$$

which is square integrable on $\mathbb{R} \times \mathbb{R}$. Using the inverse Fourier transform, we obtain

$$[\underline{R}_{\mathbf{F}_c}(t,t')] = \int_{\mathbb{R}}\int_{\mathbb{R}} e^{i(\omega t-\omega' t')}\, [\widetilde{\underline{R}}_{\mathbf{F}_c}(\omega,\omega')]\, d\omega\, d\omega' \quad . \tag{173}$$

As above, Eqs. (172) and (173) do not correspond to the Fourier transform convention used in this book and consequently, a tilda is introduced.

Expression of the autocorrelation function of the random fluctuation of the excitation based on the use of an integral representation of the nonstationary stochastic process. In most cases, matrix-valued autocorrelation function $[R_{\mathbf{F}_c}(t,t')]$ of centered nonstationary stochastic process $\mathbf{F}_c$ can be written as [11] (see Guikhman and Skorokhod, 1979; Priestley, 1981 and 1988; Kree and Soize 1986; Soize, 1993a)

$$[R_{\mathbf{F}_c}(t,t')] = \int_{\mathbb{R}} e^{i\omega(t-t')}\, [Q_{\mathbf{F}_c}(t,\omega)]\, [S(\omega)]\, [Q_{\mathbf{F}_c}(t',\omega)]^*\, d\omega \quad , \tag{174}$$

[1]1 Let W be a subset of $\mathbb{R}$, and $\mathcal{W}$ be the Borel σ-algebra of W. The general integral representation of the $\mathbb{R}^n$-valued centered stochastic process $\mathbf{F}_c$ indexed by Θ is written as $\mathbf{F}_c(t) = \int_W e^{i\omega t}[Q_{\mathbf{F}_c}(t,\omega)]\, \mathbf{H}(d\omega)$ in which $\mathbf{H}(d\omega)$ is a $\mathbb{C}^n$-valued stochastic spectral measure on $(W,\mathcal{W})$

in which $[S(\omega)]$ is an $(n \times n)$ complex positive Hermitian matrix such that

$$\int_{\mathbb{R}} |[S(\omega)]_{jk}| \, d\omega < +\infty \quad , \tag{175}$$

and such that, for all ω in $\mathbb{R}$,

$$[S(-\omega)] = \overline{[S(\omega)]} \quad . \tag{176}$$

Function $[Q_{\mathbf{F}_c}(t,\omega)]$ defined on $\Theta \times \mathbb{R}$, with values in $Mat_{\mathbb{C}}(n,n)$, satisfies the following assumptions. For all t in Θ,

$$\int_{\mathbb{R}} \operatorname{tr}\{[Q_{\mathbf{F}_c}(t,\omega)]\,[S(\omega)]\,[Q_{\mathbf{F}_c}(t,\omega)]^*\} \, d\omega < +\infty \quad . \tag{177}$$

For all ω in $\mathbb{R}$,

$$\int_{\Theta} \|Q_{\mathbf{F}_c}(t,\omega)\|_F^2 \, dt < +\infty \quad . \tag{178}$$

Finally, for all t in Θ and ω in $\mathbb{R}$,

$$[Q_{\mathbf{F}_c}(t,-\omega)] = \overline{[Q_{\mathbf{F}_c}(t,\omega)]} \quad . \tag{179}$$

In the representation defined by Eq. (174), the data are $[Q_{\mathbf{F}_c}(t,\omega)]$ and $[S(\omega)]$. From Eq. (174), we deduce that, for all t in Θ,

$$E\{\|\mathbf{F}_c(t)\|^2\} = \operatorname{tr}[R_{\mathbf{F}_c}(t,t)] = \int_{\mathbb{R}} \operatorname{tr}[S_{\mathbf{F}_c}(t,\omega)] \, d\omega \quad , \tag{180}$$

in which $[S_{\mathbf{F}_c}(t,\omega)]$ is the matrix-valued instantaneous spectral density function with respect to $d\omega$ such that

$$[S_{\mathbf{F}_c}(t,\omega)] = [Q_{\mathbf{F}_c}(t,\omega)]\,[S(\omega)]\,[Q_{\mathbf{F}_c}(t,\omega)]^* \quad . \tag{181}$$

---

such that, for all B and B' in $\mathcal{W}$, $E\{\mathbf{H}(B)\,\mathbf{H}(B')^*\}=[M(B\cap B')]$, where the so-called structural measure $[M(d\omega)]$ of $\mathbf{H}(d\omega)$ is a $Mat_{\mathbb{C}}(n,n)$-valued bounded measure on $(W,\mathcal{W})$ such that, for all B in $\mathcal{W}$, $[M(B)]$ is a positive Hermitian matrix. Function $[Q_{\mathbf{F}_c}(t,\omega)]$ defined on $\Theta\times W$ with values in $Mat_{\mathbb{C}}(n,n)$ is such that, $\forall t\in\Theta$, $\int_W tr\{[Q_{\mathbf{F}_c}(t,\omega)]\,[M(d\omega)]\,[Q_{\mathbf{F}_c}(t,\omega)]^*\}<+\infty$. We deduce that the matrix-valued autocorrelation function of stochastic process $\mathbf{F}_c$ can be written as $[R_{\mathbf{F}_c}(t,t')]=\int_W e^{i\omega(t-t')}\,[Q_{\mathbf{F}_c}(t,\omega)]\,[M(d\omega)]\,[Q_{\mathbf{F}_c}(t',\omega)]^*$. Consequently, Eq. (174) appears as a particular case for which $W=\mathbb{R}$ and $[M(d\omega)]=[S(\omega)]\,d\omega$ (see Guikhman and Skorokhod, 1979; Priestley, 1981; Kree and Soize 1986; Soize, 1993a).

For all t in Θ and ω in $\mathbb{R}$, $[S_{\mathbf{F}_c}(t,\omega)]$ is an $(n\times n)$ complex positive Hermitian matrix. Substituting Eq. (174) into Eq. (172) yields

$$[\underline{\widetilde{R}}_{\mathbf{F}_c}(\omega,\omega')] = \int_{\mathbb{R}} [\widetilde{Q}_{\mathbf{F}_c}(\omega-v,v)]\,[S(v)]\,[\widetilde{Q}_{\mathbf{F}_c}(\omega'-v,v)]^*\,dv \quad , \tag{182}$$

in which, due to Eq. (178), for all ω in $\mathbb{R}$ and t in Θ,

$$[\widetilde{Q}_{\mathbf{F}_c}(\omega,v)] = \frac{1}{2\pi}\int_{\Theta} e^{-i\omega t}\,[Q_{\mathbf{F}_c}(t,v)]\,dt \quad ,$$

$$[Q_{\mathbf{F}_c}(t,v)] = \int_{\mathbb{R}} e^{i\omega t}\,[\widetilde{Q}_{\mathbf{F}_c}(\omega,v)]\,d\omega \quad . \tag{183}$$

As an example, if we introduce an $\mathbb{R}^n$-valued second-order centered mean-square stationary stochastic process $\mathbf{\Lambda}$ indexed by $\mathbb{R}$ such that

$$[R_{\mathbf{\Lambda}}(\tau)] = E\{\mathbf{\Lambda}(t+\tau)\,\mathbf{\Lambda}(t)^T\} = \int_{\mathbb{R}} e^{i\omega\tau}\,[S_{\mathbf{\Lambda}}(\omega)]\,d\omega \quad , \tag{184}$$

then the $\mathbb{R}^n$-valued second-order centered nonstationary stochastic process $\mathbf{F}_c$ indexed by Θ, defined by

$$\mathbf{F}_c(t) = \lambda(t)\,\mathbf{\Lambda}(t) \quad , \tag{185}$$

in which $\lambda(t)$ is a bounded and square integrable real function defined on Θ, has the representation given by Eq. (174) with

$$[S(\omega)] = [S_{\mathbf{\Lambda}}(\omega)] \quad , \quad [Q_{\mathbf{F}_c}(t,\omega)] = \lambda(t)\,[\,I\,] \quad . \tag{186}$$

The instantaneous spectral density function is such that

$$[S_{\mathbf{F}_c}(t,\omega)] = \lambda(t)^2\,[S_{\mathbf{\Lambda}}(\omega)] \quad . \tag{187}$$

From Eq. (182), we deduce that

$$[\underline{\widetilde{R}}_{\mathbf{F}_c}(\omega,\omega')] = \int_{\mathbb{R}} \widetilde{\lambda}(\omega-v)\,\overline{\widetilde{\lambda}(\omega'-v)}\,[S_{\mathbf{\Lambda}}(v)]\,dv \quad , \tag{188}$$

in which

$$[\widetilde{\lambda}(\omega)] = \frac{1}{2\pi}\int_{\Theta} e^{-i\omega t}\,\lambda(t)\,dt \quad . \tag{189}$$

6.2. Theoretical formula for second-order moments of the nonstationary response

Definition of the time-variable problem with Cauchy initial conditions. We consider the time response of the master structure for $t > 0$, submitted to stochastic body and surface force fields defined in Section 6.1 and to zero Cauchy initial conditions. Using the Ritz-Galerkin approximation introduced in Section 2.2, $\mathbf{U}(t) = (U_1(t), \ldots, U_n(t))$ is such that

$$\mathbf{U}(0) = \mathbf{0} \quad , \quad \dot{\mathbf{U}}(0) = \mathbf{0} \quad . \tag{190}$$

Consequently, Eq. (31) must be replaced by

$$\mathbf{U}(t) = \int_0^t [\,h(t')\,]\,\mathbf{F}(t-t')\,dt' \quad , \quad t \in \Theta = [\,0\,,T\,] \quad . \tag{191}$$

It should be noted that the right-hand side of Eq. (191) is not a linear filtering of stochastic process $\mathbf{F}(t)$ with respect to t, contrary to the case presented in Section 5.

Decomposition of the response into its mean part and its random fluctuation. Substituting Eq. (161) into Eq. (191), we deduce that, for all t in Θ, $\mathbf{U}(t)$ can be written as

$$\mathbf{U}(t) = \mathbf{m}_{\mathbf{U}}(t) + \mathbf{U}_c(t) \quad , \tag{192}$$

in which $\mathbf{m}_{\mathbf{U}}(t)$ is the mean function of stochastic process $\mathbf{U}$, defined by

$$\begin{aligned} \mathbf{m}_{\mathbf{U}}(t) &= E\{\mathbf{U}(t)\} \\ &= \int_0^t [\,h(t')\,]\,\mathbf{m}_{\mathbf{F}}(t-t')\,dt' \\ &= \int_0^t [\,h(t-t')\,]\,\mathbf{m}_{\mathbf{F}}(t')\,dt' \quad . \end{aligned} \tag{193}$$

Random fluctuation $\mathbf{U}_c$ is given by [12]

$$\begin{aligned} \mathbf{U}_c(t) &= \int_0^t [\,h(t')\,]\,\mathbf{F}_c(t-t')\,dt' \\ &= \int_0^t [\,h(t-t')\,]\,\mathbf{F}_c(t')\,dt' \quad , \quad t \in \Theta = [\,0\,,T\,] \quad . \end{aligned} \tag{194}$$

---

12 The integral in the right-hand side of Eq. (194) is defined as a mean-square integral.

It can be shown that random fluctuation $\mathbf{U}_c$ defined by Eq. (194) is an $\mathbb{R}^n$-valued second-order [13] nonstationary stochastic process indexed by Θ. From Eq. (30), we deduce that

$$\mathbf{u}^n(\mathbf{x},t) = \mathbf{m}_{\mathbf{u}^n}(\mathbf{x},t) + \mathbf{u}_c^n(\mathbf{x},t) \quad , \tag{195}$$

in which

$$\mathbf{m}_{\mathbf{u}^n}(\mathbf{x},t) = \sum_{j=1}^{n} m_{U_j}(t)\, \mathbf{e}_j(\mathbf{x}) \quad . \tag{196}$$

Eq. (195) defines the random fluctuation displacement field $\mathbf{u}_c^n(\mathbf{x},t)$ of the displacement field $\mathbf{u}^n(\mathbf{x},t)$,

$$\mathbf{u}_c^n(\mathbf{x},t) = \sum_{j=1}^{n} U_{c,j}(t)\, \mathbf{e}_j(\mathbf{x}) \quad . \tag{197}$$

Mean part of the response in the time and frequency domains. Using Eqs. (15) and (165), Eq. (193) can be rewritten as

$$\mathbf{m}_{\mathbf{U}}(t) = \int_{\mathbb{R}} [\, h(t')\,]\; \underline{\mathbf{m}}_{\mathbf{F}}(t-t')\, dt' \quad , \quad t \in \Theta \quad . \tag{198}$$

Function $\mathbf{m}_{\mathbf{U}}$ is a square integrable function [14] on Θ. Substituting Eq. (169) into (198) and using Eq. (33) yields

$$\mathbf{m}_{\mathbf{U}}(t) = \int_{\mathbb{R}} e^{i\omega t}\, \widetilde{\underline{\mathbf{m}}}_{\mathbf{U}}(\omega)\, d\omega \quad , \tag{199}$$

in which $\widetilde{\underline{\mathbf{m}}}_{\mathbf{U}}(\omega)$ is a square integrable function on $\mathbb{R}$, such that

$$\widetilde{\underline{\mathbf{m}}}_{\mathbf{U}}(\omega) = [T(\omega)]\, \widetilde{\underline{\mathbf{m}}}_{\mathbf{F}}(\omega) \quad , \tag{200}$$

where $[T(\omega)]$ is the $(n \times n)$ complex symmetric matrix defined by Eq. (25).

---

[13] $\mathbf{U}_c(t) = \int_0^t [\, h(t')\,]\, \mathbf{F}_c(t-t')\, dt'$ is a second-order stochastic process indexed by Θ, because $(t,t') \mapsto [R_{\mathbf{F}_c}(t,t')]$ is bounded on $\Theta \times \Theta$ (see Eq. (171)) and for all t in Θ, $t' \mapsto [\, h(t-t')\,]$ is an integrable function on Θ (see for instance Soize, 1993a, p. 378 or, Soize, 1994, p. 85).

[14] Since $[\, h\,]$ is an integrable function on $\mathbb{R}$, i.e. $\int_{\mathbb{R}} |[\, h(t)\,]_{jk}|\, dt = \int_0^{+\infty} |[\, h(t)\,]_{jk}|\, dt < +\infty$, and since $\underline{\mathbf{m}}_{\mathbf{F}}$ is square integrable on $\mathbb{R}$, then $\mathbf{m}_{\mathbf{U}} = [\, h\,] * \underline{\mathbf{m}}_{\mathbf{F}}$ is a square integrable function on $\mathbb{R}$ (see for instance Soize, 1993a, p. 18).

Second-order representation of the random fluctuation part of the nonstationary response in the time and frequency domains. From Eqs. (162) and (194), we deduce that stochastic process $\mathbf{U}_c$ and thus stochastic field $\mathbf{u}_c^n$ defined by Eq. (197) are centered, i.e., for all fixed $\mathbf{x}$ in Ω and t in Θ,

$$E\{\mathbf{U}_c(t)\} = \mathbf{0} \quad , \quad E\{\mathbf{u}_c^n(\mathbf{x},t)\} = \mathbf{0} \quad . \tag{201}$$

For all t and t' in Θ, the matrix-valued autocorrelation function $[R_{\mathbf{U}_c}(t,t')]$ of stochastic process $\mathbf{U}_c$ is defined by

$$[R_{\mathbf{U}_c}(t,t')] = E\{\mathbf{U}_c(t)\,\mathbf{U}_c(t')^T\} \quad . \tag{202}$$

Eqs. (194), (163) and (202) yield

$$[R_{\mathbf{U}_c}(t,t')] = \int_0^t \int_0^{t'} [\,h(t-\tau)\,]\,[R_{\mathbf{F}_c}(\tau,\tau')]\,[\,h(t'-\tau')\,]^T\,d\tau\,d\tau' \quad . \tag{203}$$

From Eqs. (197) and (202), we deduce that, for all $\mathbf{x}$ and $\mathbf{x}'$ in Ω and t and t' in Θ, the $Mat_{\mathbb{R}}(3,3)$-valued cross-correlation function of stochastic field $\mathbf{u}_c^n$ can be written as

$$\begin{aligned}[R_{\mathbf{u}_c^n}(\mathbf{x},\mathbf{x}',t,t')] &= E\{\mathbf{u}_c^n(\mathbf{x},t)\,\mathbf{u}_c^n(\mathbf{x}',t')^T\} \\ &= \sum_{j,k=1}^{n} [R_{\mathbf{U}_c}(t,t')]_{jk}\,\mathbf{e}_j(\mathbf{x})\,\mathbf{e}_k(\mathbf{x}')^T \quad .\end{aligned} \tag{204}$$

Let $\mathbf{v}_c^n$ and $\mathbf{w}_c^n$ be the velocity and acceleration fluctuation fields such that

$$\mathbf{v}_c^n(\mathbf{x},t) = \partial_t \mathbf{u}_c^n(\mathbf{x},t) \quad , \quad \mathbf{w}_c^n(\mathbf{x},t) = \partial_t \mathbf{v}_c^n(\mathbf{x},t) \quad . \tag{205}$$

Fields $\mathbf{v}_c^n$ and $\mathbf{w}_c^n$ are second-order time-nonstationary centered stochastic fields indexed by $\Omega \times \Theta$ with values in $\mathbb{R}^3$. Their matrix-valued cross-correlation functions are such that

$$\begin{aligned}[R_{\mathbf{v}_c^n}(\mathbf{x},\mathbf{x}',t,t')] &= E\{\mathbf{v}_c^n(\mathbf{x},t)\,\mathbf{v}_c^n(\mathbf{x}',t')^T\} \\ &= \frac{\partial^2}{\partial t\,\partial t'}[R_{\mathbf{u}_c^n}(\mathbf{x},\mathbf{x}',t,t')] \quad ,\end{aligned} \tag{206}$$

$$\begin{aligned}[R_{\mathbf{w}_c^n}(\mathbf{x},\mathbf{x}',t,t')] &= E\{\mathbf{w}_c^n(\mathbf{x},t)\,\mathbf{w}_c^n(\mathbf{x}',t')^T\} \\ &= \frac{\partial^2}{\partial t\,\partial t'}[R_{\mathbf{v}_c^n}(\mathbf{x},\mathbf{x}',t,t')] \quad .\end{aligned} \tag{207}$$

1- Expression of the autocorrelation function. Using Eqs. (15) and (170), Eq. (203) can be rewritten as

$$[R_{\mathbf{U}_c}(t,t')] = \int_{\mathbb{R}}\int_{\mathbb{R}} [\,h(\tau)\,]\,[\underline{R}_{\mathbf{F}_c}(t-\tau,t'-\tau')]\,[\,h(\tau')\,]^T\,d\tau\,d\tau' \quad . \tag{208}$$

It should be noted that in Eq. (208), impulse function $[\,h(\tau)\,]$ requires calculation of the Fourier transform of frequency response function $[\,T(\omega)\,]$ which is the quantity constructed in the proposed methodology.

2- First expression of the autocorrelation function in the frequency domain. It can be proved [15] that $[R_{\mathbf{U}_c}(t,t')]$ is a square integrable function on $\Theta\times\Theta$. Substituting Eq. (173) into (208) and using Eq. (33) yields

$$[R_{\mathbf{U}_c}(t,t')] = \int_{\mathbb{R}}\int_{\mathbb{R}} e^{i(\omega t-\omega' t')}\,[\underline{\widetilde{R}}_{\mathbf{U}_c}(\omega,\omega')]\,d\omega\,d\omega' \quad , \tag{209}$$

in which $[\underline{\widetilde{R}}_{\mathbf{U}_c}(\omega,\omega')]$ is a square integrable function on $\mathbb{R}\times\mathbb{R}$, such that

$$[\underline{\widetilde{R}}_{\mathbf{U}_c}(\omega,\omega')] = [T(\omega)]\,[\underline{\widetilde{R}}_{\mathbf{F}_c}(\omega,\omega')]\,[T(\omega')]^* \quad , \tag{210}$$

where $[T(\omega)]$ is the $(n\times n)$ complex symmetric matrix defined by Eq. (25) and $[\underline{\widetilde{R}}_{\mathbf{F}_c}(\omega,\omega')]$ is given by Eq. (172) or Eq. (182).

3- Second expression in the frequency domain. Substituting Eq. (182) into Eq. (210) and using Eq. (209) yields [16]

$$[R_{\mathbf{U}_c}(t,t')] = \int_{\mathbb{R}} e^{i\omega(t-t')}\,[Q_{\mathbf{U}_c}(t,\omega)]\,[S(\omega)]\,[Q_{\mathbf{U}_c}(t',\omega)]^*\,d\omega \quad , \tag{211}$$

$$[Q_{\mathbf{U}_c}(t,\omega)] = \int_{\mathbb{R}} e^{i(v-\omega)t}\,[T(v)]\,[\widetilde{Q}_{\mathbf{F}_c}(v-\omega,\omega)]\,dv \quad . \tag{212}$$

---

[15] Since $[\,h\,]$ is an integrable function on $\mathbb{R}$, i.e. $\int_{\mathbb{R}} |[\,h(t)\,]_{jk}|\,dt = \int_0^{+\infty} |[\,h(t)\,]_{jk}|\,dt < +\infty$, and since $[\underline{R}_{\mathbf{F}_c}(t,t')]$ is square integrable on $\mathbb{R}\times\mathbb{R}$, then $[R_{\mathbf{U}_c}(t,t')]$ is a square integrable function on $\Theta\times\Theta$.

[16] Eqs. (211) and (212) can be obtained by substituting the general integral representation $\mathbf{F}_c(t) = \int_{\mathbb{R}} e^{i\omega t}[Q_{\mathbf{F}_c}(t,\omega)]\,\mathbf{H}(d\omega)$ introduced in footnote 11 into Eq. (194). We obtain $\mathbf{U}_c(t) = \int_{\mathbb{R}} e^{i\omega t}[Q_{\mathbf{U}_c}(t,\omega)]\,\mathbf{H}(d\omega)$ with $[Q_{\mathbf{U}_c}(t,\omega)] = \int_0^t e^{-i\omega t'}\,[\,h(t')\,]\,[Q_{\mathbf{F}_c}(t-t',\omega)]\,dt'$. Substituting the second Eq. (183) into this last equality yields Eq. (212). It should be noted that in this last equality, impulse function $[\,h(\tau)\,]$ requires calculation of the Fourier transform of frequency response function $[\,T(\omega)\,]$ which is the quantity constructed in the proposed methodology.

These equations, which are directly deduced from the general integral representation theory, are not suited to calculation of structures described by a large number of degrees of freedom. In effect, Eq. (212) shows that, in order to calculate $[Y(v,\omega)] = [T(v)]\,[\widetilde{Q}_{\mathbf{F}_c}(v-\omega,\omega)]$, it is necessary to solve $[A(v)]\,[Y(v,\omega)] = [\widetilde{Q}_{\mathbf{F}_c}(v-\omega,\omega)]$ for a large number of second members, due to the presence of parameter $\omega \in \mathbb{R}$.

6.3. Simplified calculation of the nonstationary random fluctuation of the response

In this section, we present a simplified calculation corresponding to two approximations. The first is that the values of function $[\underline{\widetilde{R}}_{\mathbf{F}_c}(\omega,\omega')]$ are negligible outside a domain $\mathbb{B}_{\text{LF}} \times \mathbb{B}_{\text{LF}}$ in which $\mathbb{B}_{\text{LF}}$ is a low-frequency band. Secondly, the quasistatic correction terms are not introduced. Consequently, we use the formulation introduced in Chapter VI, i.e. from Eq. (VI.58), $[T_{N,\,\text{acc}}(\omega)] = [T_N(\omega)] + [T_{N,\,\text{stat}}]$ in which $[T_{N,\,\text{stat}}]$ is neglected. We then have

$$[T_{N,\,\text{acc}}(\omega)] = [T_N(\omega)] \quad , \tag{213}$$

in which $[T_N(\omega)]$ is given by Eq. (VI.30),

$$[T_N(\omega)] = \sum_{\alpha=1}^{N}\sum_{\beta=1}^{N} [\mathcal{T}(\omega)]_{\alpha\beta}\,\mathbf{U}_\alpha\,\mathbf{U}_\beta^T \quad . \tag{214}$$

Replacing $[T(\omega)]$ by $[T_N(\omega)]$ in Eq. (210) yields

$$[\underline{\widetilde{R}}_{\mathbf{U}_c}(\omega,\omega')] \simeq \sum_{\alpha,\beta=1}^{N}\sum_{\alpha',\beta'=1}^{N} [\mathcal{T}(\omega)]_{\alpha\beta}\,\overline{[\mathcal{T}(\omega')]}_{\alpha'\beta'}\,\{\mathbf{U}_\beta^T\,[\underline{\widetilde{R}}_{\mathbf{F}_c}(\omega,\omega')]\,\mathbf{U}_{\beta'}\}\,\mathbf{U}_\alpha\,\mathbf{U}_{\alpha'}^T \quad . \tag{215}$$

Finally, Eq. (209) gives

$$[R_{\mathbf{U}_c}(t,t')] = \int_{\mathbb{R}}\int_{\mathbb{R}} e^{i(\omega t-\omega' t')}\,[\underline{\widetilde{R}}_{\mathbf{U}_c}(\omega,\omega')]\,d\omega\,d\omega' \quad . \tag{216}$$

Remark 1. If the data is $[R_{\mathbf{F}_c}(t,t')]$ on $\Theta\times\Theta$, then $[\underline{\widetilde{R}}_{\mathbf{F}_c}(\omega,\omega')]$ is calculated using Eq. (172). We then calculate $[\underline{\widetilde{R}}_{\mathbf{U}_c}(\omega,\omega')]$ using Eq. (215) and finally, $[R_{\mathbf{U}_c}(t,t')]$ is deduced from Eq. (216).

Remark 2. If $[R_{\mathbf{F}_c}(t,t')]$ is defined by Eq. (174), i.e. if the data are $[\widetilde{Q}_{\mathbf{F}_c}(\omega,v)]$ or $[Q_{\mathbf{F}_c}(t,v)]$ (see Eq. (183)) and $[S(v)]$, then $[\underline{\widetilde{R}}_{\mathbf{F}_c}(\omega,\omega')]$ can be expressed by Eq. (182) and consequently, Eq. (215) is used with

$$\{\mathbf{U}_\beta^T\,[\underline{\widetilde{R}}_{\mathbf{F}_c}(\omega,\omega')]\,\mathbf{U}_{\beta'}\} = \int_{\mathbb{R}} \mathbf{X}_\beta(\omega,v)^*\,[S(v)]\,\mathbf{X}_{\beta'}(\omega',v)\,dv \quad , \tag{217}$$

$$\mathbf{X}_\beta(\omega,v) = [\widetilde{Q}_{\mathbf{F}_c}(\omega-v,v)]^*\,\mathbf{U}_\beta \quad . \tag{218}$$

It should be noted that Eq. (215) with Eqs. (217) and (218) can also be deduced from Eqs. (211), (212) and (214).

6.4. Calculation of the nonstationary random fluctuation of the response

In this section, we consider the case for which the values of function $[\underline{\widetilde{R}}_{\mathbf{F}_c}(\omega,\omega')]$ are negligible outside a domain $\{\mathbb{B}_{\mathrm{LF}}\cup\mathbb{B}_{\mathrm{MF}}\}\times\{\mathbb{B}_{\mathrm{LF}}\cup\mathbb{B}_{\mathrm{MF}}\}$ in which $\mathbb{B}_{\mathrm{LF}}$ and $\mathbb{B}_{\mathrm{MF}}$ are low- and medium-frequency bands. In this case, a reduced representation of the random fluctuation of stochastic excitation is used.

Reduced representation of the random fluctuation of the stochastic excitation. In this subsection, we use the expansion of autocorrelation function $[R_{\mathbf{F}_c}(t,t')]$ on its eigenfunctions to construct a reduced expression $[R_{\mathbf{F}_{c,M}}(t,t')]$ of $[R_{\mathbf{F}_c}(t,t')]$. The reduced expression obtained is used to calculate an approximation $[R_{\mathbf{U}_{c,M}}(t,t')]$ of autocorrelation function $[R_{\mathbf{U}_c}(t,t')]$ of nonstationary random fluctuation $\mathbf{U}_c$ of the response.

1- Expansion of the autocorrelation function. Since $[R_{\mathbf{F}_c}(t,t')]$ is square integrable on $\Theta\times\Theta$ (Section 6.1),

$$\mu = \int_\Theta\int_\Theta \|R_{\mathbf{F}_c}(t,t')\|_F^2\,dt\,dt' < +\infty \quad , \tag{219}$$

it can be shown [17] that, for all t and t' in Θ,

$$[R_{\mathbf{F}_c}(t,t')] = \sum_{m=1}^{+\infty} \lambda_m\,\mathbf{X}_m(t)\,\mathbf{X}_m(t')^T \quad , \tag{220}$$

The linear integral operator $\mathbf{R}$ defined by the kernel $[R_{\mathbf{F}_c}(t,t')]$ on $\Theta\times\Theta$ is a symmetric and positive Hilbert-Schmidt operator on $L^2(\Theta,\mathbb{R}^n)$ (due to the general properties of a matrix-valued autocorrelation function and the square integrability of $[R_{\mathbf{F}_c}(t,t')]$). Since $\mathbf{R}$ is a Hilbert-Schmidt operator, $\mathbf{R}$ is compact, its spectrum is countable, each eigenvalue has a finite multiplicity and Eq. (223) holds. For more mathematical developments, see for instance Dautray and Lions, 1992; Guikhman and Skorokhod, 1979; Soize, 1993a, p. 151 and p. 419.

in which for each m, λ_m and $\mathbf{X}_m$ satisfy the following eigenvalue problem

$$\int_\Theta [R_{\mathbf{F}_c}(t,t')]\,\mathbf{X}_m(t')\,dt' = \lambda_m\,\mathbf{X}_m(t) \quad , \quad \forall t \in \Theta \quad . \tag{221}$$

The spectrum $\lambda_1, \lambda_2, \ldots$ is countable and the eigenvalues λ_m are positive real numbers such that

$$\lambda_1 \geq \lambda_2 \geq \ldots \to 0 \quad , \tag{222}$$

$$\mu = \sum_{m=1}^{+\infty} \lambda_m^2 < +\infty \quad . \tag{223}$$

The eigenfunctions $\{\mathbf{X}_m\}_m$ form a complete set of square integrable functions from Θ into $\mathbb{R}^n$ and satisfy the orthogonality condition [18]

$$\int_\Theta \mathbf{X}_m(t)^T\,\mathbf{X}_{m'}(t)\,dt = \delta_{mm'} \quad . \tag{224}$$

The expansion of the autocorrelation function defined by Eq. (220) allows construction of the Karhunen-Loeve expansion [19] of nonstationary stochastic process $\mathbf{F}_c$ indexed by finite interval Θ.

2- Construction of a reduced representation of the autocorrelation function. The reduced representation $[R_{\mathbf{F}_{c,M}}(t,t')]$ of $[R_{\mathbf{F}_c}(t,t')]$ is defined, for all t and t' in Θ, by

$$[R_{\mathbf{F}_{c,M}}(t,t')] = \sum_{m=1}^{M} \lambda_m\,\mathbf{X}_m(t)\,\mathbf{X}_m(t')^T \quad , \tag{225}$$

in which M is a positive integer which can be calculated as follows. For any fixed $\varepsilon > 0$, there exists an integer M such that

$$0 \leq \mu - \sum_{m=1}^{M} \lambda_m^2 \leq \varepsilon \quad . \tag{226}$$

Since $\mu = \sum_{m=1}^{+\infty} \lambda_m^2$, the difference $\mu - \sum_{m=1}^{M} \lambda_m^2$ appearing in Eq. (226) is equal to $\sum_{m=M+1}^{\infty} \lambda_m^2$ which can be estimated without calculating the

---

[18] The eigenfunctions $\{\mathbf{X}_m\}_m$ form a Hilbert basis of $L^2(\Theta,\mathbb{R}^n)$.

[19] The Karhunen-Loeve expansion is written as $\mathbf{F}_c(t)=\sum_{m=1}^{+\infty} Z_m\,\mathbf{X}_m(t)$ in which $Z_1,Z_2,\ldots$ are independent second-order centered real-valued random variables such that $E\{Z_m Z_{m'}\}=\lambda_m \delta_{mm'}$ (see for instance Guikhman and Skorokhod, 1979; Soize, 1993a, p. 417).

eigenvalues λ_m for $m \geq M+1$. Equation (226) allows order M of the representation to be calculated. It should be noted that, if for ε fixed, Eq. (226) implies that $M \ll n$, then Eq. (225) defines a reduced representation. Consequently, we have to calculate only the first M eigenvalues and eigenvectors of Eq. (221).

3- Construction of a reduced representation of the autocorrelation function in the frequency domain. From Eq. (172), the approximation $[\underline{\widetilde{R}}_{\mathbf{F}_{c,M}}(\omega,\omega')]$ of $[\underline{\widetilde{R}}_{\mathbf{F}_c}(\omega,\omega')]$ is given by

$$[\underline{\widetilde{R}}_{\mathbf{F}_{c,M}}(\omega,\omega')] = \frac{1}{4\pi^2}\int_\Theta\int_\Theta e^{-i(\omega t-\omega' t')}\,[R_{\mathbf{F}_{c,M}}(t,t')]\,dt\,dt' \quad . \tag{227}$$

Substituting Eq. (225) into Eq. (227) yields

$$[\underline{\widetilde{R}}_{\mathbf{F}_{c,M}}(\omega,\omega')] = \sum_{m=1}^{M} \lambda_m\,\widetilde{\mathbf{X}}_m(\omega)\,\widetilde{\mathbf{X}}_m(\omega')^* \quad , \tag{228}$$

in which

$$\widetilde{\mathbf{X}}_m(\omega) = \frac{1}{2\pi}\int_\Theta e^{-i\omega t}\,\mathbf{X}_m(t)\,dt \quad . \tag{229}$$

For every m, $\mathbb{C}^n$-valued function $\widetilde{\mathbf{X}}_m(\omega)$ is square integrable on $\mathbb{R}$.

Calculation of the nonstationary random fluctuation of the response. The approximation $[\underline{\widetilde{R}}_{\mathbf{U}_{c,M}}(\omega,\omega')]$ of $[\underline{\widetilde{R}}_{\mathbf{U}_c}(\omega,\omega')]$ defined by Eq. (210), is given by

$$[\underline{\widetilde{R}}_{\mathbf{U}_{c,M}}(\omega,\omega')] = [T(\omega)]\,[\underline{\widetilde{R}}_{\mathbf{F}_{c,M}}(\omega,\omega')]\,[T(\omega')]^* \quad . \tag{230}$$

Substituting Eq. (228) into Eq. (230) yields

$$[\underline{\widetilde{R}}_{\mathbf{U}_{c,M}}(\omega,\omega')] = \sum_{m=1}^{M} \lambda_m\,\widetilde{\mathbf{Y}}_m(\omega)\,\widetilde{\mathbf{Y}}_m(\omega')^* \quad , \tag{231}$$

in which $\widetilde{\mathbf{Y}}_m(\omega)$ is a $\mathbb{C}^n$-valued square integrable function on $\mathbb{R}$ such that

$$\widetilde{\mathbf{Y}}_m(\omega) = [T(\omega)]\,\widetilde{\mathbf{X}}_m(\omega) \quad . \tag{232}$$

Since $\widetilde{\mathbf{X}}_m(\omega)$ is a square integrable function on $\mathbb{R}$, $\widetilde{\mathbf{Y}}_m(\omega)$ is constructed using the methodology presented in Section 4. For all t and t' in Θ, the approximation $[R_{\mathbf{U}_{c,M}}(t,t')$ of $[R_{\mathbf{U}_c}(t,t')]$ defined by Eq. (209), is given by

$$[R_{\mathbf{U}_{c,M}}(t,t')] = \int_{\mathbb{R}}\int_{\mathbb{R}} e^{i(\omega t-\omega' t')}\,[\underline{\widetilde{R}}_{\mathbf{U}_{c,M}}(\omega,\omega')]\,d\omega\,d\omega' \quad . \tag{233}$$

Substituting Eq. (231) into Eq. (233) yields

$$[R_{\mathbf{U}_{c,M}}(t,t')] = \sum_{m=1}^{M} \lambda_m \, \mathbf{Y}_m(t) \, \mathbf{Y}_m(t')^T \quad , \quad \forall (t,t') \in \Theta \times \Theta \quad , \tag{234}$$

in which, for all t in Θ,

$$\mathbf{Y}_m(t) = \int_{\mathbb{R}} e^{i\omega t} \, \widetilde{\mathbf{Y}}_m(\omega) \, d\omega \quad . \tag{235}$$

6.5. Bibliographical comments

For further details on nonstationary stochastic processes, their transformations (linear integral transformations, derivatives, etc.), nonstationary random vibration and nonstationary stochastic dynamics, we refer the reader to Guikhman and Skorokhod, 1979; Kree and Soize, 1983 and 1986; Lin, 1967; Priestley, 1981 and 1988; Soize, 1988, 1993a, 1994; Soong, 1973.

CHAPTER X

Linear Acoustic Equations

1. Introduction

This chapter is devoted to the derivation of the basic equations of linear acoustics in the context of structural-acoustic problems. The fluid is assumed to be homogeneous and compressible. In the reference configuration, the fluid is at rest. The fluid is either a gas or a liquid and gravity effects are neglected. Such a fluid is called an *acoustic fluid.* For more detailed developments on fluid mechanics and acoustics, the reader is referred to Chorin and Marsden, 1993; Germain, 1973 and 1986; Landau and Lifchitz, 1992b; Lighthill, 1978; Morse and Ingard, 1968; Pierce, 1989.

In Section 2, we establish the equations for an *inviscid acoustic fluid.* This model will be used for the external acoustic problem (Chapter XII).

In Section 3, we construct the equations for a *dissipative acoustic fluid* introducing a small internal acoustic dissipation in the internal cavity and a wall dissipation modeled by an impedance boundary condition. This dissipative acoustic fluid model will be used for the internal acoustic problem.

2. Inviscid Acoustic Fluid

2.1. Basic assumptions

Physical assumptions. We consider low amplitude vibrations of an inviscid homogeneous compressible fluid around a static equilibrium position taken as the reference configuration. We neglect gravity effects [1] to establish the linear acoustic equations. In this reference configuration, the fluid is at rest and occupies a fixed three-dimensional domain Ω referred to a

---

[1] It should be noted that gravity effects are taken into account to determine the equilibrium configuration which defines the geometry of the fixed reference domain Ω (the horizontal free surface for a liquid subjected to the Earth's gravity field, etc.).

cartesian reference system $(\mathbf{i}, \mathbf{j}, \mathbf{k})$. We denote the generic point of $\Omega \subset \mathbb{R}^3$ as $\mathbf{x} = (x_1, x_2, x_3)$. Domain Ω has a smooth boundary $\partial\Omega$ and the outward unit normal to $\partial\Omega$ is denoted as $\mathbf{n}$ (see Fig. 1).

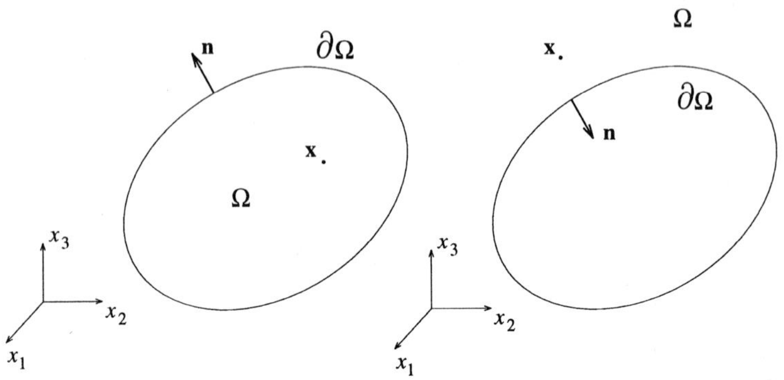

a- Bounded domain

b- Unbounded domain

Fig. 1. Geometrical configuration

Geometry of domain Ω. In this book, two cases are considered. Either domain Ω is a bounded simply connected domain [2] of $\mathbb{R}^3$ (see Fig. 2) or

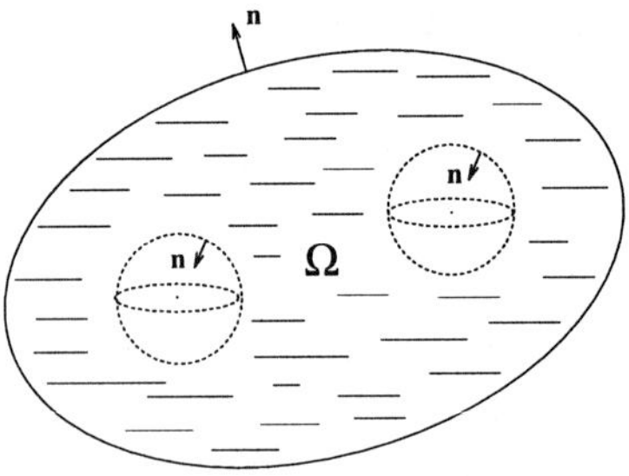

Fig. 2. Case of a bounded simply connected domain Ω of $\mathbb{R}^3$

domain Ω is an unbounded simply connected domain whose complement $\Omega' = \mathbb{R}^3 \backslash \Omega$ is a bounded domain (see Fig. 3). Cases not considered herein are shown in Fig. 4-a (torus immersed in an external unbounded fluid) and Fig. 4-b (torus containing an internal bounded fluid).

---

[2] A domain Ω is called simply connected if any continuous closed curve in Ω can be continuously shrunk to a point without leaving Ω.

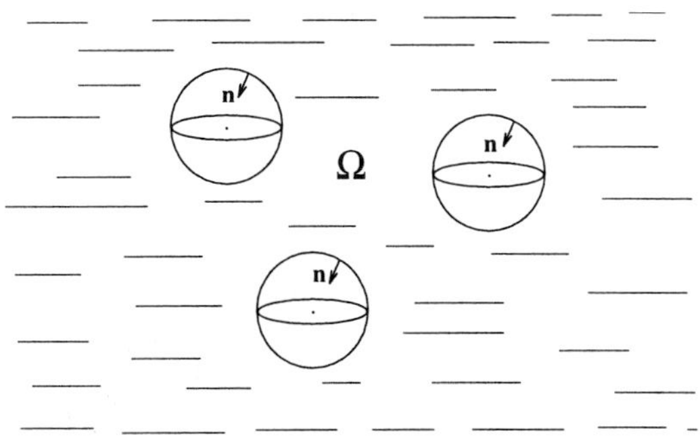

Fig. 3. Case of an unbounded simply connected domain Ω of $\mathbb{R}^3$ (with bounded complement)

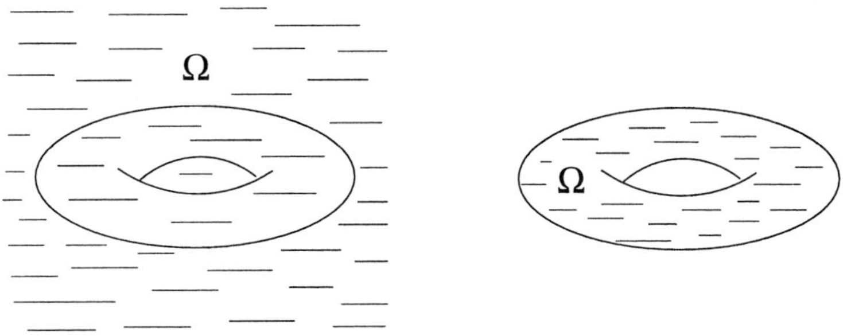

a- Unbounded domain Ω b- Bounded domain Ω

Fig. 4. Cases of connected but not simply connected domains Ω of $\mathbb{R}^3$

Hypotheses of linearization around a reference configuration of the fluid at rest. Let ρ_0 be the mass density and p_0 be the pressure at equilibrium in the reference configuration Ω. Since the fluid is homogeneous and the gravity effects are neglected to establish the linear acoustic equations, ρ_0 and p_0 are constant. Let $\rho(\mathbf{x}, t)$ and $p(\mathbf{x}, t)$ be the disturbances of the mass density and pressure [3] from their undisturbed values ρ_0 and p_0 respectively. In the context of the linearized theory, it is assumed that the mass density and pressure disturbances, the velocity and their gradients are small and of the same order (in particular, $\rho \ll \rho_0$, $p \ll p_0$). In reference system $(\mathbf{i}, \mathbf{j}, \mathbf{k})$, the velocity field and the corresponding displacement

Since gravity effects are neglected in establishing the linear acoustic equations, there is only one definition of the pressure disturbance. In the other case, see Wilcox, 1984; Morand and Ohayon, 1995, p. 33.

field of a particle located in point $\mathbf{x}$ of Ω and at time t are denoted as $\mathbf{v}(\mathbf{x},t) = (v_1(\mathbf{x},t), v_2(\mathbf{x},t), v_3(\mathbf{x},t))$ and $\mathbf{u}(\mathbf{x},t) = (u_1(\mathbf{x},t), u_2(\mathbf{x},t), u_3(\mathbf{x},t))$ respectively ($\mathbf{v}(\mathbf{x},t) = \partial_t \mathbf{u}(\mathbf{x},t)$).

2.2. Linearized equations in the time domain

Linearized mass conservation equation in the context of linear acoustics. Following Lighthill, 1978, we assume that there are distributed monopole sources located in a source region included in fluid domain Ω. These sources cause the emergence of new mass at rate $Q(\mathbf{x},t)$ per unit volume per unit time in the source region. Quantity $Q(\mathbf{x},t)$ is called the *acoustic source density*. Then $\partial_t Q(\mathbf{x},t)$ represents the source strength per unit volume. It should be noted that $Q(\mathbf{x},t)$ is equal to zero except in the source region where fluid is being created. In the context of the linear acoustic theory, the linearized mass conservation equation around the reference configuration is then written as (Lighthill, 1978)

$$\partial_t \rho(\mathbf{x},t) + \rho_0 \nabla \cdot \mathbf{v}(\mathbf{x},t) = Q(\mathbf{x},t) \quad . \tag{1}$$

It should be noted that outside the source region or if there are no distributed monopole sources, Eq. (1) gives the usual linearized mass conservation equation $\partial_t \rho(\mathbf{x},t) + \rho_0 \nabla \cdot \mathbf{v}(\mathbf{x},t) = 0$.

Linearized Euler equation. The linearized Euler equation around the reference configuration is written as

$$\rho_0 \, \partial_t \mathbf{v}(\mathbf{x},t) + \nabla p(\mathbf{x},t) = \mathbf{0} \quad . \tag{2}$$

Constitutive law. Let c_0 be the constant speed of sound in the fluid at equilibrium. In the context of the linearized theory, the constitutive law of a barotropic fluid relating the pressure disturbance to the mass density disturbance is written as

$$p(\mathbf{x},t) = c_0^2 \, \rho(\mathbf{x},t) \quad . \tag{3}$$

It should be noted that the linearization hypotheses are justified if

$$\|\mathbf{v}\| \ll c_0 \quad . \tag{4}$$

Equations in terms of the pressure and velocity fields in the time domain. These equations include Eq. (2) and the equation resulting from the elimination of $\rho(\mathbf{x},t)$ between Eqs. (1) and (3). We obtain

$$\rho_0 \, \partial_t \mathbf{v}(\mathbf{x},t) + \nabla p(\mathbf{x},t) = \mathbf{0} \quad , \tag{5}$$

$$\partial_t p(\mathbf{x},t) = -\rho_0 \, c_0^2 \; \nabla \cdot \mathbf{v}(\mathbf{x},t) + c_0^2 \, Q(\mathbf{x},t) \quad . \tag{6}$$

2.3. Linearized equations in the frequency domain

Fourier transform. To obtain the equations in the frequency domain, we introduce the Fourier transform for various quantities, using the same symbol for a quantity and its Fourier transform (see Mathematical Notations in the appendix). For all fixed $\mathbf{x}$, displacement field $\mathbf{u}(\mathbf{x}, t)$, velocity field $\mathbf{v}(\mathbf{x}, t)$, pressure field $p(\mathbf{x}, t)$ and source term $Q(\mathbf{x}, t)$ have the following Fourier transforms with respect to t

$$\mathbf{u}(\mathbf{x}, \omega) = \int_{\mathbb{R}} e^{-i\omega t}\, \mathbf{u}(\mathbf{x}, t)\, dt \quad , \quad \mathbf{v}(\mathbf{x}, \omega) = \int_{\mathbb{R}} e^{-i\omega t}\, \mathbf{v}(\mathbf{x}, t)\, dt \quad ,$$

$$p(\mathbf{x}, \omega) = \int_{\mathbb{R}} e^{-i\omega t}\, p(\mathbf{x}, t)\, dt \quad , \quad Q(\mathbf{x}, \omega) = \int_{\mathbb{R}} e^{-i\omega t}\, Q(\mathbf{x}, t)\, dt \quad . \tag{7}$$

We assume that $Q(\mathbf{x}, \omega)$, $\boldsymbol{\nabla} Q(\mathbf{x}, \omega)$ and $\boldsymbol{\nabla}^2 Q(\mathbf{x}, \omega)$ are square integrable functions on $\mathbb{R}$ with respect to ω and are such that, for all $\mathbf{x}$ in Ω,

$$\lim_{\omega \to 0} \frac{Q(\mathbf{x}, \omega)}{\omega^2} = 0 \ , \ \lim_{\omega \to 0} \frac{\boldsymbol{\nabla} Q(\mathbf{x}, \omega)}{\omega^2} = \mathbf{0} \ , \ \lim_{\omega \to 0} \frac{\boldsymbol{\nabla}^2 Q(\mathbf{x}, \omega)}{\omega^2} = 0 \ . \tag{8}$$

Equations in terms of the pressure and velocity fields in the frequency domain. Taking the Fourier transform of Eqs. (5) and (6) and using Eq. (7), we obtain

$$i\omega\, \rho_0\, \mathbf{v}(\mathbf{x}, \omega) + \boldsymbol{\nabla} p(\mathbf{x}, \omega) = \mathbf{0} \quad , \tag{9}$$

$$i\omega\, p(\mathbf{x}, \omega) = -\rho_0\, c_0^2\, \boldsymbol{\nabla} \cdot \mathbf{v}(\mathbf{x}, \omega) + c_0^2\, Q(\mathbf{x}, \omega) \quad . \tag{10}$$

Equations in terms of the pressure and displacement fields in the frequency domain. Substituting $\mathbf{v}(\mathbf{x}, \omega) = i\omega\, \mathbf{u}(\mathbf{x}, \omega)$ in Eqs. (9) and (10) yields

$$-\omega^2\, \rho_0\, \mathbf{u}(\mathbf{x}, \omega) + \boldsymbol{\nabla} p(\mathbf{x}, \omega) = \mathbf{0} \quad , \tag{11}$$

$$p(\mathbf{x}, \omega) = -\rho_0\, c_0^2\, \boldsymbol{\nabla} \cdot \mathbf{u}(\mathbf{x}, \omega) + c_0^2\, \frac{Q(\mathbf{x}, \omega)}{i\omega} \quad . \tag{12}$$

2.4. Boundary conditions in the frequency domain

Wall boundary condition. Let Γ be a part of the boundary $\partial\Omega$ on which a displacement field $\mathbf{u}_{\text{wall}}(\mathbf{x}, t)$ is prescribed. In the frequency domain this field is written as $\mathbf{u}_{\text{wall}}(\mathbf{x}, \omega)$. In the context of the linearized theory and since the fluid is inviscid, the wall boundary condition is written in the time domain as

$$\mathbf{v}(\mathbf{x}, t) \cdot \mathbf{n}(\mathbf{x}) = \partial_t\, \mathbf{u}_{\text{wall}}(\mathbf{x}, t) \cdot \mathbf{n}(\mathbf{x}) \quad , \quad \forall\, \mathbf{x} \in \Gamma \quad . \tag{13}$$

In the frequency domain, this equation is written in terms of **v** or **u** as

$$\mathbf{v}(\mathbf{x},\omega)\cdot\mathbf{n}(\mathbf{x}) = i\omega\,\mathbf{u}_{\text{wall}}(\mathbf{x},\omega)\cdot\mathbf{n}(\mathbf{x}) \quad , \quad \forall\mathbf{x}\in\Gamma \tag{14}$$

$$\mathbf{u}(\mathbf{x},\omega)\cdot\mathbf{n}(\mathbf{x}) = \mathbf{u}_{\text{wall}}(\mathbf{x},\omega)\cdot\mathbf{n}(\mathbf{x}) \quad , \quad \forall\mathbf{x}\in\Gamma \quad . \tag{15}$$

Wall impedance boundary condition. Let Γ_Z be a part of the boundary $\partial\Omega$ having acoustical properties modeled by a *wall acoustic impedance* $Z(\mathbf{x},\omega)$ defined for $\mathbf{x}\in\Gamma_Z$, with complex values. A displacement field is prescribed on Γ_Z and written in the frequency domain as $\mathbf{u}_{\text{wall}}(\mathbf{x},\omega), \mathbf{x}\in\Gamma_Z$. The wall impedance boundary condition is written as

$$p(\mathbf{x},\omega) = Z(\mathbf{x},\omega)\left\{\mathbf{v}(\mathbf{x},\omega)\cdot\mathbf{n}(\mathbf{x}) - i\omega\,\mathbf{u}_{\text{wall}}(\mathbf{x},\omega)\cdot\mathbf{n}(\mathbf{x})\right\} , \ \forall\mathbf{x}\in\Gamma_Z \ . \tag{16}$$

Wall acoustic impedance $Z(\mathbf{x},\omega)$ must satisfy appropriate conditions in order to ensure that the problem is correctly stated. For this purpose, $Z(\mathbf{x},\omega)$ is written as

$$Z(\mathbf{x},\omega) = Z_R(\mathbf{x},\omega) + i\,Z_I(\mathbf{x},\omega) \quad , \tag{17}$$

in which the real part $Z_R(\mathbf{x},\omega)$ and the imaginary part $Z_I(\mathbf{x},\omega)$ are the *acoustic resistance* and *acoustic reactance* respectively. For all $\omega\in\mathbb{R}$, function $\mathbf{x}\mapsto 1/\{i\omega\,Z(\mathbf{x},\omega)\}$ is bounded on Γ_Z,

$$\frac{1}{|i\omega\,Z(\mathbf{x},\omega)|} \leq \text{Const}(\omega) \quad , \quad \forall\mathbf{x}\in\Gamma_Z \quad , \tag{18}$$

in which $\text{Const}(\omega)$ is a finite positive constant independent of $\mathbf{x}$. For all $\mathbf{x}$ in Γ_Z and for all ω in $\mathbb{R}$,

$$Z_R(\mathbf{x},-\omega) = Z_R(\mathbf{x},\omega) \quad , \quad Z_I(\mathbf{x},-\omega) = -Z_I(\mathbf{x},\omega) \quad , \tag{19}$$

$$Z_R(\mathbf{x},\omega) > 0 \quad , \quad -\omega\,Z_I(\mathbf{x},\omega) \geq 0 \quad . \tag{20}$$

For all $\mathbf{x}$ in Γ_Z, real function $\omega\mapsto Z_I(\mathbf{x},\omega)$ is continuous on $\mathbb{R}-\{0\}$ and

$$\lim_{\omega\to 0}\{-\omega\,Z_I(\mathbf{x},\omega)\} = \alpha(\mathbf{x}) \geq \alpha_{\text{min}} > 0 \quad , \quad \forall\mathbf{x}\in\Gamma_Z \quad , \tag{21}$$

in which α_{min} is a given real positive constant and $\mathbf{x}\mapsto\alpha(\mathbf{x})$ is a positive-valued function defined on Γ_Z. Eq. (21) means that

$$Z_I(\mathbf{x},\omega) \sim -\frac{\alpha(\mathbf{x})}{\omega} \quad \text{as} \quad \omega\to 0 \quad . \tag{22}$$

For all $\mathbf{x}$ in Γ_Z, real function $\omega \mapsto Z_R(\mathbf{x}, \omega)$ is continuous on $\mathbb{R}$ and consequently,

$$\lim_{\omega \to 0} \{\omega \, Z_R(\mathbf{x}, \omega)\} = 0 \quad , \quad \forall \mathbf{x} \in \Gamma_Z \quad . \tag{23}$$

From Eq. (20), we deduce that

$$Z(\mathbf{x}, \omega) \neq 0 \quad , \quad \forall \mathbf{x} \in \Gamma_Z \quad , \quad \forall \omega \in \mathbb{R} \quad , \tag{24}$$

and from Eqs. (21) and (23), we deduce that

$$\{i\omega \, Z(\mathbf{x}, \omega)\}_{\omega=0} = \{-\omega \, Z_I(\mathbf{x}, \omega)\}_{\omega=0} = \alpha(\mathbf{x}) > 0 \quad . \tag{25}$$

The assumption defined by Eq. (20) corresponds to an acoustic impedance stiffness-controlled over the entire frequency band (see Fig. 5). This can be useful for modeling an acoustic material coating whose thickness is neglected.

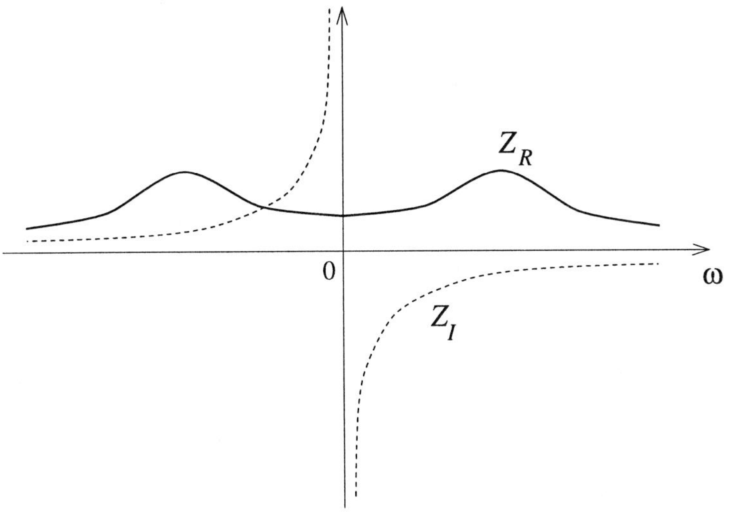

Fig. 5. Acoustic resistance Z_R (solid line) and acoustic reactance Z_I (dashed line) for a wall acoustic impedance Z stiffness-controlled over the frequency band

It should be noted that the sign of $Z_I(\mathbf{x}, \omega)$ depends on the Fourier transform convention used. With the convention defined by Eq. (7), $Z_I(\mathbf{x}, \omega) < 0$ for $\omega > 0$. If the other convention were used, then we would have $Z_I(\mathbf{x}, \omega) > 0$ for $\omega > 0$. Acoustic resistance $Z_R(\mathbf{x}, \omega)$ does not depend on the Fourier transform convention.

Zero pressure condition on a part of the boundary. Let Γ_0 be a part of the boundary $\partial\Omega$ submitted to a zero pressure field. We have

$$p(\mathbf{x},\omega) = 0 \quad , \quad \forall \mathbf{x} \in \Gamma_0 \quad . \tag{26}$$

2.5. Introduction of velocity and displacement potentials

Irrotationality and existence of a velocity potential. For $\omega \neq 0$, applying operator $(\nabla\times)$ to the two sides of Eq. (9) yields $i\omega\,\rho_0\,\nabla\times\mathbf{v}(\mathbf{x},\omega) = \mathbf{0}$ and gives the irrotationality condition

$$\nabla \times \mathbf{v}(\mathbf{x},\omega) = \mathbf{0} \quad . \tag{27}$$

Since Ω is a simply connected domain [4], then Eq. (27) shows that the velocity field can be written as

$$\mathbf{v}(\mathbf{x},\omega) = \nabla\psi(\mathbf{x},\omega) \quad , \tag{28}$$

in which $\psi(\mathbf{x},\omega)$ is the velocity potential defined to within an additive constant. It should be noted that for $\omega = 0$ there is a nonunique solution for the velocity field. Among those solutions, we select the solution satisfying Eq. (28) to ensure the continuity of the unique solution of the structural-acoustic equations for $\omega \neq 0$, as $\omega \to 0$. Consequently, Eq. (28) is used for all real ω. As ψ is defined to within an additive constant, we introduce a linear constraint on ψ,

$$\ell(\psi) = 0 \quad \text{with} \quad \ell(1) \neq 0 \quad , \tag{29}$$

in which ℓ is any linear form [5] on the space of functions ψ. For instance, we can choose $\ell(\psi)$ such that $\ell(\psi) = 0$ is written as

$$\int_\Omega \psi(\mathbf{x},\omega)\,d\mathbf{x} = 0 \quad . \tag{30}$$

Introduction of a displacement potential. We introduce the displacement potential $\varphi(\mathbf{x},\omega)$ such that

$$\psi(\mathbf{x},\omega) = i\omega\,\varphi(\mathbf{x},\omega) \quad . \tag{31}$$

---

4 It should be noted that Eqs. (27) and (28) are equivalent only for a simply connected domain (see Dautray and Lions, 1992).

5 Let C be any real constant, let $\psi+C$ and ψ satisfy Eq. (28). Then C=0 because $\ell(\psi+C)$= $\ell(\psi)+C\,\ell(1)$ implies $C\,\ell(1)$=0, i.e. C=0.

Substituting Eq. (31) into Eq. (28) yields

$$\mathbf{u}(\mathbf{x},\omega) = \boldsymbol{\nabla}\varphi(\mathbf{x},\omega) \quad . \tag{32}$$

As above, we introduce a linear constraint equation on φ

$$\ell(\varphi) = 0 \quad , \quad \text{with} \quad \ell(1) \neq 0 \quad , \tag{33}$$

in which ℓ is any linear form [6] on the space of functions φ. As above, we can choose $\ell(\varphi)$ such that $\ell(\varphi) = 0$ is written as

$$\int_\Omega \varphi(\mathbf{x},\omega)\, d\mathbf{x} = 0 \quad . \tag{34}$$

2.6. Formulation in terms of the pressure and the velocity potential

Equations in terms of p and ψ. Substituting Eq. (28) into Eqs. (9) and (10) yields

$$i\omega\, \rho_0\, \boldsymbol{\nabla}\psi + \boldsymbol{\nabla}p = \mathbf{0} \qquad \text{in} \quad \Omega \quad , \tag{35}$$

$$i\omega\, p = -\rho_0\, c_0^2\, \boldsymbol{\nabla}^2\psi + c_0^2\, Q \qquad \text{in} \quad \Omega \quad . \tag{36}$$

Eq. (35) can be rewritten as $\boldsymbol{\nabla}\{i\omega\,\rho_0\,\psi + p\} = \mathbf{0}$ whose solution is

$$p(\mathbf{x},\omega) = -i\omega\,\rho_0\,\psi(\mathbf{x},\omega) + \pi(\omega) \qquad \text{in} \quad \Omega \quad , \tag{37}$$

in which $\pi(\omega)$ depends only on ω but not on $\mathbf{x}$ (see Ohayon, 1987; Morand and Ohayon, 1992 and 1995).

Helmholtz equation in terms of ψ. Substituting Eq. (37) into Eq. (36) yields the Helmholtz equation expressed in terms of velocity potential $\psi(\mathbf{x},\omega)$ and modified by the presence of $\pi(\omega)$,

$$\boldsymbol{\nabla}^2\psi + \frac{\omega^2}{c_0^2}\,\psi + \frac{i\omega}{\rho_0 c_0^2}\,\pi(\omega) = \frac{1}{\rho_0}\,Q \qquad \text{in} \quad \Omega \quad . \tag{38}$$

Boundary conditions in terms of ψ. We express the boundary condition presented in Section 2.4 in terms of velocity potential $\psi(\mathbf{x},\omega)$.

---

6 Let C be any real constant, let $\varphi{+}C$ and φ satisfy Eq. (32). Then $C{=}0$ because $\ell(\varphi{+}C){=}\ell(\varphi){+}C\,\ell(1)$ implies $C\,\ell(1){=}0$, i.e. $C{=}0$.

1- Neumann boundary condition on Γ. Substituting Eq. (28) into Eq. (14) yields the Neumann condition on Γ,

$$\frac{\partial \psi}{\partial \mathbf{n}} = i\omega\, \mathbf{u}_{\text{wall}} \cdot \mathbf{n} \quad , \tag{39}$$

in which $\partial\psi/\partial\mathbf{n} = \boldsymbol{\nabla}\psi \cdot \mathbf{n}$.

2- Neumann boundary condition on Γ_Z with wall acoustic impedance. Substituting Eqs. (28) and (37) into Eq. (16) yields the Neumann condition on Γ_Z,

$$\frac{\partial \psi}{\partial \mathbf{n}} = i\omega\, \mathbf{u}_{\text{wall}} \cdot \mathbf{n} - \frac{i\omega\, \rho_0}{Z}\, \psi + \frac{\pi(\omega)}{Z} \quad . \tag{40}$$

3- Dirichlet boundary condition on Γ_0. From Eq. (26), since $p(\mathbf{x},\omega)$ is equal to zero on Γ_0, Eq. (37) yields $\pi(\omega) = i\omega\, \rho_0\, \psi(\mathbf{x},\omega)$ for all $\mathbf{x}$ in Γ_0, which shows that $\psi(\mathbf{x},\omega)$ is independent of $\mathbf{x}$ on Γ_0. Consequently, since $\psi(\mathbf{x},\omega)$ is defined to within an additive constant, this constant is removed by choosing

$$\psi = 0 \quad \text{on} \quad \Gamma_0 \quad . \tag{41}$$

We then deduce that $\pi(\omega) = 0$ in Eq. (37) and therefore, $p(\mathbf{x},\omega) = -i\omega\, \rho_0\, \psi(\mathbf{x},\omega)$.

2.7. Formulation in terms of the pressure and the displacement potential.

Equations in terms of p and φ. Substituting Eq. (31) into Eqs. (35) and (36) yields

$$-\omega^2\, \rho_0\, \boldsymbol{\nabla}\varphi + \boldsymbol{\nabla} p = \mathbf{0} \quad \text{in} \quad \Omega \quad , \tag{42}$$

$$p = -\rho_0\, c_0^2\, \boldsymbol{\nabla}^2 \varphi + c_0^2\, \frac{Q}{i\omega} \quad \text{in} \quad \Omega \quad . \tag{43}$$

Eq. (42) can be written as $\boldsymbol{\nabla}\{-\omega^2\, \rho_0\, \varphi + p\} = \mathbf{0}$ whose solution is

$$p(\mathbf{x},\omega) = \omega^2\, \rho_0\, \varphi(\mathbf{x},\omega) + \pi(\omega) \quad \text{in} \quad \Omega \quad , \tag{44}$$

which can also be found by substituting Eq. (31) into Eq. (37).

Helmholtz equation in terms of φ. Substituting Eq. (44) into Eq. (43) yields the Helmholtz equation in terms of displacement potential $\varphi(\mathbf{x},\omega)$ modified by the presence of $\pi(\omega)$,

$$\boldsymbol{\nabla}^2 \varphi + \frac{\omega^2}{c_0^2}\, \varphi + \frac{\pi(\omega)}{\rho_0 c_0^2} = \frac{1}{\rho_0}\, \frac{Q}{i\omega} \quad \text{in} \quad \Omega \quad . \tag{45}$$

Boundary conditions in terms of φ. We express the boundary condition presented in Section 2.4 in terms of displacement potential $\varphi(\mathbf{x}, \omega)$.

1- Neumann boundary condition on Γ. Substituting Eq. (32) into Eq. (15) yields the Neumann condition on Γ,

$$\frac{\partial \varphi}{\partial \mathbf{n}} = \mathbf{u}_{\text{wall}} \cdot \mathbf{n} \quad . \tag{46}$$

2- Neumann boundary condition on Γ_Z with wall acoustic impedance. Substituting Eqs. (32) and Eq. (44) into Eq. (16) yields the Neumann condition on Γ_Z,

$$\frac{\partial \varphi}{\partial \mathbf{n}} = \mathbf{u}_{\text{wall}} \cdot \mathbf{n} - \frac{i\omega\, \rho_0}{Z}\, \varphi + \frac{\pi(\omega)}{i\omega\, Z} \quad . \tag{47}$$

3- Dirichlet boundary condition on Γ_0. From Eq. (26), since $p(\mathbf{x}, \omega)$ is equal to zero on Γ_0, Eq. (44) yields $\pi(\omega) = -\omega^2\, \rho_0\, \varphi(\mathbf{x}, \omega)$ for all $\mathbf{x}$ in Γ_0, which shows that $\varphi(\mathbf{x}, \omega)$ is independent of $\mathbf{x}$ on Γ_0. Consequently, since $\varphi(\mathbf{x}, \omega)$ is defined to within an additive constant, this constant is removed by choosing

$$\varphi = 0 \quad \text{on} \quad \Gamma_0 \quad . \tag{48}$$

We then deduce that $\pi(\omega) = 0$ in Eq. (44) and therefore, $p(\mathbf{x}, \omega) = \omega^2\, \rho_0\, \varphi(\mathbf{x}, \omega)$.

3. Dissipative Acoustic Fluid

3.1. Basic assumptions

In the model presented in Section 2, the acoustic fluid is inviscid and consequently there is no acoustic dissipation inside the fluid. In this section, we introduce a dissipation model for the acoustic fluid. Generally, there are two main physical dissipations. The first is an internal acoustic dissipation inside the cavity due to the viscosity and thermal conduction of the fluid. These dissipation mechanisms are assumed to be small. In this model, we consider only the dissipation due to viscosity. This correction introduces an additional dissipative term in the Helmholtz equation (see Eqs. (38) and (45)) without modifying the conservative part. The second dissipation is generated inside the "wall viscothermical boundary layer" of the cavity. This dissipation is not considered in this section, but can possibly be modeled with an equivalent wall acoustic impedance (already introduced in Section 2.4). Acoustic impedance $Z(\mathbf{x}, \omega)$ on boundary Γ_Z is useful for modeling an acoustic material coating whose thickness is neglected in the model. We then consider only the acoustic mode (irrotational motion) predominant in the volume. The vorticity and entropy modes which mainly play a role in the "wall viscothermical boundary layer" are not directly modeled. For additional details concerning dissipation in acoustic fluids, we refer the reader to Landau and Lifchitz, 1992b; Lighthill, 1978; Mason, 1971; Pierce, 1989.

3.2. Linearized equations in the time domain

Concerning the linearized equations recalled in this section, we refer the reader, for instance, to Landau and Lifchitz, 1992b. We denote as p, ρ, $\mathbf{v}$, T and s the disturbances of the pressure, mass density, velocity vector, temperature and entropy per unit mass (or specific entropy) of the fluid from their undisturbed values p_0, ρ_0, $\mathbf{v}_0$, T_0 and s_0 corresponding to the fluid at equilibrium in the reference configuration (fluid at rest). The disturbances p, ρ, T and s satisfy the following classical thermodynamic equations of state

$$\rho = \frac{1}{c_0^2}\, p - \frac{\rho_0 \beta T_0}{c_p}\, s \quad , \tag{49}$$

$$T = \frac{\beta T_0}{\rho_0 c_p}\, p + \frac{T_0}{c_p}\, s \quad , \tag{50}$$

in which c_p and c_v are the specific-heat coefficients at constant pressure and constant volume respectively, and β is the coefficient of thermal expansion. The values of coefficients c_p, c_v and β correspond to the reference configuration. The linearized mass conservation equation, the linearized Navier-Stokes equation and the linearized general equation of heat propagation (also called the Kirchhoff-Fourier equation or entropy equation) are written as

$$\partial_t \rho + \rho_0\, \boldsymbol{\nabla} \cdot \mathbf{v} = Q \quad , \tag{51}$$

$$\rho_0\, \partial_t \mathbf{v} + \boldsymbol{\nabla} p = \eta\, \boldsymbol{\nabla}^2 \mathbf{v} + \left(\zeta + \frac{\eta}{3}\right) \boldsymbol{\nabla}\{\boldsymbol{\nabla} \cdot \mathbf{v}\} \quad , \tag{52}$$

$$\rho_0 T_0\, \partial_t s = \kappa\, \boldsymbol{\nabla}^2 T \quad , \tag{53}$$

in which η is the dynamic viscosity, $\nu = \eta/\rho_0$ is the kinematic viscosity, ζ is the second viscosity and κ is the coefficient of thermal conduction (or thermal conductivity). The values of coefficients η, ζ and κ correspond to the reference configuration. It should be noted that outside the source region or if there are no distributed monopole sources, Eq. (51) gives the usual linearized mass conservation equation, $\partial_t \rho + \rho_0\, \boldsymbol{\nabla} \cdot \mathbf{v} = 0$.

3.3. Helmholtz equation with viscosity and thermal conduction

As explained in Section 3.1, we are interested in the linear acoustic equations corresponding to irrotational motion

$$\boldsymbol{\nabla}\times\mathbf{v} = \mathbf{0} \quad . \tag{54}$$

From the identity $\boldsymbol{\nabla}^2\mathbf{v} = \boldsymbol{\nabla}\{\boldsymbol{\nabla}\cdot\mathbf{v}\} - \boldsymbol{\nabla}\times(\boldsymbol{\nabla}\times\mathbf{v})$ and using Eq. (54), we deduce that $\boldsymbol{\nabla}^2\mathbf{v} = \boldsymbol{\nabla}\{\boldsymbol{\nabla}\cdot\mathbf{v}\}$. Replacing $\boldsymbol{\nabla}^2\mathbf{v}$ by $\boldsymbol{\nabla}\{\boldsymbol{\nabla}\cdot\mathbf{v}\}$ in Eq. (52), and taking the Fourier transform with respect to time of Eqs. (49) to (53) yields

$$\rho = \frac{1}{c_0^2}\,p - \frac{\rho_0\beta T_0}{c_p}\,s \quad , \tag{55}$$

$$T = \frac{\beta T_0}{\rho_0 c_p}\,p + \frac{T_0}{c_p}\,s \quad , \tag{56}$$

$$i\omega\,\rho + \rho_0\,\boldsymbol{\nabla}\cdot\mathbf{v} = Q \quad , \tag{57}$$

$$i\omega\,\rho_0\,\mathbf{v} + \boldsymbol{\nabla}p = \Big(\frac{4}{3}\eta + \zeta\Big)\boldsymbol{\nabla}\{\boldsymbol{\nabla}\cdot\mathbf{v}\} \quad , \tag{58}$$

$$i\omega\,\rho_0 T_0\,s = \kappa\,\boldsymbol{\nabla}^2 T \quad . \tag{59}$$

In a first step, ρ is eliminated in Eq. (57) using Eq. (55). In a second step, T is eliminated in Eq. (59) using Eq. (56) and neglecting the term in $\boldsymbol{\nabla}^2 s$. Then, eliminating s between the two resulting equations yields

$$i\omega\,p = -\rho_0\,c_0^2\;\boldsymbol{\nabla}\cdot\mathbf{v} + \frac{\beta^2}{c_p{}^2}\,\frac{c_0^2\,\kappa T_0}{\rho_0}\,\boldsymbol{\nabla}^2 p + c_0^2\,Q \quad . \tag{60}$$

Eliminating $\boldsymbol{\nabla}\cdot\mathbf{v}$ in Eq. (58) using the equation obtained in the first step and neglecting the term in $\boldsymbol{\nabla}s$ yields

$$i\omega\,\rho_0\,\mathbf{v} + \boldsymbol{\nabla}p = \tau c_0^2\boldsymbol{\nabla}Q - i\omega\,\tau\boldsymbol{\nabla}p \quad , \tag{61}$$

in which τ is given by

$$\tau = \frac{1}{\rho_0 c_0^2}\Big(\frac{4}{3}\eta + \zeta\Big) > 0 \quad . \tag{62}$$

In general, τ can depend [7] on frequency ω. Taking the divergence of Eq. (61) and using Eq. (60), we obtain a Helmholtz equation with a dissipative term and a source term

$$\frac{\omega^2}{c_0^2}\,p + i\omega\,\tau_{\text{tot}}\,\boldsymbol{\nabla}^2 p + \boldsymbol{\nabla}^2 p = \tau\,c_0^2\,\boldsymbol{\nabla}^2 Q - i\omega\,Q \quad , \tag{63}$$

---

[7] Second viscosity ζ can depend on ω. To simplify the notation, we often write τ instead of $\tau(\omega)$.

in which τ_{tot} is defined by

$$\tau_{\text{tot}} = \tau + \tau_\kappa \quad , \tag{64}$$

where τ is the coefficient due to viscosity (defined by Eq. (62)) and τ_κ is the coefficient due to thermal conduction,

$$\tau_\kappa = \frac{\beta^2}{c_p{}^2} \frac{\kappa T_0}{\rho_0} > 0 \quad . \tag{65}$$

Various expressions for the parameter τ_{tot}. Using the classical thermodynamic identity (see for instance Landau and Lifchitz, 1992b),

$$c_p - c_v = T_0 \beta^2 c_0^2 \, \frac{c_v}{c_p} \quad , \tag{66}$$

and introducing the specific-heat ratio γ defined by

$$\gamma = \frac{c_p}{c_v} \quad , \tag{67}$$

coefficient τ_κ defined by Eq. (65) can be rewritten as

$$\tau_\kappa = \frac{\kappa}{\rho_0 c_0^2} \left(\frac{1}{c_v} - \frac{1}{c_p} \right) = \frac{\eta}{\rho_0 c_0^2} \, \frac{(\gamma - 1)}{\mathrm{P}} \quad , \tag{68}$$

in which P is the Prandtl number defined by

$$\mathrm{P} = \frac{\eta \, c_p}{\kappa} \quad . \tag{69}$$

From Eqs. (62), (64) and (68), we deduce the following two expressions,

$$\tau_{\text{tot}} = \frac{1}{\rho_0 c_0^2} \left\{ \left(\frac{4}{3} \eta + \zeta \right) + \kappa \left(\frac{1}{c_v} - \frac{1}{c_p} \right) \right\} \quad , \tag{70}$$

$$\tau_{\text{tot}} = \frac{\eta}{\rho_0 c_0^2} \left\{ \frac{4}{3} + \frac{\zeta}{\eta} + \frac{\gamma - 1}{\mathrm{P}} \right\} \quad . \tag{71}$$

The numerical values of the different coefficients for pure water and air at 20° C are

	Water	Air
ρ_0 (kg/m^3)	1000	1.2
c_0 (m/s)	1482	340
η $(Pa.s)$	0.001	0.0000181
γ	1.006	1.4
C_p $(J/(kg.K)$	4180	1004
C_v $(J/(kg.K)$	4150	717
κ $(W/(m.K))$	0.597	0.02569
P	7.00	0.707

Dispersion relation. Taking the spatial Fourier transform

$$p(\mathbf{k},\omega) = \int_{\mathbb{R}^3} p(\mathbf{x},\omega)\,\exp(i\,\mathbf{k}\cdot\mathbf{x})\,d\mathbf{x} \tag{72}$$

of the homogeneous equation associated with Eq. (63), and introducing the wave number $k = \|\mathbf{k}\|$, we obtain the classical dispersion relation

$$k^2 = \frac{1}{(1+i\omega\,\tau_{\text{tot}})}\,\frac{\omega^2}{c_0^2} \simeq \frac{\omega^2}{c_0^2} - \frac{i\omega^3}{c_0^2}\tau_{\text{tot}} \quad . \tag{73}$$

Eq. (73) shows that the dissipation coefficient varies as ω^2.

3.4. Linearized acoustic equations in the frequency domain in terms of pressure and velocity

Below, the dissipation due to thermal conduction is neglected. Eqs. (60) and (61) become

$$i\omega\,p = -\rho_0\,c_0^2\;\nabla\cdot\mathbf{v} + c_0^2\,Q \quad . \tag{74}$$

$$i\omega\,\rho_0\,\mathbf{v} + \nabla p = \tau c_0^2 \nabla Q - i\omega\,\tau\nabla p \quad , \tag{75}$$

in which τ is given by

$$\tau = \frac{1}{\rho_0 c_0^2}\left(\frac{4}{3}\eta + \zeta\right) > 0 \quad . \tag{76}$$

3.5. Introduction of the velocity and displacement potentials

Velocity potential $\widetilde{\psi}(\mathbf{x},\omega)$ and field $\psi(\mathbf{x},\omega)$. Eq. (54) allows a velocity potential $\widetilde{\psi}(\mathbf{x},\omega)$ to be introduced such that

$$\mathbf{v}(\mathbf{x},\omega) = \boldsymbol{\nabla}\widetilde{\psi}(\mathbf{x},\omega) \quad . \tag{77}$$

It is convenient to introduce a new unknown field $\psi(\mathbf{x},\omega)$ related to velocity potential $\widetilde{\psi}(\mathbf{x},\omega)$ by (Soize et al., 1986b)

$$\widetilde{\psi}(\mathbf{x},\omega) = (1 + i\omega\,\tau)\psi(\mathbf{x},\omega) + \frac{\tau c_0^2}{\rho_0}\frac{Q(\mathbf{x},\omega)}{i\omega} \quad . \tag{78 – 1}$$

Velocity field $\mathbf{v}$ given by Eq. (77) is then written as

$$\mathbf{v}(\mathbf{x},\omega) = (1 + i\omega\,\tau)\,\boldsymbol{\nabla}\psi(\mathbf{x},\omega) + \frac{\tau c_0^2}{\rho_0}\,\boldsymbol{\nabla}\left(\frac{Q(\mathbf{x},\omega)}{i\omega}\right) \quad . \tag{78 – 2}$$

Since velocity potential $\widetilde{\psi}(\mathbf{x},\omega)$ is defined to within an additive constant, field $\psi(\mathbf{x},\omega)$ is also defined to within an additive constant. As in Section 2.5, we introduce a linear constraint equation on ψ,

$$\ell(\psi) = 0 \quad , \quad \text{with} \quad \ell(1) \neq 0 \quad , \tag{79}$$

in which ℓ is any linear form on the space of functions $\mathbf{x} \mapsto \psi(\mathbf{x},\omega)$. For instance, we can choose $\ell(\psi)$ such that $\ell(\psi) = 0$ is written as

$$\int_\Omega \psi(\mathbf{x},\omega)\,d\mathbf{x} = 0 \quad . \tag{80}$$

It should be noted that the constraint introduced on ψ leads to the following constraint $\int_\Omega \widetilde{\psi}(\mathbf{x},\omega)\,d\mathbf{x} = \tau c_0^2/\rho_0 \int_\Omega Q(\mathbf{x},\omega)/i\omega\,d\mathbf{x}$ for velocity potential $\widetilde{\psi}$.

Displacement potential $\widetilde{\varphi}(\mathbf{x},\omega)$ and field $\varphi(\mathbf{x},\omega)$. As in Section 2.5, we introduce a displacement potential $\widetilde{\varphi}(\mathbf{x},\omega)$ such that $\widetilde{\psi}(\mathbf{x},\omega) = i\omega\,\widetilde{\varphi}(\mathbf{x},\omega)$. From Eqs. (77) and (78-1), we deduce that

$$\mathbf{u}(\mathbf{x},\omega) = \boldsymbol{\nabla}\widetilde{\varphi}(\mathbf{x},\omega) \quad . \tag{81}$$

We then introduce an unknown field $\varphi(\mathbf{x},\omega)$ related to displacement potential $\widetilde{\varphi}(\mathbf{x},\omega)$ by (see Eq. (78-1))

$$\widetilde{\varphi}(\mathbf{x},\omega) = (1 + i\omega\,\tau)\varphi(\mathbf{x},\omega) - \frac{\tau c_0^2}{\rho_0}\frac{Q(\mathbf{x},\omega)}{\omega^2} \quad . \tag{82 – 1}$$

Displacement field $\mathbf{u}$ given by Eq. (81) is written as

$$\mathbf{u}(\mathbf{x},\omega) = (1 + i\omega\,\tau)\,\nabla\varphi(\mathbf{x},\omega) - \frac{\tau c_0^2}{\rho_0}\,\nabla\left(\frac{Q(\mathbf{x},\omega)}{\omega^2}\right) \quad . \tag{82 -- 2}$$

Since φ is defined to within an additive constant, as above, we introduce a linear constraint equation on φ,

$$\ell(\varphi) = 0 \quad , \quad \text{with} \quad \ell(1) \neq 0 \quad , \tag{83}$$

in which ℓ is any linear form on the space of functions $\mathbf{x} \mapsto \varphi(\mathbf{x},\omega)$. For instance, we can choose $\ell(\varphi)$ such that $\ell(\varphi) = 0$ is written as

$$\int_\Omega \varphi(\mathbf{x},\omega)\, d\mathbf{x} = 0 \quad . \tag{84}$$

It should be noted that the constraint introduced on φ leads to the following constraint $\int_\Omega \widetilde{\varphi}(\mathbf{x},\omega)\, d\mathbf{x} = -\tau c_0^2/\rho_0 \int_\Omega Q(\mathbf{x},\omega)/\omega^2\, d\mathbf{x}$ for displacement potential $\widetilde{\varphi}$.

3.6. Equations in terms of ψ

Equations in terms of p and ψ. Substituting Eq. (78-2) into Eq. (75) yields $(1 + i\omega\,\tau)\nabla\{i\omega\,\rho_0\,\psi + p\} = \mathbf{0}$ whose solution is

$$p = -i\omega\,\rho_0\,\psi + \pi(\omega) \quad , \tag{85}$$

in which $\pi(\omega)$ depends only on ω but not on $\mathbf{x}$ (see Ohayon, 1987; Morand and Ohayon, 1992 and 1995). Substituting Eq. (78-2) into Eq. (74) yields

$$i\omega\, p = -\rho_0 c_0^2(1+i\omega\,\tau)\nabla^2\psi - \tau c_0^4\nabla^2\left(\frac{Q}{i\omega}\right) + c_0^2 Q \quad . \tag{86}$$

Dissipative Helmholtz equation in ψ. Substituting Eq. (85) into Eq. (86) yields a Helmholtz type equation,

$$\nabla^2\psi + i\omega\,\tau\nabla^2\psi + \frac{\omega^2}{c_0^2}\,\psi + \frac{i\omega}{\rho_0 c_0^2}\,\pi(\omega) = \frac{1}{\rho_0}\,Q - \frac{\tau c_0^2}{\rho_0}\,\nabla^2\left(\frac{Q}{i\omega}\right) \quad . \tag{87}$$

Boundary conditions in terms of ψ. We express the boundary condition presented in Section 2.4 in terms of field $\psi(\mathbf{x},\omega)$.

1- Neumann boundary condition on Γ. Substituting Eq. (78-2) into Eq. (14) yields the Neumann condition on Γ,

$$(1 + i\omega\tau)\frac{\partial\psi}{\partial\mathbf{n}} = i\omega\,\mathbf{u}_{\text{wall}}\cdot\mathbf{n} - \frac{\tau c_0^2}{\rho_0}\frac{\partial}{\partial\mathbf{n}}\left(\frac{Q}{i\omega}\right) \quad . \tag{88}$$

2- Neumann boundary condition on Γ_Z with wall acoustic impedance. Substituting Eqs. (78-2) and (85) into Eq. (16) yields the Neumann condition on Γ_Z,

$$(1 + i\omega\tau)\frac{\partial\psi}{\partial\mathbf{n}} = i\omega\,\mathbf{u}_{\text{wall}}\cdot\mathbf{n} - \frac{\tau c_0^2}{\rho_0}\frac{\partial}{\partial\mathbf{n}}\left(\frac{Q}{i\omega}\right) - \frac{i\omega\,\rho_0}{Z}\psi + \frac{\pi(\omega)}{Z} \quad . \tag{89}$$

3- Dirichlet boundary condition on Γ_0. From Eq. (26), since $p(\mathbf{x},\omega)$ is equal to zero on Γ_0, Eq. (85) yields $\pi(\omega) = i\omega\,\rho_0\,\psi(\mathbf{x},\omega)$ for all $\mathbf{x}$ in Γ_0, which shows that $\psi(\mathbf{x},\omega)$ is independent of $\mathbf{x}$ on Γ_0. Consequently, since $\psi(\mathbf{x},\omega)$ is defined to within an additive constant, this constant is removed by choosing

$$\psi = 0 \quad \text{on} \quad \Gamma_0 \quad . \tag{90}$$

We then deduce that $\pi(\omega) = 0$ in Eq. (85) and consequently, $p = -i\omega\,\rho_0\,\psi$.

3.7. Equations in terms of φ

Equations in terms of p and φ. Substituting $\psi = i\omega\,\varphi$ into Eq. (85) yields

$$p = \omega^2\,\rho_0\,\varphi + \pi(\omega) \quad , \tag{91}$$

in which $\pi(\omega)$ depends only on ω but not on $\mathbf{x}$. Substituting $\psi = i\omega\,\varphi$ into Eq. (86) yields

$$p = -\rho_0\,c_0^2(1+i\omega\tau)\boldsymbol{\nabla}^2\varphi + \tau c_0^4\boldsymbol{\nabla}^2\left(\frac{Q}{\omega^2}\right) + c_0^2\frac{Q}{i\omega} \quad . \tag{92}$$

Dissipative Helmholtz equation in φ. Substituting Eq. (91) into Eq. (92) yields the Helmholtz type equation

$$\boldsymbol{\nabla}^2\varphi + i\omega\,\tau\boldsymbol{\nabla}^2\varphi + \frac{\omega^2}{c_0^2}\varphi + \frac{1}{\rho_0 c_0^2}\pi(\omega) = \frac{1}{\rho_0}\frac{Q}{i\omega} + \frac{\tau c_0^2}{\rho_0}\boldsymbol{\nabla}^2\left(\frac{Q}{\omega^2}\right) \quad . \tag{93}$$

Boundary conditions in terms of φ. We express the boundary condition presented in Section 2.4 in terms of field $\varphi(\mathbf{x},\omega)$.

1- Neumann boundary condition on Γ. Substituting $\psi = i\omega\,\varphi$ into Eq. (88) yields the Neumann condition on Γ,

$$(1 + i\omega\tau)\frac{\partial\varphi}{\partial\mathbf{n}} = \mathbf{u}_{\text{wall}} \cdot \mathbf{n} + \frac{\tau c_0^2}{\rho_0}\frac{\partial}{\partial\mathbf{n}}\left(\frac{Q}{\omega^2}\right) \quad . \tag{94}$$

2- Neumann boundary condition on Γ_Z with wall acoustic impedance. Substituting $\psi = i\omega\,\varphi$ into Eq. (89) yields the Neumann condition on Γ_Z,

$$(1 + i\omega\tau)\frac{\partial\varphi}{\partial\mathbf{n}} = \mathbf{u}_{\text{wall}} \cdot \mathbf{n} + \frac{\tau c_0^2}{\rho_0}\frac{\partial}{\partial\mathbf{n}}\left(\frac{Q}{\omega^2}\right) - \frac{i\omega\,\rho_0}{Z}\,\varphi + \frac{\pi(\omega)}{i\omega\,Z} \quad . \tag{95}$$

3- Dirichlet boundary condition on Γ_0. As for the Dirichlet boundary condition defined by Eq. (90), we have

$$\varphi = 0 \quad \text{on} \quad \Gamma_0 \quad . \tag{96}$$

We then deduce that $\pi(\omega) = 0$ in Eq. (91) and consequently $p = \omega^2\,\rho_0\,\varphi$.

CHAPTER XI

Internal Acoustic Fluid Formulation for the LF and MF Ranges

1. Introduction

In this chapter we present the formulation for the calculation of the frequency response function of an internal acoustic fluid (gas or liquid) in the LF and MF ranges. The boundary of the acoustic cavity is submitted to a prescribed wall displacement field and there is a wall acoustic impedance on a part of this boundary. In addition, an acoustic source density is given inside the cavity.

In Sections 2 and 3, we state the internal acoustic problem and give the boundary value problem.

In Sections 4 to 6, we establish the variational formulation, the linear operator equation whose unique solution allows the frequency response function to be constructed and finally the finite element discretization.

In Sections 7 and 8, we give the formulation of the acoustic modes of the problem, the corresponding spectral properties and the finite element discretization.

Using the methodology presented in Chapter VI, Section 9 gives the method for calculating the frequency response function in the low-frequency range. The construction of the frequency response function is based on its projection on a finite set of acoustic modes including quasi-static correction terms.

In Section 10, we present a method for calculating the frequency response function in the medium-frequency range. This method is based on the frequency transform technique presented in Chapter VII.

Section 11 deals with the case of an acoustic cavity with a zero pressure condition on a part of the boundary (returning to Sections 2 to 10).

In Section 12, we discuss the particular case of an axisymmetric acoustic cavity.

Finally, Section 13 refers to Chapter IX for calculating the response of an acoustic cavity subjected to deterministic and random excitations.

2. Statement of the Internal Acoustic Problem

We consider a bounded acoustic cavity Ω filled with a dissipative acoustic fluid (gas or liquid). The boundary $\partial\Omega$ is written as $\Gamma \cup \Gamma_Z$ (see Fig. 1).

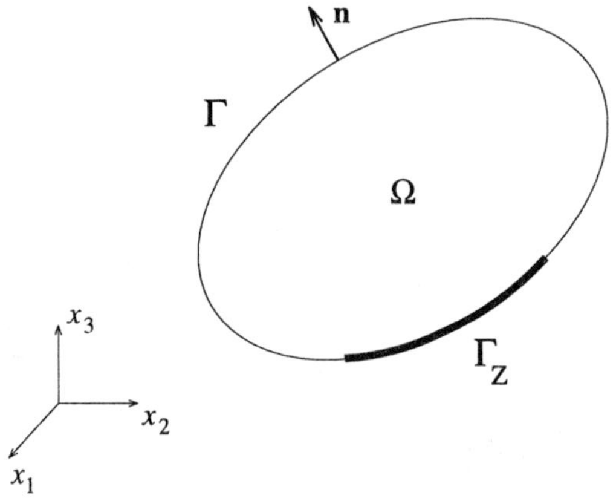

Fig. 1. Configuration of the acoustic cavity

On wall $\Gamma \cup \Gamma_Z$, a displacement field $\mathbf{u}_{\text{wall}}(\mathbf{x}, \omega)$ is prescribed. Wall Γ_Z has acoustical properties modeled by an acoustic impedance $Z(\mathbf{x}, \omega)$ satisfying the hypotheses defined by Eqs. (X.18) to (X.25). In addition, an acoustic source density $Q(\mathbf{x}, \omega)$ is given inside Ω and satisfies the hypotheses defined by Eq. (X.8). The problem consists in calculating the FRF of the system in the low- and medium-frequency ranges, whose input is due to $\mathbf{u}_{\text{wall}}(\mathbf{x}, \omega)$ and $Q(\mathbf{x}, \omega)$, and whose output is field φ. The pressure field $p(\mathbf{x}, \omega)$ in cavity Ω can then be deduced. The mathematical model of this physical case is described in Section X.3.

3. Boundary Value Problem

Equations in terms of φ. The equations of the boundary value problem are described in Section X.3.7 and are rewritten as

$$-\omega^2 \frac{\rho_0}{c_0^2}\, \varphi - i\omega\, \tau\, \rho_0 \boldsymbol{\nabla}^2 \varphi - \rho_0\, \boldsymbol{\nabla}^2 \varphi - \frac{1}{c_0^2}\, \pi(\omega) = -\frac{Q}{i\omega} - \tau c_0^2\, \boldsymbol{\nabla}^2 \left(\frac{Q}{\omega^2}\right) \quad \text{in } \Omega\,, \quad (1)$$

with the Neumann boundary condition,

$$\rho_0(1+i\omega\tau)\frac{\partial\varphi}{\partial\mathbf{n}} = \rho_0\,\mathbf{u}_{\text{wall}}\cdot\mathbf{n} + \tau c_0^2\frac{\partial}{\partial\mathbf{n}}\Big(\frac{Q}{\omega^2}\Big) \quad \text{on} \quad \Gamma\ , \tag{2}$$

the Neumann boundary condition with wall acoustic impedance,

$$\rho_0(1+i\omega\tau)\frac{\partial\varphi}{\partial\mathbf{n}} = \rho_0\,\mathbf{u}_{\text{wall}}\cdot\mathbf{n} + \tau c_0^2\frac{\partial}{\partial\mathbf{n}}\Big(\frac{Q}{\omega^2}\Big) + \rho_0{}^2\omega^2\frac{\varphi}{i\omega\,Z} + \frac{\rho_0\pi(\omega)}{i\omega\,Z} \quad \text{on} \quad \Gamma_Z\ , \tag{3}$$

and the uniqueness condition defined by Eq. (X.84),

$$\int_\Omega \varphi\, d\mathbf{x} = 0 \quad . \tag{4}$$

The pressure field is given by Eq. (X.91),

$$p(\mathbf{x},\omega) = \omega^2\,\rho_0\,\varphi(\mathbf{x},\omega) + \pi(\omega) \qquad \text{in} \quad \Omega \quad . \tag{5}$$

Expression for $\pi(\omega)$. Integrating Eq. (1) in domain Ω, using Stokes' formula and Eqs. (2)–(3) yields

$$\pi(\omega) = \kappa(\omega)\ \{-\omega^2\rho_0\,\pi_1(\omega\,;\varphi) - \pi_2(\mathbf{u}_{\text{wall}}) + \pi_Q(\omega)\} \quad , \tag{6}$$

in which

$$\pi_1(\omega\,;\varphi) = \int_{\Gamma_Z}\frac{\varphi(\mathbf{x},\omega)}{i\omega\,Z(\mathbf{x},\omega)}\,ds(\mathbf{x}) \quad , \tag{7-1}$$

$$\pi_2(\mathbf{u}) = \int_{\Gamma\cup\Gamma_Z}\mathbf{u}(\mathbf{x},\omega)\cdot\mathbf{n}(\mathbf{x})\,ds(\mathbf{x}) \quad , \tag{7-2}$$

$$\pi_Q(\omega) = \frac{1}{\rho_0}\int_\Omega\frac{Q(\mathbf{x},\omega)}{i\omega}\,d\mathbf{x} \quad , \tag{7-3}$$

$|\Omega|$ is the volume of domain Ω,

$$|\Omega| = \int_\Omega d\mathbf{x} \quad , \tag{8}$$

and $\kappa(\omega)$ is the dimensionless complex number defined by

$$\kappa(\omega) = \frac{\rho_0\,c_0^2}{|\Omega|}\times\Big\{1+\frac{\rho_0\,c_0^2}{|\Omega|}\int_{\Gamma_Z}\frac{ds(\mathbf{x})}{i\omega\,Z(\mathbf{x},\omega)}\Big\}^{-1} \quad . \tag{9}$$

From Eqs. (X.25) and (X.21), we deduce that in the limit static case $\omega = 0$, and $\kappa(0)$ is real and is such that

$$0 < \kappa(0) < +\infty \quad , \quad \lim_{\alpha_{\min} \to +\infty} \kappa(0) = \frac{\rho_0 c_0^2}{|\Omega|} \quad . \tag{10}$$

It should be noted that the general expression of $\pi(\omega)$ defined by Eq. (6) includes the term $-\kappa(\omega)\,(\omega^2/c_0^2) \int_\Omega \varphi\, d\mathbf{x}$ which is equal to zero due to Eq. (4).

Problem to be solved. For all real ω, for $\mathbf{u}_{\text{wall}}(\omega)$ and $Q(\omega)$ given, the problem consists in finding $\varphi(\omega)$ as a solution of Eqs. (1) to (4) in which $\pi(\omega)$ is expressed by Eqs. (6) to (9). The pressure is then given by Eq. (5). It should be noted that the static problem corresponding to $\omega = 0$ is correctly stated.

4. Variational Formulation and Linear Operator Equation

4.1. Admissible function space

The variational formulation is constructed using the test function method presented in Chapter II. We introduce the complex vector space $\mathcal{E}^c$ of sufficiently differentiable [1] functions φ defined on Ω with values in $\mathbb{C}$ and the admissible function space $\mathcal{E}_\varphi^c$ which is the subspace of $\mathcal{E}^c$ defined by

$$\mathcal{E}_\varphi^c = \{ \varphi \in \mathcal{E}^c \quad ; \quad \int_\Omega \varphi\, d\mathbf{x} = 0 \} \quad . \tag{11}$$

4.2. Variational formulation

Let φ be in $\mathcal{E}_\varphi^c$. Multiplying Eq. (1) by $\overline{\delta\varphi} \in \mathcal{E}_\varphi^c$, using Green's formula, taking into account the Neumann boundary condition defined by Eqs. (2) and (3), and since $\int_\Omega \overline{\delta\varphi}\, d\mathbf{x} = 0$, yields the following variational formulation.

Variational formulation. For all fixed real ω, find φ in $\mathcal{E}_\varphi^c$ such that

$$a(\omega\,; \varphi\,, \delta\varphi) = f(\omega\,; \delta\varphi) \quad , \quad \forall\, \delta\varphi \in \mathcal{E}_\varphi^c \quad , \tag{12}$$

[1] Space $\mathcal{E}^c$ is the Sobolev space $H^1(\Omega)$.

in which sesquilinear form $a(\omega\,;\varphi\,,\delta\varphi)$ on $\mathcal{E}_\varphi^c \times \mathcal{E}_\varphi^c$ is the restriction to $\mathcal{E}_\varphi^c$ of the following sesquilinear form $a(\omega\,;\varphi\,,\delta\varphi)$ on $\mathcal{E}^c \times \mathcal{E}^c$ defined by

$$a(\omega;\varphi,\delta\varphi) = -\omega^2 m(\varphi,\delta\varphi) + i\omega\, d_\tau(\omega;\varphi,\delta\varphi) + k(\varphi,\delta\varphi) + s_z(\omega;\varphi,\delta\varphi)\,. \quad (13)$$

Sesquilinear form $m(\varphi\,,\delta\varphi)$ and operator M. Positive-definite sesquilinear form $m(\varphi\,,\delta\varphi)$ on $\mathcal{E}^c \times \mathcal{E}^c$ is such that

$$m(\varphi\,,\delta\varphi) = \frac{\rho_0}{c_0^2}\int_\Omega \varphi\,\overline{\delta\varphi}\;d\mathbf{x} \quad , \quad (14)$$

and defines the linear operator $\mathbf{M}$ such that [2]

$$<\mathbf{M}\,\varphi\,,\delta\varphi> = m(\varphi\,,\delta\varphi) \quad . \quad (15)$$

Sesquilinear form $k(\varphi\,,\delta\varphi)$ and operator K. Sesquilinear form $k(\varphi\,,\delta\varphi)$ is defined on $\mathcal{E}^c \times \mathcal{E}^c$ by

$$k(\varphi\,,\delta\varphi) = \rho_0\int_\Omega \boldsymbol{\nabla}\varphi\cdot\boldsymbol{\nabla}\overline{\delta\varphi}\;d\mathbf{x} \quad , \quad (16)$$

and is positive semidefinite ($k(\varphi\,,\delta\varphi) = 0$ if φ is equal to a constant). The restriction of $k(\varphi\,,\delta\varphi)$ to the admissible function space $\mathcal{E}_\varphi^c \times \mathcal{E}_\varphi^c$ is positive-definite. Sesquilinear form $k(\varphi\,,\delta\varphi)$ defines the linear operator [3] $\mathbf{K}$ such that

$$<\mathbf{K}\,\varphi\,,\delta\varphi> = k(\varphi\,,\delta\varphi) \quad . \quad (17)$$

Sesquilinear form $d_\tau(\omega\,;\varphi\,,\delta\varphi)$ and operator $\mathbf{D}_\tau(\omega)$. Positive-semidefinite sesquilinear form $d_\tau(\omega\,;\varphi\,,\delta\varphi)$ on $\mathcal{E}^c \times \mathcal{E}^c$ is such that

$$d_\tau(\omega\,;\varphi\,,\delta\varphi) = \tau(\omega)\,k(\varphi\,,\delta\varphi) \quad , \quad (18)$$

and defines the linear operator $\mathbf{D}_\tau(\omega)$ such that

$$<\mathbf{D}_\tau(\omega)\,\varphi\,,\delta\varphi> = d_\tau(\omega\,;\varphi\,,\delta\varphi) \quad . \quad (19)$$

---

2 Sesquilinear form $m(\varphi,\delta\varphi)$ is continuous on $H^c\times H^c$ with $H^c = L^2(\Omega)$ and is also continuous on $\mathcal{E}^c\times\mathcal{E}^c$, $\mathbf{M}$ is a real continuous operator from $\mathcal{E}^c$ into its antidual space $\mathcal{E}^{c\prime}$, and the angle brackets in Eq. (15) denote the antiduality product between $\mathcal{E}^{c\prime}$ and $\mathcal{E}^c$.

3 Sesquilinear form $k(\varphi,\delta\varphi)$ is continuous on $\mathcal{E}^c\times\mathcal{E}^c$, $\mathbf{K}$ is a continuous operator from $\mathcal{E}^c$ into its antidual space $\mathcal{E}^{c\prime}$ and the angle brackets in Eq. (17) denote the antiduality product between $\mathcal{E}^{c\prime}$, and $\mathcal{E}^c$. Continuous linear operator $\mathbf{K}$ is positive semi-definite Hermitian from $\mathcal{E}^c$ into $\mathcal{E}^{c\prime}$. This operator is coercive on $\mathcal{E}_\varphi^c$.

Sesquilinear form $d_T(\omega\,;\varphi\,,\delta\varphi)$ has the same properties as $k(\varphi\,,\delta\varphi)$ and consequently is positive-definite on $\mathcal{E}^c_\varphi \times \mathcal{E}^c_\varphi$.

Sesquilinear form $s_Z(\omega\,;\varphi\,,\delta\varphi)$ and operator $\mathbf{S}_Z(\omega)$. Sesquilinear form $s_Z(\omega\,;\varphi\,,\delta\varphi)$ defined on $\mathcal{E}^c \times \mathcal{E}^c$ is such that

$$s_Z(\omega\,;\varphi\,,\delta\varphi) = -\rho_0^2\omega^2\int_{\Gamma_Z}\frac{1}{i\omega Z}\,\varphi\,\overline{\delta\varphi}\;ds + \omega^2\kappa(\omega)\,\rho_0^2\,\pi_1(\omega;\varphi)\,\pi_1(\omega;\overline{\delta\varphi}), \tag{20}$$

and defines the linear operator $\mathbf{S}_Z(\omega)$ such that [4]

$$<\mathbf{S}_Z(\omega)\,\varphi\,,\delta\varphi>= s_Z(\omega\,;\varphi\,,\delta\varphi) \quad . \tag{21}$$

Antilinear form $f(\omega\,;\delta\varphi)$ and excitation vector $\mathbf{f}(\omega)$. Antilinear form $\delta\varphi \mapsto f(\omega\,;\delta\varphi)$ on $\mathcal{E}^c$ can be written as

$$f(\omega\,;\delta\varphi) = f_Q(\omega\,;\delta\varphi) + f_{\mathbf{u}_{\text{wall}}}(\omega\,;\delta\varphi) \quad , \tag{22}$$

in which (using Green's formula for the term containing $\boldsymbol{\nabla}^2 Q$),

$$f_Q(\omega;\delta\varphi) = -\int_\Omega \frac{Q}{i\omega}\overline{\delta\varphi}\,d\mathbf{x} + \frac{\tau c_0^2}{\omega^2}\int_\Omega \boldsymbol{\nabla}Q\cdot\boldsymbol{\nabla}\overline{\delta\varphi}\,d\mathbf{x} + \rho_0\,\kappa(\omega)\,\pi_Q(\omega)\,\pi_1(\omega\,;\overline{\delta\varphi}), \tag{23}$$

and where $f_{\mathbf{u}_{\text{wall}}}(\omega\,;\delta\varphi)$ is such that

$$f_{\mathbf{u}_{\text{wall}}}(\omega\,;\delta\varphi) = \rho_0\int_{\Gamma\cup\Gamma_Z}\mathbf{u}_{\text{wall}}\cdot\mathbf{n}\;\overline{\delta\varphi}\;ds - \rho_0\,\kappa(\omega)\,\pi_2(\mathbf{u}_{\text{wall}})\;\pi_1(\omega\,;\overline{\delta\varphi})\,. \tag{24}$$

Antilinear form $f(\omega\,;\delta\varphi)$ defines $\mathbf{f}(\omega)$ such that [5]

$$<\mathbf{f}(\omega)\,,\delta\varphi>= f(\omega\,;\delta\varphi) \quad . \tag{25}$$

---

[4] For all ω fixed in $\mathbb{R}$, since function $\mathbf{x}\mapsto\{i\omega Z(\mathbf{x},\omega)\}^{-1}$ is bounded on Γ_Z (see Eq. (X.18)), we deduce that $s_Z(\omega\,;\varphi\,,\delta\varphi)$ is a continuous sesquilinear form on $\mathcal{E}^c\times\mathcal{E}^c$. This sesquilinear form defines a continuous operator $\mathbf{S}_Z(\omega)$ from $\mathcal{E}^c$ into its antidual space $\mathcal{E}^{c\,\prime}$, and the angle brackets in Eq. (21) denote the antiduality product between $\mathcal{E}^{c\,\prime}$ and $\mathcal{E}^c$. It is assumed that Z is such that $\omega^{-1}\Re e\{-is_Z(\omega;\varphi,\varphi)\}=\omega^{-1}\Im m\{s_Z(\omega;\varphi,\varphi)\}\geq 0$.

[5] For all ω fixed in $\mathbb{R}$, it is assumed that function $\mathbf{x}\mapsto Q(\mathbf{x},\omega)/\omega^2$ belongs to $\mathcal{E}^c=H^1(\Omega)$ and $\mathbf{x}\mapsto\mathbf{u}_{\text{wall}}(\mathbf{x},\omega)\cdot\mathbf{n}(\mathbf{x})$ belongs to $\mathcal{E}^c_{\partial\Omega}=H^{1/2}(\partial\Omega)\subset L^2(\partial\Omega)$. Since function $\mathbf{x}\mapsto\{i\omega Z(\mathbf{x},\omega)\}^{-1}$ is bounded on Γ_Z (see Eq. (X.18)), we then deduce that $\delta\varphi\mapsto f(\omega\,;\delta\varphi)$ is a continuous antilinear form on $\mathcal{E}^c$. This antilinear form defines an element $\mathbf{f}(\omega)$ belonging to the antidual space $\mathcal{E}^{c\,\prime}$ of $\mathcal{E}^c$ and the angle brackets in Eq. (25) denote the antiduality product between $\mathcal{E}^{c\,\prime}$ and $\mathcal{E}^c$.

4.3. Linear operator equation

Using Eqs. (15), (17), (19), (21) and (25), the linear operator equation associated with Eqs. (12) and (13) is written as [6]

$$\mathbf{A}(\omega)\,\varphi = \mathbf{f}(\omega) \quad , \quad \varphi \in \mathcal{E}_\varphi^c \quad , \tag{26}$$

in which operator $\mathbf{A}(\omega)$ is defined by [7]

$$\mathbf{A}(\omega) = -\omega^2\,\mathbf{M} + i\omega\,\mathbf{D}_\tau(\omega) + \mathbf{K} + \mathbf{S}_Z(\omega) \quad . \tag{27}$$

5. Frequency Response Function

The operator-valued frequency response function is defined by the linear operator which associates each $\mathbf{f}$ with field φ as the solution of the variational formulation defined by Eq. (12). It can be proved [8] that, for all fixed real ω and $\mathbf{f}(\omega)$ given, the variational formulation defined by Eq. (12) has a unique solution $\mathbf{x} \mapsto \varphi(\mathbf{x},\omega)$ in $\mathcal{E}_\varphi^c$. Consequently, linear operator $\mathbf{A}(\omega)$ is invertible on $\mathcal{E}_\varphi^c$ and Eq. (26) yields

$$\varphi(\omega) = \mathbf{T}(\omega)\,\mathbf{f}(\omega) \quad , \tag{28}$$

in which $\omega \mapsto \mathbf{T}(\omega) = \mathbf{A}(\omega)^{-1}$ is the operator-valued frequency response function [9].

6. Finite Element Discretization

Since the finite element method uses a real basis for constructing the finite element matrices, finite element discretization of sesquilinear forms

---

[6] Continuous sesquilinear form $a(\omega\,;\varphi,\delta\varphi)$ on $\mathcal{E}_\varphi^c \times \mathcal{E}_\varphi^c$ defines a continuous linear operator $\mathbf{A}(\omega)$ from $\mathcal{E}_\varphi^c$ into $\mathcal{E}_\varphi^{c\,\prime}$ by $<\mathbf{A}(\omega)\,\varphi,\delta\varphi>=a(\omega\,;\varphi,\delta\varphi)$. Eq. (26) is an equality in $\mathcal{E}_\varphi^{c\,\prime}$.

[7] From footnote 6, linear operator $\mathbf{A}(\omega)$ is continuous from $\mathcal{E}_\varphi^c$ into $\mathcal{E}_\varphi^{c\,\prime}$. The operator on the right-hand side of Eq. (27) is the restriction to $\mathcal{E}_\varphi^c$ of an operator defined on $\mathcal{E}^c$.

[8] For $\omega>0$, $\tau(\omega)>0$ (see Eq. (X.76)) and $\Re e\{-is_Z(\omega;\varphi,\varphi)\}\geq 0$ (see footnote 4). The existence and uniqueness of the solution in $\mathcal{E}_\varphi^c$ is then obtained by applying the Lax-Milgram theorem to $a_c(\omega;\varphi,\delta\varphi)=f_c(\omega;\delta\varphi)$ in which $a_c(\omega;\varphi,\delta\varphi)=-i\,a(\omega;\varphi,\delta\varphi)$ and $f_c(\omega;\delta\varphi)=-i\,f(\omega;\delta\varphi)$. In effect, $\Re e\,a_c(\omega;\varphi,\varphi)=\omega d_\tau(\omega;\varphi,\varphi)+\Re e\{-is_Z(\omega;\varphi,\varphi)\}$ is coercive because $d_\tau(\omega;\varphi,\delta\varphi)$ is coercive on $\mathcal{E}_\varphi^c$ and $\Re e\{-is_Z(\omega;\varphi,\varphi)\}\geq 0$. The proof is similar for $\omega<0$ considering $a_c(\omega;\varphi,\delta\varphi)= i\,a(\omega;\varphi,\delta\varphi)$ and $f_c(\omega;\delta\varphi)=i\,f(\omega;\delta\varphi)$.

[9] Since linear operator $\mathbf{A}(\omega)$ is continuous and invertible from $\mathcal{E}_\varphi^c$ into $\mathcal{E}_\varphi^{c\,\prime}$, $\mathbf{T}(\omega)=\mathbf{A}(\omega)^{-1}$ is continuous from $\mathcal{E}_\varphi^{c\,\prime}$ into $\mathcal{E}_\varphi^c$.

$m(\varphi\,,\delta\varphi)$, $k(\varphi\,,\delta\varphi)$, $d_\tau(\varphi\,,\delta\varphi)$ and $s_z(\omega\,;\varphi\,,\delta\varphi)$ on $\mathcal{E}^c \times \mathcal{E}^c$ defined by Eqs. (14), (16), (18) and (20) respectively, yields $(n\times n)$ real matrices $[M]$, $[K]$, $[D_\tau]$ and $(n\times n)$ complex matrix $[S_z(\omega)]$. Consequently, finite element discretization of Eqs. (12) and (13) yields

$$[\mathbb{A}(\omega)]\,\mathbf{\Phi} = \mathbf{F}(\omega) \quad , \tag{29}$$

with the constraint

$$\mathbf{L}^T\mathbf{\Phi} = 0 \quad . \tag{30}$$

Eq. (30) corresponds to the finite element discretization of linear constraint $\int_\Omega \varphi\, d\mathbf{x} = 0$ appearing in the admissible function space $\mathcal{E}^c_\varphi$ defined by Eq. (11) and $\mathbf{L}$ is an $\mathbb{R}^n$ vector. In Eqs. (29) and (30), $\mathbf{\Phi} = (\Phi_1, \ldots, \Phi_n)$ denotes the complex vector of the DOFs which are the values of φ at the nodes of the finite element mesh of domain Ω. Complex vector $\mathbf{F} = (F_1, \ldots, F_n)$ corresponds to the finite element discretization of antilinear form $f(\omega\,;\delta\varphi)$ defined by Eq. (22) and $[\mathbb{A}(\omega)]$ is the $(n\times n)$ symmetric complex matrix such that

$$[\mathbb{A}(\omega)] = -\omega^2\,[M] + i\omega\,[D_\tau(\omega)] + [K] + [S_z(\omega)] \quad . \tag{31}$$

The $(n\times n)$ real matrix $[M]$ is symmetric and positive definite. The $(n\times n)$ real matrices $[K]$ and $[D_\tau(\omega)]$ are symmetric and positive semidefinite, of rank $n-1$ and have the same null space spanned by the vector $c\,\mathbf{1}$ where c is any real constant and $\mathbf{1} = (1, \ldots, 1) \in \mathbb{R}^n$. The $(n\times n)$ complex matrix $[S_z(\omega)]$ is symmetric and such that $\omega^{-1}\Im m[S_z(\omega)]$ is positive semidefinite. For all fixed real ω, Eq. (29) with the constraint defined by Eq. (30) has a unique solution which is written as

$$\mathbf{\Phi}(\omega) = [T(\omega)]\,\mathbf{F}(\omega) \quad , \tag{32}$$

in which the $(n\times n)$ symmetric complex matrix $[T(\omega)]$ is the matrix-valued frequency response function. It should be noted that $[T(\omega)]$ is not the inverse of matrix $[\mathbb{A}(\omega)]$, which is not invertible. The corresponding $(n\times n)$ real matrix-valued impulse response function $t \mapsto [\,h(t)\,]$ is such that (see Eq. (V.72)),

$$[T(\omega)] = \int_{\mathbb{R}} e^{-i\omega t}\,[h(t)]\,dt \quad . \tag{33}$$

7. Acoustic Modes of the Acoustic Cavity

In this section, we introduce the acoustic modes which are the eigenfunctions of a spectral problem. The basic notions on this subject were introduced in Chapter III.

7.1. The spectral boundary value problem

The acoustic modes are obtained by considering the associated conservative system corresponding to an inviscid acoustic fluid in acoustic cavity Ω with a fixed wall, without wall acoustic impedance and with no acoustic sources. From Eqs. (1) to (4), the equations of this classical spectral problem are the Helmholtz equation

$$\boldsymbol{\nabla}^2\varphi + \frac{\omega^2}{c_0^2}\,\varphi = 0 \qquad \text{in} \quad \Omega \quad , \tag{34}$$

with the Neumann boundary condition,

$$\frac{\partial\varphi}{\partial\mathbf{n}} = 0 \qquad \text{on} \quad \partial\Omega \quad , \tag{35}$$

and the constraint

$$\int_\Omega \varphi\, d\mathbf{x} = 0 \quad . \tag{36}$$

From Eqs. (6) to (9), we deduce that $\pi(\omega) = 0$. Then Eq. (5) yields

$$p(\mathbf{x},\omega) = \omega^2\,\rho_0\,\varphi(\mathbf{x},\omega) \qquad \text{in} \quad \Omega \quad . \tag{37}$$

7.2. Variational formulation of the spectral boundary value problem

The eigenvalue problem. Multiplying Eq. (34) by ρ_0, we deduce the variational formulation of the spectral boundary value problem defined by Eqs. (34) to (36). Find $\lambda = \omega^2$ and φ in $\mathcal{E}_\varphi^c$ such that

$$k(\varphi\,,\delta\varphi) = \lambda\; m(\varphi\,,\delta\varphi) \quad , \quad \forall\,\delta\varphi \in \mathcal{E}_\varphi^c \quad . \tag{38}$$

Countable number of positive eigenvalues. Since Ω is a bounded domain and from the properties indicated in Section 4 (symmetry and positive definiteness of sesquilinear forms m and k on $\mathcal{E}_\varphi^c \times \mathcal{E}_\varphi^c$), there exists an increasing sequence of positive eigenvalues

$$0 < \lambda_1 \leq \lambda_2 \leq \ldots \leq \lambda_\alpha \leq \ldots \quad . \tag{39}$$

In addition, any multiple eigenvalue has a finite multiplicity.

Completeness of the eigenfunctions. Let φ_α be the eigenfunction associated with eigenvalue λ_α. The eigenfunctions which are in $\mathcal{E}_\varphi^c$ are real, and $\{\varphi_\alpha\}_{\alpha\geq 1}$ form a complete set in $\mathcal{E}_\varphi^c$ which means that an arbitrary function φ belonging to $\mathcal{E}_\varphi^c$ can be expanded as

$$\varphi(\mathbf{x}) = \sum_{\alpha=1}^{+\infty} q_\alpha \, \varphi_\alpha(\mathbf{x}) \quad , \tag{40}$$

in which $\{q_\alpha\}_\alpha$ is a sequence of complex numbers. Eigenfunction φ_α is called the *acoustic mode* corresponding to *eigenfrequency* ω_α such that

$$\omega_\alpha = \sqrt{\lambda_\alpha} \quad . \tag{41}$$

Orthogonality of the eigenfunctions. The sequence of eigenfunctions $\{\varphi_\alpha\}_\alpha$ satisfies the orthogonality conditions

$$m(\varphi_\alpha \, , \varphi_\beta) = \mu_\alpha \, \delta_{\alpha\beta} \quad , \tag{42}$$

$$k(\varphi_\alpha \, , \varphi_\beta) = \mu_\alpha \, \omega_\alpha^2 \, \delta_{\alpha\beta} \quad , \tag{43}$$

in which μ_α is a positive real number depending on the normalization of eigenfunction φ_α, because the eigenfunctions are defined to within a multiplication constant.

Linear operator equation for the eigenvalue problem. The linear operator equation corresponding to the variational problem defined by Eq. (38) is written as

$$\mathbf{K}\varphi = \lambda \, \mathbf{M}\varphi \quad , \quad \varphi \in \mathcal{E}_\varphi^c \quad , \tag{44}$$

in which $\mathbf{M}$ and $\mathbf{K}$ are defined by Eqs. (15) and (17). The corresponding algebraic properties defined by Eqs. (42) and (43) can be rewritten as

$$<\mathbf{M}\,\varphi_\alpha \, , \varphi_\beta> = \mu_\alpha \, \delta_{\alpha\beta} \quad , \tag{45}$$

$$<\mathbf{K}\,\varphi_\alpha \, , \varphi_\beta> = \mu_\alpha \, \omega_\alpha^2 \, \delta_{\alpha\beta} \quad . \tag{46}$$

Referring to Section 4, we recall that $\mathbf{M}$ and $\mathbf{K}$ are symmetric positive-definite operators on $\mathcal{E}_\varphi^c$. It should be noted that operator $\mathbf{K}$ is only positive semidefinite on $\mathcal{E}^c$.

8. Finite Element Discretization and Generalized Symmetric Matrix Eigenvalue Problem

We consider the eigenvalue problem defined by

$$k(\varphi\,,\delta\varphi) = \lambda\; m(\varphi\,,\delta\varphi) \quad , \quad \forall\,\delta\varphi \in \mathcal{E}^c \quad , \tag{47}$$

corresponding to Eq. (38) for which the admissible function space $\mathcal{E}^c_\varphi$ is replaced by $\mathcal{E}^c$. This means that we remove the constraint on φ. It can easily be seen that the solutions $\{\lambda_\alpha > 0, \varphi_\alpha\}_{\alpha\geq 1}$ of the eigenproblem defined by Eq. (47) are the solutions of the eigenvalue problem defined by Eq. (38) with an additional solution $\lambda_0 = 0$ corresponding to the eigenfunction $\varphi_0 = c$ in which c is any real constant. For α and β in $\{0, 1, 2, \ldots\}$, the orthogonality conditions defined by Eqs. (42) and (43) hold. In particular, applying Eq. (42) with $\beta = 0$ and $\alpha \geq 1$ yields

$$\int_\Omega \varphi_\alpha(\mathbf{x})\, d\mathbf{x} = 0 \quad , \quad \forall\,\alpha \geq 1 \quad . \tag{48}$$

Consequently, every solution of Eq. (38) is a solution of Eq. (47) and every solution $\alpha \geq 1$ of Eq. (47) is a solution of Eq. (38). Therefore, instead of solving Eq. (38) numerically which requires solving a problem with a constraint, we solve Eq. (48) without any constraint and solution $(\lambda_0 = 0, \varphi_0 = c)$ is not considered. The matrix equation of the generalized symmetric eigenvalue problem corresponding to the finite element discretization of Eq. (47) is written as

$$[\,K\,]\,\mathbf{\Phi} = \lambda\,[\,M\,]\,\mathbf{\Phi} \quad , \tag{49}$$

in which $[\,M\,]$, $[\,K\,]$ and $\mathbf{\Phi}$ are defined in Section 6. The numerical solution of the generalized symmetric eigenvalue problem defined by Eq. (49) requires special eigenvalue solvers, see Parlett, 1980; Bathe and Wilson, 1976; Golub and Van Loan, 1989; Chatelin, 1993.

9. FRF Calculation in the LF Range

As before, we denote the LF band as $\mathbb{B}_{\rm LF}$ such that

$$\mathbb{B}_{\rm LF} = [\,0\,,\,\omega_{\rm LF,\,final}\,] \quad . \tag{50}$$

We consider the low-frequency case for which ω belongs to $\mathbb{B}_{\rm LF}$. The methodology presented in Chapter VI can be applied directly.

9.1. Response to a given excitation vector

Using the Ritz-Galerkin methodology (see Section VI.3.1), we introduce the subspace $\mathcal{E}^c_{\varphi,N}$ of $\mathcal{E}^c_{\varphi}$, of dimension $N \geq 1$, spanned by the finite family $\{\varphi_1, \ldots, \varphi_N\}$ of acoustic modes φ_α defined in Section 7, for which $\lambda_\alpha > 0$. The projection φ^N on $\mathcal{E}^c_{\varphi,N}$ of the unique solution φ of Eq. (12) can be written as

$$\varphi^N(\mathbf{x}, \omega) = \sum_{\alpha=1}^{N} q_\alpha(\omega)\, \varphi_\alpha(\mathbf{x}) \quad , \tag{51}$$

in which

$$\mathbf{q} = (q_1, \ldots, q_N) \tag{52}$$

is a complex vector of the generalized coordinates. From the orthogonality conditions defined by Eqs. (42) and (43), we deduce that $\mathbf{q}$ is the solution of the following matrix equation of dimension N

$$[\mathcal{A}(\omega)]\, \mathbf{q}(\omega) = \boldsymbol{\mathcal{F}}(\omega) \quad . \tag{53}$$

From Eq. (13), we deduce that the $(N \times N)$ complex symmetric matrix $[\mathcal{A}(\omega)]$ is written as

$$[\mathcal{A}(\omega)] = -\omega^2 [\mathcal{M}] + i\omega\, [\mathcal{D}_\tau(\omega)] + [\mathcal{K}] + [\mathcal{S}_z(\omega)] \quad , \tag{54}$$

where $[\mathcal{M}]$, $[\mathcal{K}]$ and $[\mathcal{D}_\tau(\omega)]$ are $(N \times N)$ positive-definite diagonal real matrices and where $[\mathcal{S}_z(\omega)]$ is a dense $(N \times N)$ symmetric complex matrix such that

$$[\mathcal{M}]_{\alpha\beta} = m(\varphi_\beta, \varphi_\alpha) = \mu_\alpha\, \delta_{\alpha\beta} \quad , \tag{55}$$

$$[\mathcal{K}]_{\alpha\beta} = k(\varphi_\beta, \varphi_\alpha) = \mu_\alpha\, \omega_\alpha^2 \delta_{\alpha\beta} \quad , \tag{56}$$

$$[\mathcal{D}_\tau(\omega)]_{\alpha\beta} = d_\tau(\omega\,; \varphi_\beta, \varphi_\alpha) = \tau(\omega)\, \mu_\alpha\, \omega_\alpha^2 \delta_{\alpha\beta} \quad , \tag{57}$$

$$[\mathcal{S}_z(\omega)]_{\alpha\beta} = s_z(\omega\,; \varphi_\beta, \varphi_\alpha) \quad , \tag{58}$$

in which sesquilinear form s_z is defined by Eq. (20). It should be noted that dense matrix $[\mathcal{S}_z(\omega)]$ is due only to the presence of an acoustic impedance $Z(\mathbf{x}, \omega)$ on boundary Γ_Z and consequently uses only the values of the acoustic modes on Γ_Z. Component $\mathcal{F}_\alpha$ of the generalized force vector

$$\boldsymbol{\mathcal{F}} = (\mathcal{F}_1, \ldots, \mathcal{F}_N) \quad , \tag{59}$$

is written as

$$\mathcal{F}_\alpha(\omega) = f(\omega\,; \varphi_\alpha) \quad , \tag{60}$$

in which antilinear form $f(\omega\,;\delta\varphi)$ is defined by Eqs. (22) to (24). From Eq. (53) and the invertibility of matrix $[\mathcal{A}(\omega)]$, we deduce that

$$\mathbf{q}(\omega) = [\mathcal{T}(\omega)]\,\mathcal{F}(\omega) \quad , \tag{61}$$

in which $[\mathcal{T}(\omega)]$ is an $(N \times N)$ complex symmetric matrix such that

$$[\mathcal{T}(\omega)] = [\mathcal{A}(\omega)]^{-1} \quad , \quad [\mathcal{T}(\omega)]^T = [\mathcal{T}(\omega)] \quad . \tag{62}$$

From Eqs. (51) and (61), we deduce that

$$\varphi^N(\mathbf{x},\omega) = \sum_{\alpha=1}^{N}\sum_{\beta=1}^{N} [\mathcal{T}(\omega)]_{\alpha\beta}\,\mathcal{F}_\beta(\omega)\,\varphi_\alpha(\mathbf{x}) \quad . \tag{63}$$

9.2. Projection of the FRF on the acoustic modes

Operator-valued frequency response function $\mathbf{T}(\omega)$ is an intrinsic operator characterizing the dynamics of the system independently of the values of excitation vector $\mathbf{f}(\omega)$. We can then choose a particular $\mathbf{f}(\omega)$ to construct the projection of $\mathbf{T}(\omega)$, and we define element $\mathbf{f}(\omega)$ by

$$\begin{aligned} f(\omega\,;\delta\varphi) &= <\mathbf{f}(\omega)\,,\delta\varphi> \\ &= \int_\Omega g(\mathbf{x},\omega)\,\overline{\delta\varphi(\mathbf{x})}\;d\mathbf{x} \quad , \end{aligned} \tag{64}$$

in which $\mathbf{x} \mapsto g(\mathbf{x},\omega)$ is any complex-valued square integrable [10] function on Ω. Consequently, Eqs. (60) and (64) yield

$$\mathcal{F}_\beta(\omega) = \int_\Omega g(\mathbf{x},\omega)\,\varphi_\beta(\mathbf{x})\;d\mathbf{x} \quad . \tag{65}$$

Let $\mathbf{T}_N(\omega)$ be the projection on $\mathcal{E}^c_{\varphi,N}$ of operator-valued frequency response function $\mathbf{T}(\omega)$ defined by Eq. (28). From Eq. (63), we deduce that

$$\mathbf{T}_N(\omega)\,\mathbf{f}(\omega) = \sum_{\alpha=1}^{N}\sum_{\beta=1}^{N} [\mathcal{T}(\omega)]_{\alpha\beta}\,\mathcal{F}_\beta(\omega)\,\varphi_\alpha \quad . \tag{66}$$

10 Since $\mathbf{x}\mapsto g(\mathbf{x},\omega)$ belongs to $H^c = L^2(\Omega)$, the antiduality product $<.\,,.>$ between $\mathcal{E}^{c\,\prime}_\varphi$ and $\mathcal{E}^c_\varphi$ is equal to the usual inner product $(\varphi\,,\delta\varphi)_{H^c} = \int_\Omega \varphi(\mathbf{x})\,\overline{\delta\varphi(\mathbf{x})}\,d\mathbf{x}$ in H^c.

Substituting Eq. (65) into Eq. (66) yields, for all $\mathbf{x}$ in Ω,

$$\big(\mathbf{T}_N(\omega)\,\mathbf{f}(\omega)\big)(\mathbf{x}) = \int_\Omega \tau_N(\omega;\mathbf{x},\mathbf{x}')\; g(\mathbf{x}',\omega)\,d\mathbf{x}' \quad , \tag{67}$$

in which $\tau_N(\omega;\mathbf{x},\mathbf{x}')$ is such that

$$\tau_N(\omega;\mathbf{x},\mathbf{x}') = \sum_{\alpha=1}^{N}\sum_{\beta=1}^{N} [\mathcal{T}(\omega)]_{\alpha\beta}\,\varphi_\alpha(\mathbf{x})\,\varphi_\beta(\mathbf{x}') \quad . \tag{68}$$

Eq. (67) shows that linear operator [11] $\mathbf{T}_N(\omega)$ is an integral operator with kernel $\tau_N(\omega;\mathbf{x},\mathbf{x}')$ satisfying the following property

$$\tau_N(\omega;\mathbf{x},\mathbf{x}') = \tau_N(\omega;\mathbf{x}',\mathbf{x}) \quad . \tag{69}$$

Consequently, for all fixed real ω, we have [12]

$$\mathbf{T}_N(\omega)^* = \overline{\mathbf{T}_N(\omega)} \quad . \tag{70}$$

9.3. Finite element discretization of the FRF projection on acoustic modes

Consider the finite element discretization introduced in Section 6. Let $\{\boldsymbol{\Phi}_\alpha\}_{\alpha=1,\ldots,N}$ be the set of N eigenvectors defined in Section 8. The matrix $[T_N(\omega)]$ of operator $\mathbf{T}_N(\omega)$ obtained by finite element discretization can then be written as

$$[T_N(\omega)] = \sum_{\alpha=1}^{N}\sum_{\beta=1}^{N} [\mathcal{T}(\omega)]_{\alpha\beta}\,\boldsymbol{\Phi}_\alpha\,\boldsymbol{\Phi}_\beta^T \quad , \tag{71}$$

in which $[\mathcal{T}(\omega)]$ is the $(N\times N)$ symmetric complex matrix defined by the first Eq. (62). Matrix $[T_N(\omega)]$ is an $(n\times n)$ complex matrix such that

$$[T_N(\omega)]^T = [T_N(\omega)] \quad . \tag{72}$$

This property corresponds to Eq. (70).

---

[11] Integral operator $\mathbf{T}_N(\omega)$ considered as an operator in $H^c = L^2(\Omega)$ can be written as $\mathbf{T}_N(\omega) = \sum_{\alpha=1}^{N}\sum_{\beta=1}^{N} [\mathcal{T}(\omega)]_{\alpha\beta}\;\varphi_\alpha \otimes_{H^c} \varphi_\beta$.

[12] $\mathbf{T}_N(\omega)^*$ is the adjoint operator of bounded operator $\mathbf{T}_N(\omega)$ (see Mathematical Notations in the appendix).

9.4. Introduction of quasi-static correction terms

Construction of the quasi-static correction terms. Let $N \geq 1$ be the integer such that

$$0 < \omega_1 \leq \ldots \leq \omega_N < \omega_{_{\text{LF, final}}} < \omega_{N+1} \leq \cdots \quad , \tag{73}$$

in which $\mathbb{B}_{\text{LF}} = [\,0\,, \omega_{_{\text{LF, final}}}\,]$ is the low-frequency band of analysis. Projection $\mathbf{T}_{_N}(\omega)$ of operator-valued frequency response function $\mathbf{T}(\omega)$, defined by Eqs. (67) and (68), represents the approximation corresponding to the first N acoustic modes $\{\varphi_\alpha\}_{\alpha=1,\ldots,N}$ whose eigenfrequencies lie inside $\mathbb{B}_{\text{LF}}$. We introduce the operator

$$\mathbf{R}_{_N}(\omega) = \mathbf{T}(\omega) - \mathbf{T}_{_N}(\omega) \quad , \tag{74}$$

which represents the projection of $\mathbf{T}(\omega)$ on the remaining acoustic modes $\{\varphi_\alpha\}_{\alpha=N+1,\ldots,+\infty}$. We then have the identity

$$\mathbf{T}(\omega) = \mathbf{T}_{_N}(\omega) + \mathbf{R}_{_N}(\omega) \quad . \tag{75}$$

For $\omega \in \mathbb{B}_{\text{LF}}$, operator $\mathbf{R}_{_N}(\omega)$ can be approximated by $\mathbf{R}_{_N}(0)$. Operator $\mathbf{R}_{_N}(0)$ represents the quasi-static contribution of the acoustic modes whose eigenfrequencies are greater than $\omega_{_{\text{LF, final}}}$. Consequently, in order to accelerate convergence, we introduce the sequence of operators including quasi-static correction terms such that

$$\mathbf{T}_{_{N,\,\text{acc}}}(\omega) = \mathbf{T}_{_N}(\omega) + \mathbf{R}_{_N}(0) \quad , \tag{76}$$

with $\mathbf{R}_{_N}(0)$ given by

$$\mathbf{R}_{_N}(0) = \mathbf{T}(0) - \mathbf{T}_{_N}(0) \quad . \tag{77}$$

From Eqs. (20) and (X.25), we deduce that $s_z(0\,;\varphi\,,\delta\varphi) = 0$ and consequently, Eq. (21) yields $\mathbf{S}_z(0) = 0$. From Eq. (27), we deduce that

$$\mathbf{T}(0) = \mathbf{K}^{-1} \quad , \tag{78}$$

in which $\mathbf{K}^{-1}$ denotes the inverse of $\mathbf{K}$, considered as an operator defined on $\mathcal{E}_\varphi^c$. From Eqs. (67) and (68), we deduce that

$$\big(\mathbf{T}_{_N}(0)\,\mathbf{f}(\omega)\big)(\mathbf{x}) = \int_\Omega \tau_{_N}(0;\mathbf{x},\mathbf{x}')\; g(\mathbf{x}',\omega)\,d\mathbf{x}' \quad , \tag{79}$$

$$\tau_N(0;\mathbf{x},\mathbf{x}') = \sum_{\alpha=1}^{N} \frac{1}{\mu_\alpha\,\omega_\alpha^2}\,\varphi_\alpha(\mathbf{x})\,\varphi_\alpha(\mathbf{x}')^T \quad . \tag{80}$$

Finally, we then deduce that

$$\mathbf{T}_{N,\,\mathrm{acc}}(\omega) = \mathbf{T}_N(\omega) + \mathbf{T}_{N,\,\mathrm{stat}} \quad , \tag{81}$$

in which $\mathbf{T}_N(\omega)$ is defined by Eqs. (67) and (68), and $\mathbf{T}_{N,\,\mathrm{stat}} = \mathbf{R}_N(0)$ is written as

$$\mathbf{T}_{N,\,\mathrm{stat}} = \mathbf{K}^{-1} - \mathbf{T}_N(0) \quad , \tag{82}$$

where $\mathbf{T}_N(0)$ is defined by Eqs. (79) and (80). For N fixed and for all ω fixed in $\mathbb{B}_{\mathrm{LF}}$, projection $\mathbf{T}_{N,\,\mathrm{acc}}(\omega)$ is a more accurate approximation of $\mathbf{T}(\omega)$ than $\mathbf{T}_N(\omega)$.

Finite element discretization of the operator-valued FRF. We consider the finite element discretization introduced in Section 6. The finite element discretization of operator $\mathbf{K}^{-1}$ is the $(n\times n)$ real symmetric matrix denoted as $[T_0]$. This matrix is such that $\mathbf{\Phi} = [T_0]\,\mathbf{F}$ where $\mathbf{\Phi}$ is the unique solution of equation $[\,K\,]\,\mathbf{\Phi} = \mathbf{F}$ under the constraint $\mathbf{L}^T\mathbf{\Phi} = 0$. The finite element discretization of Eq. (81) is then written as

$$[T_{N,\,\mathrm{acc}}(\omega)] = [T_N(\omega)] + [T_{N,\,\mathrm{stat}}] \quad , \tag{83}$$

in which $[T_N(\omega)]$ is defined by Eq. (71) and $[T_{N,\,\mathrm{stat}}]$ is such that

$$[T_{N,\,\mathrm{stat}}] = [T_0] - \sum_{\alpha=1}^{N} \frac{1}{\mu_\alpha\,\omega_\alpha^2}\,\mathbf{\Phi}_\alpha\,\mathbf{\Phi}_\alpha^T \quad . \tag{84}$$

9.5. Frequency-by-frequency construction of the FRF

We use the numerical procedure described in Section VI.6 to construct an approximation of $\omega \mapsto \mathbf{T}(\omega)$ on low-frequency band $\mathbb{B}_{\mathrm{LF}}$ based on the use of projection $\mathbf{T}_N(\omega)$ of $\mathbf{T}(\omega)$ on the first N acoustic modes, taking into account the quasi-static correction terms (if necessary). In a first step, the acoustic modes whose eigenfrequencies lie inside $\mathbb{B}_{\mathrm{LF}}$ are calculated. In the low-frequency range, N is small i.e., in the context of finite element discretization with n DOFs, we have $N \ll n$. Consequently, a frequency-by-frequency construction can be used. In a second step, matrix $[\mathcal{A}(\omega)]$ defined by Eq. (54) is inverted numerically frequency by frequency in $\mathbb{B}_{\mathrm{LF}}$ in order to construct $[\mathcal{T}(\omega)]$. From Eq. (71), we then deduce $[T_N(\omega)]$. If the quasi-static correction terms are taken into account, $[T_{N,\,\mathrm{stat}}]$ is calculated using Eq. (84) and, from Eq. (83), we deduce $[T_{N,\,\mathrm{acc}}(\omega)]$. It should be noted that matrix $[T_0]$ is generally not constructed explicitly as explained in Sections IX.3.3, IX.4.3 and IX.5.4 for deterministic and random excitations.

10. FRF Calculation in the MF Range

Let $\mathbb{B}_{\mathrm{MF}}$ be the MF broad band defined by

$$\mathbb{B}_{\mathrm{MF}} = [\omega_{\mathrm{MF,\,init}} \,,\, \omega_{\mathrm{MF,\,final}}] \quad . \tag{85}$$

For the reasons given in Section VII.1, we use the finite element discretization introduced in Section 6 to construct the FRF in the MF range. Consequently, for all ω in $\mathbb{B}_{\mathrm{MF}}$, we have to construct $[T(\omega)]$ defined by Eq. (32). From Eqs. (29) to (31), we then have to solve for the class of MF narrow band excitations $\theta_\nu(\omega)\,\mathbf{B}$ introduced in Section VII.8,

$$\left(-\omega^2\,[M] + i\omega\,[D_{\mathcal{T}}(\omega)] + [K] + [S_{\mathcal{Z}}(\omega)]\right)\boldsymbol{\Phi}(\omega) = \theta_\nu(\omega)\,\mathbf{B} \;,\; \omega \in \mathbb{B}_\nu \;, \tag{86}$$

with the constraint

$$\mathbf{L}^T\boldsymbol{\Phi}(\omega) = 0 \quad , \quad \omega \in \mathbb{B}_\nu \quad , \tag{87}$$

in which the MF narrow band $\mathbb{B}_\nu$ is defined by

$$\mathbb{B}_\nu = [\,\Omega_\nu - \Delta\omega/2 \,,\, \Omega_\nu + \Delta\omega/2\,] \quad , \tag{88}$$

where $\Omega_\nu > 0$ is the center frequency of band $\mathbb{B}_\nu$ and $\Delta\omega$ is its bandwidth. For construction of the frequency response function, we take $\theta_\nu(\omega) = 1$ for $\omega \in \mathbb{B}_\nu$, and $\mathbf{B}$ represents the vectors of the canonical basis of $\mathbb{R}^n$. Concerning the methods of calculation of the response to various types of deterministic and random excitations, we refer the reader to Chapter IX.

10.1. Direct use of the MF method

The MF method presented in Chapter VII uses an approximation of Eq. (86) due to the frequency-dependent matrices $[D_{\mathcal{T}}(\omega)]$ and $[S_{\mathcal{Z}}(\omega)]$, consisting in replacing Eq. (86) by

$$\left(-\omega^2\,[M] + i\omega\,[D_\nu] + [K_\nu]\right)\boldsymbol{\Phi}_\nu(\omega) = \theta_\nu(\omega)\,\mathbf{B} \quad , \quad \omega \in \mathbb{B}_\nu \quad , \tag{89}$$

in which the frequency-independent $(n\times n)$ real matrices $[D_\nu]$ and $[K_\nu]$ are written as

$$[D_\nu] = [D_\tau(\Omega_\nu)] + [S_z^{\mathrm{I}}(\Omega_\nu)] \quad , \quad [K_\nu] = [K] + [S_z^{\mathrm{R}}(\Omega_\nu)] \quad , \tag{90}$$

where the $(n\times n)$ real matrices $[S_z^{\mathrm{I}}(\Omega_\nu)]$ and $[S_z^{\mathrm{R}}(\Omega_\nu)]$ are such that

$$[S_z^{\mathrm{R}}(\Omega_\nu)] = \Re e\,[S_z(\Omega_\nu)] \quad , \quad \omega\,[S_z^{\mathrm{I}}(\Omega_\nu)] = \Im m\,[S_z(\Omega_\nu)] \quad . \tag{91}$$

By inspecting Eq. (86), it can be seen that $\omega \mapsto [D_\tau(\omega)]$ and $[S_z(\omega)]$ can be considered as slowly varying functions on MF narrow band $\mathbb{B}_\nu$ (function $\omega \mapsto [D_\tau(\omega)]$ is due to the dissipative acoustic phenomenon inside the internal acoustic fluid and varies slowly on $\mathbb{B}_\nu$; functions $\omega \mapsto [S_z^{\mathrm{R}}(\omega)]$, $[S_z^{\mathrm{I}}(\omega)]$ and $\kappa(\omega)$ are related to the wall acoustic impedance $Z(\mathbf{x},\omega)$ and vary slowly on $\mathbb{B}_\nu$). We then apply the frequency transform technique presented in Sections VII.6.1 and VII.6.2 to obtain the associated LF equation which is written in the time domain as

$$[M]\,\ddot{\mathbf{\Phi}}_0(t) + [\widetilde{D}_\nu]\,\dot{\mathbf{\Phi}}_0(t) + [\widetilde{K}_\nu]\mathbf{\Phi}_0(t) = \theta_0(t)\,\mathbf{B} \quad , \quad t > t_{\mathrm{i}} \ , \tag{92}$$

$$\mathbf{\Phi}_0(t_{\mathrm{i}}) = \mathbf{0} \quad , \quad \dot{\mathbf{\Phi}}_0(t_{\mathrm{i}}) = \mathbf{0} \quad , \tag{93}$$

with the constraint

$$\mathbf{L}^T\mathbf{\Phi}_0(t) = 0 \quad , \quad t > t_{\mathrm{i}} \quad , \tag{94}$$

in which $(n\times n)$ complex symmetric matrices $[\widetilde{D}_\nu]$ and $[\widetilde{K}_\nu]$ are such that

$$[\widetilde{D}_\nu] = [D_\nu] + 2\,i\,\Omega_\nu\,[M] \quad , \tag{95}$$

$$[\widetilde{K}_\nu] = -\Omega_\nu^2\,[M] + i\,\Omega_\nu\,[D_\nu] + [K_\nu] \quad . \tag{96}$$

Once $\mathbf{\Phi}_0$ has been calculated in the time domain, the expression of $\mathbf{\Phi}_\nu(\omega)$ in MF narrow band $\mathbb{B}_\nu$ is calculated by

$$\mathbf{\Phi}_\nu(\omega) \simeq \mathbf{1}_{\mathbb{B}_\nu}(\omega)\,\Delta t \sum_{m=m_{\mathrm{i}}}^{m_{\mathrm{f}}} \mathbf{\Phi}_0(m\,\Delta t)\,e^{-im\,\Delta t\,(\omega-\Omega_\nu)} \quad , \quad \forall\omega\in\mathbb{R} \ , \tag{97}$$

in which $\mathbf{1}_{\mathbb{B}_\nu}(\omega) = 1$ if $\omega\in\mathbb{B}_\nu$ and $= 0$ if $\omega\notin\mathbb{B}_\nu$ and where integers m_{i} and m_{f} are defined in Section VII.6.4.

10.2. Special procedure

From Eq. (20), matrix $[S_z(\omega)]$ can be written as

$$[S_z(\omega)] = [S_1^c(\omega)] + [S_2^c(\omega)] \quad , \tag{98}$$

in which $[S_1^c(\omega)]$ is an $(n \times n)$ sparse complex symmetric matrix corresponding to the finite element discretization of the sesquilinear form

$$(\varphi, \delta\varphi) \mapsto -\rho_0^2 \omega^2 \int_{\Gamma_Z} \frac{1}{i\omega\, Z}\, \varphi\, \overline{\delta\varphi}\, ds \quad , \tag{99}$$

and $[S_2^c(\omega)]$ is an $(n \times n)$ complex symmetric matrix corresponding to the finite element discretization of the sesquilinear form

$$(\varphi, \delta\varphi) \mapsto \omega^2\, \kappa(\omega)\, \rho_0^2\, \pi_1(\omega\,; \varphi)\; \pi_1(\omega\,; \overline{\delta\varphi}) \quad . \tag{100}$$

This matrix is then written as

$$[S_2^c(\omega)] = \omega^2\, \kappa(\omega)\, \rho_0^2\, \mathbf{\Pi}_1(\omega)\, \mathbf{\Pi}_1(\omega)^T \quad , \tag{101}$$

where $\mathbf{\Pi}_1(\omega)$ is a $\mathbb{C}^n$ vector such that $\mathbf{\Pi}_1(\omega)^T$ corresponds to the finite element discretization of linear form $\varphi \mapsto \pi_1(\omega\,; \varphi)$ defined by Eq. (7-1). The right-hand side of Eq. (101) shows that the block submatrix corresponding to the DOFs on Γ_Z is dense, not sparse (the other elements of this matrix are zero). The algebraic property of matrix $[S_2^c(\omega)]$ can be used to solve Eqs. (86) and (87), using the following algebraic result and the MF method.

Algebraic result. Let $[\,A\,]$ be an $(n \times n)$ invertible complex matrix, $\mathbf{\Gamma}$ be a $\mathbb{C}^n$ vector, κ be a complex number and $\mathbf{F}$ be a $\mathbb{C}^n$ vector. If the linear matrix equation

$$(\,[\,A\,] + \kappa\, \mathbf{\Gamma}\mathbf{\Gamma}^T)\, \mathbf{X} = \mathbf{F} \tag{102}$$

has a unique solution $\mathbf{X}$, then this solution is given by

$$\mathbf{X} = \mathbf{V} - \frac{\kappa\, \mathbf{\Gamma}^T \mathbf{V}}{1 + \kappa\, \mathbf{\Gamma}^T \mathbf{W}}\, \mathbf{W} \quad , \tag{103}$$

in which $\mathbf{V}$ and $\mathbf{W}$ are the solutions of linear matrix equations

$$[A]\, \mathbf{V} = \mathbf{F} \quad , \tag{104}$$

$$[A]\, \mathbf{W} = \mathbf{\Gamma} \quad . \tag{105}$$

It should be noted that Eqs. (104) and (105) correspond to the same system with two different second members.

Application to the FRF calculation in the MF range. Matrix $[S_1^c(\omega)]$ appearing in Eq. (98) is written as

$$[S_1^c(\omega)] = [S_1^R(\omega)] + i\omega\,[S_1^I(\omega)] \quad , \tag{106}$$

in which sparse real symmetric matrices $[S_1^R(\omega)]$ and $[S_1^I(\omega)]$ are such that

$$[S_1^R(\omega)] = \Re e\,[S_z(\omega)] \quad , \quad \omega\,[S_1^I(\omega)] = \Im m\,[S_z(\omega)] \quad . \tag{107}$$

The approximation of Eq. (86) defined by Eq. (89) is then replaced by the new approximation

$$\left([A_\nu(\omega)] + \kappa(\omega)\,\mathbf{\Gamma}(\omega)\,\mathbf{\Gamma}(\omega)^T\right)\mathbf{\Phi}_\nu(\omega) = \theta_\nu(\omega)\,\mathbf{B} \quad , \quad \omega \in \mathbb{B}_\nu \quad , \tag{108}$$

with the constraint,

$$\mathbf{L}^T\mathbf{\Phi}_\nu(\omega) = 0 \quad , \quad \omega \in \mathbb{B}_\nu \quad . \tag{109}$$

In Eq. (108), sparse $(n\times n)$ complex symmetric matrix $[A_\nu(\omega)]$ is defined by

$$[A_\nu(\omega)] = -\omega^2\,[\,M\,] + i\omega\left([D_\tau(\Omega_\nu)] + [S_1^I(\Omega_\nu)]\right) + [\,K\,] + [S_1^R(\Omega_\nu)] \quad , \tag{110}$$

and $\mathbf{\Gamma}(\omega)$ is the vector in $\mathbb{C}^n$ such that

$$\mathbf{\Gamma}(\omega) = |\omega|\,\rho_0\,\mathbf{\Pi}_1(\omega) \quad . \tag{111}$$

Using the above algebraic result, the solution of Eqs. (108) and (109) can then be written as

$$\mathbf{\Phi}_\nu(\omega) = \mathbf{V}_\nu(\omega) - \frac{\kappa(\omega)\mathbf{\Gamma}(\omega)^T\mathbf{V}_\nu(\omega)}{1 + \kappa(\omega)\mathbf{\Gamma}(\omega)^T\mathbf{W}_\nu(\omega)}\,\mathbf{W}_\nu(\omega) \quad , \tag{112}$$

in which $\mathbf{V}(\omega)$ and $\mathbf{W}(\omega)$ are the solutions of the linear matrix equations,

$$[A_\nu(\omega)]\,\mathbf{V}_\nu(\omega) = \theta_\nu(\omega)\,\mathbf{B} \quad , \quad \omega \in \mathbb{B}_\nu \quad , \tag{113}$$

with the constraint

$$\mathbf{L}^T\mathbf{V}_\nu(\omega) = 0 \quad , \quad \omega \in \mathbb{B}_\nu \quad , \tag{114}$$

and

$$[A_\nu(\omega)]\,\mathbf{W}_\nu(\omega) = \mathbf{\Gamma}(\omega) \quad , \quad \omega \in \mathbb{B}_\nu \quad , \tag{115}$$

with the constraint,

$$\mathbf{L}^T\mathbf{W}_\nu(\omega) = 0 \quad , \quad \omega \in \mathbb{B}_\nu \quad . \tag{116}$$

Eqs. (113) and (114) and Eqs. (115) and (116) are solved using the MF method presented in Section 10.1.

11. Case of a Zero Pressure Condition on Part of the Boundary

11.1. Statement of the internal acoustic problem

We consider the problem described in Section 2 for which part Γ of the boundary is replaced by $\Gamma \cup \Gamma_0$ where Γ_0 is a part of $\partial\Omega$ submitted to a zero pressure field (see Fig. 2). There is no gravity effect.

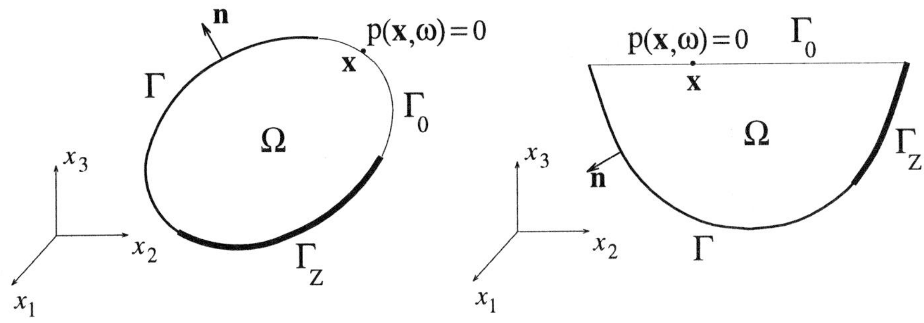

a- Gas or liquid configuration b- Liquid with free surface

Fig. 2. Configurations of the acoustic cavity

11.2. Boundary value problem

Equations in terms of φ. From Eq. (X.96), we deduce that $\pi(\omega) = 0$ in this case. In addition, the constraint defined by Eq. (4) is not necessary. Eqs. (1) to (4) of the boundary value problem are then replaced by

$$-\omega^2\frac{\rho_0}{c_0^2}\,\varphi - i\omega\,\tau\,\rho_0\boldsymbol{\nabla}^2\varphi - \rho_0\,\boldsymbol{\nabla}^2\varphi = -\frac{Q}{i\omega} - \tau c_0^2\,\boldsymbol{\nabla}^2\Big(\frac{Q}{\omega^2}\Big) \quad \text{in} \quad \Omega\,, \tag{117}$$

with the Neumann boundary condition,

$$\rho_0(1+i\omega\tau)\frac{\partial\varphi}{\partial\mathbf{n}} = \rho_0\,\mathbf{u}_{\text{wall}}\cdot\mathbf{n} + \tau c_0^2\frac{\partial}{\partial\mathbf{n}}\Big(\frac{Q}{\omega^2}\Big) \quad \text{on} \quad \Gamma\,, \tag{118}$$

the Neumann boundary condition with wall acoustic impedance,

$$\rho_0(1+i\omega\tau)\frac{\partial\varphi}{\partial\mathbf{n}} = \rho_0\,\mathbf{u}_{\text{wall}}\cdot\mathbf{n} + \tau c_0^2\frac{\partial}{\partial\mathbf{n}}\Big(\frac{Q}{\omega^2}\Big) + {\rho_0}^2\omega^2\frac{\varphi}{i\omega\, Z} \quad \text{on} \quad \Gamma_Z\ , \tag{119}$$

and the Dirichlet condition,

$$\varphi = 0 \quad \text{on} \quad \Gamma_0\ . \tag{120}$$

The pressure field is given by

$$p = \omega^2\,\rho_0\,\varphi\ . \tag{121}$$

Problem to be solved. For all real ω, for given $\mathbf{u}_{\text{wall}}(\omega)$ and $Q(\omega)$, the problem consists in finding $\varphi(\omega)$ as a solution of Eqs. (117) to (120). The pressure is then given by Eq. (121).

11.3. Variational formulation and linear operator equation

Admissible function space. We applied the test function method to Eq. (117) to (120). First, we introduce the complex vector space $\mathcal{E}^c$ of sufficiently differentiable[13] functions $\delta\varphi$ defined on Ω with values in $\mathbb{C}$ and the admissible function space $\mathcal{E}_0^c$ which is the subspace of $\mathcal{E}^c$ defined by

$$\mathcal{E}_0^c = \{\ \varphi\in\mathcal{E}^c\ \ ;\ \ \varphi(\mathbf{x}) = 0\ \ ,\ \ \forall\mathbf{x}\in\Gamma_0\}\ . \tag{122}$$

Variational formulation. For all fixed real ω, find φ in $\mathcal{E}_0^c$ such that

$$a(\omega\,;\varphi\,,\delta\varphi) = f(\omega\,;\delta\varphi)\quad ,\quad \delta\varphi\in\mathcal{E}_0^c\ , \tag{123}$$

in which sesquilinear form $a(\omega\,;\varphi\,,\delta\varphi)$ on $\mathcal{E}_0^c\times\mathcal{E}_0^c$ is the restriction to $\mathcal{E}_0^c$ of the following sesquilinear form $a(\omega\,;\varphi\,,\delta\varphi)$ on $\mathcal{E}^c\times\mathcal{E}^c$ defined by

$$a(\omega;\varphi,\delta\varphi) = -\omega^2 m(\varphi,\delta\varphi) + i\omega\, d_\tau(\omega;\varphi,\delta\varphi) + k(\varphi,\delta\varphi) + s_Z(\omega;\varphi,\delta\varphi)\,. \tag{124}$$

Sesquilinear forms m, k and d_τ on $\mathcal{E}^c\times\mathcal{E}^c$ are defined by Eqs. (14), (16) and (18) and the associated linear operators[14] $\mathbf{M}$, $\mathbf{K}$ and $\mathbf{D}_\tau(\omega)$ are defined

---

13 Space $\mathcal{E}^c$ is the Sobolev space $H^1(\Omega)$.

14 The mathematical properties of operators $\mathbf{M}$, $\mathbf{K}$ and $\mathbf{D}_\tau(\omega)$ are those given in footnotes 2 and 3, replacing $\mathcal{E}_\varphi^c$ by $\mathcal{E}_0^c$ in footnote 3.

by Eqs. (15), (17) and (19). The sesquilinear form $s_Z(\omega\,;\varphi\,,\delta\varphi)$ defined on $\mathcal{E}^c\times\mathcal{E}^c$ is such that

$$s_Z(\omega\,;\varphi\,,\delta\varphi) = -\rho_0^2\omega^2\int_{\Gamma_Z}\frac{1}{i\omega\,Z}\,\varphi\,\overline{\delta\varphi}\;ds \quad , \tag{125}$$

and defines a linear operator $\mathbf{S}_Z(\omega)$ such that [15]

$$<\mathbf{S}_Z(\omega)\,\varphi\,,\delta\varphi>= s_Z(\omega\,;\varphi\,,\delta\varphi) \quad . \tag{126}$$

Antilinear form $\delta\varphi\mapsto f(\omega\,;\delta\varphi)$ on $\mathcal{E}^c$ can be written as

$$f(\omega\,;\delta\varphi) = f_Q(\omega\,;\delta\varphi) + f_{\mathbf{u}_{\text{wall}}}(\omega\,;\delta\varphi) \quad , \tag{127}$$

in which

$$f_Q(\omega\,;\delta\varphi) = -\int_\Omega \frac{Q}{i\omega}\,\overline{\delta\varphi}\;d\mathbf{x} + \tau c_0^2\int_\Omega\frac{1}{\omega^2}\boldsymbol{\nabla}Q\cdot\boldsymbol{\nabla}\overline{\delta\varphi}\,d\mathbf{x} \quad , \tag{128}$$

$$f_{\mathbf{u}_{\text{wall}}}(\omega\,;\delta\varphi) = \rho_0\int_{\Gamma\cup\Gamma_Z}\mathbf{u}_{\text{wall}}\cdot\mathbf{n}\;\overline{\delta\varphi}\;ds \quad , \tag{129}$$

and defines the element $\mathbf{f}(\omega)$ such that [16]

$$<\mathbf{f}(\omega)\,,\delta\varphi>= f(\omega\,;\delta\varphi) \quad . \tag{130}$$

Linear operator equation. Using Eqs. (15), (17), (19), (126) and (130), the linear operator equation associated with Eqs. (123) and (124) is written as [17]

$$\mathbf{A}(\omega)\,\varphi = \mathbf{f}(\omega) \quad , \quad \varphi\in\mathcal{E}_0^c \quad , \tag{131}$$

in which operator $\mathbf{A}(\omega)$ is defined by [18]

$$\mathbf{A}(\omega) = -\omega^2\,\mathbf{M} + i\omega\,\mathbf{D}_\tau(\omega) + \mathbf{K} + \mathbf{S}_Z(\omega) \quad . \tag{132}$$

The mathematical properties of operator $\mathbf{S}_Z(\omega)$ are those given in footnote 4.

The mathematical properties of $\mathbf{f}(\omega)$ are those given in footnote 5.

Sesquilinear form $\{\varphi\,,\delta\varphi\}\mapsto a(\omega\,;\varphi,\delta\varphi)$ is continuous on $\mathcal{E}_0^c\times\mathcal{E}_0^c$ and defines a linear operator $\mathbf{A}(\omega)$ continuous from $\mathcal{E}_0^c$ into $\mathcal{E}_0^{c\,\prime}$ and Eq. (131) is an equality in $\mathcal{E}_0^{c\,\prime}$.

Since linear operator $\mathbf{A}(\omega)$ is continuous from $\mathcal{E}_0^c$ into $\mathcal{E}_0^{c\,\prime}$, then the operator in the right-hand side of Eq. (132) is the restriction to $\mathcal{E}_0^c$ of the operator defined on $\mathcal{E}^c$.

11.4. Frequency response function (FRF)

The operator-valued frequency response function is defined by the linear operator which associates each $\mathbf{f}$ with field φ as the solution of the variational formulation defined by Eq. (123). It can be proved [19] that, for all fixed real ω and $\mathbf{f}(\omega)$ given, the variational formulation defined by Eq. (123) has a unique solution $\mathbf{x} \mapsto \varphi(\mathbf{x}, \omega)$ in $\mathcal{E}_0^c$. Consequently, linear operator $\mathbf{A}(\omega)$ defined on $\mathcal{E}_0^c$ is invertible and Eq. (131) yields

$$\varphi(\omega) = \mathbf{T}(\omega)\,\mathbf{f}(\omega) \quad , \tag{133}$$

in which $\omega \mapsto \mathbf{T}(\omega) = \mathbf{A}(\omega)^{-1}$ is the operator-valued frequency response function [20].

11.5. Finite element discretization

The finite element discretization of Eq. (123) yields

$$[\mathbb{A}(\omega)]\,\mathbf{\Phi} = \mathbf{F}(\omega) \quad , \tag{134}$$

in which $\mathbf{\Phi} = (\Phi_1, \ldots, \Phi_n)$ denotes the complex vector of the DOFs which are the values of φ at the nodes of the finite element mesh of domain Ω. Vector $\mathbf{F} = (F_1, \ldots, F_n)$ is the complex vector corresponding to the discretization in $\mathcal{E}_0^c$ of antilinear form $f(\omega\,; \delta\varphi)$ defined by Eq. (130). The $(n \times n)$ complex matrix $[\mathbb{A}(\omega)]$ is symmetric and is defined by

$$[\mathbb{A}(\omega)] = -\omega^2\,[M] + i\omega\,[D_\tau(\omega)] + [K] + [S_z(\omega)] \quad , \tag{135}$$

where the real symmetric positive-definite matrices $[M]$, $[K]$, $[D_\tau(\omega)]$ correspond to the finite element discretization in $\mathcal{E}_0^c$ and the symmetric complex matrix $[S_z(\omega)]$ is such that $\omega^{-1}\Im m[S_z(\omega)]$ is positive semidefinite. For all fixed real ω, Eq. (134) has a unique solution which is written as

$$\mathbf{\Phi}(\omega) = [T(\omega)]\,\mathbf{F}(\omega) \quad , \tag{136}$$

in which the $(n \times n)$ symmetric complex matrix $[T(\omega)] = [\mathbb{A}(\omega)]^{-1}$ is the matrix-valued frequency response function. The corresponding $(n \times n)$ real matrix-valued impulse response function $t \mapsto [\,h(t)\,]$ is such that (see Eq. (V.72)),

$$[T(\omega)] = \int_{\mathbb{R}} e^{-i\omega t}\,[h(t)]\,dt \quad . \tag{137}$$

[19] The proof of existence and uniqueness is similar to the proof given in footnote 8.

[20] Since linear operator $\mathbf{A}(\omega)$ is continuous and invertible from $\mathcal{E}_0^c$ into $\mathcal{E}_0^{c\,\prime}$, $\mathbf{T}(\omega)$=$\mathbf{A}(\omega)^{-1}$ is continuous from $\mathcal{E}_0^{c\,\prime}$ into $\mathcal{E}_0^c$.

11.6. Acoustic modes of the acoustic cavity

The spectral boundary value problem. As in Section 7 and from Eqs. (117) to (120), the equations of this classical spectral problem are the Helmholtz equation

$$\nabla^2\varphi + \frac{\omega^2}{c_0^2}\,\varphi = 0 \qquad \text{in} \quad \Omega \quad , \tag{138}$$

with the Neumann boundary condition,

$$\frac{\partial\varphi}{\partial\mathbf{n}} = 0 \quad \text{on} \quad \partial\Omega\backslash\Gamma_0 \quad , \tag{139}$$

and the Dirichlet condition,

$$\varphi = 0 \quad \text{on} \quad \Gamma_0 \quad . \tag{140}$$

Variational formulation of the spectral boundary value problem. From Section 11.3, we deduce the variational formulation of the spectral boundary value problem defined by Eqs. (138) to (140). Find $\lambda = \omega^2$ and φ in $\mathcal{E}_0^c$ such that

$$k(\varphi\,,\delta\varphi) = \lambda\; m(\varphi\,,\delta\varphi) \quad , \quad \forall\,\delta\varphi \in \mathcal{E}_0^c \quad . \tag{141}$$

Properties of the eigenvalues and eigenfunctions. They are similar to those given in Section 7.2. There is a countable number of positive eigenvalues, the family of eigenfunctions is a complete set in $\mathcal{E}_0^c$ (see Eq. (40)) and the orthogonality conditions with respect to m and k are satisfied (see Eqs. (42) and (43)).

Finite element discretization and generalized symmetric matrix eigenvalue problem. The matrix equation of the generalized symmetric eigenvalue problem corresponding to the finite element discretization of Eq. (141) is written as

$$[\,K\,]\,\mathbf{\Phi} = \lambda\,[\,M\,]\,\mathbf{\Phi} \quad , \tag{142}$$

in which $[\,M\,]$, $[\,K\,]$ and $\mathbf{\Phi}$ are defined in Section 11.5.

11.7. FRF calculation in the LF and MF ranges

The results are similar to those presented in Sections 9 and 10. It is sufficient to replace the admissible function space $\mathcal{E}_\varphi^c$ by $\mathcal{E}_0^c$. It should be noted that the constraint defined by Eq. (87) must be removed.

12. Particular Case of an Axisymmetric Acoustic Cavity

We consider the boundary value problem defined in Sections 2 and 3 for which Ω is an axisymmetric domain and Γ and Γ_Z are axisymmetric parts of boundary $\partial\Omega$. The wall acoustic impedance $Z(\mathbf{x},\omega)$ on Γ_Z is assumed to have the same axisymmetry property. Consequently, using the analysis presented in Section III.8.2, $p(\mathbf{x},\omega)$ and $\varphi(\mathbf{x},\omega)$ can be written as

$$p(r,\theta,z) = p_0(r,z) + \sum_{n=1}^{+\infty}\{p_n^+(r,z)\,\cos n\theta + p_n^-(r,z)\,\sin n\theta\} \quad , \tag{143}$$

$$\varphi(r,\theta,z) = \varphi_0(r,z) + \sum_{n=1}^{+\infty}\{\varphi_n^+(r,z)\,\cos n\theta + \varphi_n^-(r,z)\,\sin n\theta\} \quad . \tag{144}$$

From Eq. (5), we deduce that

$$p_0(r,z) = \omega^2\,\rho_0\,\varphi_0(r,z) + \pi(\omega) \quad , \tag{145}$$

and for all $n \geq 1$,

$$p_n^\pm(r,z) = \omega^2\,\rho_0\,\varphi_n^\pm(r,z) \quad . \tag{146}$$

These equations show that $\pi(\omega) = 0$ for $n \geq 1$. Consequently, in the boundary value problem, for $n \geq 1$, we set $\pi(\omega) = 0$ in Eqs. (1) and (3), Eq. (4) must be suppressed, admissible function space $\mathcal{E}_\varphi^c$ is replaced by subspaces $\mathcal{E}_n^{c+}$ and $\mathcal{E}_n^{c-}$ of $\mathcal{E}^c$, the second term in the right-hand side of Eq. (20) must be suppressed and finally, in Eqs. (23) and (24), the terms containing $\pi_Q(\omega)$ and $\pi_2(\mathbf{u}_{\text{wall}})$ must also be suppressed.

13. Response to Deterministic and Random Excitations

In the boundary value problem defined in Section 3 or 11, the excitations are wall displacement field $\mathbf{u}_{\text{wall}}(\mathbf{x},t)$ and internal acoustic source density $Q(\mathbf{x},t)$. For calculation of the reponse of the acoustic cavity subjected to these deterministic or random excitations, we refer the reader to Chapter IX.

CHAPTER XII

External Acoustic Fluid. Boundary Integral Formulation for the LF and MF Ranges

1. Introduction

This chapter is devoted to the external acoustic problem in the frequency domain related to the Helmholtz equation. We present an appropriate boundary element method to model the external acoustic fluid in structural-acoustic problems. Various methods can be used for solving this external acoustic problem. For simple shapes and geometries, we refer the reader to Bowman et al., 1969; Crighton et al., 1992; Junger and Feit, 1993. For methods based on the use of a finite element mesh of the external fluid, see Everstine and Yang, 1984; Givoli and Keller, 1989; Pinsky and Abboud, 1989; Harari and Hughes, 1994, and for the infinite element application, see Bettess, 1993 and Zienkiewicz et al., 1985. For methods based on the use of asymptotic approximations, concerning the slender body theory, see Coupry and Soize, 1984; Chabas and Soize, 1986; concerning the doubly asymptotic approximation theory (DAA), see Belytschko and Geers, 1977; Geers and Felippa, 1983; Geers and Zhang, 1994.

Herein, the external acoustic fluid problem is solved using integral equation techniques. This kind of approach is well suited to this problem and various methodologies are proposed in the literature (Maue, 1949; Panich, 1965; Mitzner, 1966; Schenck, 1968; Burton and Miller, 1971; Jones, 1974; Belytschko and Geers, 1977; Brebbia, 1978; Angelini and Hutin, 1983; Everstine and Yang, 1984; Brebbia et al. 1984; Mathews, 1986; Kress, 1989; Amini and Harris, 1990; Chen et al., 1990; Amini et al., 1992; Brebbia and Dominguez, 1992; Chen and Zhou, 1992; Colton and Kress, 1992; Dautray and Lions, Vol 4, 1992). In this chapter, we present an appropriate symmetric boundary integral formulation valid for all real values of the frequency, and the corresponding boundary element method for constructing the acoustic impedance boundary operator and the radiation impedance operator in the LF and MF ranges.

In Section 2, we introduce the external acoustic problem in the frequency domain, consisting in finding the pressure field in the external domain Ω_E and on its boundary Γ_E for a given velocity field on Γ_E, an acoustic source density in Ω_E and an incident plane wave.

In Section 3, we present the basic exterior Neumann problem related to the Helmholtz equation expressed in terms of the velocity potential field, for a given normal velocity field on boundary Γ_E. This problem, which includes the outward Sommerfeld radiation condition at infinity, has a unique solution for all real values of the frequency.

In Section 4, we introduce the *acoustic impedance boundary operator* which relates the normal velocity field on Γ_E to the pressure field on Γ_E. This fundamental operator is related to the basic exterior Neumann problem related to the Helmholtz equation and will be used for coupling an external acoustic fluid with the structure in the LF and MF ranges. The symmetry and positivity properties of this operator are proved. We then introduce another fundamental operator, the so-called *radiation impedance operator*, which allows calculation of the near field and far field pressure.

Section 5 deals with the free-space Green's function, used for constructing the integral boundary representations on Γ_E.

In Section 6, due to the linearity of the problem, we show that the solution of the above basic exterior Neumann problem allows the construction of the equations for radiation by an elastic boundary, scattering of an acoustic source density by a rigid boundary and scattering of an incident plane wave by a rigid boundary.

Section 7 contains a brief review of the potential theory introducing the single- and double-layer potentials used for constructing the boundary integral equations related to the solution of the basic exterior Neumann problem.

Section 8 is devoted to Helmholtz integral representations and their normal derivatives for the external domain and the internal domain.

In Section 9, we introduce two basic boundary integral equations and their variational formulation for the exterior Neumann problem related to the Helmholtz equation. These basic formulations do not allow construction of a unique solution of the physical problem in the case of a sequence of real frequencies called *spurious* or *irregular frequencies*. Various methods are proposed in the literature to overcome this mathematical difficulty (Panich, 1965; Schenck, 1968; Burton and Miller, 1971; Mathews, 1986; Amini et al., 1990 and 1992; Colton and Kress, 1992). It appears that the most popular and robust formulation is due to Burton and Miller, 1971. Nevertheless, it should be noted that this formulation is not symmetric.

In Section 10, we present a method initially developed by Angelini and Hutin, 1983, based on an appropriate use of the two basic boundary integral equations presented in Section 9, leading to an appropriate symmetric boundary integral method valid for all real values of the frequency. Finite element discretization based on a variational formulation leads to a boundary element method which is numerically stable and very efficient. This method does not require introducing additional degrees of freedom in the numerical discretization for treatment of irregular frequencies. This method was validated by comparisons with analytical and experimental results in the LF and MF ranges (Angelini and Hutin, 1983; Soize et al., 1986b and 1992; Petitjean, 1992) and then extended to the Maxwell equations (Angelini and Soize, 1989 and 1993).

In Section 11, we construct the radiation impedance operator and we give an asymptotic formula for the radiated pressure field.

Section 12 is devoted to the boundary element method corresponding to finite element discretization of the method presented in Section 10 and 11.

Finally, in Section 13, using the image method, we show that the case of an unbounded domain with a free surface is equivalent to the case of an unbounded domain without a free surface.

2. Statement of the External Acoustic Problem

We consider an external acoustic fluid (gas or liquid) occupying an unbounded open three-dimensional simply connected domain [1] Ω_E whose complement $\Omega_i = \mathbb{R}^3 \backslash (\Omega_E \cup \Gamma_E)$ is an open bounded domain (see Fig. 1). The acoustic fluid is assumed to be inviscid. A displacement field $\mathbf{u}_{\text{wall}}(\mathbf{y}, \omega)$ is prescribed on the boundary $\Gamma_E = \partial\Omega_E = \partial\Omega_i$ which is assumed to be smooth. We denote the outward unit normal to Γ_E as $\mathbf{n}$ (see Fig. 1). In addition, an acoustic source density $Q(\mathbf{x}, \omega)$ and an incident plane wave defined by the velocity potential $\psi_{\text{inc}}(\mathbf{x}, \omega) = \psi_0(\omega) \exp\{-i\,\mathbf{k}\cdot\mathbf{x}\}$ are given inside Ω_E. The problem consists in calculating the pressure field $p_E(\mathbf{y}, \omega)$ on boundary Γ_E and the pressure field $p_E(\mathbf{x}, \omega)$ in the unbounded

[1] A domain Ω is called simply connected if any continuous closed curve in Ω can be continuously shrunk to a point without leaving Ω.

domain Ω_E in the low- and medium-frequency ranges, for given $\mathbf{u}_{\text{wall}}(\mathbf{y},\omega)$, $Q(\mathbf{x},\omega)$ and $\psi_{\text{inc}}(\mathbf{x},\omega)$. For this problem, we need to solve the basic problem corresponding to the Helmholtz equation in external unbounded domain Ω_E with an arbitrary Neumann boundary condition on Γ_E. This problem is called the *exterior Neumann problem related to the Helmholtz equation.* The linearity of the equations then allows us to construct the solution for a given acoustic source density and a given incident plane wave.

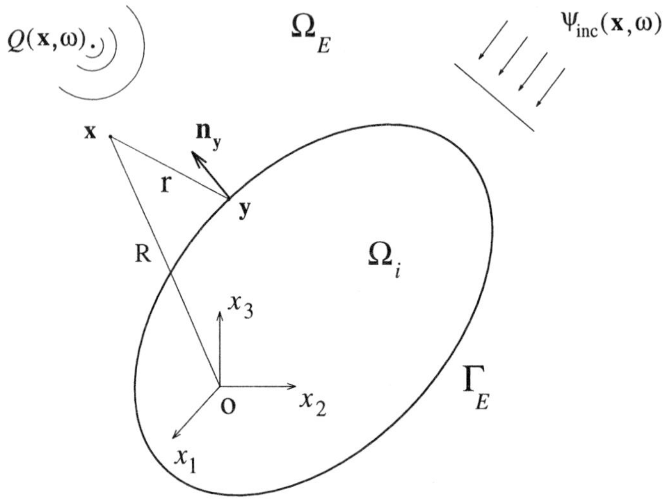

Fig. 1. Geometry of the external unbounded domain

3. Exterior Neumann Problem Related to the Helmholtz Equation

3.1. Helmholtz equation and Neumann boundary condition

This basic problem corresponds to the case for which ω is a given real number and there is neither a source nor an incident plane wave inside Ω_E. Using Section X.2, we deduce the Helmholtz equation for velocity potential $\psi(\mathbf{x},\omega)$ and we write the Neumann boundary condition for a given normal velocity field $v(\mathbf{y})$ defined on Γ_E with values in $\mathbb{C}$.

Expression of the pressure in terms of the velocity potential. The region at infinity is at rest, which means that for all fixed real ω, the pressure field $p(\mathbf{x},\omega)$ and the velocity field $\mathbf{v}(\mathbf{x},\omega)$ converge to zero as the distance $\|\mathbf{x}\|$

from the origin approaches infinity [2]. We then have

$$p(\mathbf{x},\omega) \longrightarrow 0 \quad \text{as} \quad \|\mathbf{x}\| \longrightarrow +\infty \quad , \tag{1}$$

$$\psi(\mathbf{x},\omega) \longrightarrow 0 \quad \text{as} \quad \|\mathbf{x}\| \longrightarrow +\infty \quad . \tag{2}$$

From Eqs. (1), (2) and (X.37), we deduce that $\pi(\omega)$ is equal to zero and consequently,

$$p(\mathbf{x},\omega) = -i\omega\,\rho_E\,\psi(\mathbf{x},\omega) \qquad \text{in} \quad \Omega_E \quad , \tag{3}$$

in which ρ_E is the constant mass density of the external fluid at equilibrium.

Helmholtz equation in terms of the velocity potential. For all real ω, we introduce k defined by

$$k = \frac{\omega}{c_E} \quad , \tag{4}$$

in which c_E is the constant speed of sound in the external fluid at equilibrium. Since Q and $\pi(\omega)$ are equal to zero, Eq. (X.38) yields the Helmholtz equation in terms of velocity potential $\psi(\mathbf{x},\omega)$

$$\boldsymbol{\nabla}^2\psi(\mathbf{x},\omega) + k^2\,\psi(\mathbf{x},\omega) = 0 \qquad \text{in} \quad \Omega_E \quad . \tag{5}$$

Neumann boundary condition on Γ_E. This condition is written as

$$\frac{\partial\psi(\mathbf{y},\omega)}{\partial\mathbf{n}_\mathbf{y}} = v(\mathbf{y}) \qquad \text{on} \quad \Gamma_E \quad , \tag{6}$$

in which the normal derivative of ψ on Γ_E is defined by

$$\frac{\partial\psi(\mathbf{y},\omega)}{\partial\mathbf{n}_\mathbf{y}} = \{\boldsymbol{\nabla}_\mathbf{x}\psi(\mathbf{x},\omega)\}_{\mathbf{x}=\mathbf{y}}\cdot\mathbf{n}_\mathbf{y} \quad , \quad \mathbf{y}\in\Gamma_E \quad . \tag{7}$$

---

[2] We will see below that in the frequency domain, the uniqueness of the exterior Neumann problem related to the Helmholtz equation is obtained by introducing the Sommerfeld conditions at infinity. These conditions imply that $\psi(\mathbf{x},\omega)$, $\boldsymbol{\nabla}\psi(\mathbf{x},\omega)$ and $p(\mathbf{x},\omega)\to 0$ as $\|\mathbf{x}\|\to+\infty$.

3.2. Introduction of the outward Sommerfeld radiation condition at infinity

For all fixed real $k \neq 0$ and for v given on Γ_E, the problem defined by Eqs. (5) and (6) does not have a unique solution. In order to make the solution unique, it is necessary to introduce the *outward Sommerfeld radiation condition at infinity* which ensures that we only consider outward traveling waves at infinity,

$$|\psi| = O(\frac{1}{R}) \quad , \quad \left| \frac{\partial \psi}{\partial R} + i\,k\,\psi \right| = O(\frac{1}{R^2}) \quad , \tag{8}$$

as $R = \|\mathbf{x}\| \to +\infty$, in which $\partial/\partial R$ denotes the derivative in the radial direction from the origin (see Fig. 1).

3.3. Existence and uniqueness of the boundary value problem

Admissible function spaces. Let $\mathcal{C}_{\Omega_E}$ be the admissible function space [3] of functions $\mathbf{x} \mapsto \psi(\mathbf{x})$ defined on Ω_E with values in $\mathbb{C}$. Let $\mathcal{C}_{\Gamma_E}$ be the function space constituted by the traces [4] $\mathbf{y} \mapsto \psi_{\Gamma_E}(\mathbf{y}) = \psi(\mathbf{y})$ on Γ_E of functions $\mathbf{x} \mapsto \psi(\mathbf{x})$ belonging to $\mathcal{C}_{\Omega_E}$. We introduce the space [5] $\mathcal{C}'_{\Gamma_E}$ of the normal derivative $\mathbf{y} \mapsto \partial\psi(\mathbf{y})/\partial\mathbf{n}_\mathbf{y}$ on Γ_E of functions $\mathbf{x} \mapsto \psi(\mathbf{x})$ belonging to $\mathcal{C}_{\Omega_E}$. Finally, we introduce the space $H_{\Gamma_E} = L^2(\Gamma_E)$ of the square integrable complex functions on Γ_E and we have $\mathcal{C}_{\Gamma_E} \subset H_{\Gamma_E} \subset \mathcal{C}'_{\Gamma_E}$.

Existence and uniqueness of the boundary value problem. For all real $k \neq 0$ and for all $\mathbf{y} \mapsto v(\mathbf{y})$ in $\mathcal{C}'_{\Gamma_E}$, the exterior Neumann problem related to the Helmholtz equation defined by Eqs. (5), (6) and (8), has a unique solution $\mathbf{x} \mapsto \psi(\mathbf{x}, \omega)$ in $\mathcal{C}_{\Omega_E}$ whose trace $\mathbf{x} \mapsto \psi_{\Gamma_E}(\mathbf{x}, \omega)$ on boundary Γ_E belongs to $\mathcal{C}_{\Gamma_E}$ (see for instance Dautray and Lions, 1992).

---

[3] Space $\mathcal{C}_{\Omega_E} = H^1_{\text{loc}}(\overline{\Omega}_E)$; $H^1_{\text{loc}}(\overline{\Omega}_E)$ is the set of functions from $\overline{\Omega}_E$ into $\mathbb{C}$ such that, for all bounded subsets $\mathcal{O} \subset \overline{\Omega}_E$, the restriction of ψ to $\mathcal{O}$ belongs to $H^1(\mathcal{O}, \mathbb{C})$.

[4] Vector space $\mathcal{C}_{\Gamma_E} = H^{1/2}(\Gamma_E)$.

[5] Vector space $\mathcal{C}'_{\Gamma_E}$ is the dual space of $\mathcal{C}_{\Gamma_E}$ for the dual brackets $<v, \psi_{\Gamma_E}>_{\mathcal{C}'_{\Gamma_E}, \mathcal{C}_{\Gamma_E}}$ in which $v \in \mathcal{C}'_{\Gamma_E}$ and $\psi_{\Gamma_E} \in \mathcal{C}_{\Gamma_E}$ and which is linear in v and linear in ψ_{Γ_E} (and not antilinear in ψ_{Γ_E}). We have $\mathcal{C}'_{\Gamma_E} = H^{-1/2}(\Gamma_E)$. Here we use the dual space instead of the antidual space as is done in all the other chapters of the book.

4. Acoustic Impedance Boundary Operator and Radiation Impedance Operator

Acoustic impedance boundary operator. From Section 3.3, there exists a unique linear operator $\mathbf{B}_{\Gamma_E}(\omega/c_E)$ such that [6]

$$\psi_{\Gamma_E}(\omega) = \mathbf{B}_{\Gamma_E}(\omega/c_E)\, v \quad \text{on} \quad \Gamma_E \quad . \tag{9}$$

From Eqs. (3) and (9), we deduce that

$$p_{\Gamma_E}(\omega) = \mathbf{Z}_{\Gamma_E}(\omega)\, v \quad \text{on} \quad \Gamma_E \quad , \tag{10}$$

in which $p_{\Gamma_E}(\omega)$ is the trace on Γ_E of $p(\omega)$, and $\mathbf{Z}_{\Gamma_E}(\omega)$ is the *acoustic impedance boundary operator* defined by [7]

$$\mathbf{Z}_{\Gamma_E}(\omega) = -i\,\omega\,\rho_E\,\mathbf{B}_{\Gamma_E}(\omega/c_E) \quad . \tag{11}$$

Radiation impedance operator in the external domain. From Section 3.3, there exists a unique linear operator $\mathbf{R}(\mathbf{x}, \omega/c_E)$ such that [8]

$$\psi(\mathbf{x},\omega) = \mathbf{R}(\mathbf{x},\omega/c_E)\, v \quad , \quad \mathbf{x} \in \Omega_E \quad . \tag{12}$$

From Eqs. (3) and (12), we deduce that

$$p(\mathbf{x},\omega) = \mathbf{Z}_{\text{rad}}(\mathbf{x},\omega)\, v \quad , \quad \mathbf{x} \in \Omega_E \quad , \tag{13}$$

in which $\mathbf{Z}_{\text{rad}}(\mathbf{x},\omega)$ is the *radiation impedance operator* defined by [9]

$$\mathbf{Z}_{\text{rad}}(\mathbf{x},\omega) = -i\,\omega\,\rho_E\,\mathbf{R}(\mathbf{x},\omega/c_E) \quad , \quad \mathbf{x} \in \Omega_E \quad . \tag{14}$$

Symmetry property of the acoustic impedance boundary operator. The transpose [10] of operator $\mathbf{B}_{\Gamma_E}(\omega/c_E)$ is the linear operator ${}^t\mathbf{B}_{\Gamma_E}(\omega/c_E)$ such that, for all v and δv in $\mathcal{C}'_{\Gamma_E}$,

$$< \delta v, \mathbf{B}_{\Gamma_E}(\omega/c_E)\, v >_{\mathcal{C}'_{\Gamma_E}, \mathcal{C}_{\Gamma_E}} = < {}^t\mathbf{B}_{\Gamma_E}(\omega/c_E)\, \delta v\, , v >_{\mathcal{C}_{\Gamma_E}, \mathcal{C}'_{\Gamma_E}} \quad . \tag{15}$$

Mapping $v \mapsto \psi_{\Gamma_E}$ is linear and defines a unique continuous linear operator $\mathbf{B}_{\Gamma_E}(\omega/c_E)$ from $\mathcal{C}'_{\Gamma_E}$ into $\mathcal{C}_{\Gamma_E}$.

Linear operator $\mathbf{Z}_{\Gamma_E}(\omega)$ is continuous from $\mathcal{C}'_{\Gamma_E}$ into $\mathcal{C}_{\Gamma_E}$.

For all $\mathbf{x}$ fixed in Ω_E, mapping $v \mapsto \psi(\mathbf{x},\omega)$ is linear and defines a unique linear operator $\mathbf{R}(\mathbf{x},\omega/c_E)$ from $\mathcal{C}'_{\Gamma_E}$ into $\mathbb{C}$.

Linear operator $\mathbf{Z}_{\text{rad}}(\mathbf{x},\omega)$ is defined from $\mathcal{C}'_{\Gamma_E}$ into $\mathbb{C}$.

Transpose ${}^t\mathbf{B}_{\Gamma_E}(\omega/c_E)$ is a continuous linear operator from $\mathcal{C}'_{\Gamma_E}$ into $(\mathcal{C}'_{\Gamma_E})' = \mathcal{C}_{\Gamma_E}$ (see footnote 5).

We then have the following symmetry property

$$ {}^t\mathbf{B}_{\Gamma_E}(\omega/c_E) = \mathbf{B}_{\Gamma_E}(\omega/c_E) \quad , \tag{16} $$

and from Eq. (11), we deduce that, for $\omega \neq 0$,

$$ {}^t\mathbf{Z}_{\Gamma_E}(\omega) = \mathbf{Z}_{\Gamma_E}(\omega) \quad . \tag{17} $$

Let us prove Eq. (16) for v and δv in the space $H_{\Gamma_E} \subset C'_{\Gamma_E}$ of the square integrable complex function on Γ_E. We introduce ψ_{Γ_E} and $\delta\psi_{\Gamma_E}$ in $C_{\Gamma_E} \subset H_{\Gamma_E}$ such that $\psi_{\Gamma_E} = \mathbf{B}_{\Gamma_E}(\omega/c_E)\, v$ and $\delta\psi_{\Gamma_E} = \mathbf{B}_{\Gamma_E}(\omega/c_E)\, \delta v$. Consequently, we can write [11],

$$ <\delta v\, ,\, \mathbf{B}_{\Gamma_E}(\omega/c_E)\, v>_{C'_{\Gamma_E}, C_{\Gamma_E}} =<\delta v\, ,\, \psi_{\Gamma_E}>= \int_{\Gamma_E} \delta v\; \psi_{\Gamma_E}\; ds \quad , \tag{18} $$

$$ <\mathbf{B}_{\Gamma_E}(\omega/c_E)\, \delta v\, ,\, v>_{C_{\Gamma_E}, C'_{\Gamma_E}} =<\delta\psi_{\Gamma_E}\, , v>= \int_{\Gamma_E} \delta\psi_{\Gamma_E}\; v\; ds \quad , \tag{19} $$

in which ds denotes the surface element. From Eqs. (15), (18), (19), we deduce that Eq. (16) holds if and only if

$$ \int_{\Gamma_E} (\delta v\; \psi_{\Gamma_E} - \delta\psi_{\Gamma_E}\; v)\; ds = 0 \quad . \tag{20} $$

Let us introduce ball $B(0,R)$ centered at the origin with radius R such that $\Omega_i \subset B(0,R) \subset \Omega_E$. Let Ω_R be the bounded domain of $\mathbb{R}^3$ such that $\Omega_R = B(0,R)\backslash\Omega_i$ (see Fig. 2).

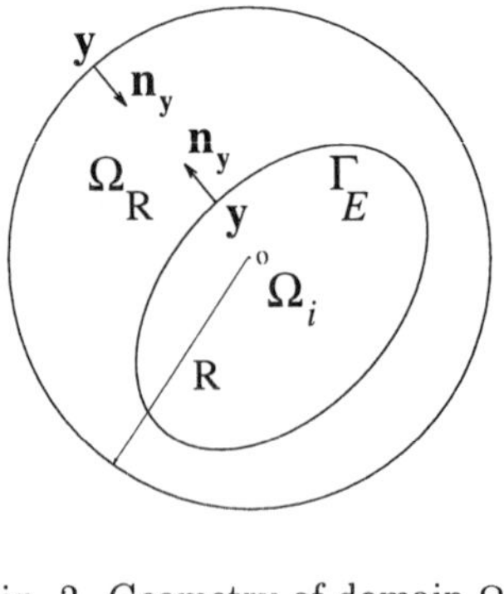

Fig. 2. Geometry of domain Ω_R

[11] Since $C_{\Gamma_E} \subset H_{\Gamma_E} = L^2(\Gamma_E)$, the second pair of angle brackets in Eqs. (18) and (19) are such that $<v\, ,w>=\int_{\Gamma_E} v\, w\; ds$ where v and w are in H_{Γ_E}. The inner product in H_{Γ_E} is $(v,w) \mapsto\; <\overline{v}\, ,w>$.

Since ψ and $\delta\psi$ satisfy the Helmholtz Eq. (5) and using Green's formula in domain Ω_R, we obtain for $R > R_0$,

$$\int_{\Gamma_E} (\delta v \; \psi_{\Gamma_E} - \delta\psi_{\Gamma_E} \; v)\, ds = - \int_{\partial B(0,R)} (\frac{\partial \delta\psi}{\partial \mathbf{n}} \; \psi - \delta\psi \; \frac{\partial \psi}{\partial \mathbf{n}})\, ds \quad . \tag{21}$$

Considering $R \to +\infty$ and using the Sommerfeld condition defined by Eq. (8), the right-hand side of Eq. (21) converges to zero and yields Eq. (20).

Positivity of the real part of the acoustic impedance boundary operator. Operator $i\,\omega\,\mathbf{Z}_{\Gamma_E}(\omega)$ can be written as

$$i\,\omega\,\mathbf{Z}_{\Gamma_E}(\omega) = -\omega^2\,\mathbf{M}_{\Gamma_E}(\omega/c_E) + i\,\omega\,\mathbf{D}_{\Gamma_E}(\omega/c_E) \quad , \tag{22}$$

in which $\mathbf{M}_{\Gamma_E}(\omega/c_E)$ and $\mathbf{D}_{\Gamma_E}(\omega/c_E)$ are two linear operators such that

$$\omega\,\mathbf{M}_{\Gamma_E}(\omega/c_E) = \Im m\,\mathbf{Z}_{\Gamma_E}(\omega) \quad , \tag{23}$$

$$\mathbf{D}_{\Gamma_E}(\omega/c_E) = \Re e\,\mathbf{Z}_{\Gamma_E}(\omega) \quad . \tag{24}$$

Let v be in H_{Γ_E}. Trace ψ_{Γ_E} on Γ_E is associated with v using Eq. (9). We have the following positivity property of the real part $\mathbf{D}_{\Gamma_E}(\omega/c_E)$ of the acoustic impedance boundary operator [12],

$$\forall \omega \in \mathbb{R} \quad , \quad <\overline{v}\,,\mathbf{D}_{\Gamma_E}(\omega/c_E)\,v> \; \geq 0 \quad . \tag{25}$$

The proof is carried out as follows. First, we have

$$<\overline{v}\,,\mathbf{D}_{\Gamma_E}(\omega/c_E)\,v>= \omega\,\rho_E\,\Im m \int_{\Gamma_E} \overline{v}\;\psi_{\Gamma_E}\; ds \quad , \tag{26}$$

because Eqs. (9), (11) and (22) yield

$$<\overline{v}\,,\psi_{\Gamma_E}>= \frac{i}{\omega\rho_E} <\overline{v}\,,\mathbf{D}_{\Gamma_E}(\omega/c_E)\,v> - \frac{1}{\rho_E} <\overline{v}\,,\mathbf{M}_{\Gamma_E}(\omega/c_E)\,v> \quad , \tag{27}$$

and because $\mathbf{M}_{\Gamma_E}(\omega/c_E)$ and $\mathbf{D}_{\Gamma_E}(\omega/c_E)$ are linear operators which have symmetry properties due to Eq. (17). Secondly, from Eq. (26), we deduce that Eq.(25) holds if and only if

$$\forall \omega \in \mathbb{R} \quad , \quad \omega\,\Im m \int_{\Gamma_E} \overline{v}\;\psi_{\Gamma_E}\; ds \geq 0 \quad . \tag{28}$$

---

[12] Since $C_{\Gamma_E} \subset H_{\Gamma_E} = L^2(\Gamma_E)$, the angle brackets in Eq. (25) are such that $<v\,,w>=\int_{\Gamma_E} v\,w\,ds$ where v and w are in H_{Γ_E}. The inner product in H_{Γ_E} is $(v,w) \mapsto <\overline{v}\,,w>$.

We then have to prove Eq. (28). Introducing the bounded domain Ω_R (see Fig. 2) and from Eq. (5), we deduce that

$$\int_{\Omega_R} (\boldsymbol{\nabla}^2\overline{\psi} + k^2\,\overline{\psi})\,\psi\,d\mathbf{x} = 0 \quad . \tag{29}$$

Using Green's formula in domain Ω_R and taking the imaginary part, we deduce that

$$\Im m \int_{\partial\Omega_R} \frac{\partial\overline{\psi}}{\partial\mathbf{n}}\,\psi\,ds = 0 \quad . \tag{30}$$

Since $\partial\Omega_R = \Gamma_E \cup \partial B(0,R)$ and using Eq. (6), Eq. (30) yields

$$\forall R > R_0 \quad , \quad \Im m \int_{\Gamma_E} \overline{v}\,\psi_{\Gamma_E}\,ds = \Im m \int_{\partial B(0,R)} \overline{\psi}\,\frac{\partial\psi}{\partial\mathbf{n}}\,ds \quad . \tag{31}$$

The right-hand side of Eq. (31) is transformed using the identity

$$2k\,\Im m \int_{\partial B(0,R)} \overline{\psi}\,\frac{\partial\psi}{\partial\mathbf{n}}\,ds = \int_{\partial B(0,R)} \left(\left|\frac{\partial\psi}{\partial\mathbf{n}}\right|^2 + k^2\,|\psi|^2 \right) ds - \int_{\partial B(0,R)} \left|\frac{\partial\psi}{\partial\mathbf{n}} + i\,k\,\psi\right|^2 ds \quad . \tag{32}$$

The first term on the right-hand side of Eq. (32) is positive or equal to zero and the second term converges to zero as $R \to +\infty$ due to the Sommerfeld condition defined by Eq. (8). From Eqs. (31) and (32), we deduce Eq. (28).

5. Free-Space Green's Function and Solution of the Inhomogeneous Helmholtz Equation

Free-space Green's function. The free-space Green's function $G(\mathbf{x}-\mathbf{x}')$ is a *fundamental solution* of the Helmholtz equation

$$\boldsymbol{\nabla}^2 G(\mathbf{x}-\mathbf{x}') + k^2\,G(\mathbf{x}-\mathbf{x}') = \delta_{\mathbf{x}'}(\mathbf{x}) \quad , \quad \forall\mathbf{x} \in \mathbb{R}^3 \quad , \tag{33}$$

in which $\boldsymbol{\nabla}^2 = \Delta$ is the Laplacian operator with respect to $\mathbf{x}$, $\mathbf{x}'$ is any point fixed in $\mathbb{R}^3$, k is defined by Eq. (4) and $\delta_{\mathbf{x}'}(\mathbf{x})$ is the Dirac generalized function at point $\mathbf{x}'$. The *outward fundamental solution* of the Helmholtz equation in $\mathbb{R}^3$ is the fundamental solution satisfying Eq. (33) and the outward Sommerfeld radiation condition at infinity

$$\left|\frac{\partial G}{\partial R} + i\,k\,G\right| = O(\frac{1}{R^2}) \quad , \tag{34}$$

as $R = \|\mathbf{x}\| \rightarrow +\infty$, in which $\partial/\partial R$ denotes the derivative in the radial direction from the origin. It can be proved (Morse and Feshbach, 1953; Schwartz, 1965) that, for all $\mathbf{x}'$ fixed in $\mathbb{R}^3$, Eqs. (33) and (34) have a unique solution which is written as

$$G(\mathbf{x}-\mathbf{x}') = g(\|\mathbf{x}-\mathbf{x}'\|) \quad , \tag{35}$$

$$g(r) = -\frac{1}{4\pi}\,\frac{e^{-ikr}}{r} \quad , \quad r = \|\mathbf{x}-\mathbf{x}'\| \quad . \tag{36}$$

In addition, it can easily be proved that

$$\nabla_{\mathbf{x}} G(\mathbf{x}-\mathbf{x}') = -\nabla_{\mathbf{x}'} G(\mathbf{x}-\mathbf{x}') \quad . \tag{37}$$

Solution of the inhomogeneous Helmholtz equation. Consider the inhomogeneous Helmholtz equation

$$\boldsymbol{\nabla}^2 \psi_{\mathrm{inc,Q}}(\mathbf{x},\omega) + \frac{\omega^2}{c_E^2}\,\psi_{\mathrm{inc,Q}}(\mathbf{x},\omega) = \frac{1}{\rho_E}\,Q(\mathbf{x},\omega) \qquad \text{in} \quad \mathbb{R}^3 \quad , \tag{38}$$

in which, for all fixed real ω, the inhomogeneous part $\mathbf{x} \mapsto Q(\mathbf{x},\omega)$ is a bounded function with compact support [13] K_Q included in Ω_E. Consequently, we have

$$\int_{\mathbb{R}^3} |Q(\mathbf{x},\omega)|\,d\mathbf{x} = \int_{K_Q} |Q(\mathbf{x},\omega)|\,d\mathbf{x} < +\infty \quad . \tag{39}$$

Introducing the outward Sommerfeld radiation condition at infinity,

$$\left| \frac{\partial \psi_{\mathrm{inc,Q}}}{\partial R} + i\,k\,\psi_{\mathrm{inc,Q}} \right| = O(\frac{1}{R^2}) \quad , \tag{40}$$

the unique solution [14] of Eqs. (38) and (40) is written as

$$\psi_{\mathrm{inc,Q}}(\mathbf{x},\omega) = \int_{\mathbb{R}^3} G(\mathbf{x}-\mathbf{x}')\,\frac{1}{\rho_E}\,Q(\mathbf{x}',\omega)\,d\mathbf{x}' = \int_{K_Q} G(\mathbf{x}-\mathbf{x}')\,\frac{1}{\rho_E}\,Q(\mathbf{x}',\omega)\,d\mathbf{x}' \,. \tag{41}$$

---

[13] Compact support $K_Q \subset \Omega_E$ means that $Q(\mathbf{x},\omega)=0$ for all $\mathbf{x} \notin K_Q$.

[14] This unique solution satisfies Eq. (38) in the generalized function sense.

6. Response to Prescribed Wall Displacement, Acoustic Source Density and Incident Plane Wave

6.1. Radiation by an elastic boundary

We consider the boundary value problem defined by the Helmholtz equation in Ω_E

$$\nabla^2 \psi_{\text{rad}}(\mathbf{x}, \omega) + \frac{\omega^2}{c_E^2} \psi_{\text{rad}}(\mathbf{x}, \omega) = 0 \quad , \tag{42}$$

with the Neumann condition on Γ_E, considered as an elastic boundary,

$$\frac{\partial \psi_{\text{rad}}(\mathbf{y}, \omega)}{\partial \mathbf{n_y}} = i\omega \, \mathbf{u}_{\text{wall}}(\mathbf{y}, \omega) \cdot \mathbf{n}(\mathbf{y}) \quad , \tag{43}$$

in which $\mathbf{u}_{\text{wall}}$ is a given wall displacement field and with the outward Sommerfeld radiation condition at infinity,

$$|\,\psi_{\text{rad}}\,| = O(\frac{1}{R}) \quad , \quad \left| \frac{\partial \psi_{\text{rad}}}{\partial R} + i\,k\,\psi_{\text{rad}} \right| = O(\frac{1}{R^2}) \quad . \tag{44}$$

Referring to the basic problem defined in Sections 3 and 4, for all real ω, the boundary value problem defined by Eqs. (42) to (44) has a unique solution. The velocity potential on Γ_E is written as (see Eqs. (9)),

$$\psi_{\text{rad}}|_{\Gamma_E}(\omega) = i\omega \, \mathbf{B}_{\Gamma_E}(\omega/c_E)\{\mathbf{u}_{\text{wall}}(\omega) \cdot \mathbf{n}\} \quad , \tag{45}$$

and the pressure field on Γ_E is such that (see Eq. (10))

$$p_{\text{rad}}|_{\Gamma_E}(\omega) = i\omega \, \mathbf{Z}_{\Gamma_E}(\omega)\{\mathbf{u}_{\text{wall}}(\omega) \cdot \mathbf{n}\} \quad , \tag{46}$$

in which $p_{\text{rad}}|_{\Gamma_E}(\omega)$ denotes the trace of $p_{\text{rad}}(\omega)$ on Γ_E, and $\mathbf{Z}_{\Gamma_E}(\omega)$ is the acoustic impedance boundary operator related to $\mathbf{B}_{\Gamma_E}(\omega/c_E)$ by Eq. (11). At any point $\mathbf{x}$ fixed in Ω_E, the radiated pressure is given by (see Eq. (13)),

$$p_{\text{rad}}(\mathbf{x}, \omega) = i\omega \, \mathbf{Z}_{\text{rad}}(\mathbf{x}, \omega)\{\mathbf{u}_{\text{wall}}(\omega) \cdot \mathbf{n}\} \quad , \tag{47}$$

in which $\mathbf{Z}_{\text{rad}}(\mathbf{x}, \omega)$ is the radiation impedance operator defined by Eq. (14). The linear mappings $\mathbf{u}_{\text{wall}}(\omega) \mapsto p_{\text{rad}}|_{\Gamma_E}(\omega)$ and $\mathbf{u}_{\text{wall}}(\omega) \mapsto p_{\text{rad}}(\mathbf{x}, \omega)$ define the frequency response functions to be calculated.

6.2. Scattering of an acoustic source density by a rigid boundary

We consider the boundary value problem defined by the Helmholtz equation in Ω_E

$$\nabla^2\psi(\mathbf{x},\omega) + \frac{\omega^2}{c_E^2}\,\psi(\mathbf{x},\omega) = \frac{1}{\rho_E}\,Q(\mathbf{x},\omega) \quad , \tag{48}$$

in which, for all fixed real ω, acoustic source density $\mathbf{x} \mapsto Q(\mathbf{x},\omega)$ is a bounded function with compact support [15] K_Q included in Ω_E, with the Neumann boundary condition on Γ_E, considered as a rigid boundary,

$$\frac{\partial\psi(\mathbf{y},\omega)}{\partial\mathbf{n}_\mathbf{y}} = 0 \quad , \tag{49}$$

and with the outward Sommerfeld radiation condition at infinity,

$$|\,\psi\,| = O(\frac{1}{R}) \quad , \quad \left|\frac{\partial\psi}{\partial R} + i\,k\,\psi\right| = O(\frac{1}{R^2}) \quad . \tag{50}$$

The solution of Eqs. (48) to (50) can then be written as

$$\psi = \psi_{\text{inc,Q}} + \psi_{\text{rig}} \quad . \tag{51}$$

The field $\psi_{\text{inc,Q}}$ is the incident velocity potential induced by the acoustic source density. This field satisfies Eqs. (38) and (40) whose solution is given by (see Eq. (41)),

$$\psi_{\text{inc,Q}}(\mathbf{x},\omega) = \int_{K_Q} G(\mathbf{x}-\mathbf{x}')\,\frac{1}{\rho_E}\,Q(\mathbf{x}',\omega)\,d\mathbf{x}' \quad . \tag{52}$$

The incident pressure field in $\Omega_E \cup \Gamma_E$ is then given by

$$p_{\text{inc,Q}}(\mathbf{x},\omega) = -i\omega\int_{K_Q} G(\mathbf{x}-\mathbf{x}')\,Q(\mathbf{x}',\omega)\,d\mathbf{x}' \quad . \tag{53}$$

The field ψ_{rig} is the velocity potential scattered by Γ_E considered as a rigid boundary. This field is a solution of the boundary-value problem

$$\nabla^2\psi_{\text{rig}}(\mathbf{x},\omega) + \frac{\omega^2}{c_E^2}\,\psi_{\text{rig}}(\mathbf{x},\omega) = 0 \quad \text{in} \quad \Omega_E \quad , \tag{54}$$

---

[15] Compact support $K_Q \subset \Omega_E$ means that $Q(\mathbf{x},\omega)=0$ for all $\mathbf{x} \notin K_Q$. Consequently, we have $\int_{\mathbb{R}^3} |Q(\mathbf{x},\omega)|\,d\mathbf{x} = \int_{K_Q} |Q(\mathbf{x},\omega)|\,d\mathbf{x} < +\infty$.

$$\frac{\partial \psi_{\text{rig}}(\mathbf{y},\omega)}{\partial \mathbf{n}_{\mathbf{y}}} = -\frac{\partial \psi_{\text{inc,Q}}(\mathbf{y},\omega)}{\partial \mathbf{n}_{\mathbf{y}}} \quad \text{on} \quad \Gamma_E \quad , \tag{55}$$

$$|\,\psi_{\text{rig}}\,| = O(\frac{1}{R}) \quad , \quad \left| \frac{\partial \psi_{\text{rig}}}{\partial R} + i\,k\,\psi_{\text{rig}} \right| = O(\frac{1}{R^2}) \quad . \tag{56}$$

Since $K_Q \cap \Omega_i = \emptyset$, the normal derivative on Γ_E of $\psi_{\text{inc,Q}}$ defined by Eq. (52) is such that, for all $\mathbf{y}$ in Γ_E,

$$\frac{\partial \psi_{\text{inc,Q}}(\mathbf{y},\omega)}{\partial \mathbf{n}_{\mathbf{y}}} = \int_{K_Q} \frac{\partial G(\mathbf{y}-\mathbf{x}')}{\partial \mathbf{n}_{\mathbf{y}}}\,\frac{1}{\rho_E}\,Q(\mathbf{x}',\omega)\,d\mathbf{x}' \quad . \tag{57}$$

Referring to the basic problem defined in Sections 3 and 4, for all real ω, the boundary value problem defined by Eqs. (54) to (56) has a unique solution. Velocity potential ψ_{rig} on Γ_E is written as (see Eqs. (9)),

$$\psi_{\text{rig}}|_{\Gamma_E}(\omega) = -\mathbf{B}_{\Gamma_E}(\omega/c_E)\{\frac{\partial \psi_{\text{inc,Q}}}{\partial \mathbf{n}}\} \quad , \tag{58}$$

and pressure field p_{rig} on Γ_E is such that (see Eq. (10))

$$p_{\text{rig}}|_{\Gamma_E}(\omega) = -\mathbf{Z}_{\Gamma_E}(\omega)\{\frac{\partial \psi_{\text{inc,Q}}}{\partial \mathbf{n}}\} \quad . \tag{59}$$

At any point $\mathbf{x}$ fixed in Ω_E, the pressure field scattered by Γ_E, considered as a rigid boundary, is given by (see Eq. (13)),

$$p_{\text{rig}}(\mathbf{x},\omega) = -\mathbf{Z}_{\text{rad}}(\mathbf{x},\omega)\{\frac{\partial \psi_{\text{inc,Q}}}{\partial \mathbf{n}}\} \quad . \tag{60}$$

The resultant pressure $p_{\Gamma_E}(\omega) = p_{\text{inc,Q}}|_{\Gamma_E}(\omega) + p_{\text{rig}}|_{\Gamma_E}(\omega)$ on Γ_E is written as

$$p_{\Gamma_E}(\omega) = p_{\text{inc,Q}}|_{\Gamma_E}(\omega) - \mathbf{Z}_{\Gamma_E}(\omega)\{\frac{\partial \psi_{\text{inc,Q}}}{\partial \mathbf{n}}\} \quad , \tag{61}$$

in which $p_{\text{inc,Q}}|_{\Gamma_E}(\omega)$, $\mathbf{Z}_{\Gamma_E}(\omega)$ and $\partial \psi_{\text{inc,Q}}/\partial \mathbf{n}$ are given by Eqs. (53), (11) and (57) respectively. At any point $\mathbf{x}$ fixed in Ω_E, the resultant pressure $p(\mathbf{x},\omega) = p_{\text{inc,Q}}(\mathbf{x},\omega) + p_{\text{rig}}(\mathbf{x},\omega)$ is such that

$$p(\mathbf{x},\omega) = p_{\text{inc,Q}}(\mathbf{x},\omega) - \mathbf{Z}_{\text{rad}}(\mathbf{x},\omega)\{\frac{\partial \psi_{\text{inc,Q}}}{\partial \mathbf{n}}\} \quad , \tag{62}$$

in which $p_{\text{inc,Q}}(\mathbf{x},\omega)$, $\mathbf{Z}_{\text{rad}}(\mathbf{x},\omega)$ and $\partial \psi_{\text{inc,Q}}/\partial \mathbf{n}$ are given by Eqs. (53), (14) and (57) respectively. The linear mappings $Q(\omega) \mapsto p_{\Gamma_E}(\omega)$ and $Q(\omega) \mapsto p(\mathbf{x},\omega)$ define the frequency response functions to be calculated.

6.3. Scattering of an incident plane wave by a rigid boundary

We consider an incident plane wave defined by the incident velocity potential

$$\psi_{\rm inc}(\mathbf{x},\omega) = \psi_0(\omega)\exp\{-i\,\mathbf{k}\cdot\mathbf{x}\} \quad , \quad \mathbf{x}\in\mathbb{R}^3 \quad , \tag{63}$$

in which $\mathbf{k}\in\mathbb{R}^3$ is a given wave vector such that $\|\mathbf{k}\| = |k| = \omega/c_E$ and $\psi_0(\omega)\in\mathbb{C}$ is a given complex amplitude. This incident velocity potential satisfies the Helmholtz equation in $\mathbb{R}^3$

$$\nabla^2\psi_{\rm inc}(\mathbf{x},\omega) + k^2\,\psi_{\rm inc}(\mathbf{x},\omega) = 0 \quad . \tag{64}$$

It should be noted that potential $\psi_{\rm inc}$ does not satisfy the Sommerfeld condition at infinity. The corresponding pressure $p_{\rm inc}$ is given by

$$p_{\rm inc}(\mathbf{x},\omega) = -i\omega\,\rho_E\,\psi_{\rm inc}(\mathbf{x},\omega) \quad . \tag{65}$$

We consider scattering of this incident plane wave by a rigid boundary Γ_E. The resultant velocity potential ψ is then written as

$$\psi = \psi_{\rm inc} + \psi_{\rm rig} \quad , \tag{66}$$

in which $\psi_{\rm rig}$ is the velocity potential scattered by Γ_E considered as a rigid boundary. This field is a solution of the boundary-value problem

$$\nabla^2\psi_{\rm rig}(\mathbf{x},\omega) + \frac{\omega^2}{c_E^2}\,\psi_{\rm rig}(\mathbf{x},\omega) = 0 \quad \text{in} \quad \Omega_E \quad , \tag{67}$$

$$\frac{\partial\psi_{\rm rig}(\mathbf{y},\omega)}{\partial\mathbf{n_y}} = -\frac{\partial\psi_{\rm inc}(\mathbf{y},\omega)}{\partial\mathbf{n_y}} \quad \text{on} \quad \Gamma_E \quad , \tag{68}$$

$$|\,\psi_{\rm rig}\,| = O(\frac{1}{R}) \quad , \quad \left|\frac{\partial\psi_{\rm rig}}{\partial R} + i\,k\,\psi_{\rm rig}\right| = O(\frac{1}{R^2}) \quad . \tag{69}$$

For all $\mathbf{y}$ in Γ_E, we have

$$\frac{\partial\psi_{\rm inc}(\mathbf{y},\omega)}{\partial\mathbf{n_y}} = -i\,\mathbf{k}\cdot\mathbf{n}(\mathbf{y})\,\psi_{\rm inc}(\mathbf{y},\omega) \quad . \tag{70}$$

Referring to the basic problem defined in Sections 3 and 4, for all real ω the boundary value problem defined by Eqs. (67) to (69) has a unique solution. Velocity potential $\psi_{\rm rig}$ on Γ_E is written as (see Eqs. (9)),

$$\psi_{\rm rig}|_{\Gamma_E}(\omega) = -\mathbf{B}_{\Gamma_E}(\omega/c_E)\{\frac{\partial\psi_{\rm inc}}{\partial\mathbf{n}}\} \quad , \tag{71}$$

and pressure field $p_{\rm rig}$ on Γ_E is such that (see Eq. (10))

$$p_{\rm rig}|_{\Gamma_E}(\omega) = -\mathbf{Z}_{\Gamma_E}(\omega)\{\frac{\partial \psi_{\rm inc}}{\partial \mathbf{n}}\} \quad . \tag{72}$$

At any point $\mathbf{x}$ fixed in Ω_E, the pressure field scattered by Γ_E, considered as a rigid boundary, is given by (see Eq. (13)),

$$p_{\rm rig}(\mathbf{x},\omega) = -\mathbf{Z}_{\rm rad}(\mathbf{x},\omega)\{\frac{\partial \psi_{\rm inc}}{\partial \mathbf{n}}\} \quad . \tag{73}$$

The resultant pressure $p_{\Gamma_E}(\omega) = p_{\rm inc}|_{\Gamma_E}(\omega) + p_{\rm rig}|_{\Gamma_E}(\omega)$ on Γ_E is written as

$$p_{\Gamma_E}(\omega) = p_{\rm inc}|_{\Gamma_E}(\omega) - \mathbf{Z}_{\Gamma_E}(\omega)\{\frac{\partial \psi_{\rm inc}}{\partial \mathbf{n}}\} \quad , \tag{74}$$

in which $p_{\rm inc}|_{\Gamma_E}(\omega)$, $\mathbf{Z}_{\Gamma_E}(\omega)$ and $\partial\psi_{\rm inc}/\partial\mathbf{n}$ are given by Eqs. (65), (11) and (70) respectively. At any point $\mathbf{x}$ fixed in Ω_E, the resultant pressure $p(\mathbf{x},\omega)$ is such that

$$p(\mathbf{x},\omega) = p_{\rm inc}(\mathbf{x},\omega) - \mathbf{Z}_{\rm rad}(\mathbf{x},\omega)\{\frac{\partial \psi_{\rm inc}}{\partial \mathbf{n}}\} \quad , \tag{75}$$

in which $p_{\rm inc}(\mathbf{x},\omega)$, $\mathbf{Z}_{\rm rad}(\mathbf{x},\omega)$ and $\partial\psi_{\rm inc}/\partial\mathbf{n}$ are given by Eqs. (65), (14) and (70) respectively. The linear mappings $\psi_0(\omega) \mapsto p_{\Gamma_E}(\omega)$ and $\psi_0(\omega) \mapsto p(\mathbf{x},\omega)$ define the frequency response functions to be calculated.

6.4. Calculation of the response to prescribed wall displacement, acoustic source density and incident plane wave

We consider the general case corrresponding to a prescribed wall displacement $\mathbf{u}_{\rm wall}$ on Γ_E, a given acoustic source density $\mathbf{Q}$ and a given incident plane wave $\psi_{\rm inc}$ as defined in Sections 6.1, 6.2 and 6.3 respectively. Due to the linearity of the problem, the resultant pressure on Γ_E is then written as

$$p_E|_{\Gamma_E}(\omega) = p_{\rm rad}|_{\Gamma_E}(\omega) + p_{\rm given}|_{\Gamma_E}(\omega) \quad , \tag{76}$$

in which $p_{\rm rad}|_{\Gamma_E}(\omega)$ is written as

$$p_{\rm rad}|_{\Gamma_E}(\omega) = i\omega\, \mathbf{Z}_{\Gamma_E}(\omega)\{\mathbf{u}_{\rm wall}(\omega)\cdot\mathbf{n}\} \quad , \tag{77}$$

and $p_{\rm given}|_{\Gamma_E}(\omega)$ is such that

$$p_{\rm given}|_{\Gamma_E}(\omega) = p_{\rm inc,Q}|_{\Gamma_E}(\omega) + p_{\rm inc}|_{\Gamma_E}(\omega) - \mathbf{Z}_{\Gamma_E}(\omega)\{\frac{\partial \psi_{\rm inc,Q}}{\partial \mathbf{n}} + \frac{\partial \psi_{\rm inc}}{\partial \mathbf{n}}\}, \tag{78}$$

where $p_{\text{inc,Q}}|_{\Gamma_E}(\omega)$, $p_{\text{inc}}|_{\Gamma_E}(\omega)$, $\partial\psi_{\text{inc,Q}}/\partial\mathbf{n}$ and $\partial\psi_{\text{inc}}/\partial\mathbf{n}$ are defined by Eqs. (53), (65), (57) and (70) respectively. At any point $\mathbf{x}$ fixed in Ω_E, the resultant pressure $p_E(\mathbf{x},\omega)$ is written as

$$p_E(\mathbf{x},\omega) = p_{\text{rad}}(\mathbf{x},\omega) + p_{\text{given}}(\mathbf{x},\omega) \quad , \tag{79}$$

in which $p_{\text{rad}}(\mathbf{x},\omega)$ is written as

$$p_{\text{rad}}(\mathbf{x},\omega) = i\omega\, \mathbf{Z}_{\text{rad}}(\mathbf{x},\omega)\{\mathbf{u}_{\text{wall}}(\omega)\cdot\mathbf{n}\} \quad , \tag{80}$$

and $p_{\text{given}}(\mathbf{x},\omega)$ is such that

$$p_{\text{given}}(\mathbf{x},\omega) = p_{\text{inc,Q}}(\mathbf{x},\omega) + p_{\text{inc}}(\mathbf{x},\omega) - \mathbf{Z}_{\text{rad}}(\mathbf{x},\omega)\{\frac{\partial\psi_{\text{inc,Q}}}{\partial\mathbf{n}} + \frac{\partial\psi_{\text{inc}}}{\partial\mathbf{n}}\} , \tag{81}$$

where $p_{\text{inc,Q}}(\mathbf{x},\omega)$, $p_{\text{inc}}(\mathbf{x},\omega)$, $\partial\psi_{\text{inc,Q}}/\partial\mathbf{n}$ and $\partial\psi_{\text{inc}}/\partial\mathbf{n}$ are defined by Eqs. (53), (65), (57) and (70) respectively. Construction of the response defined by Eqs. (76) to (81) requires calculating acoustic impedance boundary operator $\mathbf{Z}_{\Gamma_E}(\omega)$ and radiation impedance operator $\mathbf{Z}_{\text{rad}}(\mathbf{x},\omega)$. From Eqs. (11) and (14), we then deduce that we have to construct operators $\mathbf{B}_{\Gamma_E}(\omega/c_E)$ and $\mathbf{R}(\mathbf{x},\omega/c_E)$. This construction is carried out below using an integral boundary formulation related to Γ_E.

7. Results of Potential Theory: Single- and Double-Layer Potentials

In this section we recall the fundamental results of potential theory (Mikhlin, 1970; Kress, 1989; Colton and Kress, 1992) which is useful for constructing the boundary integral equations related to the solution of the exterior Neumann problem introduced in Section 3.

7.1. Single- and double-layer potentials

Consider the external unbounded domain Ω_E of $\mathbb{R}^3$ introduced in Section 2, with boundary $\partial\Omega_E = \Gamma_E$ and $\Omega_i = \mathbb{R}^3\backslash\Omega_E$. Let $\mathbf{x}$, $\mathbf{y}$ and $\mathbf{n}_\mathbf{x}$, $\mathbf{n}_\mathbf{y}$ be two points in Γ_E and their outward unit normals to Γ_E respectively (see Fig. 3). Let $\mathbf{x}'$ be a point in Ω_E (see Fig. 3-a) or in Ω_i (see Fig. 3-b) such that

$$\mathbf{x}' = \mathbf{x} + \eta\,\mathbf{n}_\mathbf{x} \quad , \quad \eta\in\mathbb{R} \quad , \quad \mathbf{x}\in\Gamma_E \quad . \tag{82}$$

Below, $f(\mathbf{y})$ denotes a continuous function from Γ_E into $\mathbb{C}$.

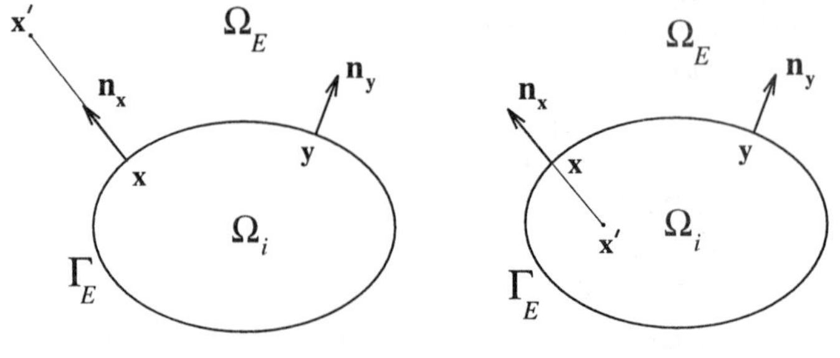

Fig. 3-a Fig. 3-b

Fig. 3. Geometry of configurations

Single-layer potential. The single-layer potential ψ_S of density f is defined by [16]

$$\psi_S(\mathbf{x}') = \int_{\Gamma_E} f(\mathbf{y})\, G(\mathbf{x}' - \mathbf{y})\, ds_\mathbf{y} \quad , \tag{83}$$

in which the integral in the right-hand side of Eq. (83) is defined in the usual sense. Then, function ψ_S is continuous in $\mathbb{R}^3$ but the normal derivatives are discontinuous across Γ_E and we have, for $\mathbf{x}'$ defined by Eq. (82) and in Ω_E (see Fig. 3-a),

$$\lim_{\eta \to 0_+} \frac{\partial \psi_S(\mathbf{x}')}{\partial \mathbf{n}_\mathbf{x}} = \frac{1}{2} f(\mathbf{x}) + \int_{\Gamma_E} f(\mathbf{y})\, \frac{\partial G(\mathbf{x} - \mathbf{y})}{\partial \mathbf{n}_\mathbf{x}}\, ds_\mathbf{y} \quad , \tag{84}$$

and for $\mathbf{x}'$ defined by Eq. (82) and in Ω_i (see Fig. 3-b),

$$\lim_{\eta \to 0_-} \frac{\partial \psi_S(\mathbf{x}')}{\partial \mathbf{n}_\mathbf{x}} = -\frac{1}{2} f(\mathbf{x}) + \int_{\Gamma_E} f(\mathbf{y})\, \frac{\partial G(\mathbf{x} - \mathbf{y})}{\partial \mathbf{n}_\mathbf{x}}\, ds_\mathbf{y} \quad . \tag{85}$$

The integrals [17] in the right-hand sides of Eqs. (84) and (85) are defined in the usual sense.

Double-layer potential. The double-layer potential ψ_D of density f is defined by [18]

$$\psi_D(\mathbf{x}') = \int_{\Gamma_E} f(\mathbf{y})\, \frac{\partial G(\mathbf{x}' - \mathbf{y})}{\partial \mathbf{n}_\mathbf{y}}\, ds_\mathbf{y} \quad , \tag{86}$$

[16] The mapping $f \mapsto \psi_S|_{\Gamma_E}$ is continuous from $H^{-1/2}(\Gamma_E)$ into $H^{1/2}(\Gamma_E)$ (see Costabel, 1988).

[17] For $\mathbf{x} \in \Gamma_E$, the mapping $f \mapsto \int_{\Gamma_E} f(\mathbf{y})\, \partial G(\mathbf{x}-\mathbf{y})/\partial \mathbf{n}_\mathbf{x}\, ds_\mathbf{y}$ is continuous from $H^{-1/2}(\Gamma_E)$ into $H^{1/2}(\Gamma_E)$ (see Costabel, 1988).

[18] The mapping $f \mapsto \psi_D|_{\Gamma_E}$ is continuous from $H^{1/2}(\Gamma_E)$ into $H^{3/2}(\Gamma_E)$ (see Costabel, 1988).

in which the integral on the right-hand sides of Eq. (86) is defined in the usual sense. Then, function ψ_{D} is continuous in Ω_i and in Ω_E, but is discontinuous across Γ_E and we have, for $\mathbf{x}'$ defined by Eq. (82) and in Ω_E (see Fig. 3-a),

$$\lim_{\eta\rightarrow 0_+} \psi_{\mathrm{D}}(\mathbf{x}') = -\frac{1}{2} f(\mathbf{x}) + \int_{\Gamma_E} f(\mathbf{y}) \frac{\partial G(\mathbf{x}-\mathbf{y})}{\partial \mathbf{n}_{\mathbf{y}}} \, ds_{\mathbf{y}} \quad , \tag{87}$$

and for $\mathbf{x}'$ defined by Eq. (82) and in Ω_i (see Fig. 3-b),

$$\lim_{\eta\rightarrow 0_-} \psi_{\mathrm{D}}(\mathbf{x}') = \frac{1}{2} f(\mathbf{x}) + \int_{\Gamma_E} f(\mathbf{y}) \frac{\partial G(\mathbf{x}-\mathbf{y})}{\partial \mathbf{n}_{\mathbf{y}}} \, ds_{\mathbf{y}} \quad . \tag{88}$$

The integrals in the right-hand sides of Eqs. (87) and (88) are defined in the usual sense. The normal derivative with respect to $\mathbf{n}_{\mathbf{x}}$ is continuous in $\mathbb{R}^3$ and for $\mathbf{x}'$ defined by Eq. (82) and in $\mathbb{R}^3$, we have [19]

$$\frac{\partial \psi_{\mathrm{D}}(\mathbf{x}')}{\partial \mathbf{n}_{\mathbf{x}}} = \int_{\Gamma_E} f(\mathbf{y}) \frac{\partial^2 G(\mathbf{x}'-\mathbf{y})}{\partial \mathbf{n}_{\mathbf{x}} \partial \mathbf{n}_{\mathbf{y}}} \, ds_{\mathbf{y}} \quad . \tag{89}$$

If we consider Eq. (89) on Γ_E, i.e. $\mathbf{x}' = \mathbf{x}$, then the integral in the right-hand side of Eq. (89) exists as a Cauchy principal value [20].

8. Helmholtz Integral Representations and their Normal Derivatives

In this section, we recall the fundamental Helmholtz integral representations for the external domain and the internal domain related to the Helmholtz equation. For the mathematical proof we refer the reader to Mikhlin, 1970 for the particular case of the Laplace equation ($k = 0$), and to Kellogg, 1953 and Colton and Kress, 1992 for the Helmholtz equation. In addition, we construct the normal derivative of each Helmholtz integral representation using the results of Section 7 concerning potential theory.

---

[19] For $\mathbf{x}\in\Gamma_E$, the mapping $f \mapsto \int_{\Gamma_E} f(\mathbf{y})\, \partial^2 G(\mathbf{x}-\mathbf{y})/\partial\mathbf{n}_{\mathbf{x}}\partial\mathbf{n}_{\mathbf{y}}\, ds_{\mathbf{y}}$ is continuous from $H^{1/2}(\Gamma_E)$ into $H^{-1/2}(\Gamma_E)$ (see Costabel, 1988).

[20] Denoting the ball with center $\mathbf{x}$ and radius η as $B(\mathbf{x},\eta)$, the Cauchy principal value is defined by $\int_{\Gamma_E} f(\mathbf{y})\, \partial^2 G(\mathbf{x}-\mathbf{y})/\partial\mathbf{n}_{\mathbf{x}}\partial\mathbf{n}_{\mathbf{y}}\, ds_{\mathbf{y}} = \lim_{\eta\rightarrow 0} \int_{\Gamma_E \setminus B(\mathbf{x},\eta)} f(\mathbf{y})\, \partial^2 G(\mathbf{x}-\mathbf{y})/\partial\mathbf{n}_{\mathbf{x}}\partial\mathbf{n}_{\mathbf{y}}\, ds_{\mathbf{y}}$.

8.1. Integral representations for the Helmholtz equation in an external domain

Let us consider the geometric configuration defined in Fig. 4 in which $\Omega_E \cap \Gamma_E \cap \Omega_i = \emptyset$.

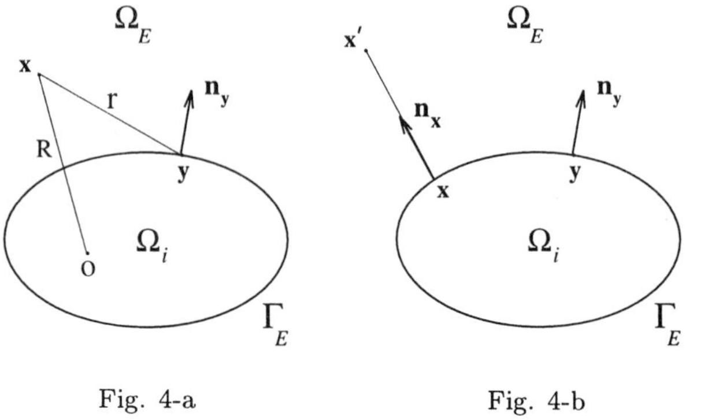

Fig. 4. Geometric configuration for the Helmholtz equation in an external domain

Let $\psi(\mathbf{x})$ be a twice continuously differentiable function on Ω_E satisfying the Helmholtz equation in the external domain Ω_E,

$$\boldsymbol{\nabla}^2\psi(\mathbf{x},\omega) + k^2\,\psi(\mathbf{x},\omega) = 0 \qquad \text{in} \quad \Omega_E \quad , \tag{90}$$

and satisfying the outward Sommerfeld radiation condition at infinity

$$|\,\psi\,| = O(\frac{1}{R}) \quad , \quad \left|\frac{\partial\psi}{\partial R} + i\,k\,\psi\right| = O(\frac{1}{R^2}) \quad , \tag{91}$$

as $R = \|\mathbf{x}\| \rightarrow +\infty$, in which $\partial/\partial R$ denotes the derivative in the radial direction from the origin (see Fig. 4).

Helmholtz integral representation for the Helmholtz equation in an external domain. Velocity potential ψ satisfying Eqs. (90) and (91) has the following integral representation related to the Helmholtz equation in external domain $\Omega_E \cup \Gamma_E$,

$$\varepsilon\,\psi(\mathbf{x}) = \int_{\Gamma_E} \left\{ G(\mathbf{x}-\mathbf{y})\,\frac{\partial\psi(\mathbf{y})}{\partial\mathbf{n}_\mathbf{y}} - \psi(\mathbf{y})\,\frac{\partial G(\mathbf{x}-\mathbf{y})}{\partial\mathbf{n}_\mathbf{y}} \right\} ds_\mathbf{y} \quad , \tag{92}$$

in which G is the free-space Green's function defined by Eqs. (35)–(36) and ε is such that

$$\varepsilon = 1 \quad \text{for} \quad \mathbf{x} \in \Omega_E \quad , \tag{93}$$

$$\varepsilon = \frac{1}{2} \quad \text{for} \quad \mathbf{x} \in \Gamma_E \quad . \tag{94}$$

For the internal open domain Ω_i, this integral representation becomes

$$0 = \int_{\Gamma_E} \left\{ G(\mathbf{x}-\mathbf{y}) \frac{\partial \psi(\mathbf{y})}{\partial \mathbf{n_y}} - \psi(\mathbf{y}) \frac{\partial G(\mathbf{x}-\mathbf{y})}{\partial \mathbf{n_y}} \right\} ds_{\mathbf{y}} \quad , \quad \mathbf{x} \in \Omega_i \quad . \qquad (95)$$

The integrals in the right-hand side of Eqs. (92) or (95) are defined in the usual sense for $\mathbf{x} \in \Gamma_E$ or $\mathbf{x} \notin \Gamma_E$.

Normal derivative of the Helmholtz integral representation for the Helmholtz equation in an external domain. For all $\mathbf{x} \in \Gamma_E$, velocity potential ψ satisfying Eqs. (90) and (91) has the following integral representation on Γ_E related to the Helmholtz equation in external domain Ω_E,

$$-\int_{\Gamma_E} \psi(\mathbf{y}) \frac{\partial^2 G(\mathbf{x}-\mathbf{y})}{\partial \mathbf{n_x} \partial \mathbf{n_y}} ds_{\mathbf{y}} = \frac{1}{2} \frac{\partial \psi(\mathbf{x})}{\partial \mathbf{n_x}} - \int_{\Gamma_E} \frac{\partial G(\mathbf{x}-\mathbf{y})}{\partial \mathbf{n_x}} \frac{\partial \psi(\mathbf{y})}{\partial \mathbf{n_y}} ds_{\mathbf{y}} \, , \qquad (96)$$

in which the integral in the left-hand side of Eq. (96) exists as a Cauchy principal value [21]. For the proof of Eq. (96), we consider Fig. 4-b in which x' is a point in Ω_E such that, for all $\mathbf{x}$ in Γ_E, $\mathbf{x}' = \mathbf{x} + \eta \, \mathbf{n_x}$, $\eta > 0$. Using the Helmholtz integral Eq. (92) with $\varepsilon = 1$ at point $x' \in \Omega_E$ and taking the normal derivative with respect to $\mathbf{n}(\mathbf{x}') = \mathbf{n}(\mathbf{x})$ yields an equation in which, using Eqs. (84) and (89), the limit is calculated when $\mathbf{x}' \in \Omega_E$ approaches $\mathbf{x} \in \Gamma_E$. We therefore obtain Eq. (96).

8.2. Integral representations for the Helmholtz equation in an internal domain

Let us consider the geometric configuration defined in Fig. 5 in which $\Omega_E \cap \Gamma_E \cap \Omega_i = \emptyset$.

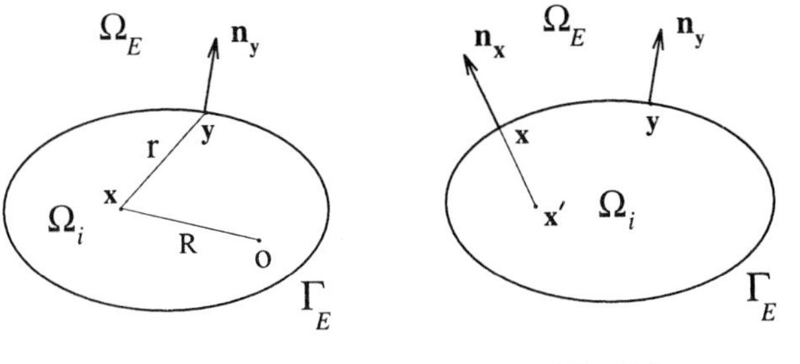

Fig. 5-a Fig. 5-b

Fig. 5. Geometric configuration for the Helmholtz equation in an internal domain

---

Denoting the ball with center $\mathbf{x}$ and radius η as $B(\mathbf{x},\eta)$, the Cauchy principal value is defined by $\int_{\Gamma_E} \psi(\mathbf{y}) \, \partial^2 G(\mathbf{x}-\mathbf{y}) / \partial \mathbf{n_x} \partial \mathbf{n_y} \, ds_{\mathbf{y}} = \lim_{\eta \to 0} \int_{\Gamma_E \setminus B(\mathbf{x},\eta)} \psi(\mathbf{y}) \, \partial^2 G(\mathbf{x}-\mathbf{y}) / \partial \mathbf{n_x} \partial \mathbf{n_y} \, ds_{\mathbf{y}}$.

Let $\psi_i(\mathbf{x})$ be a twice continuously differentiable function on Ω_i satisfying the Helmholtz equation in the internal domain Ω_i,

$$\nabla^2\psi_i(\mathbf{x},\omega) + k^2\,\psi_i(\mathbf{x},\omega) = 0 \qquad \text{in} \quad \Omega_i \quad . \tag{97}$$

Helmholtz integral representation for the Helmholtz equation in an internal domain. Velocity potential ψ_i satisfying Eq. (97) has the following integral equation related to the Helmholtz equation in internal domain $\Omega_i \cup \Gamma_E$,

$$\varepsilon_i\,\psi_i(\mathbf{x}) = -\int_{\Gamma_E}\left\{G(\mathbf{x}-\mathbf{y})\,\frac{\partial\psi_i(\mathbf{y})}{\partial\mathbf{n}_\mathbf{y}} - \psi_i(\mathbf{y})\,\frac{\partial G(\mathbf{x}-\mathbf{y})}{\partial\mathbf{n}_\mathbf{y}}\right\} ds_\mathbf{y} \quad , \tag{98}$$

in which G is the free-space Green's function defined by Eqs. (35)–(36) and ε is such that

$$\varepsilon_i = 1 \quad \text{for} \quad \mathbf{x}\in\Omega_i \quad . \tag{99}$$

$$\varepsilon_i = \frac{1}{2} \quad \text{for} \quad \mathbf{x}\in\Gamma_E \quad . \tag{100}$$

For the external open domain Ω_E, this integral representation becomes

$$0 = -\int_{\Gamma_E}\left\{G(\mathbf{x}-\mathbf{y})\,\frac{\partial\psi_i(\mathbf{y})}{\partial\mathbf{n}_\mathbf{y}} - \psi_i(\mathbf{y})\,\frac{\partial G(\mathbf{x}-\mathbf{y})}{\partial\mathbf{n}_\mathbf{y}}\right\} ds_\mathbf{y}\ ,\ \mathbf{x}\in\Omega_E \quad . \tag{101}$$

The integrals in the right-hand side of Eqs. (98) and (101) are defined in the usual sense for $\mathbf{x}\in\Gamma_E$ or $\mathbf{x}\notin\Gamma_E$.

Normal derivative of the Helmholtz integral representation for the Helmholtz equation in an internal domain. For all $\mathbf{x}\in\Gamma_E$, velocity potential ψ_i satisfying Eq. (97) has the following integral representation on Γ_E related to the Helmholtz equation in internal domain Ω_i,

$$-\int_{\Gamma_E}\psi_i(\mathbf{y})\frac{\partial^2 G(\mathbf{x}-\mathbf{y})}{\partial\mathbf{n}_\mathbf{x}\partial\mathbf{n}_\mathbf{y}}\,ds_\mathbf{y} = -\frac{1}{2}\frac{\partial\psi_i(\mathbf{x})}{\partial\mathbf{n}_\mathbf{x}} - \int_{\Gamma_E}\frac{\partial G(\mathbf{x}-\mathbf{y})}{\partial\mathbf{n}_\mathbf{x}}\,\frac{\partial\psi_i(\mathbf{y})}{\partial\mathbf{n}_\mathbf{y}}\,ds_\mathbf{y}\,, \tag{102}$$

in which the integral on the left-hand side of Eq. (102) exists as a Cauchy principal value[22]. For the proof of Eq. (102), we consider Fig. 5-b in which x' is a point in Ω_i such that, for all $\mathbf{x}$ in Γ_E, $\mathbf{x}' = \mathbf{x}+\eta\,\mathbf{n}_\mathbf{x}$, $\eta<0$. Using the Helmholtz integral Eq. (98) with $\varepsilon_i = 1$ at point $x'\in\Omega_i$ and taking the normal derivative with respect to $\mathbf{n}(\mathbf{x}') = \mathbf{n}(\mathbf{x})$ yields an equation in which, using Eqs. (85) and (89), the limit is calculated when $\mathbf{x}'\in\Omega_i$ approaches $\mathbf{x}\in\Gamma_E$. We therefore obtain Eq. (102).

22 Denoting the ball with center $\mathbf{x}$ and radius η as $B(\mathbf{x},\eta)$, the Cauchy principal value is defined by $\int_{\Gamma_E}\psi_i(\mathbf{y})\,\partial^2 G(\mathbf{x}-\mathbf{y})/\partial\mathbf{n}_\mathbf{x}\partial\mathbf{n}_\mathbf{y}\,ds_\mathbf{y} = \lim_{\eta\to 0}\int_{\Gamma_E\setminus B(\mathbf{x},\eta)}\psi_i(\mathbf{y})\,\partial^2 G(\mathbf{x}-\mathbf{y})/\partial\mathbf{n}_\mathbf{x}\partial\mathbf{n}_\mathbf{y}\,ds_\mathbf{y}$.

9. Boundary Integral Equations for the Exterior Neumann Problem Related to the Helmholtz Equation

An important problem is the construction of operator $\mathbf{B}_{\Gamma_E}(\omega/c_E)$ introduced in Section 4, which allows the calculation of acoustic impedance boundary operator $\mathbf{Z}_{\Gamma_E}(\omega)$ (see Eq. (11)). We recall that, for all real ω, this operator is such that

$$\psi^{\text{sol}}_{\Gamma_E}(\omega) = \mathbf{B}_{\Gamma_E}(\omega/c_E)\, v \quad \text{on} \quad \Gamma_E \quad , \tag{103}$$

in which $\psi^{\text{sol}}_{\Gamma_E}$ is the trace on Γ_E of the unique solution ψ^{sol} of the basic exterior Neumann problem related to the Helmholtz equation (see Section 3),

$$\nabla^2 \psi^{\text{sol}}(\mathbf{x},\omega) + k^2\, \psi^{\text{sol}}(\mathbf{x},\omega) = 0 \qquad \text{in} \quad \Omega_E \quad , \tag{104}$$

$$\frac{\partial \psi^{\text{sol}}(\mathbf{y},\omega)}{\partial \mathbf{n}_{\mathbf{y}}} = v(\mathbf{y}) \qquad \text{on} \quad \Gamma_E \quad , \tag{105}$$

$$|\,\psi^{\text{sol}}\,| = O(\frac{1}{R}) \quad , \quad \left| \frac{\partial \psi^{\text{sol}}}{\partial R} + i\,k\,\psi^{\text{sol}} \right| = O(\frac{1}{R^2}) \quad . \tag{106}$$

Since the boundary value problem defined by Eqs. (104) to (106) is related to an *unbounded* external domain, the use of a boundary integral equation method is well suited and allows us to make a boundary integral formulation of operator $\mathbf{B}_{\Gamma_E}(\omega/c_E)$ on Γ_E . In this section, we introduce two boundary integral equations which will be used together in Section 10 to construct the boundary integral formulation of operator $\mathbf{B}_{\Gamma_E}(\omega/c_E)$.

9.1. First boundary integral equation and its variational formulation

First integral equation on Γ_E. The first boundary integral equation is deduced directly from the Helmholtz integral Eq. (92) for $\mathbf{x}$ in Γ_E ($\varepsilon = 1/2$), related to the unique solution ψ^{sol} of Eqs. (104) to (106). Using Eq. (105), we deduce that, for all $\mathbf{x} \in \Gamma_E$,

$$\int_{\Gamma_E} G(\mathbf{x}-\mathbf{y})\, v(\mathbf{y})\, ds_{\mathbf{y}} - \left\{ \frac{1}{2}\psi^{\text{sol}}_{\Gamma_E}(\mathbf{x}) + \int_{\Gamma_E} \psi^{\text{sol}}_{\Gamma_E}(\mathbf{y})\, \frac{\partial G(\mathbf{x}-\mathbf{y})}{\partial \mathbf{n}_{\mathbf{y}}}\, ds_{\mathbf{y}} \right\} = 0\,. \tag{107}$$

Definition of integral operators $\mathbf{S}_{\text{S}}(\omega/c_E)$ and $\mathbf{S}_{\text{D}}(\omega/c_E)$. Let $\mathbf{S}_{\text{S}}(\omega/c_E)$ and $\mathbf{S}_{\text{D}}(\omega/c_E)$ be the linear integral operators defined, for all $\mathbf{x} \in \Gamma_E$, by [23]

$$\{\mathbf{S}_{\text{S}}(\omega/c_E)\, v\}(\mathbf{x}) = \int_{\Gamma_E} G(\mathbf{x}-\mathbf{y})\, v(\mathbf{y})\, ds_{\mathbf{y}} \quad , \tag{108}$$

The integrals in the right-hand side of Eqs. (108) and (109) are defined for sufficiently smooth functions. According to Costabel and Stephan, 1985; Costabel, 1988, $\text{S}_{\text{S}}(\omega/c_E)$ is a linear continuous operator from $\mathcal{C}'_{\Gamma_E} = H^{-1/2}(\Gamma_E)$ into $\mathcal{C}_{\Gamma_E} = H^{1/2}(\Gamma_E)$, and $\text{S}_{\text{D}}(\omega/c_E)$ is a linear continuous operator from $\mathcal{C}_{\Gamma_E} = H^{1/2}(\Gamma_E)$ into $H^{3/2}(\Gamma_E) \subset H^{1/2}(\Gamma_E) = \mathcal{C}_{\Gamma_E}$.

$$\{\mathbf{S}_{\mathrm{D}}(\omega/c_E)\,\psi_{\Gamma_E}\}(\mathbf{x}) = \int_{\Gamma_E} \psi_{\Gamma_E}(\mathbf{y})\,\frac{\partial G(\mathbf{x}-\mathbf{y})}{\partial \mathbf{n}_{\mathbf{y}}}\,ds_{\mathbf{y}} \quad . \tag{109}$$

Symmetry property of integral operator $\mathbf{S}_{\mathrm{S}}(\omega/c_E)$. The transpose [24] of operator $\mathbf{S}_{\mathrm{S}}(\omega/c_E)$ is the linear operator ${}^t\mathbf{S}_{\mathrm{S}}(\omega/c_E)$ such that, for all v and δv in $\mathcal{C}'_{\Gamma_E}$,

$$<\mathbf{S}_{\mathrm{S}}(\omega/c_E)\,v\,,\delta v>_{\mathcal{C}_{\Gamma_E},\mathcal{C}'_{\Gamma_E}} =<v\,,{}^t\mathbf{S}_{\mathrm{S}}(\omega/c_E)\,\delta v>_{\mathcal{C}'_{\Gamma_E},\mathcal{C}_{\Gamma_E}} \quad . \tag{110}$$

We then have the following symmetry property

$${}^t\mathbf{S}_{\mathrm{S}}(\omega/c_E) = \mathbf{S}_{\mathrm{S}}(\omega/c_E) \quad . \tag{111}$$

Let us prove Eq. (111) for v and δv in the space $H_{\Gamma_E} \subset \mathcal{C}'_{\Gamma_E}$ of the square integrable complex function on Γ_E. Consequently, we can write

$$<\mathbf{S}_{\mathrm{S}}(\omega/c_E)\,v\,,\delta v>_{\mathcal{C}_{\Gamma_E},\mathcal{C}'_{\Gamma_E}} = \int_{\Gamma_E}\int_{\Gamma_E} G(\mathbf{x}-\mathbf{y})\,v(\mathbf{y})\,\delta v(\mathbf{x})\,ds_{\mathbf{y}}\,ds_{\mathbf{x}}\,. \tag{112}$$

Since $G(\mathbf{x}-\mathbf{y}) = G(\mathbf{y}-\mathbf{x})$ (see Section 5), Eq. (112) yields Eq. (111).

Calculation of the transpose ${}^t\mathbf{S}_{\mathrm{D}}(\omega/c_E)$ of operator $\mathbf{S}_{\mathrm{D}}(\omega/c_E)$. The transpose of $\mathbf{S}_{\mathrm{D}}(\omega/c_E)$ is the linear operator ${}^t\mathbf{S}_{\mathrm{D}}(\omega/c_E)$ such that [25], for all ψ_{Γ_E} in $\mathcal{C}_{\Gamma_E}$ and for all δv in $\mathcal{C}'_{\Gamma_E}$,

$$<\mathbf{S}_{\mathrm{D}}(\omega/c_E)\,\psi_{\Gamma_E}\,,\delta v>_{\mathcal{C}_{\Gamma_E},\mathcal{C}'_{\Gamma_E}} =<\psi_{\Gamma_E}\,,{}^t\mathbf{S}_{\mathrm{D}}(\omega/c_E)\,\delta v>_{\mathcal{C}_{\Gamma_E},\mathcal{C}'_{\Gamma_E}} \quad . \tag{113}$$

Let us consider δv in the space $H_{\Gamma_E} \subset \mathcal{C}'_{\Gamma_E}$. Consequently, we can write

$$<\mathbf{S}_{\mathrm{D}}(\omega/c_E)\,\psi_{\Gamma_E}\,,\delta v>_{\mathcal{C}_{\Gamma_E},\mathcal{C}'_{\Gamma_E}} = \int_{\Gamma_E}\int_{\Gamma_E} \frac{\partial G(\mathbf{x}-\mathbf{y})}{\partial \mathbf{n}_{\mathbf{y}}}\,\psi_{\Gamma_E}(\mathbf{y})\,\delta v(\mathbf{x})\,ds_{\mathbf{y}}\,ds_{\mathbf{x}}\,. \tag{114}$$

Since $\partial G(\mathbf{x}-\mathbf{y})/\partial \mathbf{n}_{\mathbf{y}} = \partial G(\mathbf{y}-\mathbf{x})/\partial \mathbf{n}_{\mathbf{y}}$ (see Section 5), from Eqs. (113) and (114), we deduce that

$$<\psi_{\Gamma_E},{}^t\mathbf{S}_{\mathrm{D}}(\omega/c_E)\,\delta v>_{\mathcal{C}_{\Gamma_E},\mathcal{C}'_{\Gamma_E}} = \int_{\Gamma_E}\int_{\Gamma_E} \frac{\partial G(\mathbf{x}-\mathbf{y})}{\partial \mathbf{n}_{\mathbf{x}}}\,\delta v(\mathbf{y})\,\psi_{\Gamma_E}(\mathbf{x})\,ds_{\mathbf{y}}\,ds_{\mathbf{x}}\,, \tag{115}$$

---

24 Space $\mathcal{C}'_{\Gamma_E}$ is the dual space of $\mathcal{C}_{\Gamma_E}$ for the dual brackets $<v,\psi_{\Gamma_E}>_{\mathcal{C}'_{\Gamma_E},\mathcal{C}_{\Gamma_E}}$ in which $v\in\mathcal{C}'_{\Gamma_E}$ and $\psi_{\Gamma_E}\in\mathcal{C}_{\Gamma_E}$ and which is linear in v and linear in ψ_{Γ_E} (and not antilinear in ψ_{Γ_E}). The transpose ${}^t\mathrm{S}_{\mathrm{S}}(\omega/c_E)$ is an operator from $\mathcal{C}'_{\Gamma_E}$ into $(\mathcal{C}'_{\Gamma_E})'=\mathcal{C}_{\Gamma_E}$.

25 The transpose of $\mathrm{S}_{\mathrm{D}}(\omega/c_E)$ from $\mathcal{C}_{\Gamma_E}$ into $\mathcal{C}_{\Gamma_E}$ is the linear operator ${}^t\mathrm{S}_{\mathrm{D}}(\omega/c_E)$ from $\mathcal{C}'_{\Gamma_E}$ into $\mathcal{C}'_{\Gamma_E}$.

which yields, for all $\mathbf{x} \in \Gamma_E$,

$$\{{}^t\mathbf{S}_{\mathrm{D}}(\omega/c_E)\, v\}(\mathbf{x}) = \int_{\Gamma_E} v(\mathbf{y})\, \frac{\partial G(\mathbf{x}-\mathbf{y})}{\partial \mathbf{n}_{\mathbf{x}}}\, ds_{\mathbf{y}} \quad . \tag{116}$$

Variational formulation in v of the first integral equation on Γ_E. For all ω fixed in $\mathbb{R}$ and for all $v \in \mathcal{C}'_{\Gamma_E}$, let $\psi^{\mathrm{sol}}_{\Gamma_E} \in \mathcal{C}_{\Gamma_E}$ be the trace on Γ_E of the unique solution ψ^{sol} of the boundary value problem defined by Eqs. (104) to (106). From Eqs. (107), (108) and (109), we deduce that, for all δv in $\mathcal{C}'_{\Gamma_E}$, v and $\psi^{\mathrm{sol}}_{\Gamma_E}$ satisfy the following variational equation in v,

$$< \mathbf{S}_{\mathrm{S}}(\omega/c_E)\, v \,, \delta v >_{\mathcal{C}_{\Gamma_E}, \mathcal{C}'_{\Gamma_E}} - < (\frac{1}{2}\,\mathbf{I} + \mathbf{S}_{\mathrm{D}}(\omega/c_E))\, \psi^{\mathrm{sol}}_{\Gamma_E} \,, \delta v >_{\mathcal{C}_{\Gamma_E}, \mathcal{C}'_{\Gamma_E}} = 0 \,, \tag{117}$$

in which $\mathbf{I}$ denotes the identity operator [26]. The corresponding operator equation between v and $\psi^{\mathrm{sol}}_{\Gamma_E}$ on Γ_E is [27]

$$\mathbf{S}_{\mathrm{S}}(\omega/c_E)\, v - (\frac{1}{2}\,\mathbf{I} + \mathbf{S}_{\mathrm{D}}(\omega/c_E))\, \psi^{\mathrm{sol}}_{\Gamma_E} = 0 \quad . \tag{118}$$

In the methodology presented below, Eq. (118) will not be directly used to solve the problem in $\psi^{\mathrm{sol}}_{\Gamma_E}$.

9.2. Second boundary integral equation and its variational formulation

Second integral equation on Γ_E. The second boundary integral equation is deduced directly from the normal derivative on Γ_E of the Helmholtz integral representation given by Eq. (96), related to the unique solution ψ^{sol} of Eqs. (104) to (106). Using Eq. (105), we deduce that for all $\mathbf{x} \in \Gamma_E$,

$$-\int_{\Gamma_E} \psi^{\mathrm{sol}}_{\Gamma_E}(\mathbf{y}) \frac{\partial^2 G(\mathbf{x}-\mathbf{y})}{\partial \mathbf{n}_{\mathbf{x}} \partial \mathbf{n}_{\mathbf{y}}}\, ds_{\mathbf{y}} = \frac{1}{2}\, v(\mathbf{x}) - \int_{\Gamma_E} v(\mathbf{y})\, \frac{\partial G(\mathbf{x}-\mathbf{y})}{\partial \mathbf{n}_{\mathbf{x}}}\, ds_{\mathbf{y}} \,. \tag{119}$$

Definition of integral operator $\mathbf{T}(\omega/c_E)$. We introduce the linear integral operator $\mathbf{S}_{\mathrm{T}}(\omega/c_E)$ such that [28], for all $\mathbf{x} \in \Gamma_E$,

$$\{\mathbf{S}_{\mathrm{T}}(\omega/c_E)\, \psi_{\Gamma_E}\}(\mathbf{x}) = -\int_{\Gamma_E} \psi_{\Gamma_E}(\mathbf{y}) \frac{\partial^2 G(\mathbf{x}-\mathbf{y})}{\partial \mathbf{n}_{\mathbf{x}} \partial \mathbf{n}_{\mathbf{y}}}\, ds_{\mathbf{y}} \quad , \tag{120}$$

---

[26] The identity operator $\mathbf{I}$ from $\mathcal{C}_{\Gamma_E}$ into $\mathcal{C}_{\Gamma_E}$ is defined by the bilinear form $(\psi_{\Gamma_E}, \delta v) \mapsto <\psi_{\Gamma_E}, \delta v>_{\mathcal{C}_{\Gamma_E}, \mathcal{C}'_{\Gamma_E}}$.

[27] Eq. (118) is an equality in $H^{1/2}(\Gamma_E)$.

[28] According to Costabel and Stephan, 1985; Costabel, 1988; Kirsch, 1989, $\mathbf{S}_{\mathrm{T}}(\omega/c_E)$ is a linear continuous operator from $\mathcal{C}_{\Gamma_E} = H^{1/2}(\Gamma_E)$ into $\mathcal{C}'_{\Gamma_E} = H^{-1/2}(\Gamma_E)$.

in which the integral in right-hand side of Eq. (120) exists as a Cauchy principal value [29].

Symmetry property of integral operator $\mathbf{S}_{\mathrm{T}}(\omega/c_E)$. The transpose [30] of operator $\mathbf{S}_{\mathrm{T}}(\omega/c_E)$ is the linear operator ${}^t\mathbf{S}_{\mathrm{T}}(\omega/c_E)$ such that, for all ψ_{Γ_E} and $\delta\psi_{\Gamma_E}$ in $\mathcal{C}_{\Gamma_E}$,

$$<\mathbf{S}_{\mathrm{T}}(\omega/c_E)\,\psi_{\Gamma_E}\,,\delta\psi_{\Gamma_E}>_{\mathcal{C}'_{\Gamma_E},\mathcal{C}_{\Gamma_E}}=<\psi_{\Gamma_E}\,,{}^t\mathbf{S}_{\mathrm{T}}(\omega/c_E)\,\delta\psi_{\Gamma_E}>_{\mathcal{C}_{\Gamma_E},\mathcal{C}'_{\Gamma_E}}\quad . \tag{121}$$

We then have the following symmetry property

$${}^t\mathbf{S}_{\mathrm{T}}(\omega/c_E)=\mathbf{S}_{\mathrm{T}}(\omega/c_E)\quad . \tag{122}$$

Let us prove Eq. (122). From Eq. (120), we deduce that

$$<\mathbf{S}_{\mathrm{T}}(\omega/c_E)\psi_{\Gamma_E},\delta\psi_{\Gamma_E}>_{\mathcal{C}'_{\Gamma_E},\mathcal{C}_{\Gamma_E}}=-\int_{\Gamma_E}\int_{\Gamma_E}\frac{\partial^2 G(\mathbf{x}-\mathbf{y})}{\partial\mathbf{n}_{\mathbf{x}}\partial\mathbf{n}_{\mathbf{y}}}\psi_{\Gamma_E}(\mathbf{y})\,\delta\psi_{\Gamma_E}(\mathbf{x})ds_{\mathbf{y}}ds_{\mathbf{x}}. \tag{123}$$

From the equality $\partial^2 G(\mathbf{x}-\mathbf{y})/\partial\mathbf{n}(\mathbf{x})\partial\mathbf{n}(\mathbf{y})=\partial^2 G(\mathbf{y}-\mathbf{x})/\partial\mathbf{n}(\mathbf{y})\partial\mathbf{n}(\mathbf{x})$ (see Section 5), we deduce Eq. (121).

Variational formulation in ψ_{Γ_E} of the second integral equation on Γ_E. Using Eqs. (116) and (120), the variational formulation of the second boundary integral equation (119) is written as follows. For all ω fixed in $\mathbb{R}$ and for v given in $\mathcal{C}'_{\Gamma_E}$, the trace $\psi^{\mathrm{sol}}_{\Gamma_E}$ in $\mathcal{C}_{\Gamma_E}$ of the unique solution ψ^{sol} of Eqs. (104) to (106) is such that, for all $\delta\psi_{\Gamma_E}$ in $\mathcal{C}_{\Gamma_E}$,

$$<\mathbf{S}_{\mathrm{T}}(\omega/c_E)\,\psi^{\mathrm{sol}}_{\Gamma_E},\delta\psi_{\Gamma_E}>_{\mathcal{C}'_{\Gamma_E},\mathcal{C}_{\Gamma_E}}=<\big(\frac{1}{2}\,{}^t\mathbf{I}-{}^t\mathbf{S}_{\mathrm{D}}(\omega/c_E)\big)\,v\,,\delta\psi_{\Gamma_E}>_{\mathcal{C}'_{\Gamma_E},\mathcal{C}_{\Gamma_E}} \tag{124}$$

in which ${}^t\mathbf{I}$ denotes the identity operator [31]. The corresponding operator equation is [32]

$$\mathbf{S}_{\mathrm{T}}(\omega/c_E)\,\psi^{\mathrm{sol}}_{\Gamma_E}=\big(\frac{1}{2}\,{}^t\mathbf{I}-{}^t\mathbf{S}_{\mathrm{D}}(\omega/c_E)\big)\,v\quad . \tag{125}$$

---

29 Denoting the ball with center $\mathbf{x}$ and radius η as $B(\mathbf{x},\eta)$, the Cauchy principal value is defined by $\int_{\Gamma_E}\psi_{\Gamma_E}(\mathbf{y})\,\partial^2 G(\mathbf{x}-\mathbf{y})/\partial\mathbf{n}_{\mathbf{x}}\partial\mathbf{n}_{\mathbf{y}}\,ds_{\mathbf{y}}=\lim_{\eta\to 0}\int_{\Gamma_E\setminus B(\mathbf{x},\eta)}\psi_{\Gamma_E}(\mathbf{y})\,\partial^2 G(\mathbf{x}-\mathbf{y})/\partial\mathbf{n}_{\mathbf{x}}\partial\mathbf{n}_{\mathbf{y}}\,ds_{\mathbf{y}}$.

30 Space $\mathcal{C}'_{\Gamma_E}$ is the dual space of $\mathcal{C}_{\Gamma_E}$ for the dual brackets $<v,\psi_{\Gamma_E}>_{\mathcal{C}'_{\Gamma_E},\mathcal{C}_{\Gamma_E}}$ in which $v\in\mathcal{C}'_{\Gamma_E}$ and $\psi_{\Gamma_E}\in\mathcal{C}_{\Gamma_E}$ and which is linear in v and linear in ψ_{Γ_E} (and not antilinear in ψ_{Γ_E}). The transpose ${}^t\mathbf{S}_{\mathrm{T}}(\omega/c_E)$ is an operator from $(\mathcal{C}'_{\Gamma_E})'=\mathcal{C}_{\Gamma_E}$ into $\mathcal{C}'_{\Gamma_E}$.

31 The identity operator ${}^t\mathbf{I}$ from $\mathcal{C}'_{\Gamma_E}$ into $\mathcal{C}'_{\Gamma_E}$ is defined by the bilinear form $(v\,,\delta\psi_{\Gamma_E})\mapsto\,<v\,,\delta\psi_{\Gamma_E}>_{\mathcal{C}'_{\Gamma_E},\mathcal{C}_{\Gamma_E}}$.

32 Eq. (125) is an equality in $H^{-1/2}(\Gamma_E)$.

From Section 3.3, for all real ω, the boundary value problem defined by Eqs. (104) to (106) has a unique solution $\psi^{\rm sol}(\omega)$. Below, we will see that operator $\mathbf{S}_{\rm T}(\omega/c_E)$ is singular for a countable set of values of ω (called the spurious or irregular frequencies). Consequently, for these spurious frequencies, the trace $\psi^{\rm sol}_{\Gamma_E}(\omega)$ on Γ_E of the solution $\psi^{\rm sol}(\omega)$ cannot be constructed by Eq. (125) and therefore does not allow the construction of $\mathbf{B}_{\Gamma_E}(\omega/c_E)$ for all real ω.

Transformation of the bilinear form associated with $\mathbf{S}_{\rm T}(\omega/c_E)$. The bilinear form $< \mathbf{S}_{\rm T}(\omega/c_E)\,\psi_{\Gamma_E}, \delta\psi_{\Gamma_E} >_{\mathcal{C}'_{\Gamma_E}, \mathcal{C}_{\Gamma_E}}$ defined by Eq. (123) contains the non-integrable singular term $\partial^2 G(\mathbf{x}-\mathbf{y})/\partial\mathbf{n}_{\mathbf{x}}\partial\mathbf{n}_{\mathbf{y}}$. This term can be transformed to eliminate this second derivative. It can be proved (see Maue, 1949; Mitzner, 1966; Nedelec, 1978) that, for sufficiently differentiable complex functions ψ_{Γ_E} and $\delta\psi_{\Gamma_E}$ belonging to $\mathcal{C}_{\Gamma_E}$,

$$< \mathbf{S}_{\rm T}(\omega/c_E)\psi_{\Gamma_E}, \delta\psi_{\Gamma_E} >_{\mathcal{C}'_{\Gamma_E}, \mathcal{C}_{\Gamma_E}} = -k^2 \int_{\Gamma_E}\int_{\Gamma_E} G(\mathbf{x}-\mathbf{y})\, \mathbf{n}_{\mathbf{x}} \cdot \mathbf{n}_{\mathbf{y}}\, \psi_{\Gamma_E}(\mathbf{y})\, \delta\psi_{\Gamma_E}(\mathbf{x})\, ds_{\mathbf{y}}\, ds_{\mathbf{x}}$$

$$+ \int_{\Gamma_E}\int_{\Gamma_E} G(\mathbf{x}-\mathbf{y})\, \{\mathbf{n}_{\mathbf{y}} \times \boldsymbol{\nabla}_{\mathbf{y}}\psi_{\Gamma_E}(\mathbf{y})\} \cdot \{\mathbf{n}_{\mathbf{x}} \times \boldsymbol{\nabla}_{\mathbf{x}}\delta\psi_{\Gamma_E}(\mathbf{x})\}\, ds_{\mathbf{y}}\, ds_{\mathbf{x}} \quad . \tag{126}$$

9.3. Null spaces of integral operators $\mathbf{S}_{\rm T}(\omega/c_E)$ and $\frac{1}{2}\,\mathbf{I} - \mathbf{S}_{\rm D}(\omega/c_E)$

To discuss the singularity of operator $\mathbf{S}_{\rm T}(\omega/c_E)$ used in Eq. (125), we have to study the eigenvalues and eigenvectors of the interior Neumann boundary value problem related to bounded domain Ω_i (see Fig. 5-b) for operator $-\boldsymbol{\nabla}^2$,

$$-\boldsymbol{\nabla}^2\psi = \lambda\,\psi \qquad \text{in} \quad \Omega_i \quad , \tag{127}$$

$$\frac{\partial\psi}{\partial\mathbf{n}_{\mathbf{y}}} = 0 \qquad \text{on} \quad \Gamma_E \quad , \tag{128}$$

$$\int_{\Omega_i} \psi(\mathbf{x})\, d\mathbf{x} = 0 \quad . \tag{129}$$

The last equation implies that $\lambda = 0$ is not an eigenvalue. This classical eigenvalue problem has an increasing sequence of positive eigenvalues $0 < \lambda_1 < \lambda_2 < \ldots < \lambda_\alpha < \ldots$. Each eigenvalue λ_α has a finite multiplicity n_α and we denote the corresponding eigenvectors as $\psi^1_\alpha, \ldots, \psi^{n_\alpha}_\alpha$. Finally, we define the countable set $\mathcal{S}^N$ such that

$$\mathcal{S}^N = \{\, k_\alpha \ , \ \alpha = 1, 2, \ldots \,\} \quad , \tag{130}$$

in which

$$k_\alpha = \omega_\alpha / c_E = \sqrt{\lambda_\alpha} \quad . \tag{131}$$

For all j in $\{1, \ldots, n_\alpha\}$, the pair $\{k_\alpha, \psi_\alpha^j\}$ satisfies

$$\nabla^2 \psi_\alpha^j + k_\alpha^2 \, \psi_\alpha^j = 0 \qquad \text{in} \quad \Omega_i \quad , \tag{132}$$

$$\frac{\partial \psi_\alpha^j}{\partial \mathbf{n}_\mathbf{y}} = 0 \qquad \text{on} \quad \Gamma_E \quad . \tag{133}$$

From Eqs. (97), (98), (100) and (102), we deduce that, for all $\mathbf{x}$ in Γ_E, the trace ψ_{α,Γ_E}^j of ψ_α^j on Γ_E verifies

$$\frac{1}{2}\psi_{\alpha,\Gamma_E}^j(\mathbf{x}) = \int_{\Gamma_E} \psi_{\alpha,\Gamma_E}^j(\mathbf{y}) \, \frac{\partial G(\mathbf{x}-\mathbf{y})}{\partial \mathbf{n}_\mathbf{y}} \, ds_\mathbf{y} \quad , \tag{134}$$

$$-\int_{\Gamma_E} \psi_{\alpha,\Gamma_E}^j(\mathbf{y}) \frac{\partial^2 G(\mathbf{x}-\mathbf{y})}{\partial \mathbf{n}_\mathbf{x} \partial \mathbf{n}_\mathbf{y}} \, ds_\mathbf{y} = 0 \quad , \tag{135}$$

in which $G(\mathbf{x}-\mathbf{y})$ is given by Eqs. (35)-(36) with $k = k_\alpha$. Using the definition of operators $\mathbf{S}_\mathrm{D}(\omega/c_E)$ and $\mathbf{S}_\mathrm{T}(\omega/c_E)$ given by Eqs. (109) and (120), Eqs. (134) and (135) can be rewritten as

$$\Big(\frac{1}{2}\mathbf{I} - \mathbf{S}_\mathrm{D}(\omega_\alpha/c_E)\Big) \, \psi_{\alpha,\Gamma_E}^j = 0 \quad , \tag{136}$$

$$\mathbf{S}_\mathrm{T}(\omega_\alpha/c_E) \, \psi_{\alpha,\Gamma_E}^j = 0 \quad . \tag{137}$$

Consequently, for all ω/c_E in $\mathcal{S}^N$, the null spaces of operators $\mathbf{S}_\mathrm{T}(\omega/c_E)$ and $\frac{1}{2}\mathbf{I} - \mathbf{S}_\mathrm{D}(\omega/c_E)$ are not reduced to zero. In addition, for all ω_α/c_E in $\mathcal{S}^N$, operators $\mathbf{S}_\mathrm{T}(\omega_\alpha/c_E)$ and $\frac{1}{2}\mathbf{I} - \mathbf{S}_\mathrm{D}(\omega_\alpha/c_E)$ have the same null space which is the real vector space $\mathcal{C}_{\alpha,\Gamma_E}^N$ of finite dimension n_α,

$$\mathcal{C}_{\alpha,\Gamma_E}^N = \text{null}\,\{\mathbf{S}_\mathrm{T}(\omega_\alpha/c_E)\} = \text{null}\,\{\frac{1}{2}\mathbf{I} - \mathbf{S}_\mathrm{D}(\omega_\alpha/c_E)\} \quad , \tag{138}$$

defined by

$$\mathcal{C}_{\alpha,\Gamma_E}^N = \text{span}\{\psi_{\alpha,\Gamma_E}^1, \ldots, \psi_{\alpha,\Gamma_E}^{n_\alpha}\} \quad . \tag{139}$$

9.4. Existence of a solution of the second boundary integral equation

For all real ω, the trace $\psi^{\rm sol}_{\Gamma_E}(\omega)$ on Γ_E of the unique solution $\psi^{\rm sol}$ of Eqs. (104) to (106) satisfies Eq. (125). Conversely, we discuss below the existence and the expression of solutions $\psi_{\Gamma_E}(\omega)$ of

$$\mathbf{S}_{\rm T}(\omega/c_E)\,\psi_{\Gamma_E}(\omega) = \big(\frac{1}{2}\,{}^t\mathbf{I} - {}^t\mathbf{S}_{\rm D}(\omega/c_E)\big)\,v \quad . \tag{140}$$

First case: $\{\omega/c_E\} \notin \mathcal{S}^N$. From Section 9.3, the null space of operator $\mathbf{S}_{\rm T}(\omega/c_E)$ is reduced to zero and consequently $\mathbf{S}_{\rm T}(\omega/c_E)$ is invertible. Therefore, Eq. (140) has a unique solution $\psi_{\Gamma_E}(\omega) = \psi^{\rm sol}_{\Gamma_E}(\omega)$,

$$\forall\omega \in \mathbb{R} \quad \text{and} \quad \omega \notin \mathcal{S}^N \quad , \quad \psi_{\Gamma_E}(\omega) = \psi^{\rm sol}_{\Gamma_E}(\omega) \quad . \tag{141}$$

Second case: $\{\omega/c_E\} \in \mathcal{S}^N$. Consider $\omega/c_E = \omega_\alpha/c_E$ in $\mathcal{S}^N$. From Section 9.3, the null space $\mathcal{C}^N_{\alpha,\Gamma_E}$ of operator $\mathbf{S}_{\rm T}(\omega_\alpha/c_E)$, defined by Eq. (138) or (139), is not reduced to zero and is of finite dimension n_α. Consequently, Eq. (140) will have at least one solution if and only if the right-hand side $b(\omega_\alpha)$ of Eq. (140),

$$b(\omega_\alpha) = \big(\frac{1}{2}\,{}^t\mathbf{I} - {}^t\mathbf{S}_{\rm D}(\omega_\alpha/c_E)\big)\,v \quad , \tag{142}$$

is orthogonal to null space $\mathcal{C}^N_{\alpha,\Gamma_E}$,

$$< b(\omega_\alpha)\,,\psi^j_{\alpha,\Gamma_E} >_{C'_{\Gamma_E},C_{\Gamma_E}} = 0 \quad , \quad j = 1,\ldots,n_\alpha \quad , \tag{143}$$

which is the case[33] due to Eq. (136). Consequently, Eq. (140) has a family of solutions which are written as

$$\{\omega_\alpha/c_E\} \in \mathcal{S}^N \quad , \quad \psi_{\Gamma_E}(\omega_\alpha) = \psi^{\rm sol}_{\Gamma_E}(\omega_\alpha) + \sum_{j=1}^{n_\alpha} \nu_j\,\psi^j_{\alpha,\Gamma_E} \quad , \tag{144}$$

in which $\nu_1,\ldots,\nu_{n_\alpha}$ are arbitrary complex numbers and where $\psi^{\rm sol}_{\Gamma_E}(\omega_\alpha)$ is the trace on Γ_E of the unique solution $\psi^{\rm sol}(\omega_\alpha)$ of Eqs. (104) to (106) with $\omega = \omega_\alpha$.

[33] We have $<b(\omega_\alpha),\psi^j_{\alpha,\Gamma_E}>_{C'_{\Gamma_E},C_{\Gamma_E}} = <(\frac{1}{2}{}^t\mathbf{I}-{}^t\mathbf{S}_{\rm D}(\omega_\alpha/c_E))v,\psi^j_{\alpha,\Gamma_E}> = <v,(\frac{1}{2}\mathbf{I}-\mathbf{S}_{\rm D}(\omega_\alpha/c_E))\psi^j_{\alpha,\Gamma_E}>$ $=0$ (see Mathematical Notations in the appendix).

Comments. The above analysis shows that Eq. (140) does not allow construction of the trace $\psi^{\rm sol}_{\Gamma_E}(\omega)$ on Γ_E of the unique solution $\psi^{\rm sol}(\omega)$ of Eqs. (104) to (106) for a countable set $\mathcal{S}^N$ of values of ω/c_E (called the spurious or irregular frequencies). In Section 10, we present a method which is valid for all real values of the frequency.

10. Acoustic Impedance Boundary Operator Construction: Appropriate Symmetric Boundary Integral Method Valid for all Real Values of the Frequency

In this section, we present an appropriate method developed by Angelini and Hutin, 1983 (and extended by Angelini and Soize, 1989 and 1993 for the Maxwell equations). This method allows construction of the trace $\psi^{\rm sol}_{\Gamma_E}(\omega)$ on Γ_E of the unique solution $\psi^{\rm sol}(\omega)$ of the basic exterior Neumann problem related to the Helmholtz equation defined by Eqs. (104) to (106) and valid for all real values of ω. We can then deduce the construction of operator $\mathbf{B}_{\Gamma_E}(\omega/c_E)$ for all real values of ω and consequently, we can construct acoustic impedance boundary operator $\mathbf{Z}_{\Gamma_E}(\omega)$ using Eq. (11).

Construction of $\psi^{\rm sol}_{\Gamma_E}(\omega)$ for v given and for all real values of ω. For v given in $\mathcal{C}'_{\Gamma_E}$, we consider the following equation

$$\mathbf{S}_{\rm T}(\omega/c_E)\,\psi_{\Gamma_E}(\omega) = \Big(\frac{1}{2}\,{}^t\mathbf{I} - {}^t\mathbf{S}_{\rm D}(\omega/c_E)\Big)\,v \quad , \tag{145}$$

in which $\psi_{\Gamma_E}(\omega)$ is the unknown field (it is recalled that the solution of Eq. (145) is not necessarily unique as was seen in Section 9.4). For all ω fixed in $\mathbb{R}$, let $\psi_{\Gamma_E}(\omega) \in \mathcal{C}_{\Gamma_E}$ be any solution of Eq. (145). Then, the trace $\psi^{\rm sol}_{\Gamma_E}(\omega)$ on Γ_E of the unique solution $\psi^{\rm sol}(\omega)$ of Eqs. (104) to (106) is given by

$$\psi^{\rm sol}_{\Gamma_E}(\omega) = \Big(\frac{1}{2}\,\mathbf{I} - \mathbf{S}_{\rm D}(\omega/c_E)\Big)\,\psi_{\Gamma_E}(\omega) + \mathbf{S}_{\rm S}(\omega/c_E)\,v \quad . \tag{146}$$

Proof of the construction. As in Section 9.4, we must consider two cases.

1- First case. Let us consider $\{\omega/c_E\} \notin \mathcal{S}^N$. From Eq. (141), we deduce that Eq. (145) has a unique solution $\psi_{\Gamma_E}(\omega) = \psi^{\rm sol}_{\Gamma_E}(\omega)$. Then the right-hand side of Eq. (146) yields

$$
\begin{aligned}
&\left(\frac{1}{2}\mathbf{I} - \mathbf{S}_{\mathrm{D}}(\omega/c_E)\right)\psi_{\Gamma_E}(\omega) + \mathbf{S}_{\mathrm{S}}(\omega/c_E)\,v \\
&\qquad\qquad = \left(\frac{1}{2}\mathbf{I} - \mathbf{S}_{\mathrm{D}}(\omega/c_E)\right)\psi^{\mathrm{sol}}_{\Gamma_E}(\omega) + \mathbf{S}_{\mathrm{S}}(\omega/c_E)\,v \\
&\qquad\qquad = \psi^{\mathrm{sol}}_{\Gamma_E}(\omega) + \left\{\mathbf{S}_{\mathrm{S}}(\omega/c_E)\,v - \left(\frac{1}{2}\mathbf{I} + \mathbf{S}_{\mathrm{D}}(\omega/c_E)\right)\psi^{\mathrm{sol}}_{\Gamma_E}(\omega)\right\} \\
&\qquad\qquad = \psi^{\mathrm{sol}}_{\Gamma_E}(\omega) \quad ,
\end{aligned}
$$

the last equality being due to Eq. (118).

2- Second case. Let us consider $\omega/c_E = \omega_\alpha/c_E$ in $\mathcal{S}^N$. Using the second case of Section 9.4, we deduce that Eq. (145) has a family of solutions $\psi_{\Gamma_E}(\omega_\alpha)$ given by Eq. (144). Then the right-hand side of Eq. (146) yields

$$
\begin{aligned}
&\left(\frac{1}{2}\mathbf{I} - \mathbf{S}_{\mathrm{D}}(\omega_\alpha/c_E)\right)\psi_{\Gamma_E}(\omega_\alpha) + \mathbf{S}_{\mathrm{S}}(\omega_\alpha/c_E)\,v \\
&\qquad = \left(\frac{1}{2}\mathbf{I} - \mathbf{S}_{\mathrm{D}}(\omega_\alpha/c_E)\right)\left(\psi^{\mathrm{sol}}_{\Gamma_E}(\omega_\alpha) + \sum_{j=1}^{n_\alpha} \nu_j\,\psi^j_{\alpha,\Gamma_E}\right) + \mathbf{S}_{\mathrm{S}}(\omega_\alpha/c_E)\,v \\
&\qquad = \left(\frac{1}{2}\mathbf{I} - \mathbf{S}_{\mathrm{D}}(\omega_\alpha/c_E)\right)\psi^{\mathrm{sol}}_{\Gamma_E}(\omega_\alpha) + \mathbf{S}_{\mathrm{S}}(\omega_\alpha/c_E)\,v \\
&\qquad\qquad + \sum_{j=1}^{n_\alpha} \nu_j \left(\frac{1}{2}\mathbf{I} - \mathbf{S}_{\mathrm{D}}(\omega_\alpha/c_E)\right)\psi^j_{\alpha,\Gamma_E} \\
&\qquad = \psi^{\mathrm{sol}}_{\Gamma_E}(\omega_\alpha) + \left\{\mathbf{S}_{\mathrm{S}}(\omega_\alpha/c_E)\,v - \left(\frac{1}{2}\mathbf{I} + \mathbf{S}_{\mathrm{D}}(\omega_\alpha/c_E)\right)\psi^{\mathrm{sol}}_{\Gamma_E}(\omega_\alpha)\right\} \\
&\qquad = \psi^{\mathrm{sol}}_{\Gamma_E}(\omega_\alpha) \quad ,
\end{aligned}
$$

the two last equalities being due to Eqs. (136) and (118) respectively.

Important remark. It should be noted that the construction of $\psi^{\mathrm{sol}}_{\Gamma_E}(\omega)$ for v given, as defined by Eqs. (145) and (146), is valid for all real ω and does not require effective construction of set $\mathcal{S}^N$. The methodology consists first in calculating a solution $\psi_{\Gamma_E}(\omega)$ of Eq. (145) then, in a second step, in replacing $\psi_{\Gamma_E}(\omega)$ in the right-hand side of Eq. (146), which then furnishes the unique solution $\psi^{\mathrm{sol}}_{\Gamma_E}(\omega)$ sought. The first step requires solving a linear complex system with a symmetric operator and the second step is performed directly without solving any linear system. Consequently, the method proposed solves the basic exterior Neumann problem related to the Helmholtz equation defined by Eqs. (104) to (106) for all real values of ω, without any problems caused by irregular frequencies. More details concerning the practical construction of a solution of Eq. (145) are given in Section 12 related to the finite element discretization of the problem.

Construction of operator $\mathbf{B}_{\Gamma_E}(\omega/c_E)$ for all real values of ω. Eqs. (145) and (146) can be rewritten as

$$\begin{bmatrix} 0 \\ \psi^{\text{sol}}_{\Gamma_E} \end{bmatrix} = \begin{bmatrix} -\mathbf{S}_{\text{T}}(\omega/c_E) & \frac{1}{2}\,{}^t\mathbf{I} - {}^t\mathbf{S}_{\text{D}}(\omega/c_E) \\ \frac{1}{2}\,\mathbf{I} - \mathbf{S}_{\text{D}}(\omega/c_E) & \mathbf{S}_{\text{S}}(\omega/c_E) \end{bmatrix} \begin{bmatrix} \psi_{\Gamma_E} \\ v \end{bmatrix} \quad . \tag{147}$$

The block operator [34] $\mathbf{H}(\omega/c_E)$ appearing in the right-hand side of Eq. (147) has the symmetry property ${}^t\mathbf{H}(\omega/c_E) = \mathbf{H}(\omega/c_E)$. The elimination of ψ_{Γ_E} in Eq. (147) yields a linear equation between $\psi^{\text{sol}}_{\Gamma_E}$ and v which defines operator $\mathbf{B}_{\Gamma_E}(\omega/c_E)$ introduced in Eq. (103). The practical procedure for eliminating of ψ_{Γ_E} is described in Section 12.

Construction of acoustic impedance boundary operator $\mathbf{Z}_{\Gamma_E}(\omega)$ for all real values of ω. From the construction of $\mathbf{B}_{\Gamma_E}(\omega/c_E)$, acoustic impedance boundary operator $\mathbf{Z}_{\Gamma_E}(\omega)$ is calculated using Eq. (11).

11. Construction of the Radiation Impedance Operator and Asymptotic Formula for the Radiated Pressure Field

11.1. Construction of the radiation impedance operator

Integral representation of $\psi^{\text{sol}}(\mathbf{x},\omega)$ in external domain Ω_E. Let $\psi^{\text{sol}}(\mathbf{x},\omega)$ be the unique solution of Eqs. (104) to (106) and $\psi^{\text{sol}}_{\Gamma_E}(\omega)$ be its trace on Γ_E. From Eqs. (92) and (93), we deduce that, for all $\mathbf{x}$ in external domain Ω_E,

$$\psi^{\text{sol}}(\mathbf{x},\omega) = \int_{\Gamma_E} \left\{ G(\mathbf{x}-\mathbf{y})\, v(\mathbf{y}) - \psi^{\text{sol}}_{\Gamma_E}(\mathbf{y},\omega)\, \frac{\partial G(\mathbf{x}-\mathbf{y})}{\partial \mathbf{n}_{\mathbf{y}}} \right\} ds_{\mathbf{y}} \quad . \tag{148}$$

Definition of integral operators $\mathbf{R}_{\text{S}}(\mathbf{x},\omega/c_E)$ and $\mathbf{R}_{\text{D}}(\mathbf{x},\omega/c_E)$. For all $\mathbf{x}$ fixed in external domain Ω_E, we define the linear integral operators $\mathbf{R}_{\text{S}}(\mathbf{x},\omega/c_E)$ and $\mathbf{R}_{\text{D}}(\mathbf{x},\omega/c_E)$ by [35]

$$\mathbf{R}_{\text{S}}(\mathbf{x},\omega/c_E)\, v = \int_{\Gamma_E} G(\mathbf{x}-\mathbf{y})\, v(\mathbf{y})\, ds_{\mathbf{y}} \quad , \tag{149}$$

[34] $\mathbf{H}(\omega/c_E)$ is a linear operator from $\mathcal{C}_{\Gamma_E} \times \mathcal{C}'_{\Gamma_E}$ into $\mathcal{C}'_{\Gamma_E} \times \mathcal{C}_{\Gamma_E}$.

[35] For all $\mathbf{x}$ fixed in Ω_E, $\mathbf{R}_{\text{S}}(\mathbf{x},\omega/c_E)$ is a linear operator from $\mathcal{C}'_{\Gamma_E}$ into $\mathbb{C}$ and $\mathbf{R}_{\text{D}}(\mathbf{x},\omega/c_E)$ is a linear operator from $\mathcal{C}_{\Gamma_E}$ into $\mathbb{C}$.

$$\mathbf{R}_{\mathrm{D}}(\mathbf{x}, \omega/c_E)\, \psi_{\Gamma_E} = \int_{\Gamma_E} \psi_{\Gamma_E}(\mathbf{y})\, \frac{\partial G(\mathbf{x}-\mathbf{y})}{\partial \mathbf{n}_{\mathbf{y}}}\, ds_{\mathbf{y}} \quad . \tag{150}$$

Expression of the radiation impedance operator $\mathbf{Z}_{\mathrm{rad}}(\mathbf{x}, \omega)$. Using Eqs. (103), (149) and (150), Eq. (148) yields

$$\psi^{\mathrm{sol}}(\mathbf{x}, \omega) = \{\mathbf{R}_{\mathrm{S}}(\mathbf{x}, \omega/c_E) - \mathbf{R}_{\mathrm{D}}(\mathbf{x}, \omega/c_E)\, \mathbf{B}_{\Gamma_E}(\omega/c_E)\}\, v \quad . \tag{151}$$

From Eqs. (151) and (12), we deduce that, for all $\mathbf{x}$ fixed in Ω_E,

$$\mathbf{R}(\mathbf{x}, \omega/c_E) = \mathbf{R}_{\mathrm{S}}(\mathbf{x}, \omega/c_E) - \mathbf{R}_{\mathrm{D}}(\mathbf{x}, \omega/c_E)\, \mathbf{B}_{\Gamma_E}(\omega/c_E) \quad . \tag{152}$$

For all $\mathbf{x}$ in Ω_E, the radiation impedance operator $\mathbf{Z}_{\mathrm{rad}}(\mathbf{x}, \omega)$ is calculated using Eqs. (14) and (152),

$$\mathbf{Z}_{\mathrm{rad}}(\mathbf{x}, \omega) = -i\, \omega\, \rho_E \{\mathbf{R}_{\mathrm{S}}(\mathbf{x}, \omega/c_E) - \mathbf{R}_{\mathrm{D}}(\mathbf{x}, \omega/c_E)\, \mathbf{B}_{\Gamma_E}(\omega/c_E)\} \quad . \tag{153}$$

11.2. Asymptotic formula for the radiated pressure field $p(\mathbf{x}, \omega)$

From Eq. (13), the radiated pressure field $p(\mathbf{x}, \omega)$ in external domain Ω_E is given by $p(\mathbf{x}, \omega) = \mathbf{Z}_{\mathrm{rad}}(\mathbf{x}, \omega)\, v$. Let R and $\mathbf{e}$ be such that (see Fig. 6)

$$\mathbf{x} = R\, \mathbf{e} \quad \text{with} \quad R = \|\mathbf{x}\| \quad . \tag{154}$$

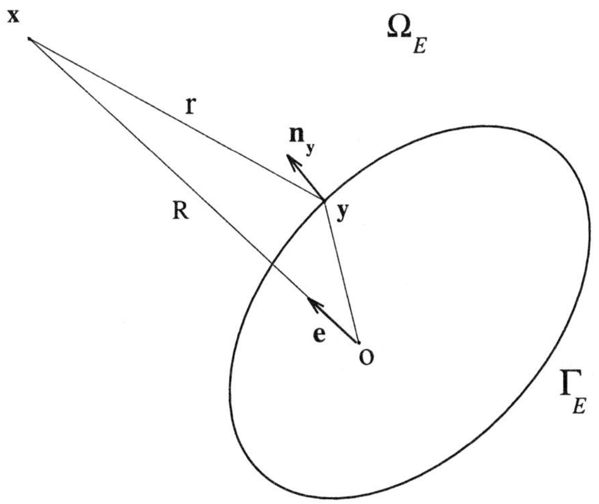

Fig. 6. Geometrical configuration

Definition of integral operators $\mathbf{R}_{\mathrm{S}}^{\infty}(\mathbf{x},\omega/c_E)$ and $\mathbf{R}_{\mathrm{D}}^{\infty}(\mathbf{x},\omega/c_E)$. For all $\mathbf{x} = R\,\mathbf{e}$ fixed in external domain Ω_E, we define the linear integral operators $\mathbf{R}_{\mathrm{S}}^{\infty}(\mathbf{x},\omega/c_E)$ and $\mathbf{R}_{\mathrm{D}}^{\infty}(\mathbf{x},\omega/c_E)$ by [36]

$$\mathbf{R}_{\mathrm{S}}^{\infty}(\mathbf{x},\omega/c_E)\, v = \frac{1}{R} e^{-i\omega R/c_E} \int_{\Gamma_E} N_{\mathbf{e}}(\mathbf{y})\, v(\mathbf{y})\, ds_{\mathbf{y}} \quad , \tag{155}$$

$$\mathbf{R}_{\mathrm{D}}^{\infty}(\mathbf{x},\omega/c_E)\, \psi_{\Gamma_E} = \frac{i\omega}{c_E}\frac{1}{R} e^{-i\omega R/c_E} \int_{\Gamma_E} \mathbf{e}\cdot\mathbf{n}_{\mathbf{y}}\, N_{\mathbf{e}}(\mathbf{y})\psi_{\Gamma_E}(\mathbf{y})\, ds_{\mathbf{y}} \quad , \tag{156}$$

in which $N_{\mathbf{e}}(\mathbf{y})$ is defined by

$$N_{\mathbf{e}}(\mathbf{y}) = -\frac{1}{4\pi} \exp(i\,\mathbf{e}\cdot\mathbf{y}\,\omega/c_E) \quad . \tag{157}$$

Asymptotic formula for radiation impedance operator $\mathbf{Z}_{\mathrm{rad}}(\mathbf{x},\omega)$. We have the following asymptotic formulas

$$\lim_{R\to+\infty} \mathbf{R}_{\mathrm{S}}(R\,\mathbf{e},\omega/c_E) = \mathbf{R}_{\mathrm{S}}^{\infty}(R\,\mathbf{e},\omega/c_E) \quad , \tag{158}$$

$$\lim_{R\to+\infty} \mathbf{R}_{\mathrm{D}}(R\,\mathbf{e},\omega/c_E) = \mathbf{R}_{\mathrm{D}}^{\infty}(R\,\mathbf{e},\omega/c_E) \quad . \tag{159}$$

From Eq. (153), we deduce the asymptotic formula for the radiation impedance operator

$$\lim_{R\to+\infty} \mathbf{Z}_{\mathrm{rad}}(R\mathbf{e},\omega) = -i\omega\rho_E\{\mathbf{R}_{\mathrm{S}}^{\infty}(R\mathbf{e},\omega/c_E) - \mathbf{R}_{\mathrm{D}}^{\infty}(R\mathbf{e},\omega/c_E)\,\mathbf{B}_{\Gamma_E}(\omega/c_E)\}. \tag{160}$$

Proof of the asymptotic formula. Let $k = \omega/c_E$, $r = \|\mathbf{x}-\mathbf{y}\|$ and $\cos\theta$ such that $\mathbf{x}\cdot\mathbf{y} = R\,\|\mathbf{y}\|\,\cos\theta$. We have $r = (R^2 + \|\mathbf{y}\|^2 - 2\,R\,\|\mathbf{y}\|\,\cos\theta)^{1/2}$. From Eqs. (35) and (36), we deduce that $R\,e^{ikR}\,G(\mathbf{x}-\mathbf{y}) = -e^{-ikR(r/R-1)}R/(4\pi r)$. Therefore, for all θ, $-ikR(r/R-1) \to ik\,\|\mathbf{y}\|\cos\theta$ as $R\to+\infty$. Using Eqs. (154), (155) and (157), we then deduce Eq. (158). It can easily be verified that $Re^{ikR}\partial G(\mathbf{x}-\mathbf{y})/\partial\mathbf{n}_{\mathbf{y}} \to \mathbf{n}_{\mathbf{y}}\cdot\boldsymbol{\nabla}_{\mathbf{y}}N_{\mathbf{e}}(\mathbf{y})$ as $R\to+\infty$ and that $\boldsymbol{\nabla}_{\mathbf{y}}N_{\mathbf{e}}(\mathbf{y}) = i\,k\,\mathbf{e}\,N_{\mathbf{e}}(\mathbf{y})$. Using Eqs. (156) and (157) we deduce Eq. (159).

---

36 For all $\mathbf{x}=R\,\mathbf{e}$ fixed in Ω_E, linear integral operator $\mathbf{R}_{\mathrm{S}}^{\infty}(\mathbf{x},\omega/c_E)$ is defined from $\mathcal{C}'_{\Gamma_E}$ into $\mathbb{C}$ and linear integral operator $\mathbf{R}_{\mathrm{D}}^{\infty}(\mathbf{x},\omega/c_E)$ is defined from $\mathcal{C}_{\Gamma_E}$ into $\mathbb{C}$.

12. Symmetric Boundary Element Method Valid for all Real Values of the Frequency

In this section, we present the finite element discretization (so-called *boundary element method*) of the method presented in Section 10 (see Eq. (147)) for calculation of acoustic impedance boundary operator $\mathbf{Z}_{\Gamma_E}(\omega)$. In addition, we present the corresponding discretization of radiation impedance operator $\mathbf{Z}_{\text{rad}}(\mathbf{x},\omega)$ constructed in Section 11.

12.1. Finite element discretization of the boundary operators of the problem

We use the finite element method to construct the discretization of boundary integral operators $\mathbf{S}_{\text{S}}(\omega/c_E)$, $\mathbf{S}_{\text{D}}(\omega/c_E)$ and $\mathbf{S}_{\text{T}}(\omega/c_E)$ related to acoustic impedance boundary operator $\mathbf{Z}_{\Gamma_E}(\omega)$. We consider a finite element mesh of surface Γ_E and subspace $\mathcal{C}_{n_E} \subset \mathcal{C}_{\Gamma_E} \subset \mathcal{C}'_{\Gamma_E}$ of finite dimension n_E. Let $\mathbf{V} = (V_1,\dots,V_{n_E})$ and $\mathbf{\Psi}_{\Gamma_E} = (\Psi_{\Gamma_E,1},\dots,\Psi_{\Gamma_E,n_E})$ be the complex vectors of the DOFs which are the values of v and ψ_{Γ_E} at the nodes of the finite element mesh of surface Γ_E. We denote as $[S_{\text{S}}(\omega/c_E)]$, $[S_{\text{D}}(\omega/c_E)]$ and $[S_{\text{T}}(\omega/c_E)]$ the $(n_E\times n_E)$ dense complex matrices corresponding to the discretization of the bilinear forms defined by Eqs. (112), (114) and (126) respectively. From Eqs. (111) and (122), we deduce that complex matrices $[S_{\text{S}}(\omega/c_E)]$ and $[S_{\text{T}}(\omega/c_E)]$ are symmetric,

$$[S_{\text{S}}(\omega/c_E)]^T = [S_{\text{S}}(\omega/c_E)] \quad , \quad [S_{\text{T}}(\omega/c_E)]^T = [S_{\text{T}}(\omega/c_E)] \quad . \tag{161}$$

Concerning the calculation of matrix $[S_{\text{T}}(\omega/c_E)]$, it should be noted that the regularization of operator $\mathbf{S}_{\text{T}}(\omega/c_E)$ defined by Eq. (126) is used instead of its initial definition given by Eq. (123). For the choice of the type of finite elements and the practical construction of the above matrices, we refer the reader to Amini et al., 1992; Angelini and Soize, 1989 and 1992; Brebbia et al., 1984; Brebbia and Dominguez, 1992; Chen and Zhou, 1992; Dautray and Lions, 1992; Hackbusch, 1995; Nedelec, 1976.

12.2. Construction of the approximation of the acoustic impedance boundary operator

The methodology for constructing acoustic impedance boundary operator $\mathbf{Z}_{\Gamma_E}(\omega)$ is presented in Section 10. From Section 12.1, we deduce the finite element discretization of Eq. (147),

$$\begin{bmatrix} \mathbf{0} \\ \mathbf{\Psi}^{\text{sol}}_{\Gamma_E} \end{bmatrix} = \begin{bmatrix} -[S_{\text{T}}(\omega/c_E)] & \frac{1}{2}[E]^T - [S_{\text{D}}(\omega/c_E)]^T \\ \frac{1}{2}[E] - [S_{\text{D}}(\omega/c_E)] & [S_{\text{S}}(\omega/c_E)] \end{bmatrix} \begin{bmatrix} \mathbf{\Psi}_{\Gamma_E} \\ \mathbf{V} \end{bmatrix} , \tag{162}$$

in which $\mathbf{\Psi}^{\text{sol}}_{\Gamma_E}$ is the complex vector of the generalized coordinates corresponding to the finite element discretization of linear form $\delta v \mapsto \int_{\Gamma_E} \psi^{\text{sol}}_{\Gamma_E}(\mathbf{y})\, \delta v(\mathbf{y})\, ds_{\mathbf{y}}$. It should be noted that $\mathbf{\Psi}^{\text{sol}}_{\Gamma_E}$ does not represent the values of $\psi^{\text{sol}}_{\Gamma_E}$ at the nodes of the finite element mesh of surface Γ_E. Matrix $[E]$ is the $(n_E \times n_E)$ real nondiagonal matrix corresponding to the finite element discretization of identity operator $\mathbf{I}$ introduced in Section 9.1. The $(2n_E \times 2n_E)$ complex matrix $[H(\omega/c_E)]$ appearing in the right-hand side of Eq. (162) is symmetric,

$$[H(\omega/c_E)]^T = [H(\omega/c_E)] \quad . \tag{163}$$

The elimination of $\mathbf{\Psi}_{\Gamma_E}$ in Eq. (162) yields a linear equation between $\mathbf{\Psi}^{\text{sol}}_{\Gamma_E}$ and $\mathbf{V}$ which defines the $(n_E \times n_E)$ complex symmetric matrix $[B_{\Gamma_E}(\omega/c_E)]$ of operator $\mathbf{B}_{\Gamma_E}(\omega/c_E)$ and such that

$$\mathbf{\Psi}^{\text{sol}}_{\Gamma_E} = [B_{\Gamma_E}(\omega/c_E)]\, \mathbf{V} \quad . \tag{164}$$

Vector $\mathbf{\Psi}_{\Gamma_E}$ can be eliminated using a Gauss elimination with a partial pivoting algorithm (Golub and Van Loan, 1989). If $\{\omega/c_E\} \notin \mathcal{S}^N$, then complex symmetric matrix $[S_{\mathrm{T}}(\omega/c_E)]$ is invertible (see the first case of the proof of the construction in Section 10) and the elimination in Eq. (162) is performed up to row number n_E. If $\omega/c_E = \omega_\alpha/c_E \in \mathcal{S}^N$, then complex symmetric matrix $[S_{\mathrm{T}}(\omega_\alpha/c_E)]$ is not invertible (see the second case of the proof of the construction in Section 10) and the null space of matrix $[S_{\mathrm{T}}(\omega_\alpha/c_E)]$ is a real subspace of $\mathbb{C}^{n_E}$ of dimension $n_\alpha < n_E$. In this case, the elimination in Eq. (162) is performed up to row number $n_E - n_\alpha$. In practice, n_α is unknown. During Gauss elimination with a partial pivoting algorithm, the elimination process is stopped when a "zero" pivot is encountered. It should be noted that when the elimination is stopped, the equations corresponding to row numbers $n_E - n_\alpha + 1, \ldots, n_E$ are automatically satisfied due to the orthogonality property defined by Eq. (143). From Eq. (11), we deduce that the $(n_E \times n_E)$ complex symmetric matrix $[Z_{\Gamma_E}(\omega)]$ of operator $\mathbf{Z}_{\Gamma_E}(\omega)$ is such that

$$[Z_{\Gamma_E}(\omega)] = -i\,\omega\,\rho_E\,[B_{\Gamma_E}(\omega/c_E)] \quad . \tag{165}$$

Remark concerning the case of cyclic symmetry of boundary Γ_E. Let us assume that boundary Γ_E has a cyclic symmetry of order N (see this concept in Section III.8.3). In addition, we assume that the finite element discretization of the integral boundary operators corresponds to this cyclic symmetry. Consequently, dense matrix $[H(\omega/c_E)]$ appears as the assembly of $N\times N$ square submatrices of dimension $(m\times m)$ with $m = n_E/N$ and is a circulant matrix (see Golub and Van Loan, 1989, for the notion of circulant matrix). The numerical cost of eliminating $\boldsymbol{\Psi}_{\Gamma_E}$ can then be reduced using this property and a Discrete Fourier Transform (DFT) technique. Instead of eliminating $\boldsymbol{\Psi}_{\Gamma_E}$ of length n_E in $(2n_E\times 2n_E)$ complex matrix $[H(\omega/c_E)]$, we only have to perform N smaller eliminations of a vector of length m in a complex matrix of dimension $(2m\times 2m)$. It should be noted that the method presented in Section III.8.3 cannot be used directly but must be adapted to take into account the fact that the integral operators of the problem are not local (all the degrees of freedom on boundary Γ_E are coupled).

12.3. Reduced matrix model of $[B_{\Gamma_E}(\omega/c_E)]$

Let us consider the projection defined by the transformation

$$\mathbf{V}(\omega) = [U_{\Gamma_E}]\,\mathbf{q}^S(\omega) \quad , \tag{166}$$

in which $[U_{\Gamma_E}]$ is an $(n_E\times N_S)$ real matrix with $N_S < n_E$ and $\mathbf{q}^S(\omega)$ is a vector in $\mathbb{C}^{N_S}$ depending on ω. The corresponding reduced matrix $[\mathcal{B}_{\Gamma_E}(\omega/c_E)]$ of $[B_{\Gamma_E}(\omega/c_E)]$ is the $(N_S\times N_S)$ complex symmetric matrix defined by

$$[\mathcal{B}_{\Gamma_E}(\omega/c_E)] = [U_{\Gamma_E}]^T\,[B_{\Gamma_E}(\omega/c_E)]\,[U_{\Gamma_E}] \quad . \tag{167}$$

In practice, instead of calculating $[\mathcal{B}_{\Gamma_E}(\omega/c_E)]$ by Eq. (167), we construct the projection $[\mathcal{H}(\omega/c_E)]$ of matrix $[H(\omega/c_E)]$ defined by Eq. (162),

$$[\mathcal{H}(\omega/c_E)]=\begin{bmatrix} -[S_{\mathrm{T}}(\omega/c_E)] & \left(\tfrac{1}{2}\,[E]^T-[S_{\mathrm{D}}(\omega/c_E)]^T\right)[U_{\Gamma_E}] \\ [U_{\Gamma_E}]^T\left(\tfrac{1}{2}\,[E]-[S_{\mathrm{D}}(\omega/c_E)]\right) & [U_{\Gamma_E}]^T[S_{\mathrm{S}}(\omega/c_E)]\,[U_{\Gamma_E}] \end{bmatrix} \tag{168}$$

and we then proceed to Gauss elimination with a partial pivoting as explained in Section 12.2.

12.4. Construction of the approximation of the acoustic radiation impedance operator

We use the same finite element mesh of boundary Γ_E for constructing the finite element discretization of integral operators $\mathbf{R}_{\mathrm{S}}(\mathbf{x},\omega/c_E)$ and $\mathbf{R}_{\mathrm{D}}(\mathbf{x},\omega/c_E)$ related to radiation impedance operator $\mathbf{Z}_{\mathrm{rad}}(\mathbf{x},\omega)$. For all fixed $\mathbf{x}$ in external domain Ω_E, we denote as $[R_{\mathrm{S}}(\mathbf{x},\omega/c_E)]$ and $[R_{\mathrm{D}}(\mathbf{x},\omega/c_E)]$ the $(1\times n_E)$

complex matrices corresponding to the finite element discretization of integral operators $\mathbf{R}_S(\mathbf{x},\omega/c_E)$ and $\mathbf{R}_D(\mathbf{x},\omega/c_E)$ defined by Eqs. (149) and (150) respectively. The methodology is similar for the finite element discretization of the corresponding asymptotic operators defined by Eqs. (155) and (156). From Eq. (153), we deduce the $(1\times n_E)$ complex matrix $[Z_{\text{rad}}(\mathbf{x},\omega)]$ of the finite element discretization of operator $\mathbf{Z}_{\text{rad}}(\mathbf{x},\omega)$ which is such that

$$[Z_{\text{rad}}(\mathbf{x},\omega)] = -i\,\omega\,\rho_E\,\{[R_S(\mathbf{x},\omega/c_E)] - [R_D(\mathbf{x},\omega/c_E)]\,[B_{\Gamma_E}(\omega/c_E)]\} \quad . \tag{169}$$

12.5. Calculation of the pressure field on Γ_E and of the pressure in any point of Ω_E

Calculation of the pressure field on boundary Γ_E. We introduce the linear form on C'_{Γ_E} defined by

$$\ell_{p_E}(\omega\,;\delta v) = \int_{\Gamma_E} p_E|_{\Gamma_E}(\mathbf{x},\omega)\,\delta v(\mathbf{x})\,ds_{\mathbf{x}} \quad , \tag{170}$$

related to resultant pressure field $p_E|_{\Gamma_E}(\omega)$ on Γ_E. From Eq. (76), (77) and (170), we deduce that

$$\ell_{p_E}(\omega\,;\delta v) = \int_{\Gamma_E} \{\mathbf{Z}_{\Gamma_E}(\omega)v(\omega)\}(\mathbf{x})\,\delta v(\mathbf{x})\,ds_{\mathbf{x}} + \ell_{p_{\text{given}}}(\omega\,;\delta v) \quad , \tag{171}$$

in which $\ell_{p_{\text{given}}}(\omega\,;\delta v)$ is such that

$$\ell_{p_{\text{given}}}(\omega\,;\delta v) = \int_{\Gamma_E} p_{\text{given}}|_{\Gamma_E}(\mathbf{x},\omega)\,\delta v(\mathbf{x})\,ds_{\mathbf{x}} \quad , \tag{172}$$

$v(\omega) = i\omega\,\mathbf{u}_{\text{wall}}(\omega)\cdot\mathbf{n}$ and $p_{\text{given}}|_{\Gamma_E}(\mathbf{x},\omega)$ is given by Eq. (78). The finite element discretization of Eq. (171) is

$$\mathbf{P}_{\Gamma_E}(\omega) = [Z_{\Gamma_E}(\omega)]\,\mathbf{V}(\omega) + \mathbf{P}_{\text{given}}(\omega) \quad , \tag{173}$$

in which $\mathbf{P}_{\Gamma_E}(\omega)$ is a $\mathbb{C}^{n_E}$-vector of the generalized pressure at the nodes of the finite element mesh of Γ_E and where $\mathbf{P}_{\text{given}}(\omega)$ is the finite element approximation of the linear form defined by Eq. (172).

Calculation of the pressure in any point of external domain Ω_E. From Eqs. (79) and (80), we deduce the corresponding approximation of resultant pressure $p_E(\mathbf{x},\omega)$ at a point $\mathbf{x}$ of external domain Ω_E,

$$p_E(\mathbf{x},\omega) = [Z_{\text{rad}}(\mathbf{x},\omega)]\,\mathbf{V}(\omega) + p_{\text{given}}(\mathbf{x},\omega) \quad , \tag{174}$$

in which $p_{\text{given}}(\mathbf{x}, \omega)$ is given by Eq. (81).

13. Case of a Free Surface

We consider an external acoustic fluid occupying an unbounded open three-dimensional domain $\widetilde{\Omega}_E$ with a free surface denoted as Γ_0 (see Fig. 7-a).

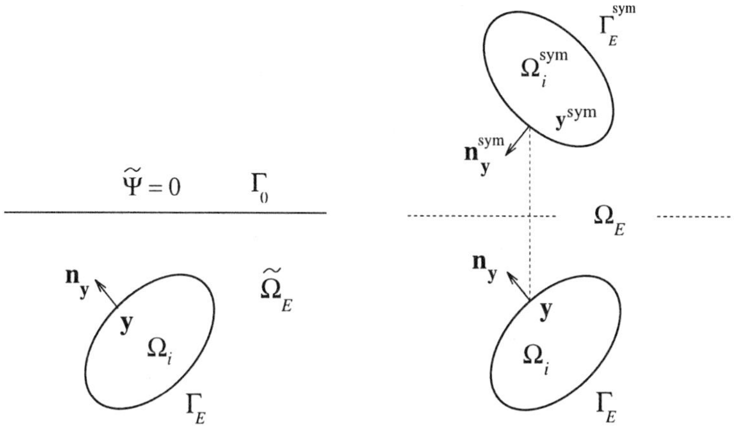

Fig. 7-a Fig. 7-b

Fig. 7. Geometrical configuration in the case of a free surface $\widetilde{\Gamma}_0$

It is assumed that Ω_i does not cross Γ_0. The boundary of the bounded domain Ω_i is denoted as Γ_E. The mechanical hypotheses are those introduced in Sections 2 and 3. We denote the outward unit normal to Γ_E as $\mathbf{n}$. On free surface Γ_0, we have the Dirichlet boundary condition $\widetilde{\psi} = 0$ for the velocity potential $\widetilde{\psi}$ (see Eq. X.41). The exterior Neumann problem related to the Helmholtz equation with a free surface can then be written as (see Section 3 for the case with no free surface),

$$\nabla^2 \widetilde{\psi}(\mathbf{x}, \omega) + k^2 \, \widetilde{\psi}(\mathbf{x}, \omega) = 0 \qquad \text{in} \quad \widetilde{\Omega}_E \quad , \tag{175}$$

$$\frac{\partial \widetilde{\psi}(\mathbf{y}, \omega)}{\partial \mathbf{n}_\mathbf{y}} = v(\mathbf{y}) \qquad \text{on} \quad \Gamma_E \quad , \tag{176}$$

$$\widetilde{\psi}(\mathbf{x}, \omega) = 0 \qquad \text{on} \quad \Gamma_0 \quad , \tag{177}$$

with the outward Sommerfeld condition at infinity,

$$|\, \widetilde{\psi} \,| = O(\frac{1}{R}) \quad , \quad \left| \frac{\partial \widetilde{\psi}}{\partial R} + i\, k \, \widetilde{\psi} \right| = O(\frac{1}{R^2}) \quad . \tag{178}$$

Let Ω_i^{sym} be the symmetrical domain of Ω_i with respect to the plane defined by boundary Γ_0 (see Fig. 7-b). The boundary of Ω_i^{sym} is denoted as Γ_E^{sym} and the outward unit normal to Γ_E^{sym} at point $\mathbf{y}^{\text{sym}}$ corresponding to $\mathbf{n}$ at point $\mathbf{y}$ is denoted as $\mathbf{n}^{\text{sym}}$. Let $\Omega_E = \mathbb{R}^3 \backslash \{\Omega_i \cup \Omega_i^{\text{sym}}\}$ be the external unbounded domain which is assumed to be simply connected [37]. Solution $\widetilde{\psi}$ of Eqs. (175) to (178) is such that

$$\widetilde{\psi} = \psi \qquad \text{in} \quad \widetilde{\Omega}_E \quad , \tag{179}$$

in which velocity potential ψ defined in Ω_E is the solution of the following exterior Neumann problem related to the Helmholtz equation in unbounded domain Ω_E with no free surface,

$$\boldsymbol{\nabla}^2 \psi(\mathbf{x}, \omega) + k^2\, \psi(\mathbf{x}, \omega) = 0 \qquad \text{in} \quad \Omega_E \quad , \tag{180}$$

$$\frac{\partial \psi(\mathbf{y}, \omega)}{\partial \mathbf{n}_{\mathbf{y}}} = v(\mathbf{y}) \qquad \text{on} \quad \Gamma_E \quad , \tag{181}$$

$$\frac{\partial \psi(\mathbf{y}^{\text{sym}}, \omega)}{\partial \mathbf{n}_{\mathbf{y}}^{\text{sym}}} = v(\mathbf{y}^{\text{sym}}) \qquad \text{on} \quad \Gamma_E^{\text{sym}} \quad , \tag{182}$$

with the outward Sommerfeld condition at infinity,

$$|\,\psi\,| = O(\frac{1}{R}) \quad , \quad \left| \frac{\partial \psi}{\partial R} + i\, k\, \psi \right| = O(\frac{1}{R^2}) \quad . \tag{183}$$

The method presented in Sections 9 to 12 for an unbounded domain without free surface can then be used.

[37] If Ω_E is simply connected then $\widetilde{\Omega}_E$ is simply connected.

CHAPTER XIII

Structural-Acoustic Master System in the LF Range

1. Introduction

In this chapter, we present the construction of a reduced model and the associated method for solving the vibration problem in the LF range of a master structure containing an internal acoustic fluid and surrounded by an external acoustic fluid. Modeling of the master structure is developed in Chapters V and VI, modeling of the internal fluid in Chapters X and XI and modeling of the external fluid in Chapter XII.

In Section 2, we state a general structural-acoustic problem in the frequency domain with prescribed internal acoustic source density, mechanical forces, external acoustic source density and external acoustic plane wave. A part of the internal fluid-structure interface has acoustical properties modeled by a wall acoustic impedance.

In Section 3, we establish the boundary value problem of the structural-acoustic master system. The external acoustic fluid is represented by the boundary impedance operator defined on the external fluid-structure interface. This boundary operator was introduced and constructed in Chapter XII. The master structure is described by the local equations introduced in Chapter V. The internal acoustic fluid is described by a variant of the local equations introduced in Section XI.3, expressed in terms of pressure field p and field φ.

In Section 4, we construct the variational formulation of the boundary value problem.

Section 5 deals with the finite element discretization of the various operators which are used for the implementation of the reduced model constructed in Section 6 (it should be noted that the solving method in the LF range uses a reduced model instead of directly solving the resulting matrix equation).

In Section 6, we construct the symmetric reduced matrix model in the LF range using the structural modes of the master structure in vacuo and the acoustic modes of the internal acoustic cavity with a rigid wall.

In Section 7, we described the solving method for the frequency-by-frequency calculation of the FRF in the LF range using the symmetric reduced matrix model.

In Section 8, we discuss the location of the resonant frequencies for the case of a weakly dissipative structural-acoustic master system.

Section 9 deals with the calculation of the structural-acoustic modes of the master structure coupled with an internal acoustic fluid (without an external acoustic fluid). A generalized symmetric matrix eigenvalue problem based on the symmetric reduced matrix model is introduced for this calculation.

In Section 10, we consider the particular case of a master structure coupled with an external acoustic fluid (without an internal acoustic fluid) which is a simplification of the general case presented in Sections 2 to 8. This section is introduced in order to have a self-contained presentation of this standard case.

In Section 11, we consider the general structural-acoustic problem introduced in Section 2 for which a zero pressure condition is imposed on a part of the boundary of the internal acoustic domain.

Finally, in Section 12 we discuss the axisymmetric case and in Section 13 the response to deterministic and random excitations.

2. Statement of the Structural-Acoustic Problem in the Frequency Domain

2.1. Geometry and mechanical assumptions

The physical space $\mathbb{R}^3$ is referred to a cartesian reference system and we denote the generic point of $\mathbb{R}^3$ as $\mathbf{x} = (x_1, x_2, x_3)$. We study the linear vibrations of a structural-acoustic master system around a static equilibrium state taken as a natural state at rest.

Master structure Ω_S. The master structure at equilibrium occupies a three-dimensional bounded domain Ω_S of $\mathbb{R}^3$ with a sufficiently smooth boundary $\partial\Omega_S = \Gamma \cup \Gamma_Z \cup \Gamma_E$ with $\Gamma \cap \Gamma_Z \cap \Gamma_E = \emptyset$. The outward unit normal to $\partial\Omega_S$ is denoted as $\mathbf{n}^S = (n_1^S, n_2^S, n_3^S)$ (see Fig. 1). The displacement field in Ω_S is denoted as $\mathbf{u}(\mathbf{x},\omega) = (u_1(\mathbf{x},\omega), u_2(\mathbf{x},\omega), u_3(\mathbf{x},\omega))$. The master structure is assumed to be free, i.e. not fixed on any part of boundary $\partial\Omega_S$. A surface force field $\mathbf{G}(\mathbf{x},\omega) = (G_1(\mathbf{x},\omega), G_2(\mathbf{x},\omega), G_3(\mathbf{x},\omega))$ is given on $\partial\Omega_S$ and a body force field $\mathbf{g}(\mathbf{x},\omega) = (g_1(\mathbf{x},\omega), g_2(\mathbf{x},\omega), g_3(\mathbf{x},\omega))$ is given in Ω_S. The master structure is a dissipative medium whose viscoelastic constitutive equation is defined in Section IV.5.1.

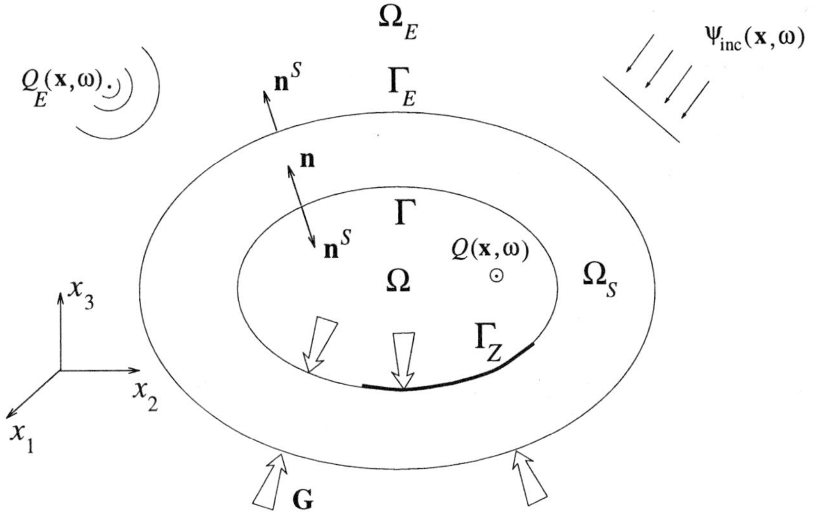

Fig. 1. Configuration of the structural-acoustic master system

Internal acoustic fluid Ω. Let Ω be the internal bounded domain filled with a dissipative acoustic fluid (gas or liquid) as described in Section X.3. The boundary $\partial\Omega$ of Ω is $\Gamma \cup \Gamma_Z$. The outward unit normal to $\partial\Omega$ is denoted as $\mathbf{n} = (n_1, n_2, n_3)$ and we have $\mathbf{n} = -\mathbf{n}^S$ on $\partial\Omega$ (see Fig. 1). Part Γ_Z of the boundary has acoustical properties modeled by a wall acoustic impedance $Z(\mathbf{x},\omega)$ satisfying the hypotheses defined by Eqs. (X.18) to (X.25). We denote the pressure field in Ω as $p(\mathbf{x},\omega)$. We assume that there is no Dirichlet boundary condition on any part of $\partial\Omega$. An acoustic source density $Q(\mathbf{x},\omega)$ is given inside Ω and satisfies the hypotheses defined by Eq. (X.8).

External acoustic fluid Ω_E. The master structure is surrounded by an external inviscid acoustic fluid (gas or liquid) as described in Chapter XII. The fluid occupies the unbounded open three-dimensional domain Ω_E whose boundary $\partial\Omega_E$ is Γ_E. We introduce the bounded open domain Ω_i defined by $\Omega_i = \mathbb{R}^3 \backslash (\Omega_E \cup \Gamma_E)$. The boundary $\partial\Omega_i$ of Ω_i is then Γ_E. The outward unit normal to $\partial\Omega_i$ is $\mathbf{n}^S$ defined above (see Fig. 1). We denote the pressure field in Ω_E as $p_E(\mathbf{x},\omega)$. We assume that there is no

Dirichlet boundary condition on any part of Γ_E. An acoustic source density $Q_E(\mathbf{x},\omega)$ and an incident plane wave defined by the velocity potential $\psi_{\text{inc}}(\mathbf{x},\omega) = \psi_0(\omega)\exp\{-i\,\mathbf{k}\cdot\mathbf{x}\}$ are given inside Ω_E. As seen in Section XII.6.4, this acoustic source density and incident plane wave induce a pressure field $p_{\text{given}}(\omega)$ on Γ_E defined by Eq. (XII.78).

2.2 Structural-acoustic problem to be solved

Definition of the FRF of the structural-acoustic master system. The problem consists in calculating the operator-valued FRF of the structural-acoustic master system in the low-frequency range, whose input is $\{\mathbf{f}_S(\omega), \mathbf{f}_Q(\omega)\}$ and whose output is $\{\mathbf{u}(\omega), \varphi(\omega)\}$ in which $\mathbf{u}(\omega)$ is the displacement field of the master structure and $\varphi(\omega)$ is the field describing the internal fluid. Concerning the input, $\mathbf{f}_S(\omega)$ represents the force vector induced by $\mathbf{g}$ in Ω_S, $\mathbf{G}$ on $\partial\Omega_S$, p_{given} on Γ_E, and $\mathbf{f}_Q(\omega)$ represents Q in Ω. The main objective of this chapter is to calculate this FRF in the low-frequency range.

Calculation of pressure field $p(\mathbf{x},\omega)$ in internal acoustic fluid Ω. Pressure field $p(\mathbf{x},\omega)$ is given by Eqs. (XI.5) to (XI.9),

$$p(\mathbf{x},\omega) = \omega^2\,\rho_0\,\varphi(\mathbf{x},\omega) + \pi(\omega\,;\mathbf{u}\,;\varphi) \quad , \tag{1}$$

in which $\pi(\omega\,;\mathbf{u}\,;\varphi)$ is written as

$$\pi(\omega\,;\mathbf{u}\,;\varphi) = \kappa(\omega)\,\left\{-\omega^2\,\rho_0\,\pi_1(\omega\,;\varphi) - \pi_2(\mathbf{u}) + \pi_Q(\omega)\right\} \quad , \tag{2}$$

where $\pi_1(\omega\,;\varphi)$, $\pi_2(\mathbf{u})$ and $\pi_Q(\omega)$ are given by

$$\pi_1(\omega\,;\varphi) = \int_{\Gamma_Z} \frac{\varphi(\mathbf{x},\omega)}{i\omega\,Z(\mathbf{x},\omega)}\,ds(\mathbf{x}) \quad , \tag{3}$$

$$\pi_2(\mathbf{u}) = \int_{\Gamma\cup\Gamma_Z} \mathbf{u}(\mathbf{x},\omega)\cdot\mathbf{n}(\mathbf{x})\,ds(\mathbf{x}) \quad , \tag{4}$$

$$\pi_Q(\omega) = \frac{1}{\rho_0}\int_\Omega \frac{Q(\mathbf{x},\omega)}{i\omega}\,d\mathbf{x} \quad , \tag{5}$$

in which $|\Omega|$ is the volume of domain Ω and $\kappa(\omega)$ is a complex number such that

$$\kappa(\omega) = \frac{\rho_0\,c_0^2}{|\Omega|}\times\left\{1 + \frac{\rho_0\,c_0^2}{|\Omega|}\int_{\Gamma_Z}\frac{ds(\mathbf{x})}{i\omega\,Z(\mathbf{x},\omega)}\right\}^{-1} \quad . \tag{6-1}$$

Real constant $\kappa(0)$ is such that

$$0 < \kappa(0) < +\infty \quad , \quad \lim_{\alpha_{\min} \to +\infty} \kappa(0) = \frac{\rho_0 c_0^2}{|\Omega|} \quad . \qquad (6-2)$$

Calculation of pressure field $p_E|_{\Gamma_E}(\omega)$ on Γ_E and resultant pressure field $p_E(\mathbf{x}, \omega)$ in any point of unbounded external acoustic fluid Ω_E. Once the trace on Γ_E of structural displacement field $\mathbf{u}(\omega)$ is known, pressure field $p_E|_{\Gamma_E}(\omega)$ is given by Eqs. (XII.76) and (XII.77),

$$p_E|_{\Gamma_E}(\omega) = i\omega\, \mathbf{Z}_{\Gamma_E}(\omega)\{\mathbf{u}(\omega) \cdot \mathbf{n}^S\} + p_{\text{given}}|_{\Gamma_E}(\omega) \quad , \qquad (7)$$

in which $p_{\text{given}}|_{\Gamma_E}(\omega)$ is defined by Eq. (XII.78) and $\mathbf{Z}_{\Gamma_E}(\omega)$ is the acoustic impedance boundary operator whose construction is given in Sections XII.10 and XII.12. At any point $\mathbf{x}$ of unbounded external acoustic fluid Ω_E, resultant pressure field $p_E(\mathbf{x}, \omega)$ is calculated using Eqs. (XII.79) and (XII.80),

$$p_E(\mathbf{x}, \omega) = i\omega\, \mathbf{Z}_{\text{rad}}(\mathbf{x}, \omega)\{\mathbf{u}(\omega) \cdot \mathbf{n}^S\} + p_{\text{given}}(\mathbf{x}, \omega) \quad , \qquad (8)$$

in which $p_{\text{given}}(\mathbf{x}, \omega)$ is given by Eq. (XII.81) and $\mathbf{Z}_{\text{rad}}(\mathbf{x}, \omega)$ is the radiation impedance operator whose construction is given in Sections XII.11 and XII.12.

3. Boundary Value Problem of the Structural-Acoustic Master System

In this section, we present the boundary value problem of the structural-acoustic master system consisting of the master structure coupled with the internal fluid and the external fluid. In this formulation, the velocity potential field of the external fluid is eliminated in terms of the structural displacement field using the acoustic impedance boundary operator.

Equations for the master structure. The equation of the master structure occupying domain Ω_S is written as

$$-\omega^2 \rho_s\, u_i - \sigma_{ij,j}(\mathbf{u}) = g_i \quad \text{in} \quad \Omega_S \quad , \qquad (9)$$

in which $\rho_s(\mathbf{x})$ is the mass density of the master structure. In the LF range, the constitutive equation is defined by Eqs. (IV-46) to (IV-50) in which tensor a_{ijkh} is independent of ω and tensor $b_{ijkh}(\omega)$ depends on ω

(see Section IV.5.1). Using Eq. (7), the boundary condition $\sigma_{ij}(\mathbf{u})n_j^S = G_i - p_E\, n_i^S$ on Γ_E can be written as

$$\sigma_{ij}(\mathbf{u})\, n_j^S = G_i - p_{\text{given}}|_{\Gamma_E}\, n_i^S - i\omega\, \mathbf{Z}_{\Gamma_E}(\omega)\{\mathbf{u}\cdot\mathbf{n}^S\}\, n_i^S \quad \text{on} \quad \Gamma_E\,. \tag{10}$$

Using Eqs. (1) and (2) and $\mathbf{n}^S = -\mathbf{n}$, the boundary condition $\sigma_{ij}(\mathbf{u})n_j^S = G_i - p(\omega)\, n_i^S = G_i + p(\omega)\, n_i$ on $\Gamma \cup \Gamma_Z$ can be written as

$$\begin{aligned}\sigma_{ij}(\mathbf{u})\, n_j^S &= G_i + \kappa\, \pi_{\mathrm{Q}}\, n_i \\ &+ \omega^2 \rho_0\, \varphi\, n_i - \omega^2 \rho_0\, \kappa\, \pi_1(\omega;\varphi) n_i - \kappa\, \pi_2(\mathbf{u}) n_i \quad \text{on} \quad \Gamma \cup \Gamma_Z\,,\end{aligned} \tag{11}$$

in which $\kappa(\omega)$ is given by Eq. (6-1), and $\pi_1(\omega\,;\varphi)$, $\pi_2(\mathbf{u})$ and $\pi_{\mathrm{Q}}(\omega)$ are given by Eqs. (3), (4) and (5) respectively.

Equations for the internal acoustic fluid. Eq. (1) can be written as $p(\mathbf{x},\omega) = p'(\mathbf{x},\omega) + \pi(\omega\,;\mathbf{u}\,;\varphi)$ in which $p'(\mathbf{x},\omega) = \omega^2\, \rho_0\, \varphi(\mathbf{x},\omega)$, which we can rewrite as

$$\frac{1}{\rho_0 c_0^2}\, p' = \frac{\omega^2}{c_0^2}\, \varphi \quad \text{in} \quad \Omega \quad . \tag{12}$$

Field p' must satisfy the constraint equation

$$\int_\Omega p'\, d\mathbf{x} = 0 \quad , \tag{13}$$

due to the constraint Eq. (XI.4),

$$\int_\Omega \varphi\, d\mathbf{x} = 0 \quad . \tag{14}$$

Using Eq. (12), the equation of the internal acoustic fluid occupying domain Ω can be rewritten as (see Eq.(XI.1)),

$$-\frac{1}{c_0^2} p' - i\omega\tau\rho_0 \nabla^2\varphi - \rho_0 \nabla^2\varphi = g_{\mathrm{Q}} - \frac{\omega^2 \rho_0 \kappa}{c_0^2}\, \pi_1(\omega;\varphi) - \frac{1}{c_0^2}\kappa\, \pi_2(\mathbf{u}) \;\text{ in } \Omega\,, \tag{15}$$

in which g_{Q} is defined by Eq. (18) below. The Neumann boundary condition on Γ is given by Eq. (XI.2),

$$\rho_0(1+i\omega\tau)\frac{\partial\varphi}{\partial\mathbf{n}} = \rho_0\, \mathbf{u}\cdot\mathbf{n} + G_{\mathrm{Q}} \quad \text{on} \quad \Gamma\,, \tag{16}$$

in which G_Q is defined by Eq. (19) below, and the Neumann boundary condition on Γ_Z with wall acoustic impedance is given by Eq. (XI.3),

$$\rho_0(1+i\omega\tau)\frac{\partial\varphi}{\partial\mathbf{n}} = \rho_0\,\mathbf{u}\cdot\mathbf{n} + \frac{\omega^2\rho_0^2}{i\omega Z}\varphi - \frac{\omega^2\rho_0^2\kappa}{i\omega Z}\pi_1(\omega;\varphi) - \frac{\kappa\rho_0}{i\omega Z}\pi_2(\mathbf{u}) + G_Q + G_{Q,Z} \quad \text{on } \Gamma_Z\,, \tag{17}$$

in which $G_{Q,Z}$ is defined by Eq. (20) below. Function g_Q is defined on Ω by

$$g_Q(\mathbf{x},\omega) = \frac{1}{c_0^2}\,\kappa(\omega)\,\pi_Q(\omega) - \frac{1}{i\omega}Q(\mathbf{x},\omega) - \tau c_0^2\,\frac{1}{\omega^2}\nabla^2 Q(\mathbf{x},\omega) \quad , \tag{18}$$

G_Q is defined on $\Gamma \cup \Gamma_Z$ by

$$G_Q(\mathbf{x},\omega) = \tau c_0^2\frac{1}{\omega^2}\frac{\partial Q(\mathbf{x},\omega)}{\partial\mathbf{n}(\mathbf{x})} \quad , \tag{19}$$

and $G_{Q,Z}$ is defined on Γ_Z by

$$G_{Q,Z}(\mathbf{x},\omega) = \frac{\rho_0}{i\omega Z(\mathbf{x},\omega)}\kappa(\omega)\,\pi_Q(\omega) \quad . \tag{20}$$

4. Variational Formulation of the Structural-Acoustic Problem

In this section we construct the variational formulation of the boundary value problem defined by Eqs. (9) to (17) in terms of $\mathbf{u}(\omega)$ and $\varphi(\omega)$. We use the results presented in Chapters V and VI for the master structure and in Chapter XI for the internal acoustic fluid.

4.1. Admissible function spaces for the structural-acoustic problem

The admissible function space of displacement field $\mathbf{u}$ of the master structure Ω_S is the complex vector space $\mathcal{C}^c$ of "sufficiently differentiable" functions [1] defined on Ω_S with values in $\mathbb{C}^3$. Concerning the internal acoustic fluid Ω, we introduce the complex vector spaces $\mathcal{E}_\varphi^c$ for field φ and $\mathcal{E}_{p'}^c$ for field p' such that

$$\mathcal{E}_\varphi^c = \{\, \varphi \in \mathcal{E}^c \quad ; \quad \int_\Omega \varphi\, d\mathbf{x} = 0 \,\} \quad , \tag{21}$$

$$\mathcal{E}_{p'}^c = \{\, p' \in H^c \quad ; \quad \int_\Omega p'\, d\mathbf{x} = 0 \,\} \quad , \tag{22}$$

[1] Complex vector space $\mathcal{C}^c$ is the Sobolev space $(H^1(\Omega_S))^3$.

in which $\mathcal{E}^c$ is the complex vector space of the "sufficiently differentiable" functions [2] defined on Ω with values in $\mathbb{C}$ and $H^c = L^2(\Omega)$. Consequently, we have $\mathcal{E}^c_\varphi \subset \mathcal{E}^c_{p'}$.

4.2. Variational formulation in terms of $\{\mathbf{u}, \varphi, p'\}$

Variational equation in $\delta\mathbf{u}$ related to the master structure. Let $\mathbf{u}$ be in $\mathcal{C}^c$ and φ in $\mathcal{E}^c_\varphi$. Multiplying Eq. (9) by $\overline{\delta\mathbf{u}} \in \mathcal{C}^c$ and integrating over domain Ω_S, using Green's formula and Eqs. (10) and (11) yields

$$a^S(\omega;\mathbf{u},\delta\mathbf{u}) - \omega^2 a^E(\omega/c_E;\mathbf{u},\delta\mathbf{u}) + \kappa\, j(\mathbf{u},\delta\mathbf{u}) - \omega^2 c(\omega;\varphi,\delta\mathbf{u}) = f_S(\omega;\delta\mathbf{u}), \quad (23)$$

in which $a^S(\omega\,;\mathbf{u}\,,\delta\mathbf{u})$ is the dynamic stiffness sesquilinear form of the master structure defined on $\mathcal{C}^c \times \mathcal{C}^c$ by

$$a^S(\omega\,;\mathbf{u}\,,\delta\mathbf{u}) = -\omega^2 m^S(\mathbf{u}\,,\delta\mathbf{u}) + i\omega\, d^S(\omega\,;\mathbf{u}\,,\delta\mathbf{u}) + k^S(\mathbf{u}\,,\delta\mathbf{u}) \quad , \quad (24-1)$$

where mass, damping and stiffness structural sesquilinear forms m^S, d^S and k^S are defined on $\mathcal{C}^c \times \mathcal{C}^c$ and are such that

$$m^S(\mathbf{u}\,,\delta\mathbf{u}) = \int_{\Omega_S} \rho_S\, \mathbf{u}\cdot\overline{\delta\mathbf{u}}\; d\mathbf{x} \quad , \quad (24-2)$$

$$d^S(\omega\,;\mathbf{u}\,,\delta\mathbf{u}) = \int_{\Omega_S} b_{ijkh}(\omega)\, \varepsilon_{kh}(\mathbf{u})\, \varepsilon_{ij}(\overline{\delta\mathbf{u}})\; d\mathbf{x} \quad , \quad (24-3)$$

$$k^S(\mathbf{u}\,,\delta\mathbf{u}) = \int_{\Omega_S} a_{ijkh}\, \varepsilon_{kh}(\mathbf{u})\, \varepsilon_{ij}(\overline{\delta\mathbf{u}})\; d\mathbf{x} \quad . \quad (24-4)$$

In Eq. (23), $a^E(\omega/c_E;\mathbf{u}\,,\delta\mathbf{u})$ is the sesquilinear form on $\mathcal{C}^c \times \mathcal{C}^c$ such that [3]

$$\begin{aligned} -\omega^2 a^E(\omega/c_E\,;\,\mathbf{u}\,,\delta\mathbf{u}) &= i\omega \int_{\Gamma_E} (\overline{\delta\mathbf{u}}\cdot\mathbf{n}^S)\; \mathbf{Z}_{\Gamma_E}(\omega)\{\mathbf{u}\cdot\mathbf{n}^S\}\; ds \\ &= \omega^2 \rho_E \int_{\Gamma_E} (\overline{\delta\mathbf{u}}\cdot\mathbf{n}^S)\; \mathbf{B}_{\Gamma_E}(\omega/c_E)\{\mathbf{u}\cdot\mathbf{n}^S\}\; ds \quad , (25) \end{aligned}$$

---

2 Space $\mathcal{E}^c$ is the Sobolev space $H^1(\Omega)$.

3 From Section XII.4, linear operator $\mathbf{B}_{\Gamma_E}(\omega/c_E)$ is continuous from $\mathcal{C}'_{\Gamma_E}$ into $\mathcal{C}_{\Gamma_E}$ where $\mathcal{C}_{\Gamma_E} = H^{1/2}(\Gamma_E)$ and $\mathcal{C}'_{\Gamma_E}$ is the dual space of $\mathcal{C}_{\Gamma_E}$ (see footnote 5 of Chapter XII). Since the vector space of the trace on Γ_E of functions of $\mathcal{C}^c$ is $\mathcal{C}_{\Gamma_E}$, the right-hand side of Eq. (25) is defined and sesquilinear form $a^E(\omega/c_E;\mathbf{u}\,,\delta\mathbf{u})$ is continuous on $\mathcal{C}^c \times \mathcal{C}^c$.

in which $\mathbf{B}_{\Gamma_E}(\omega/c_E)$ is related to operator $\mathbf{Z}_{\Gamma_E}(\omega)$ by Eq. (XII.11). Sesquilinear form $j(\mathbf{u}, \delta\mathbf{u})$ is defined on $\mathcal{C}^c \times \mathcal{C}^c$ and is such that

$$j(\mathbf{u}, \delta\mathbf{u}) = \pi_2(\mathbf{u})\, \pi_2(\overline{\delta\mathbf{u}}) \quad . \tag{26}$$

Coupling sesquilinear form $c(\omega\,; \varphi, \delta\mathbf{u})$ is defined on $\mathcal{E}^c \times \mathcal{C}^c$ and is such that

$$c(\omega\,; \varphi, \delta\mathbf{u}) = c_0(\varphi, \delta\mathbf{u}) - \rho_0\, \kappa(\omega)\, \pi_1(\omega; \varphi)\, \pi_2(\overline{\delta\mathbf{u}}) \quad , \tag{27-1}$$

$$c_0(\varphi, \delta\mathbf{u}) = \rho_0 \int_{\Gamma \cup \Gamma_Z} \varphi\, \mathbf{n} \cdot \overline{\delta\mathbf{u}}\, ds \quad , \tag{27-2}$$

in which coupling sesquilinear form $c_0(\varphi, \delta\mathbf{u})$ is defined on $\mathcal{E}^c \times \mathcal{C}^c$ and linear forms $\pi_1(\omega; \varphi)$ and $\pi_2(\mathbf{u})$ are defined on $\mathcal{E}^c$ and $\mathcal{C}^c$ respectively. Finally, $f_S(\omega\,; \delta\mathbf{u})$ in the right-hand side of Eq. (23) is the antilinear form on $\mathcal{C}^c$ defined by

$$f_S(\omega\,; \delta\mathbf{u}) = f(\omega\,; \delta\mathbf{u}) + f_{p_{\text{given}}}(\omega\,; \delta\mathbf{u}) + \kappa(\omega)\, \pi_Q(\omega)\, \pi_2(\overline{\delta\mathbf{u}}) \quad , \tag{28}$$

in which antilinear form $f(\omega\,; \delta\mathbf{u})$ is defined by

$$f(\omega\,; \delta\mathbf{u}) = \int_{\Omega_S} \mathbf{g}(\mathbf{x}, \omega) \cdot \overline{\delta\mathbf{u}(\mathbf{x})}\, d\mathbf{x} + \int_{\partial\Omega_S} \mathbf{G}(\mathbf{x}, \omega) \cdot \overline{\delta\mathbf{u}(\mathbf{x})}\, ds(\mathbf{x}) \quad , \tag{29}$$

and antilinear form $f_{p_{\text{given}}}(\omega\,; \delta\mathbf{u})$ is defined by

$$f_{p_{\text{given}}}(\omega\,; \delta\mathbf{u}) = -\int_{\Gamma_E} p_{\text{given}}|_{\Gamma_E}\, \mathbf{n}^S \cdot \overline{\delta\mathbf{u}}\, ds \quad . \tag{30}$$

Variational equation in $\delta\varphi$ related to the internal acoustic fluid. Let $\mathbf{u}$ be in $\mathcal{C}^c$, φ in $\mathcal{E}^c_\varphi$ and p' in $\mathcal{E}^c_{p'}$. Multiplying Eq. (15) by $\overline{\delta\varphi} \in \mathcal{E}^c_\varphi$, integrating over domain Ω, using Green's formula, taking into account the Neumann boundary conditions defined by Eqs. (16) and (17), and since $\int_\Omega \delta\varphi\, d\mathbf{x} = 0$, we obtain

$$-\widetilde{c}(\omega\,; \mathbf{u}, \delta\varphi) + i\omega\, d_T(\omega; \varphi, \delta\varphi) + k(\varphi, \delta\varphi) + s_Z(\omega; \varphi, \delta\varphi) - \frac{1}{c_0^2} c_1(p', \delta\varphi)$$
$$= f_Q(\omega\,; \delta\varphi) \quad , \tag{31}$$

in which coupling sesquilinear form $\widetilde{c}(\omega\,; \mathbf{u}, \delta\varphi)$ is defined on $\mathcal{C}^c \times \mathcal{E}^c$ and is such that

$$\widetilde{c}(\omega\,; \mathbf{u}, \delta\varphi) = \overline{c_0(\delta\varphi\,, \mathbf{u})} - \rho_0\, \kappa(\omega)\, \pi_1(\omega; \overline{\delta\varphi})\, \pi_2(\mathbf{u}) \quad . \tag{32}$$

It should be noted that $\widetilde{c}(\omega\,;\mathbf{u},\delta\varphi)$ is not equal to $\overline{c(\omega\,;\delta\varphi,\mathbf{u})}$ due to the presence of complex term $i\omega\, Z(\mathbf{x},\omega)$ in $\pi_1(\omega\,;\varphi)$ defined by Eq. (3). Sesquilinear forms $d_T(\omega;\varphi,\delta\varphi)$, $k(\varphi,\delta\varphi)$ and $s_Z(\omega;\varphi,\delta\varphi)$ are defined on $\mathcal{E}^c \times \mathcal{E}^c$ by Eqs. (XI.18), (XI.16) and (XI.20) respectively. Sesquilinear form $c_1(p',\delta\varphi)$ is defined on $H^c \times \mathcal{E}^c$ by

$$c_1(p',\delta\varphi) = \int_\Omega p'\,\overline{\delta\varphi}\,d\mathbf{x} \quad . \tag{33}$$

Finally, $f_Q(\omega\,;\delta\varphi)$ in the right-hand side of Eq. (31) is the antilinear form on $\mathcal{E}^c$ defined by

$$f_Q(\omega\,;\delta\varphi) = \int_\Omega g_Q\,\overline{\delta\varphi}\,d\mathbf{x} + \int_{\Gamma\cup\Gamma_Z} G_Q\,\overline{\delta\varphi}\,ds + \int_{\Gamma_Z} G_{Q,Z}\,\overline{\delta\varphi}\,ds \quad . \tag{34}$$

Using Eqs. (18) to (20) and Green's formula for the term containing $\boldsymbol{\nabla}^2 Q$, we obtain

$$f_Q(\omega;\delta\varphi) = -\int_\Omega \frac{Q}{i\omega}\overline{\delta\varphi}\,d\mathbf{x} + \frac{\tau c_0^2}{\omega^2}\int_\Omega \boldsymbol{\nabla} Q\cdot\boldsymbol{\nabla}\overline{\delta\varphi}\,d\mathbf{x} + \rho_0\,\kappa(\omega)\,\pi_Q(\omega)\,\pi_1(\omega\,;\overline{\delta\varphi}). \tag{35}$$

Variational equation in $\delta p'$ related to the internal acoustic fluid. Let φ be in $\mathcal{E}^c_\varphi$ and p' be in $\mathcal{E}^c_{p'}$. Multiplying Eq. (12) by $\overline{\delta p'} \in \mathcal{E}^c_{p'}$ and integrating over domain Ω yields

$$-\omega^2 \frac{1}{c_0^2}\,\overline{c_1(\delta p'\,,\varphi)} + \frac{1}{\rho_0^2} m(p',\delta p') = 0 \quad , \tag{36}$$

in which sesquilinear form $c_1(p'\,,\delta\varphi)$ is defined by Eq. (33) and sesquilinear form $m(p',\delta p')$ on $H^c \times H^c$ is defined by

$$m(p',\delta p') = \frac{\rho_0}{c_0^2}\int_\Omega p'\,\overline{\delta p'}\,d\mathbf{x} \quad . \tag{37}$$

Variational formulation for $\omega \neq 0$. For all fixed real $\omega \neq 0$, the variational formulation is written as follows. Find $\mathbf{u}$ in $\mathcal{C}^c$, φ in $\mathcal{E}^c_\varphi$ and p' in $\mathcal{E}^c_{p'}$ such that, for all $\delta\mathbf{u}$ in $\mathcal{C}^c$, $\delta\varphi$ in $\mathcal{E}^c_\varphi$ and $\delta p'$ in $\mathcal{E}^c_{p'}$, Eqs. (23), (31) and (36) are satisfied.

Solution for $\omega = 0$. For $\omega = 0$, $\mathbf{u}(0)$ must be searched for in $\mathcal{C}^c_{\text{elas}} = \mathcal{C}^c \setminus \mathcal{C}_{\text{rig}} \subset \mathcal{C}^c$ in which $\mathcal{C}_{\text{rig}}$ is the set of rigid body displacements. The restriction to $\mathcal{C}^c_{\text{elas}} \times \mathcal{E}^c_{\varphi} \times \mathcal{E}^c_{p'}$ of the variational formulation defined by Eqs. (23), (31) and (36) yields

$$k^S(\mathbf{u}(0), \delta\mathbf{u}) + \kappa(0)\, j(\mathbf{u}(0), \delta\mathbf{u}) = f_S(0; \delta\mathbf{u}) \quad , \quad \forall \delta\mathbf{u} \in \mathcal{C}^c_{\text{elas}} \quad , \tag{38}$$

$$-\widetilde{c}(0; \mathbf{u}(0), \delta\varphi) + k(\varphi(0), \delta\varphi) - \frac{1}{c_0^2} c_1(p'(0), \delta\varphi) = 0 \quad , \quad \forall \delta\varphi \in \mathcal{E}^c_{\varphi} \quad , \tag{39}$$

$$m(p'(0), \delta p') = 0 \quad , \quad \forall \delta p' \in \mathcal{E}^c_{p'} \quad , \tag{40}$$

in which

$$\widetilde{c}(0; \mathbf{u}(0), \delta\varphi) = c_0(\delta\varphi, \mathbf{u}(0)) - \rho_0\, \kappa(0)\, \pi_1(0; \overline{\delta\varphi})\, \pi_2(\mathbf{u}(0)) \quad . \tag{41}$$

Eq. (39) holds because $f_Q(0; \delta\varphi) = 0$ and [4] $s_Z(0; \varphi, \delta\varphi) = 0$. Since Hermitian forms [5] $k^S(\mathbf{u}(0), \mathbf{u}(0)) > 0$ and $j(\mathbf{u}(0), \mathbf{u}(0)) > 0$ for all $\mathbf{u}(0) \neq 0$ in $\mathcal{C}^c_{\text{elas}}$, and since $\kappa(0) > 0$ (see Eq. (6-2)), we deduce that $k^S(\mathbf{u}(0), \mathbf{u}(0)) + \kappa(0)\, j(\mathbf{u}(0), \mathbf{u}(0)) > 0$ for all $\mathbf{u}(0) \neq 0$ in $\mathcal{C}^c_{\text{elas}}$. Consequently, Eq. (38) has a unique solution $\mathbf{u}(0)$ in $\mathcal{C}^c_{\text{elas}}$. Since $m(p'(0), p'(0)) > 0$ for all $p'(0) \neq 0$ in $\mathcal{E}^c_{p'}$, we deduce that Eq. (40) has a unique solution

$$p'(0) = 0 \quad . \tag{42}$$

Substituting Eq. (42) into Eq. (39) yields

$$k(\varphi(0), \delta\varphi) = \widetilde{c}(0; \mathbf{u}(0), \delta\varphi) \quad , \quad \forall \delta\varphi \in \mathcal{E}^c_{\varphi} \quad , \tag{43}$$

which has a unique solution in $\mathcal{E}^c_{\varphi}$ because $k(\varphi(0), \varphi(0)) > 0$ for all $\varphi(0) \neq 0$ in $\mathcal{E}^c_{\varphi}$. From Eqs. (1) to (6-1), (X.8) and (X.25), we deduce that

$$p(\mathbf{x}, 0) = -\kappa(0) \int_{\Gamma \cup \Gamma_Z} \mathbf{u}(\mathbf{x}, 0) \cdot \mathbf{n}(\mathbf{x})\, ds(\mathbf{x}) \quad , \quad \forall \mathbf{x} \in \Omega \quad . \tag{44}$$

Symmetric variational formulation for $\omega \neq 0$. In order to obtain a symmetric matrix finite element model or a symmetric reduced matrix model, valid for all $\omega \neq 0$, Eq. (31) is multiplied by ω^2 and Eqs. (23) and (36)

---

[4] From Eqs. (XI.20), (XI.10), (X.21) and (X.25), we deduce that $s_Z(0; \varphi, \delta\varphi) = 0$ and consequently, Eq. (XI.21) yields $\mathbf{S}_Z(0) = 0$.

[5] $k^S(\mathbf{u}(0), \mathbf{u}(0))$ and $j(\mathbf{u}(0), \mathbf{u}(0))$ have the same null space which is $\mathcal{C}^c_{\text{rig}}$ (space spanned by the set of rigid body displacements).

are preserved. The symmetric variational formulation is written as follows. For all fixed real $\omega \neq 0$, find $\mathbf{u}$ in $\mathcal{C}^c$, φ in $\mathcal{E}_\varphi^c$ and p' in $\mathcal{E}_{p'}^c$ such that,

$$a^S(\omega;\mathbf{u},\delta\mathbf{u}) - \omega^2 a^E(\omega/c_E;\mathbf{u},\delta\mathbf{u}) + \kappa\, j(\mathbf{u},\delta\mathbf{u}) - \omega^2 c(\omega;\varphi,\delta\mathbf{u}) = f_S(\omega;\delta\mathbf{u}) \quad , \quad \forall\delta\mathbf{u}\in\mathcal{C}^c \quad , \tag{45}$$

$$-\omega^2\widetilde{c}(\omega\,;\mathbf{u},\delta\varphi) + i\omega^3\, d_T(\omega;\varphi,\delta\varphi) + \omega^2 k(\varphi,\delta\varphi) + \omega^2 s_Z(\omega;\varphi,\delta\varphi) - \omega^2\frac{1}{c_0^2}c_1(p',\delta\varphi) = \omega^2 f_Q(\omega\,;\delta\varphi) \;,\; \forall\delta\varphi\in\mathcal{E}_\varphi^c \;, \tag{46}$$

$$-\omega^2\frac{1}{c_0^2}\,\overline{c_1(\delta p'\,,\varphi)} + \frac{1}{\rho_0^2}m(p',\delta p') = 0 \quad , \quad \forall\delta p'\in\mathcal{E}_{p'}^c \quad . \tag{47}$$

4.3. Linear operator equation

For $\omega \neq 0$, the linear operator equation corresponding to the symmetric variational formulation defined by Eqs. (45) to (47) is

$$\mathbf{A}(\omega)\begin{bmatrix}\mathbf{u}\\ \varphi\\ p'\end{bmatrix} = \begin{bmatrix}\mathbf{f}_S(\omega)\\ \omega^2\,\mathbf{f}_Q(\omega)\\ \mathbf{0}\end{bmatrix} \quad , \quad \{\mathbf{u},\varphi,p'\}\in\mathcal{C}^c\times\mathcal{E}_\varphi^c\times\mathcal{E}_{p'}^c \quad , \tag{48}$$

in which $\mathbf{A}(\omega)$ is the linear operator [6] defined by

$$\mathbf{A}(\omega) = \begin{bmatrix}\mathbf{A}^S(\omega) - \omega^2\mathbf{A}^E(\omega/c_E) + \kappa(\omega)\mathbf{J} & -\omega^2\mathbf{C}(\omega) & \mathbf{0}\\ -\omega^2\,{}^t\overline{\mathbf{C}}(\omega) & i\omega^3\mathbf{D}_T(\omega) + \omega^2\mathbf{K} + \omega^2\mathbf{S}_Z(\omega) & -\frac{\omega^2}{c_0^2}\mathbf{C}_1\\ \mathbf{0} & -\frac{\omega^2}{c_0^2}\,{}^t\mathbf{C}_1 & \frac{1}{\rho_0^2}\mathbf{M}\end{bmatrix} . \tag{49}$$

The dynamic stiffness linear operator $\mathbf{A}^S(\omega)$ is such that [7]

$$<\mathbf{A}^S(\omega)\,\mathbf{u}\,,\delta\mathbf{u}> = a^S(\omega\,;\mathbf{u}\,,\delta\mathbf{u}) \quad , \tag{50}$$

---

6 Linear operator $\mathbf{A}(\omega)$ defined by Eq. (49) is continuous from $\mathcal{C}^c\times\mathcal{E}^c\times H^c$ into $\mathcal{C}^{c\prime}\times\mathcal{E}^{c\prime}\times H^{c\prime}$ in which $H^{c\prime} = H^c = L^2(\Omega)$. The restriction of this linear operator to $\mathcal{C}^c\times\mathcal{E}_\varphi^c\times\mathcal{E}_{p'}^c$ is continuous from $\mathcal{C}^c\times\mathcal{E}_\varphi^c\times\mathcal{E}_{p'}^c$ into $\mathcal{C}^{c\prime}\times\mathcal{E}_\varphi^{c\prime}\times\mathcal{E}_{p'}^{c\prime}$.

7 Sesquilinear form $a^S(\omega;\mathbf{u},\delta\mathbf{u})$ is continuous on $\mathcal{C}^c\times\mathcal{C}^c$, $\mathbf{A}^S(\omega)$ is a continuous operator from $\mathcal{C}^c$ into its antidual space $\mathcal{C}^{c\prime}$ and the angle brackets in Eqs. (50) to (51-4) denote the antiduality product between $\mathcal{C}^{c\prime}$ and $\mathcal{C}^c$. For all real $\omega\neq 0$, linear operator $\mathbf{A}^S(\omega)$ is invertible. For $\omega = 0$, the restriction of continuous linear operator $\mathbf{A}^S(\omega)$ to $\mathcal{C}_S^c$ is invertible (see Eq. (V.56) for the definition of $\mathcal{C}_S^c$).

and can then be written as

$$\mathbf{A}^S(\omega) = -\omega^2\mathbf{M}^S + i\omega\,\mathbf{D}^S(\omega) + \mathbf{K}^S \quad , \qquad (51-1)$$

where $\mathbf{M}^S$, $\mathbf{D}^S(\omega)$ and $\mathbf{K}^S$ are the mass, damping and stiffness operators of the master structure defined by

$$<\mathbf{M}^S\,\mathbf{u}\,,\delta\mathbf{u}> = m^S(\mathbf{u}\,,\delta\mathbf{u}) \quad , \qquad (51-2)$$

$$<\mathbf{D}^S(\omega)\,\mathbf{u}\,,\delta\mathbf{u}> = d^S(\omega\,;\mathbf{u}\,,\delta\mathbf{u}) \quad , \qquad (51-3)$$

$$<\mathbf{K}^S\,\mathbf{u}\,,\delta\mathbf{u}> = k^S(\mathbf{u}\,,\delta\mathbf{u}) \quad . \qquad (51-4)$$

Linear operator $\mathbf{A}^E(\omega/c_E)$ is such that [8]

$$<\mathbf{A}^E(\omega/c_E)\,\mathbf{u}\,,\delta\mathbf{u}> = a^E(\omega/c_E\,;\,\mathbf{u}\,,\delta\mathbf{u}) \quad . \qquad (52)$$

Linear operator $\mathbf{J}$ is such that [9]

$$<\mathbf{J}\,\mathbf{u}\,,\delta\mathbf{u}> = j(\mathbf{u},\delta\mathbf{u}) \quad . \qquad (53)$$

Coupling operator $\mathbf{C}(\omega)$ is such that [10]

$$<\mathbf{C}(\omega)\,\varphi\,,\delta\mathbf{u}> = c(\omega\,;\varphi,\delta\mathbf{u}) \quad , \qquad (54-1)$$

and can be written as

$$\mathbf{C}(\omega) = \mathbf{C}_0 + \mathbf{C}_\kappa(\omega) \quad , \qquad (54-2)$$

in which operators $\mathbf{C}_0$ and $\mathbf{C}_\kappa(\omega)$ are defined by

$$<\mathbf{C}_0\,\varphi\,,\delta\mathbf{u}> = c_0(\varphi,\delta\mathbf{u}) \quad , \qquad (54-3)$$

$$<\mathbf{C}_\kappa(\omega)\,\varphi\,,\delta\mathbf{u}> = -\rho_0\,\kappa(\omega)\,\pi_1(\omega;\varphi)\,\pi_2(\overline{\delta\mathbf{u}}) \quad . \qquad (54-4)$$

---

8 For all ω fixed in $\mathbb{R}$, sesquilinear form $a^E(\omega/c_0\,;\mathbf{u},\delta\mathbf{u})$ is continuous on $\mathcal{C}^c\times\mathcal{C}^c$ (see footnote 3) and then $\mathbf{A}^E(\omega/c_E)$ is a continuous linear operator from $\mathcal{C}^c$ into its antidual space $\mathcal{C}^{c\,\prime}$. The angle brackets in Eq. (52) denote the antiduality product between $\mathcal{C}^{c\,\prime}$ and $\mathcal{C}^c$.

9 Sesquilinear form $j(\mathbf{u},\delta\mathbf{u})$ is continuous on $\mathcal{C}^c\times\mathcal{C}^c$ and then $\mathbf{J}$ is a continuous linear operator from $\mathcal{C}^c$ into $\mathcal{C}^{c\,\prime}$. The angle brackets in Eq. (53) denote the antiduality product between $\mathcal{C}^{c\,\prime}$ and $\mathcal{C}^c$.

10 For all ω fixed in $\mathbb{R}$, sesquilinear form $c(\omega\,;\varphi,\delta\mathbf{u})$ is continuous on $\mathcal{E}^c\times\mathcal{C}^c$ and then $\mathbf{C}(\omega)$ is a continuous linear operator from $\mathcal{E}^c$ into $\mathcal{C}^{c\,\prime}$. The angle brackets in Eqs. (54-1) to (54-4) denote the antiduality product between $\mathcal{C}^{c\,\prime}$ and $\mathcal{C}^c$.

Coupling operator ${}^t\overline{\mathbf{C}}(\omega)$ is such that [11]

$$<{}^t\overline{\mathbf{C}}(\omega)\,\mathbf{u}\,,\delta\varphi>=\widetilde{c}(\omega\,;\mathbf{u},\delta\varphi)\quad , \qquad (55)$$

and we have $<{}^t\overline{\mathbf{C}}(\omega)\,\mathbf{u}\,,\delta\varphi>=<\mathbf{u}\,,\overline{\mathbf{C}}(\omega)\,\delta\varphi>$. Linear operators $\mathbf{K}$, $\mathbf{D}_T(\omega)$ and $\mathbf{S}_Z(\omega)$ are defined by Eqs. (XI.17), (XI.19) and (XI.21) respectively. Linear operator $\mathbf{C}_1$ is such that [12]

$$<\mathbf{C}_1 p'\,,\delta\varphi>=c_1(p'\,,\delta\varphi)\quad , \qquad (56)$$

and linear operator ${}^t\mathbf{C}_1$ is such that [13]

$$<{}^t\mathbf{C}_1\varphi\,,\delta p'>=\overline{c_1(\delta p'\,,\varphi)}\quad . \qquad (57)$$

Linear operator $\mathbf{M}$ is such that [14]

$$<\mathbf{M}p'\,,\delta p'>=m(p'\,,\delta p')\quad , \qquad (58)$$

where m is defined by Eq. (37). Finally, elements $\mathbf{f}_S(\omega)$ and $\mathbf{f}_Q(\omega)$ are such that [15]

$$<\mathbf{f}_S(\omega)\,,\delta\mathbf{u}>=f_S(\omega\,;\delta\mathbf{u})\quad , \qquad (59)$$

$$<\mathbf{f}_Q(\omega)\,,\delta\varphi>=f_Q(\omega\,;\delta\varphi)\quad . \qquad (60)$$

Once $\{\mathbf{u},\varphi,p'\}$ is known, internal pressure field $p(\mathbf{x},\omega)$ in Ω is deduced from the variational formulation of Eq. (1) which is such that, for all δp in H^c,

$$\int_\Omega p\,\overline{\delta p}\,d\mathbf{x}=\omega^2\rho_0\int_\Omega \varphi\,\overline{\delta p}\,d\mathbf{x}+\pi(\omega\,;\mathbf{u}\,;\varphi)\int_\Omega\overline{\delta p}\,d\mathbf{x}\quad , \qquad (61-1)$$

---

11 For all ω fixed in $\mathbb{R}$, sesquilinear form $\widetilde{c}(\omega\,;\delta\varphi,\mathbf{u})$ is continuous on $\mathcal{E}^c\times\mathcal{C}^c$ and then ${}^t\overline{\mathbf{C}}(\omega)$ is a continuous linear operator from $\mathcal{C}^c$ into $\mathcal{E}^{c\prime}$. The angle brackets in Eq. (55) denote the antiduality product between $\mathcal{E}^{c\prime}$ and $\mathcal{E}^c$.

12 Sesquilinear form $c_1(p'\,,\delta\varphi)$ is continuous on $H^c\times\mathcal{E}^c$ and then $\mathbf{C}_1$ is a continuous linear operator from H^c into $\mathcal{E}^{c\prime}$. The angle brackets in Eq. (56) denote the antiduality product between $\mathcal{E}^{c\prime}$ and $\mathcal{E}^c$.

13 The angle brackets in Eq. (57) denote the inner product in H^c.

14 Sesquilinear form $m(p'\,,\delta p')$ is continuous on $H^c\times H^c$ and consequently defines a continuous linear operator $\mathbf{M}$ from H^c into H^c. The angle brackets in Eq. (58) denote the inner product in H^c.

15 For all ω fixed in $\mathbb{R}$, antilinear forms $f_S(\omega;\delta\mathbf{u})$ and $f_Q(\omega;\delta\varphi)$ are continuous on $\mathcal{C}^c$ and $\mathcal{E}^c$ respectively. These continuous antilinear forms define elements $\mathbf{f}_S(\omega)$ in $\mathcal{C}^{c\prime}$ and $\mathbf{f}_Q(\omega)$ in $\mathcal{E}^{c\prime}$ respectively. The angle brackets in Eq. (59) denote the antiduality product between $\mathcal{C}^{c\prime}$ and $\mathcal{C}^c$. The angle brackets in Eq. (60) denote the antiduality product between $\mathcal{E}^{c\prime}$ and $\mathcal{E}^c$.

which can be rewritten as

$$\frac{c_0^2}{\rho_0}\, m(p\,,\delta p) = \omega^2 \rho_0 \overline{c_1(\delta p\,,\varphi)} + \pi(\omega\,;\mathbf{u}\,;\varphi) \int_\Omega \overline{\delta p}\, d\mathbf{x} \quad . \tag{61 – 2}$$

We then deduce that internal pressure field $p(\mathbf{x},\omega)$ is the solution of the linear operator equation

$$\frac{c_0^2}{\rho_0}\,\mathbf{M}p = \omega^2 \rho_0\, {}^t\mathbf{C}_1 \varphi + \pi(\omega\,;\mathbf{u}\,;\varphi)\,\mathbf{l} \quad , \tag{62}$$

in which $\mathbf{l}$ is the antilinear form defined on H^c such that

$$<\mathbf{l}\,,\delta p>= \int_\Omega \overline{\delta p}\, d\mathbf{x} \quad . \tag{63}$$

Pressure field $p_E|_{\Gamma_E}(\omega)$ on Γ_E due to the external acoustic fluid is given by Eq. (7) and, at any point $\mathbf{x}$ of unbounded external acoustic fluid Ω_E, resultant pressure field $p_E(\mathbf{x},\omega)$ is given by Eq. (8).

5. Finite Element Discretization

We consider a finite element mesh of master structure Ω_S and a finite element mesh of internal acoustic fluid Ω. We assume that the two finite element meshes are compatible on interface $\Gamma \cup \Gamma_Z$. The finite element mesh of surface Γ_E is the trace of the mesh of Ω_S (see Fig. 2).

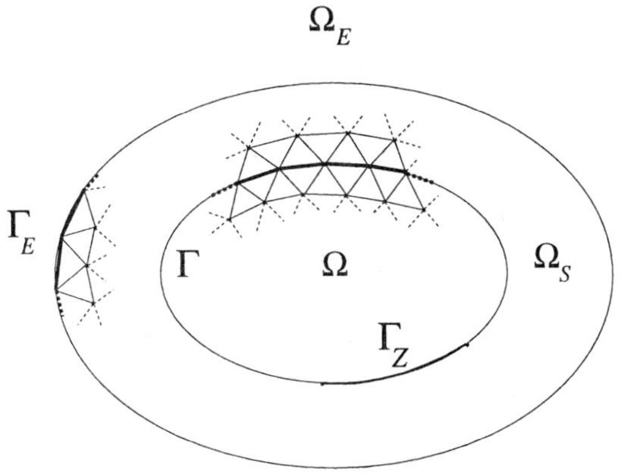

Fig. 2. Example of structure and internal fluid finite element meshes

We used the finite element method to construct the discretization of the various sesquilinear forms of the problem. For the master structure, we introduce subspace $\mathcal{C}_{n_S} \subset \mathcal{C}^c$ of finite dimension n_S. Let $\mathbf{U} = (U_1, \ldots, U_{n_S})$ be the complex vector of the DOFs which are the values of $\mathbf{u}$ at the nodes of the finite element mesh of domain Ω_S. For the internal acoustic fluid, we introduce subspace [16] $\mathcal{E}_n \subset \mathcal{E}^c \subset H^c$ of finite dimension n. Let $\mathbf{\Phi} = (\Phi_1, \ldots, \Phi_n)$ and $\mathbf{P}' = (p'_1, \ldots, p'_n)$ be the complex vectors of the DOFs which are the values of φ and p' at the nodes of the finite element mesh of domain Ω. Since the finite element method uses a real basis for constructing the finite element matrices, the finite element discretization of the symmetric variational formulation defined by Eqs. (45) to (47) yields the symmetric complex matrix equation (corresponding to the operator equation defined by Eqs. (48) and (49)),

$$[A(\omega)] \begin{bmatrix} \mathbf{U} \\ \mathbf{\Phi} \\ \mathbf{P}' \end{bmatrix} = \begin{bmatrix} \mathbf{F}_S(\omega) \\ \omega^2 \mathbf{F}_Q(\omega) \\ \mathbf{0} \end{bmatrix} \quad , \tag{64}$$

with the constraints

$$\mathbf{L}^T \mathbf{\Phi} = 0 \quad , \quad \mathbf{L}^T \mathbf{P}' = 0 \quad . \tag{65}$$

Eq. (65) corresponds to the finite element discretization in $\mathcal{E}_n$ of the linear constraints $\int_\Omega \varphi \, d\mathbf{x} = 0$ and $\int_\Omega p' \, d\mathbf{x} = 0$ appearing in admissible function spaces $\mathcal{E}^c_\varphi$ and $\mathcal{E}^c_{p'}$ defined by Eqs. (21) and (22) respectively, in which $\mathbf{L}$ is a vector in $\mathbb{R}^n$. Complex matrix $[A(\omega)]$ is symmetric and is defined by

$$\begin{bmatrix} [A^S(\omega)] - \omega^2 [A^E(\omega/c_E)] + \kappa(\omega)[J] & -\omega^2 [C(\omega)] & [0] \\ -\omega^2 [C(\omega)]^T & i\omega^3 [D_\tau(\omega)] + \omega^2 [K] + \omega^2 [S_z(\omega)] & -\frac{\omega^2}{c_0^2} [C_1] \\ [0] & -\frac{\omega^2}{c_0^2} [C_1]^T & \frac{1}{\rho_0^2} [M] \end{bmatrix} , \tag{66}$$

where $[A^S(\omega)]$ is the dynamic stiffness matrix of the master structure which is an $(n_S \times n_S)$ symmetric complex matrix (invertible for all real $\omega \neq 0$ and not invertible for $\omega = 0$ due to the presence of the rigid body modes) written as

$$[A^S(\omega)] = -\omega^2 [M^S] + i\omega \, [D^S(\omega)] + [K^S] \quad . \tag{67}$$

Mass, damping and stiffness matrices $[M^S]$, $[D^S(\omega)]$ and $[K^S]$ are $(n_S \times n_S)$ symmetric real matrices. Matrix $[M^S]$ is positive definite and matrices

[16] Since $\mathcal{E}^c = H^1(\Omega)$ and $H^c = L^2(\Omega)$, we have $\mathcal{E}^c \subset H^c$.

$[D^S(\omega)]$ and $[K^S]$ are positive semidefinite. The finite element discretization of sesquilinear form $a^E(\omega/c_E\,;\mathbf{u}\,,\delta\mathbf{u})$ yields the $(n_S \times n_S)$ symmetric complex matrix $[A^E(\omega/c_E)]$ related to the nodal values of $\mathbf{u}$, written as

$$[A^E(\omega/c_E)] = -\rho_E\,[\,\Theta\,]^T\,[B_{\Gamma_E}(\omega/c_E)]\,[\,\Theta\,] \quad , \tag{68}$$

where $[B_{\Gamma_E}(\omega/c_E)]$ is an $(n_E\times n_E)$ symmetric dense complex matrix defined by Eq. (XII.164) and related to the nodal values of the normal displacement field on Γ_E. Rectangular matrix $[\,\Theta\,]$ is a $(n_E \times n_S)$ sparse real matrix corresponding to the finite element discretization $\overline{\delta\mathbf{W}}^T\,[\,\Theta\,]\,\mathbf{U}$ of the sesquilinar form

$$(\mathbf{u}, \delta w) \mapsto \theta(\mathbf{u}, \delta w) = \int_{\Gamma_E} \overline{\delta w}\; \mathbf{u}\cdot\mathbf{n}^S\, ds \tag{69}$$

defined on $\mathcal{C}^c \times \mathcal{C}_{\Gamma_E}$. Eqs. (68) and (69) show that matrix $[A^E(\omega/c_E)]$ has zero rows and zero columns corresponding to the structural nodes which are not located on interface Γ_E. Matrix $[\,J\,]$ is an $(n_S \times n_S)$ symmetric real matrix which can be written as

$$[\,J\,] = \mathbf{\Pi}_2\,\mathbf{\Pi}_2^T \quad , \tag{70}$$

in which $\mathbf{\Pi}_2$ is an $\mathbb{R}^{n_S}$ vector such that $\mathbf{\Pi}_2^T$ corresponds to the finite element discretization of the linear form $\pi_2(\mathbf{u})$ defined by Eq. (4). Matrix $[\,J\,]$ has zero rows and zero columns corresponding to the structural nodes which are not located on interface $\Gamma \cup \Gamma_Z$. Referring to Section XI.6, the $(n\times n)$ real matrices $[\,K\,]$ and $[D_{\tau}(\omega)]$ are symmetric positive semidefinite and $[S_z(\omega)]$ is an $(n\times n)$ symmetric complex matrix such that $\omega^{-1}\Im m[S_z(\omega)]$ is positive semidefinite. Matrix $[\,M\,]$ is an $(n\times n)$ symmetric positive-definite matrix corresponding to the finite element discretization of sesquilinear form $m(p', \delta p')$ defined by Eq. (37). Coupling rectangular matrix $[C(\omega)]$ is an $(n_S\times n)$ complex matrix which is written as

$$[C(\omega)] = [C_0] - \rho_0\,\kappa(\omega)\,\mathbf{\Pi}_2\,\mathbf{\Pi}_1(\omega)^T \quad , \tag{71}$$

in which $[C_0]$ is the $(n_S\times n)$ real matrix corresponding to the finite element discretization of sesquilinear form $c_0(\varphi\,,\delta\mathbf{u})$ defined by Eq. (27-2), and $\mathbf{\Pi}_1(\omega)$ is a $\mathbb{C}^n$ vector such that $\mathbf{\Pi}_1(\omega)^T$ corresponds to the finite element discretization of the linear form $\pi_1(\omega\,;\varphi)$ defined by Eq. (3). Matrix $[C(\omega)]$ has zero rows and zero columns corresponding to the structural and fluid nodes which are not located on interface $\Gamma\cup\Gamma_Z$. Since we have chosen the same subspace $\mathcal{E}_n$ for the finite element approximation of p' and φ, then the $(n\times n)$ symmetric real matrix $[C_1]$ corresponding to the finite element discretization of sesquilinear form $c_1(p', \delta\varphi)$ defined by Eq. (33) is written as

$$[C_1] = \frac{c_0^2}{\rho_0}[\,M\,] \quad . \tag{72}$$

Finally, the finite element discretization of antilinear forms $f_{\text{S}}(\omega\,; \delta\mathbf{u})$ and $f_{\text{Q}}(\omega\,; \delta\varphi)$ defined by Eqs. (28) and (35) yields the complex vectors $\mathbf{F}_{\text{S}}(\omega)$ in $\mathbb{C}^{n_S}$ and $\mathbf{F}_{\text{Q}}(\omega)$ in $\mathbb{C}^n$ respectively.

Let $\mathbf{P} = (p_1, \ldots, p_n)$ be the complex vector of the DOFs which are the values of p at the nodes of the finite element mesh of domain Ω. The finite element discretization of Eq. (61-2) is written as

$$\frac{c_0^2}{\rho_0}[\,M\,]\,\mathbf{P} = \omega^2 c_0^2[\,M\,]\mathbf{\Phi} + \kappa(\omega)\{-\omega^2\rho_0\mathbf{\Pi}_1(\omega)^T\mathbf{\Phi} - \mathbf{\Pi}_2^T\mathbf{U} + \pi_{\text{Q}}(\omega)\}\;\mathbf{L}, \tag{73}$$

in which $\mathbf{L} \in \mathbb{R}^n$ is the finite element discretization of antilinear form $\mathbf{l}$ defined by Eq. (63) (which is also used in Eq. (65)) and where $\pi_{\text{Q}}(\omega)$ is defined by Eq. (5). It should be noted that the construction of $\mathbf{P}$ requires solving a linear matrix equation with nondiagonal matrix $(c_0^2/\rho_0)\,[\,M\,]$. An alternative discretization of Eq. (61-2) can be used to recover p locally element by element. Since $p(\omega)$ belongs to H^c, the pressure can be taken, for instance, as a constant over each element. In this case, no linear matrix equation of order n has to be solved.

6. Symmetric Reduced Matrix Model in the LF Range

As previously, we denote the LF band as

$$\mathbb{B}_{\text{LF}} =]\,0\,,\,\omega_{\text{LF},\,\text{final}}\,] \quad . \tag{74}$$

It should be noted that $\omega = 0$ does not belong to $\mathbb{B}_{\text{LF}}$. We consider the low-frequency case for which ω belongs to $\mathbb{B}_{\text{LF}}$. The methodology for constructing the symmetric reduced matrix model has two steps. The first step consists in using the Ritz-Galerkin projection of the variational formulation in $(\mathbf{u}, \varphi, p')$ on a finite dimension subspace spanned by a set of structural modes of the master structure in vacuo and a set of acoustic modes of the internal acoustic cavity with rigid wall. The second step consists in eliminating the generalized coordinates associated with field p'.

6.1. Structural modes of the master structure in vacuo

The structural modes $\mathbf{u}_\alpha$ of master structure Ω_S in vacuo satisfy the spectral problem,

$$k^S(\mathbf{u}_\alpha\,,\delta\mathbf{u}) = \lambda_\alpha^S\, m^S(\mathbf{u}_\alpha\,,\delta\mathbf{u}) \quad , \quad \forall \delta\mathbf{u} \in \mathcal{C}^c \quad , \tag{75}$$

and satisfy the orthogonality conditions

$$m^S(\mathbf{u}_\alpha, \mathbf{u}_\beta) = \mu_\alpha^S\, \delta_{\alpha\beta} \quad , \tag{76 – 1}$$

$$k^S(\mathbf{u}_\alpha, \mathbf{u}_\beta) = \mu_\alpha^S\, \lambda_\alpha^S\, \delta_{\alpha\beta} \quad , \tag{76 – 2}$$

in which eigenvalues $\lambda_\alpha^S = \{\omega_\alpha^S\}^2$ are such that $0 = \lambda_1^S = \ldots = \lambda_6^S < \lambda_7^S \leq \lambda_8^S \ldots$. Eigenfunctions $\mathbf{u}_1, \ldots, \mathbf{u}_6$ associated with the solution $\lambda_\alpha^S = 0$ are the rigid body modes and eigenfunctions $\{\mathbf{u}_\alpha, \alpha \geq 7\}$ are the elastic structural modes. Concerning the practical calculation of the structural modes by the finite element method, we refer the reader to Section III.9.

6.2. Acoustic modes of the internal acoustic cavity with a rigid wall

The acoustic modes φ_α of internal acoustic cavity Ω with a fixed wall and without wall acoustic impedance satisfy the spectral problem

$$k(\varphi_\alpha\,,\delta\varphi) = \lambda_\alpha\, m(\varphi_\alpha\,,\delta\varphi) \quad , \quad \forall\, \delta\varphi \in \mathcal{E}_\varphi^c \quad , \tag{77}$$

and satisfy the orthogonality conditions

$$m(\varphi_\alpha\,,\varphi_\beta) = \mu_\alpha\, \delta_{\alpha\beta} \quad , \tag{78}$$

$$k(\varphi_\alpha\,,\varphi_\beta) = \mu_\alpha\, \lambda_\alpha\, \delta_{\alpha\beta} \quad , \tag{79}$$

in which eigenvalues $\lambda_\alpha = \omega_\alpha^2$ are such that $0 < \lambda_1 \leq \lambda_2 \leq \ldots$. It should be noted that since φ_α belongs to $\mathcal{E}_\varphi^c$, the constraint defined by Eq. (14) is automatically satisfied. For the practical calculation of the acoustic modes by the finite element method, see Section XI.8.

6.3. Finite dimension subspaces for the projection

We introduce subspace $\mathcal{C}_{N_S}^c$ of $\mathcal{C}^c$, of dimension $N_S \geq 1$, spanned by the finite family $\{\mathbf{u}_1, \ldots, \mathbf{u}_{N_S}\}$ of structural modes. Let $\mathcal{E}_{\varphi,N}^c$ be the subspace of $\mathcal{E}_\varphi^c$, of dimension $N \geq 1$, spanned by the finite family $\{\varphi_1, \ldots, \varphi_N\}$ of acoustic modes. Since $\mathcal{E}_\varphi^c \subset \mathcal{E}_{p'}^c$, we can choose $\mathcal{E}_{\varphi,N}^c$ as the finite approximation subspace of $\mathcal{E}_{p'}^c$ (which implies that the constraint on p' defined by Eq. (13) is automatically satisfied). Consequently, the finite dimension subspace used for constructing the reduced model is $\mathcal{C}^c_{N_S} \times \mathcal{E}^c_{\varphi,N} \times \mathcal{E}^c_{\varphi,N}$. The projection $\mathbf{u}^{N_S}$ on $\mathcal{C}^c_{N_S}$ of $\mathbf{u} \in \mathcal{C}^c$ is written as

$$\mathbf{u}^{N_S}(\mathbf{x},\omega) = \sum_{\alpha=1}^{N_S} q^S_\alpha(\omega)\,\mathbf{u}_\alpha(\mathbf{x}) \quad , \tag{80}$$

in which

$$\mathbf{q}^S = (q^S_1, \ldots, q^S_{N_S}) \tag{81}$$

is a complex vector of generalized coordinates. The projection φ^N on $\mathcal{E}^c_{\varphi,N}$ of $\varphi \in \mathcal{E}^c_\varphi$ is written as

$$\varphi^N(\mathbf{x},\omega) = \sum_{\alpha=1}^{N} q_\alpha(\omega)\,\varphi_\alpha(\mathbf{x}) \quad , \tag{82}$$

in which

$$\mathbf{q} = (q_1, \ldots, q_N) \tag{83}$$

is a complex vector of generalized coordinates. Finally, the projection p'_N on $\mathcal{E}^c_{\varphi,N}$ of $p' \in \mathcal{E}^c_{p'}$ is written as

$$p'_N(\mathbf{x},\omega) = \sum_{\alpha=1}^{N} r_\alpha(\omega)\,\varphi_\alpha(\mathbf{x}) \quad , \tag{84}$$

in which

$$\mathbf{r} = (r_1, \ldots, r_N) \tag{85}$$

is a complex vector of generalized coordinates.

6.4. Symmetric reduced matrix model

Using Eqs. (80), (82) and (84), the projection of the variational formulation defined by Eqs. (45) to (47) yields

$$[\mathcal{A}(\omega)] \begin{bmatrix} \mathbf{q}^S \\ \mathbf{q} \\ \mathbf{r} \end{bmatrix} = \begin{bmatrix} \boldsymbol{\mathcal{F}}_{\mathrm{S}}(\omega) \\ \omega^2\,\boldsymbol{\mathcal{F}}_{\mathrm{Q}}(\omega) \\ \mathbf{0} \end{bmatrix} \quad , \tag{86}$$

in which complex matrix $[\mathcal{A}(\omega)]$ is symmetric and is defined by

$$\begin{bmatrix} [\mathcal{A}^S(\omega)] - \omega^2[\mathcal{A}^E(\omega/c_E)] + \kappa(\omega)[\mathcal{J}] & -\omega^2[\mathcal{C}(\omega)] & [0] \\ -\omega^2\,[\mathcal{C}(\omega)]^T & i\omega^3[\mathcal{D}_T(\omega)] + \omega^2[\mathcal{K}] + \omega^2[\mathbb{S}_z(\omega)] & -\frac{\omega^2}{c_0^2}[\mathcal{C}_1] \\ [0] & -\frac{\omega^2}{c_0^2}\,[\mathcal{C}_1]^T & \frac{1}{\rho_0^2}[\mathcal{M}] \end{bmatrix} \tag{87}$$

in which $[\mathcal{A}^S(\omega)]$ is an $(N_S \times N_S)$ symmetric complex matrix written as

$$[\mathcal{A}^S(\omega)] = -\omega^2[\mathcal{M}^S] + i\omega\,[\mathcal{D}^S(\omega)] + [\mathcal{K}^S] \quad , \tag{88}$$

where $[\mathcal{M}^S]$ is an $(N_S \times N_S)$ symmetric positive-definite real matrix and $[\mathcal{D}^S(\omega)]$ and $[\mathcal{K}^S]$ are $(N_S \times N_S)$ symmetric positive-semidefinite real matrices. Generalized mass and stiffness matrices $[\mathcal{M}^S]$ and $[\mathcal{K}^S]$ are diagonal and are such that

$$[\mathcal{M}^S]_{\alpha\beta} = \mu_\alpha^S\, \delta_{\alpha\beta} \quad , \tag{89}$$

$$[\mathcal{K}^S]_{\alpha\beta} = \mu_\alpha^S\, \lambda_\alpha^S\, \delta_{\alpha\beta} \quad . \tag{90}$$

Generalized damping matrix $[\mathcal{D}^S(\omega)]$ is dense and is such that

$$[\mathcal{D}^S(\omega)]_{\alpha\beta} = d^S(\omega\,; \mathbf{u}_\beta\,, \mathbf{u}_\alpha) \quad . \tag{91}$$

The $(N_S \times N_S)$ symmetric complex matrix $[\mathcal{A}^E(\omega/c_E)]$ is such that

$$\begin{aligned}[\mathcal{A}^E(\omega/c_E)]_{\alpha\beta} &= a^E(\omega/c_E\,; \mathbf{u}_\beta\,, \mathbf{u}_\alpha) \\ &= -\rho_E \int_{\Gamma_E} (\mathbf{u}_\beta \cdot \mathbf{n}^S)\ \mathbf{B}_{\Gamma_E}(\omega/c_E)\{\mathbf{u}_\alpha \cdot \mathbf{n}^S\}\, ds \quad .\end{aligned} \tag{92}$$

Matrix $[\,\mathcal{J}\,]$ is an $(N_S \times N_S)$ symmetric positive-semidefinite real matrix such that

$$[\,\mathcal{J}\,]_{\alpha\beta} = \pi_2(\mathbf{u}_\alpha)\,\pi_2(\mathbf{u}_\beta) \quad , \tag{93}$$

in which $\pi_2(\mathbf{u})$ is defined by Eq. (4). Matrices $[\mathcal{M}]$, $[\mathcal{K}]$ and $[\mathcal{D}_\tau(\omega)]$ are $(N \times N)$ diagonal positive-definite real matrices and $[\mathcal{S}_z(\omega)]$ is a dense $(N \times N)$ symmetric complex matrix such that

$$[\mathcal{M}\,]_{\alpha\beta} = \mu_\alpha\, \delta_{\alpha\beta} \quad , \tag{94}$$

$$[\mathcal{K}\,]_{\alpha\beta} = \mu_\alpha\, \lambda_\alpha \delta_{\alpha\beta} \quad , \tag{95}$$

$$[\mathcal{D}_\tau(\omega)]_{\alpha\beta} = d_\tau(\omega\,; \varphi_\beta, \varphi_\alpha) = \tau(\omega)\,\mu_\alpha\,\lambda_\alpha \delta_{\alpha\beta} \quad , \tag{96}$$

$$[\mathcal{S}_z(\omega)]_{\alpha\beta} = s_z(\omega\,; \varphi_\beta, \varphi_\alpha) \quad , \tag{97}$$

in which sesquilinear form s_z is defined by Eq. (XI.20). It should be noted that dense matrix $[\mathcal{S}_z(\omega)]$ is due only to the presence of acoustic impedance $Z(\mathbf{x},\omega)$ on boundary Γ_Z. Coupling rectangular matrix $[\,\mathcal{C}(\omega)]$ is an $(N_S \times N)$ complex matrix which can be written as

$$[\,\mathcal{C}(\omega)]_{\alpha\beta} = [\,\mathcal{C}_0]_{\alpha\beta} + [\,\mathcal{C}_K(\omega)]_{\alpha\beta} \quad , \tag{98}$$

where $[\mathcal{C}_0]$ is the $(N_S \times N)$ real matrix such that

$$[\mathcal{C}_0]_{\alpha\beta} = c_0(\varphi_\beta\,, \mathbf{u}_\alpha) = \rho_0 \int_{\Gamma \cup \Gamma_Z} \varphi_\beta\, \mathbf{n}\cdot\mathbf{u}_\alpha\, ds \quad , \tag{99 – 1}$$

and where matrix $[\mathcal{C}_\kappa(\omega)]$ is an $(N_S \times N)$ complex matrix such that

$$[\mathcal{C}_\kappa(\omega)]_{\alpha\beta} = -\rho_0\, \kappa(\omega)\, \pi_2(\mathbf{u}_\alpha)\, \pi_1(\omega\,; \varphi_\beta) \quad . \tag{99 – 2}$$

Matrix $[\mathcal{C}_1]$ is an $(N \times N)$ diagonal positive-definite real matrix such that

$$[\mathcal{C}_1]_{\alpha\beta} = \frac{c_0^2}{\rho_0} [\,\mathcal{M}\,]_{\alpha\beta} = \frac{c_0^2}{\rho_0} \mu_\alpha\, \delta_{\alpha\beta} \quad . \tag{100}$$

Generalized forces $\mathcal{F}_S(\omega)$ and $\mathcal{F}_Q(\omega)$ are $\mathbb{C}^{N_S}$- and $\mathbb{C}^N$-vectors respectively, such that

$$\mathcal{F}_S(\omega)_\alpha = f_S(\omega\,; \mathbf{u}_\alpha) \quad , \tag{101}$$

$$\mathcal{F}_Q(\omega)_\alpha = f_Q(\omega; \varphi_\alpha) \quad . \tag{102}$$

6.5. Construction of the reduced model using the finite element approximation

Step 1. This step concerns the finite element calculation of the structural modes of master structure Ω_S in vacuo (see Section III.9). The finite element discretization $\mathbf{U}_\alpha$ of structural mode $\mathbf{u}_\alpha$ is a solution of the $(n_s \times n_S)$ generalized symmetric real matrix eigenvalue problem

$$[K^S]\,\mathbf{U} = \lambda^S\,[M^S]\,\mathbf{U} \quad , \tag{103}$$

which gives $\{\mu_\alpha^S, \lambda_\alpha^S, \mathbf{U}_\alpha\}_{\alpha=1,\ldots,N_S}$ and consequently which allows the construction of diagonal matrices $[\mathcal{M}^S]$ and $[\mathcal{K}^S]$. We introduce the $(n_S \times N_S)$ real matrix of the N_S structural modes $\mathbf{U}_\alpha$,

$$[\,U\,] = [\,\mathbf{U}_1 \ldots \mathbf{U}_\alpha \ldots \mathbf{U}_{N_S}\,] \quad . \tag{104}$$

Solution $\lambda^S = 0$ corresponds to the six rigid body modes $\mathbf{U}_1, \ldots, \mathbf{U}_6$.

Step 2. This step corresponds to the finite element calculation of the acoustic modes of internal acoustic cavity Ω with a rigid wall (see Section XI.8). The finite element approximation $\mathbf{\Phi}_\alpha$ of acoustic mode φ_α is a solution of the $(n \times n)$ generalized symmetric real matrix eigenvalue problem

$$[\,K\,]\,\mathbf{\Phi} = \lambda\,[\,M\,]\,\mathbf{\Phi} \quad , \tag{105}$$

which gives $\{\mu_\alpha, \lambda_\alpha, \mathbf{\Phi}_\alpha\}_{\alpha=1,\ldots,N}$ and consequently which allows the construction of diagonal matrices $[\mathcal{M}]$, $[\mathcal{K}]$, $[\mathcal{D}_T(\omega)]$ and $[\mathcal{C}_1]$. We introduce the $(n\times N)$ real matrix of the N acoustic modes $\mathbf{\Phi}_\alpha$,

$$[\,\Phi\,] = [\,\mathbf{\Phi}_1 \ldots \mathbf{\Phi}_\alpha \ldots \mathbf{\Phi}_N\,] \quad . \tag{106}$$

Step 3. The third step is the finite element calculation of the nondiagonal matrices of the reduced model. For matrix $[\mathcal{D}^S(\omega)]$, the finite element discretization leads to

$$[\mathcal{D}^S(\omega)] \simeq [\,U\,]^T\,[D^S(\omega)]\,[\,U\,] \quad . \tag{107}$$

For matrix $[\,\mathcal{J}\,]$, we obtain

$$[\,\mathcal{J}\,] \simeq (\mathbf{\Pi}_2^T[\,U\,])^T\,(\mathbf{\Pi}_2^T[\,U\,]) \quad , \tag{108}$$

which uses only the values of the structural modes at the nodes belonging to boundary $\Gamma \cup \Gamma_Z$. For matrix $[\mathcal{S}_z(\omega)]$, we have

$$[\mathcal{S}_z(\omega)] \simeq [\,\Phi\,]^T\,[S_z(\omega)]\,[\,\Phi\,] \quad , \tag{109}$$

which uses only the values of the acoustic modes at the nodes belonging to boundary Γ_Z, and for matrix $[\mathcal{C}(\omega)]$ we have

$$[\mathcal{C}(\omega)] \simeq [\,U\,]^T\,[C_0(\omega)]\,[\,\Phi\,] - \rho_0\,\kappa(\omega)\,(\mathbf{\Pi}_2^T[\,U\,])^T\,(\mathbf{\Pi}_1(\omega)^T[\,\Phi\,]) \quad , \tag{110}$$

which uses only the values of the structural modes and acoustic modes at the nodes belonging to boundary $\Gamma \cup \Gamma_Z$. For matrix $[\mathcal{A}^E(\omega/c_E)]$, the finite element discretization yields

$$[\mathcal{A}^E(\omega/c_E)] \simeq -\rho_E\,[\mathcal{B}_{\Gamma_E}(\omega/c_E)] \quad , \tag{111}$$

in which the $(N_S\times N_S)$ symmetric complex matrix $[\mathcal{B}_{\Gamma_E}(\omega/c_E)]$ is given by

$$[\mathcal{B}_{\Gamma_E}(\omega/c_E)] = [\,U_{\Gamma_E}]^T\,[B_{\Gamma_E}(\omega/c_E)]\,[\,U_{\Gamma_E}] \quad , \tag{112}$$

with $[\,U_{\Gamma_E}]$ the $(n_E\times N_S)$ real matrix such that

$$[\,U_{\Gamma_E}] = [\,\Theta\,]\,[\,U\,] \quad . \tag{113}$$

The $(n_E\times n_E)$ symmetric dense complex matrix $[B_{\Gamma_E}(\omega/c_E)]$ is defined by Eq. (XII.164) and the $(n_E\times n_S)$ real matrix $[\,\Theta\,]$ is defined by Eq. (69).

The right-hand side of Eq. (112) is calculated as presented in Section XII.12.3.

7. FRF Calculation in the LF Range

In this section, for $\omega \in \mathbb{B}_{\rm LF}$ $(\omega \neq 0)$, we construct the projection of the frequency response function using the symmetric reduced model presented in Section 6.

7.1. Definition of the FRF in the LF range

For all $\omega \in \mathbb{B}_{\rm LF}$, since $\mathbf{M}$ is an invertible operator, we eliminate field p' in Eq. (48) using the third equation. For all $\omega \in \mathbb{B}_{\rm LF}$, the resulting operator is invertible and gives the operator-valued frequency response function $\mathbf{T}(\omega)$ such that [17]

$$\begin{bmatrix} \mathbf{u} \\ \varphi \end{bmatrix} = \mathbf{T}(\omega) \begin{bmatrix} \mathbf{f}_{\rm S}(\omega) \\ \omega^2\, \mathbf{f}_{\rm Q}(\omega) \end{bmatrix} \quad . \tag{114}$$

Once $\varphi(\omega)$ is known, pressure field $p(\omega)$ is calculated using Eq. (62).

7.2. Response to an excitation vector and projection of the FRF in the LF range

The projection of FRF $\mathbf{T}(\omega)$ is then obtained by eliminating $\mathbf{r}$ in the symmetric reduced model defined by Eq. (86). We then have

$$[\widetilde{\mathcal{A}}(\omega)] \begin{bmatrix} \mathbf{q}^S \\ \mathbf{q} \end{bmatrix} = \begin{bmatrix} \mathcal{F}_{\rm S}(\omega) \\ \omega^2\, \mathcal{F}_{\rm Q}(\omega) \end{bmatrix} \quad , \tag{115}$$

in which complex matrix $[\widetilde{\mathcal{A}}(\omega)]$ is symmetric and is defined by

$$\begin{bmatrix} [\mathcal{A}^S(\omega)] - \omega^2 [\mathcal{A}^E(\omega/c_E)] + \kappa(\omega)[\mathcal{J}] & -\omega^2 [\mathcal{C}(\omega)] \\ -\omega^2\, [\mathcal{C}(\omega)]^T & i\omega^3 [\mathcal{D}_T(\omega)] + \omega^2 [\mathcal{K}] + \omega^2 [\mathbb{S}_Z(\omega)] - \omega^4 [\mathcal{M}] \end{bmatrix} . \tag{116}$$

For all $\omega \in \mathbb{B}_{\rm LF}$ $(\omega \neq 0)$, matrix $[\widetilde{\mathcal{A}}(\omega)]$ is invertible and consequently,

$$\begin{bmatrix} \mathbf{q}^S \\ \mathbf{q} \end{bmatrix} = [\mathcal{T}(\omega)] \begin{bmatrix} \mathcal{F}_{\rm S}(\omega) \\ \omega^2\, \mathcal{F}_{\rm Q}(\omega) \end{bmatrix} \quad , \tag{117}$$

[17] For all $\omega \in \mathbb{B}_{\rm LF}$ $(\omega \neq 0)$, the operator resulting from the elimination of p' in Eq. (48) is a continuous and invertible operator from $\mathcal{C}^c \times \mathcal{E}_\varphi^c$ into $\mathcal{C}^{c\,\prime} \times \mathcal{E}_\varphi^{c\,\prime}$ and consequently, $\mathbf{T}(\omega) = \mathbf{A}(\omega)^{-1}$ is a continuous operator from $\mathcal{C}^{c\,\prime} \times \mathcal{E}_\varphi^{c\,\prime}$ into $\mathcal{C}^c \times \mathcal{E}_\varphi^c$.

in which $[\mathcal{T}(\omega)]$ is an $((N_S+N)\times(N_S+N))$ symmetric complex matrix such that

$$[\mathcal{T}(\omega)] = [\widetilde{\mathcal{A}}(\omega)]^{-1} \quad , \quad [\mathcal{T}(\omega)]^T = [\mathcal{T}(\omega)] \quad . \tag{118}$$

Introducing the appropriate partition of matrix $[\mathcal{T}(\omega)]$, Eq. (117) is rewritten as

$$\begin{bmatrix} \mathbf{q}^S \\ \mathbf{q} \end{bmatrix} = \begin{bmatrix} [\mathcal{T}_{11}(\omega)] & [\mathcal{T}_{12}(\omega)] \\ [\mathcal{T}_{12}(\omega)]^T & [\mathcal{T}_{22}(\omega)] \end{bmatrix} \begin{bmatrix} \mathcal{F}_{\mathrm{S}}(\omega) \\ \omega^2\, \mathcal{F}_{\mathrm{Q}}(\omega) \end{bmatrix} \quad . \tag{119}$$

Substituting Eq. (119) into Eqs. (80) and (82) yields the response of the structural-acoustic master system,

$$\mathbf{u}^{N_S}(\mathbf{x},\omega) = \sum_{\alpha=1}^{N_S} \{\, [\mathcal{T}_{11}(\omega)]\, \mathcal{F}_{\mathrm{S}}(\omega) + \omega^2 [\mathcal{T}_{12}(\omega)]\, \mathcal{F}_{\mathrm{Q}}(\omega)\, \}_\alpha\ \mathbf{u}_\alpha(\mathbf{x}) \quad , \tag{120}$$

$$\varphi^N(\mathbf{x},\omega) = \sum_{\beta=1}^{N} \{\, [\mathcal{T}_{12}(\omega)]^T\, \mathcal{F}_{\mathrm{S}}(\omega) + \omega^2 [\mathcal{T}_{22}(\omega)]\, \mathcal{F}_{\mathrm{Q}}(\omega)\, \}_\beta\ \varphi_\beta(\mathbf{x}) \quad . \tag{121}$$

Consequently, the projection $\mathbf{T}^{N_S,N}(\omega)$ of operator-valued frequency response function $\mathbf{T}(\omega)$ is such that

$$\begin{bmatrix} \mathbf{u}^{N_S} \\ \varphi^N \end{bmatrix} = \mathbf{T}^{N_S,N}(\omega) \begin{bmatrix} \mathbf{f}_{\mathrm{S}}(\omega) \\ \omega^2\, \mathbf{f}_{\mathrm{Q}}(\omega) \end{bmatrix} \quad . \tag{122}$$

We introduce the following partition of operator $\mathbf{T}^{N_S,N}(\omega)$,

$$\mathbf{T}^{N_S,N}(\omega) = \begin{bmatrix} \mathbf{T}_{11}^{N_S,N}(\omega) & \mathbf{T}_{12}^{N_S,N}(\omega) \\ \mathbf{T}_{21}^{N_S,N}(\omega) & \mathbf{T}_{22}^{N_S,N}(\omega) \end{bmatrix} \quad , \tag{123}$$

which is such that

$$\mathbf{T}_{11}^{N_S,N}(\omega)\, \mathbf{f}_{\mathrm{S}}(\omega) = \sum_{\alpha=1}^{N_S} \sum_{\alpha'=1}^{N_S} [\mathcal{T}_{11}(\omega)]_{\alpha\alpha'}\, f_{\mathrm{S}}(\omega\,; \mathbf{u}_{\alpha'})\, \mathbf{u}_\alpha \quad , \tag{124}$$

$$\mathbf{T}_{12}^{N_S,N}(\omega)\, \mathbf{f}_{\mathrm{Q}}(\omega) = \sum_{\alpha=1}^{N_S} \sum_{\beta=1}^{N} [\mathcal{T}_{12}(\omega)]_{\alpha\beta}\, f_{\mathrm{Q}}(\omega; \varphi_\beta)\, \mathbf{u}_\alpha \quad , \tag{125}$$

$$\mathbf{T}_{21}^{N_S,N}(\omega)\, \mathbf{f}_{\mathrm{S}}(\omega) = \sum_{\alpha=1}^{N_S} \sum_{\beta=1}^{N} [\mathcal{T}_{12}(\omega)]_{\alpha\beta}\, f_{\mathrm{S}}(\omega\,; \mathbf{u}_\alpha)\, \varphi_\beta \quad , \tag{126}$$

$$\mathbf{T}_{22}^{N_S,N}(\omega)\, \mathbf{f}_{\mathrm{Q}}(\omega) = \sum_{\beta=1}^{N} \sum_{\beta'=1}^{N} [\mathcal{T}_{22}(\omega)]_{\beta\beta'}\, f_{\mathrm{Q}}(\omega; \varphi_{\beta'})\, \varphi_\beta \quad . \tag{127}$$

7.3. Finite element discretization of the projection of the FRF

In order to calculate $[\mathcal{T}(\omega)]$, the effective construction of matrix $[\widetilde{\mathcal{A}}(\omega)]$ is performed using the expression of the submatrices introduced in Section 6.5. The finite element discretization of Eq. (122) is then written as

$$\begin{bmatrix} \mathbf{U}^{N_S} \\ \mathbf{\Phi}^N \end{bmatrix} = \begin{bmatrix} [T_{11}^{N_S,N}(\omega)] & [T_{12}^{N_S,N}(\omega)] \\ [T_{12}^{N_S,N}(\omega)]^T & [T_{22}^{N_S,N}(\omega)] \end{bmatrix} \begin{bmatrix} \mathbf{F}_S(\omega) \\ \omega^2\, \mathbf{F}_Q(\omega) \end{bmatrix} \quad , \tag{128}$$

in which $\mathbf{U}^{N_S} = (U_1^{N_S}, \ldots, U_{n_S}^{N_S})$ is the complex vector of the DOFs which are the values of $\mathbf{u}^{N_S}$ at the nodes of the finite element mesh of domain Ω_S, $\mathbf{\Phi}^N = (\Phi_1^N, \ldots, \Phi_n^N)$ is the complex vector of the DOFs which are the values of φ^N at the nodes of the finite element mesh of domain Ω and, vectors $\mathbf{F}_S(\omega) \in \mathbb{C}^{n_S}$ and $\mathbf{F}_Q(\omega) \in \mathbb{C}^n$ are defined in Section 5. Matrix $[T_{11}^{N_S,N}(\omega)]$ is an $(n_S \times n_S)$ symmetric complex matrix, $[T_{22}^{N_S,N}(\omega)]$ is an $(n \times n)$ symmetric complex matrix and $[T_{12}^{N_S,N}(\omega)]$ is an $(n_S \times n)$ complex matrix such that,

$$[T_{11}^{N_S,N}(\omega)] = \sum_{\alpha=1}^{N_S} \sum_{\alpha'=1}^{N_S} [\mathcal{T}_{11}(\omega)]_{\alpha\alpha'}\, \mathbf{U}_\alpha\, \mathbf{U}_{\alpha'}^T \quad , \tag{129}$$

$$[T_{12}^{N_S,N}(\omega)] = \sum_{\alpha=1}^{N_S} \sum_{\beta=1}^{N} [\mathcal{T}_{12}(\omega)]_{\alpha\beta}\, \mathbf{U}_\alpha\, \mathbf{\Phi}_\beta^T \quad , \tag{130}$$

$$[T_{22}^{N_S,N}(\omega)] = \sum_{\beta=1}^{N} \sum_{\beta'=1}^{N} [\mathcal{T}_{22}(\omega)]_{\beta\beta'}\, \mathbf{\Phi}_\beta\, \mathbf{\Phi}_{\beta'}^T \quad . \tag{131}$$

Calculation of the pressure field $p(\mathbf{x}, \omega)$ inside Ω. The complex vector $\mathbf{P}^{N_S,N}$ of the values of pressure field $p(\mathbf{x}, \omega)$ at the nodes of the finite element mesh of domain Ω is calculated using Eq. (73),

$$\frac{c_0^2}{\rho_0}[M]\mathbf{P}^{N_S,N} = \omega^2\left\{c_0^2[M]\mathbf{\Phi}^N - \kappa(\omega)\rho_0\mathbf{\Pi}_1(\omega)^T\mathbf{\Phi}^N\,\mathbf{L}\right\}$$
$$- \kappa(\omega)\mathbf{\Pi}_2^T\,\mathbf{U}^{N_S}\,\mathbf{L} + \kappa(\omega)\pi_Q(\omega)\,\mathbf{L} \quad , \tag{132}$$

in which $\mathbf{U}^{N_S}$ and $\mathbf{\Phi}^N$ are given by Eqs. (128) to (131). As explained at the end of Section 5, an alternative to Eq. (132) can be used in order to avoid having to solve a linear matrix equation of order n.

Calculation of the pressure field on boundary Γ_E and of the resultant pressure at any point of external domain Ω_E. Finite element approximation $\mathbf{P}_{\Gamma_E}(\omega)$ of pressure field $p_E|_{\Gamma_E}(\omega)$ on Γ_E is calculated by Eq. (XII.173) in which $\mathbf{V}(\omega) = [U_{\Gamma_E}]\,\mathbf{q}^S(\omega)$ where $[U_{\Gamma_E}]$ is defined by Eq. (113). At any point of external domain Ω_E, the finite element discretization of resultant pressure $p_E(\mathbf{x}, \omega)$ is given by Eq. (XII.174) in which $\mathbf{V}(\omega)$ is defined above.

7.4. Frequency-by-frequency construction of the FRF in the LF range

The numerical procedure for constructing an approximation of $\omega \mapsto \mathbf{T}(\omega)$ on low-frequency band $\mathbb{B}_{\rm LF}$ is based on the use of projection $\mathbf{T}^{N_S,N}(\omega)$ of $\mathbf{T}(\omega)$ introduced in Sections 7.2 and 7.3. In this case, no quasi-static correction terms are introduced. Convergence must be controlled by increasing the number N_S of structural modes of the master structure in vacuo and the number N of the acoustic modes of the internal acoustic cavity with a rigid wall. This procedure requires the construction of matrix $[\mathcal{T}(\omega)]$ (see Eqs. (117) and (118)). In the low-frequency range, N_S and N are small. In the context of a finite element discretization with n_S DOFs for the master structure and n DOFs for the internal fluid, we therefore have $N_S \ll n_S$ and $N \ll n$. Consequently, a frequency-by-frequency construction can be used and matrix $[\widetilde{\mathcal{A}}(\omega)]$ defined by Eq. (116) is numerically inverted frequency by frequency in $\mathbb{B}_{\rm LF}$ to construct $[\mathcal{T}(\omega)]$.

7.5. Approximation of matrix $[\mathcal{A}^E(\omega/c_E)]$ using spline functions

Since matrix $[\mathcal{A}^E(\omega/c_E)]$ appears in Eq. (116), it has to be calculated for a large number of sampling frequency points ω in frequency band $\mathbb{B}_{\rm LF}$. This generalized matrix requires the calculation of matrix $[A^E(\omega/c_E)]$, related to matrix $[B_{\Gamma_E}(\omega/c_E)]$ by Eq. (68). This last matrix is calculated by the method presented in Section XII.12.2, devoted to the construction of the acoustic impedance boundary operator related to the external acoustic fluid. Using Eq. (XII.22), matrix $[\mathcal{A}^E(\omega/c_E)]$ can be written as

$$-\omega^2[\mathcal{A}^E(\omega/c_E)] = -\omega^2[\mathcal{M}^E(\omega/c_E)] + i\omega[\mathcal{D}^E(\omega/c_E)] \quad , \tag{133}$$

in which

$$[\mathcal{M}^E(\omega/c_E)] = \Re e[\mathcal{A}^E(\omega/c_E)] \quad , \quad [\mathcal{D}^E(\omega/c_E)] = -\omega\,\Im m[\mathcal{A}^E(\omega/c_E)] \quad . \tag{134}$$

Function $\omega \mapsto [\mathcal{A}^E(\omega/c_E)]$ has a nonresonant behavior and varies slowly on frequency band $\mathbb{B}_{\rm LF}$. Consequently, this function can be approximated using spline functions to reduce the computational effort of the frequency-by-frequency method. For instance, Fig. 3 shows an example of function $\omega \mapsto \mathrm{tr}[\mathcal{M}^E(\omega/c_E)]$ ("added mass" expressed in terms of the generalized coordinates) and function $\omega \mapsto \mathrm{tr}[\mathcal{D}^E(\omega/c_E)]$ ("damping" due to the radiation at infinity expressed in terms of the generalized coordinates).

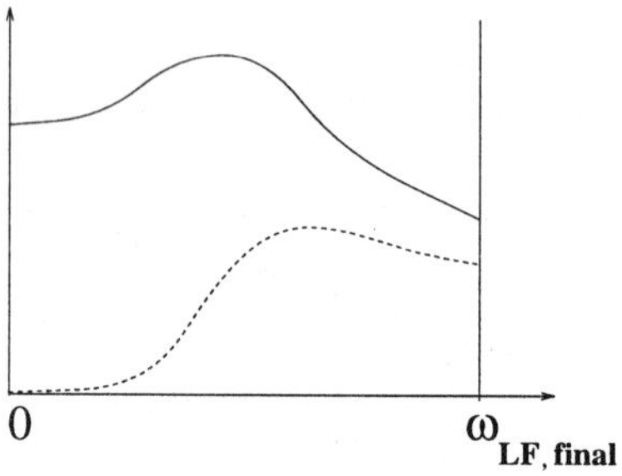

Fig. 3. Example of slowly varying "added mass" (——) and "damping" (- - -) functions for the external acoustic fluid

8. Location of the Resonant Frequencies of the Coupled System

In the case of a weakly dissipative coupled system, due to the resonances of the master structure coupled with the internal acoustic cavity, the structural-acoustic master system exhibits peaks (resonant behavior) in LF range $\mathbb{B}_{\mathrm{LF}}$. The frequency-by-frequency method presented in Section 7.4 allows the calculation of FRF $\omega \mapsto \mathbf{T}^{N_S,N}(\omega)$ defined on $\mathbb{B}_{\mathrm{LF}}$. Once the graph of the FRF is known on $\mathbb{B}_{\mathrm{LF}}$, the local maxima (the peaks) define the resonances and their frequency locations define the resonant frequencies denoted as $\omega_\nu^{\mathrm{reson}}$, $\nu = 1, 2, \ldots$ (see Fig. 4).

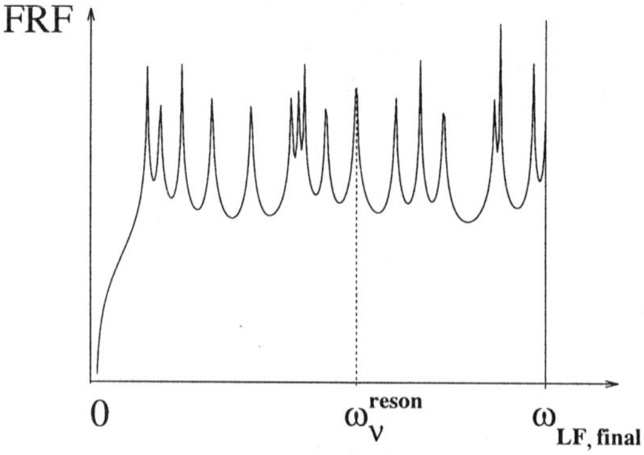

Fig. 4. FRF in the LF range

For a weakly dissipative coupled system, resonant frequencies $\omega_\nu^{\text{reson}}$, $\nu = 1, 2, \ldots$ can be estimated by

$$\omega_\nu^{\text{reson}} \simeq \omega_\nu^{\text{conserv}} \quad , \quad \nu = 1, 2, \ldots \quad , \tag{135}$$

in which $\omega_\nu^{\text{conserv}}$, $\nu = 1, 2, \ldots$, are the solutions of the conservative system deduced from Eq. (48) by considering the homogeneous equation and cancelling the dissipative terms, i.e. considering the real part of operator $\mathbf{A}(\omega)$,

$$\Re e\left\{\mathbf{A}(\omega_\nu^{\text{conserv}})\right\} \begin{bmatrix} \mathbf{u}_\nu \\ \varphi_\nu \\ p'_\nu \end{bmatrix} = \begin{bmatrix} 0 \\ 0 \\ 0 \end{bmatrix} \quad , \quad \{\mathbf{u}_\nu, \varphi_\nu, p'_\nu\} \in \mathcal{C}^c \times \mathcal{E}_\varphi^c \times \mathcal{E}_{p'}^c \quad . \tag{136}$$

It should be noted that $\mathbf{u}_\nu$ is a field with values in $\mathbb{R}^3$ and φ_ν and p'_ν are fields with values in $\mathbb{R}$. As explained in Section 7 for the construction of the FRF, instead of solving Eq. (136), the reduced symmetric matrix model $[\widetilde{\mathcal{A}}(\omega)]$ defined by Eq. (116) is used. We have then to calculate the solutions $\omega_\nu^{\text{conserv}}$ of the following matrix equation

$$\Re e\left\{[\widetilde{\mathcal{A}}(\omega_\nu^{\text{conserv}})]\right\} \begin{bmatrix} \mathbf{q}_\nu^S \\ \mathbf{q}_\nu \end{bmatrix} = \begin{bmatrix} 0 \\ 0 \end{bmatrix} \quad , \quad \mathbf{q}_\nu^S \in \mathbb{R}^{N_S} \quad , \quad \mathbf{q}_\nu \in \mathbb{R}^N \quad . \tag{137}$$

The $\omega_\nu^{\text{conserv}}$ terms are then the zeros, in $\mathbb{B}_{\text{LF}}$, of function $\omega \mapsto \det[\widetilde{\mathcal{A}}(\omega)]$. For the mathematical properties related to the problem defined by Eq. (137), we refer the reader to Lancaster, 1966 and Markus, 1986.

9. Structural-Acoustic Modes of the Master Structure Coupled with an Internal Acoustic Fluid

9.1. Definition of the structural-acoustic modes

Consider Eq. (136) without an external acoustic fluid, which we rewrite as

$$\begin{bmatrix} \mathbf{K}^S + \Re e\{\kappa(\omega)\}\mathbf{J} & 0 & 0 \\ 0 & 0 & 0 \\ 0 & 0 & \frac{1}{\rho_0^2}\mathbf{M} \end{bmatrix} \begin{bmatrix} \mathbf{u} \\ \varphi \\ p' \end{bmatrix}$$

$$= \omega^2 \begin{bmatrix} \mathbf{M}^S & \mathbf{C}_0 + \Re e\{\mathbf{C}_\kappa(\omega)\} & 0 \\ {}^t(\mathbf{C}_0 + \Re e\{\mathbf{C}_\kappa(\omega)\}) & -\mathbf{K} - \Re e\{\mathbf{S}_z(\omega)\} & \frac{1}{c_0^2}\mathbf{C}_1 \\ 0 & \frac{1}{c_0^2}{}^t\mathbf{C}_1 & 0 \end{bmatrix} \begin{bmatrix} \mathbf{u} \\ \varphi \\ p' \end{bmatrix} \quad . \tag{138}$$

Eq. (138) can be rewritten as $\mathfrak{M}(\omega)\,\mathbf{v} = \omega^2 \mathfrak{N}(\omega)\,\mathbf{v}$ and does not appear as a classical spectral problem because $\mathfrak{M}(\omega)$ and $\mathfrak{N}(\omega)$ depend on ω. By definition, the structural-acoustic modes are such that $\mathfrak{M}(0)\,\mathbf{v} = \omega^2 \mathfrak{N}(0)\,\mathbf{v}$. Since $\Re e\{\mathbf{S}_Z(0)\} = 0$ (see footnote 4), this classical spectral problem can be rewritten as

$$\begin{bmatrix} \mathbf{K}^S + \kappa(0)\mathbf{J} & \mathbf{0} & \mathbf{0} \\ \mathbf{0} & \mathbf{0} & \mathbf{0} \\ \mathbf{0} & \mathbf{0} & \frac{1}{\rho_0^2}\mathbf{M} \end{bmatrix} \begin{bmatrix} \mathbf{u} \\ \varphi \\ p' \end{bmatrix} = \omega^2 \begin{bmatrix} \mathbf{M}^S & \mathbf{C}_0 + \mathbf{C}_\kappa(0) & \mathbf{0} \\ {}^t(\mathbf{C}_0 + \mathbf{C}_\kappa(0)) & -\mathbf{K} & \frac{1}{c_0^2}\mathbf{C}_1 \\ \mathbf{0} & \frac{1}{c_0^2}{}^t\mathbf{C}_1 & \mathbf{0} \end{bmatrix} \begin{bmatrix} \mathbf{u} \\ \varphi \\ p' \end{bmatrix} , \quad (139)$$

in which $\kappa(0)$ is deduced from Eqs. (6-1) and (X.25),

$$\kappa(0) = \frac{\rho_0\, c_0^2}{|\Omega|} \times \left\{ 1 - \frac{\rho_0\, c_0^2}{|\Omega|} \int_{\Gamma_Z} \frac{ds(\mathbf{x})}{\alpha(\mathbf{x})} \right\}^{-1} , \quad (140)$$

and $\mathbf{C}_\kappa(0)$ is deduced from Eq. (54-4) by using Eq. (3) with the property defined by Eq. (X.25),

$$< \mathbf{C}_\kappa(0)\,\varphi \,, \delta\mathbf{u} >= -\rho_0\, \kappa(0)\, \pi_2(\overline{\delta\mathbf{u}}) \int_{\Gamma_Z} \frac{\varphi(\mathbf{x})}{\alpha(\mathbf{x})}\, ds(\mathbf{x}) \quad . \quad (141)$$

From Eq. (6-2), if parameter $\alpha_{\min} \to +\infty$, then $\kappa(0) \to \rho_0 c_0^2/|\Omega|$, $1/\alpha(\mathbf{x}) \to 0$ and then $\mathbf{C}_\kappa(0) \to 0$. Associated real-valued pressure field p is deduced from Eq. (62) by applying the methodology leading to Eq. (139) and can then be written as

$$\frac{c_0^2}{\rho_0}\,\mathbf{M}p = \omega^2 \rho_0 \left\{ {}^t\mathbf{C}_1 \varphi - \kappa(0) \left\{ \int_{\Gamma_Z} \frac{\varphi(\mathbf{x})}{\alpha(\mathbf{x})}\, ds(\mathbf{x}) \right\} \mathbf{l} \right\} - \kappa(0)\, \pi_2(\mathbf{u})\, \mathbf{l} \quad . \quad (142)$$

9.2. Reduced spectral problem and calculation of the structural-acoustic modes

From Section 6.4, we deduce that the symmetric reduced matrix model of the spectral problem defined by Eq. (139) is written as

$$\begin{bmatrix} [\mathcal{K}^S] + \kappa(0)[\mathcal{J}] & [0] & [0] \\ [0] & [0] & [0] \\ [0] & [0] & \frac{1}{\rho_0^2}[\mathcal{M}] \end{bmatrix} \begin{bmatrix} \mathbf{q}^S \\ \mathbf{q} \\ \mathbf{r} \end{bmatrix}$$

$$= \omega^2 \begin{bmatrix} [\mathcal{M}^S] & [\mathcal{C}(0)] & [0] \\ [\mathcal{C}(0)]^T & -[\mathcal{K}] & \frac{1}{c_0^2}[\mathcal{C}_1] \\ [0] & \frac{1}{c_0^2}[\mathcal{C}_1]^T & [0] \end{bmatrix} \begin{bmatrix} \mathbf{q}^S \\ \mathbf{q} \\ \mathbf{r} \end{bmatrix} , \quad (143)$$

in which

$$[\mathcal{C}(0)] = [\mathcal{C}_0] + [\mathcal{C}_\kappa(0)] \quad . \tag{144}$$

The second row of Eq. (143) yields

$$\mathbf{q} = [\mathcal{K}]^{-1}[\mathcal{C}(0)]^T\mathbf{q}^S + \frac{1}{c_0^2}[\mathcal{K}]^{-1}[\mathcal{C}_1]\,\mathbf{r} \quad , \tag{145}$$

which can be rewritten as

$$q_\alpha = \frac{1}{\mu_\alpha\lambda_\alpha}\left\{[\mathcal{C}(0)]^T\mathbf{q}^S\right\}_\alpha + \frac{1}{\rho_0\lambda_\alpha}r_\alpha \quad . \tag{146}$$

Eliminating $\mathbf{q}$ in Eq.(143) then yields the symmetric generalized eigenvalue problem with the positive-definite "mass matrix",

$$\begin{bmatrix} [\mathcal{K}_1] & [0] \\ [0] & [\mathcal{K}_2] \end{bmatrix}\begin{bmatrix} \mathbf{q}^S \\ \mathbf{r} \end{bmatrix} = \omega^2\begin{bmatrix} [\mathcal{M}_1] & [\mathcal{M}_c] \\ [\mathcal{M}_c]^T & [\mathcal{M}_2] \end{bmatrix}\begin{bmatrix} \mathbf{q}^S \\ \mathbf{r} \end{bmatrix} \quad , \tag{147}$$

in which

$$[\mathcal{K}_1]_{\alpha\beta} = \mu_\alpha^S\,\lambda_\alpha^S\,\delta_{\alpha\beta} + \kappa(0)\,[\mathcal{J}]_{\alpha\beta} \quad , \tag{148}$$

$$[\mathcal{K}_2]_{\alpha\beta} = \frac{\mu_\alpha}{\rho_0^2}\,\delta_{\alpha\beta} \quad , \tag{149}$$

$$[\mathcal{M}_1]_{\alpha\beta} = \mu_\alpha^S\,\delta_{\alpha\beta} + \sum_{\gamma=1}^{N}\frac{1}{\mu_\gamma\lambda_\gamma}[\mathcal{C}(0)]_{\alpha\gamma}\,[\mathcal{C}(0)]_{\beta\gamma} \quad , \tag{150}$$

$$[\mathcal{M}_2]_{\alpha\beta} = \frac{1}{\rho_0^2}\frac{\mu_\alpha}{\lambda_\alpha}\,\delta_{\alpha\beta} \quad , \tag{151}$$

$$[\mathcal{M}_c]_{\alpha\beta} = \frac{1}{\rho_0\lambda_\alpha}[\mathcal{C}(0)]_{\alpha\beta} \quad . \tag{152}$$

The structural-acoustic eigenfrequencies are a sequence of real numbers $\{\omega_j\}_j$ such that

$$0 = \omega_1 = \ldots = \omega_6 < \omega_7 \leq \omega_8 \leq \ldots \quad . \tag{153}$$

It should be noted that the solution $\omega = 0$ is associated with the six rigid body modes of the master structure. The approximation of the structural-acoustic mode associated with ω_j is given by

$$\mathbf{u}_j^{N_S}(\mathbf{x}) = \sum_{\alpha=1}^{N_S} q_{j,\alpha}^S\,\mathbf{u}_\alpha(\mathbf{x}) \quad , \tag{154}$$

$$p'_{N,j}(\mathbf{x}) = \sum_{\alpha=1}^{N} r_{j,\alpha}\,\varphi_\alpha(\mathbf{x}) \quad , \tag{155}$$

in which $\{\mathbf{q}_j^S, \mathbf{r}_j\}$ is the eigenvector in $\mathbb{R}^{N_S} \times \mathbb{R}^N$ associated with the eigenfrequency ω_j. Introducing the vectors $\mathbf{q}_j^S = (q_{j,1}^S, \ldots q_{j,N_S}^S)$ and $\mathbf{r}_j = (r_{j,1}, \ldots r_{j,N})$, we have

$$\begin{bmatrix} [\mathcal{K}_1] & [0] \\ [0] & [\mathcal{K}_2] \end{bmatrix} \begin{bmatrix} \mathbf{q}_j^S \\ \mathbf{r}_j \end{bmatrix} = \omega_j^2 \begin{bmatrix} [\mathcal{M}_1] & [\mathcal{M}_c] \\ [\mathcal{M}_c]^T & [\mathcal{M}_2] \end{bmatrix} \begin{bmatrix} \mathbf{q}_j^S \\ \mathbf{r}_j \end{bmatrix} \quad . \tag{156}$$

The corresponding field φ_j^N is given by Eq. (82),

$$\varphi_j^N(\mathbf{x}) = \sum_{\alpha=1}^{N} q_{j,\alpha}\, \varphi_\alpha(\mathbf{x}) \quad , \tag{157}$$

in which $q_{j,\alpha}$ is given by Eq. (146),

$$q_{j,\alpha} = \frac{1}{\mu_\alpha \lambda_\alpha} \left\{ [\mathcal{C}(0)]^T \mathbf{q}_j^S \right\}_\alpha + \frac{1}{\rho_0 \lambda_\alpha} r_{j,\alpha} \quad . \tag{158}$$

The corresponding pressure field $p_j^{N_S,N}(\mathbf{x})$ is then given by Eq. (142),

$$\frac{c_0^2}{\rho_0} \mathbf{M} p_j^{N_S,N} = \omega_j^2 \rho_0 \Big\{ {}^t\mathbf{C}_1 \varphi_j^N - \kappa(0) \Big\{ \int_{\Gamma_Z} \frac{\varphi_j^N(\mathbf{x})}{\alpha(\mathbf{x})}\, ds(\mathbf{x}) \Big\} \mathbf{1} \Big\} - \kappa(0) \pi_2(\mathbf{u}_j^{N_S}) \mathbf{1} . \tag{159}$$

9.3. Construction of the spectral problem using finite element discretization

Let $[\,U\,]$ and $[\,\Phi\,]$ be the matrices of the structural modes and acoustic modes defined by Eqs. (104) and (106). The submatrices in Eq. (147) are calculated using Eqs. (148) to (152) in which matrix $[\,\mathcal{J}\,]$ is calculated by Eq. (108) and matrix $[\mathcal{C}(0)]$ is deduced from Eq. (110),

$$[\mathcal{C}(0)] \simeq [\,U\,]^T [C_0(0)] [\,\Phi\,] - \rho_0\, \kappa(0)\, (\mathbf{\Pi}_2^T [\,U\,])^T\, (\mathbf{\Pi}_1(0)^T [\,\Phi\,]) \quad , \tag{160}$$

in which $\mathbf{\Pi}_1(0)^T$ corresponds to the finite element discretization of the linear form

$$\pi_1(0\,;\varphi) = \int_{\Gamma_Z} \frac{\varphi(\mathbf{x})}{\alpha(\mathbf{x})}\, ds(\mathbf{x}) \quad . \tag{161}$$

The finite element discretizations of Eqs. (154) and (157) allowing calculation of the jth structural-acoustic mode are

$$\mathbf{U}_j^{N_S} = \sum_{\alpha=1}^{N_S} q_{j,\alpha}^S\, \mathbf{U}_\alpha \quad , \tag{162}$$

$$\mathbf{\Phi}_j^N = \sum_{\alpha=1}^{N} q_{j,\alpha}\, \mathbf{\Phi}_\alpha \quad , \tag{163}$$

in which $q_{j,\alpha}$ is calculated by Eq. (158). From Eq. (132), we deduce that finite element approximation $\mathbf{P}_j^{N_S,N}$ of $p_j^{N_S,N}(\mathbf{x})$ defined by Eq. (159) is given by

$$\frac{c_0^2}{\rho_0}[\,M\,]\,\mathbf{P}_j^{N_S,N} = \omega_j^2 \left\{ c_0^2[\,M\,]\mathbf{\Phi}_j^N - \kappa(0)\rho_0 \mathbf{\Pi}_1(0)^T \mathbf{\Phi}_j^N\, \mathbf{L} \right\}$$
$$- \kappa(0)\, \mathbf{\Pi}_2^T \mathbf{U}_j^{N_S}\, \mathbf{L} \quad . \tag{164}$$

As explained at the end of Section 5, an alternative to Eq. (164) can be used in order to avoid having to solve a linear matrix equation of order n.

9.4. Bibliographical comments

Formulations devoted to the calculation of structural-acoustic modes have been extensively covered in the literature. A general overview on this topic can be found in Morand and Ohayon, 1995 (Chapter 8 and Section 9.3 of Chapter 9). A distinction can be made between formulations using a vector-valued field for the internal fluid such as a displacement field (see for instance Bermúdez and Rodriguez, 1994 and, for slender bodies, Ohayon, 1986) and formulations using a scalar-valued field such as the pressure or the displacement potential field for the internal fluid. The latter formulations introduce nonsymmetric eigenvalue problems of the type $(\mathbf{A} - \omega^2\mathbf{C})\mathbf{w} = 0$ with $\mathbf{A}$ and/or $\mathbf{C}$ nonsymmetric, see for instance Gladwell, 1966; Zienkiewicz and Newton, 1969; Craggs, 1971; Zienkiewicz and Bettess, 1978; Daniel, 1980; Sung and Nefske, 1986. Using the velocity potential field for the internal fluid, symmetric formulations of the type $(\mathbf{A} + i\omega\mathbf{B} - \omega^2\mathbf{C})\mathbf{w} = 0$ with $\mathbf{A}$, $\mathbf{B}$ and $\mathbf{C}$ symmetric, have been constructed (see for instance Everstine, 1981). In addition, symmetric formulations of the type $(\mathbf{A} - \omega^2\mathbf{C})\mathbf{w} = 0$ with $\mathbf{A}$ and $\mathbf{C}$ symmetric have been developed (see for instance Irons, 1970; Morand and Ohayon, 1976 and 1979; Ohayon, 1979 and 1984; Ohayon and Valid, 1983 and 1984; Felippa, 1985 and 1988; Kanarachos and Antoniadis, 1988). It should be noted that all these formulations use a scalar-valued field for the internal fluid and do not take into account the variation in volume of the internal acoustic cavity represented by operator $\mathbf{J}$ which is generated by the term $\pi_2(\mathbf{u})$ appearing in the expression of the pressure (see Eqs. (1) to (5)). As explained in Chapters X and XI, this term is taken into account in this chapter and is based on initial work by Ohayon, 1987 and 1989, followed by Morand and Ohayon, 1989, Felippa and Ohayon, 1990; Ohayon and Felippa, 1990;

Kehr-Candille and Ohayon, 1992; Morand and Ohayon, 1992 and 1995 (Chapter 8). Finally, there are alternative formulations using a stiffness coupling (see Ohayon et al., 1987; Sandberg and Göransson, 1988; Morand and Ohayon, 1992 and 1995).

10. Case of a Master Structure Coupled with an External Acoustic Fluid

In this section, we summarize the model, equations and solving methods for a master structure coupled with an external acoustic fluid and without internal acoustic fluid. The following results are deduced directly from Sections 2 to 9 in which the internal acoustic fluid has been eliminated.

10.1. Variational formulation

For all fixed real ω, the variational equation is

$$a^S(\omega; \mathbf{u}, \delta\mathbf{u}) - \omega^2 a^E(\omega/c_E; \mathbf{u}, \delta\mathbf{u}) = f_S(\omega; \delta\mathbf{u}) \quad , \quad \forall \delta\mathbf{u} \in \mathcal{C}^c , \tag{165}$$

in which $a^S(\omega\,;\mathbf{u}\,,\delta\mathbf{u})$ is the dynamic stiffness sesquilinear form of the master structure defined on $\mathcal{C}^c \times \mathcal{C}^c$ by

$$a^S(\omega\,;\mathbf{u}\,,\delta\mathbf{u}) = -\omega^2 m^S(\mathbf{u}\,,\delta\mathbf{u}) + i\omega\, d^S(\omega\,;\mathbf{u}\,,\delta\mathbf{u}) + k^S(\mathbf{u}\,,\delta\mathbf{u}) \quad , \tag{166}$$

where mass, damping and stiffness structural sesquilinear forms m^S, d^S and k^S defined on $\mathcal{C}^c \times \mathcal{C}^c$ are given by Eqs. (24-1), (24-2) and (24-3) respectively. Sesquilinear form $a^E(\omega/c_E; \mathbf{u}\,,\delta\mathbf{u})$ defined on $\mathcal{C}^c \times \mathcal{C}^c$ and due to the external acoustic fluid is written as

$$-\omega^2 a^E(\omega/c_E\,;\, \mathbf{u}\,,\delta\mathbf{u}) = \omega^2 \rho_E \int_{\Gamma_E} (\overline{\delta\mathbf{u}}\cdot \mathbf{n}^S)\ \mathbf{B}_{\Gamma_E}(\omega/c_E)\{\mathbf{u}\cdot \mathbf{n}^S\}\, ds \quad . \tag{167}$$

Antilinear form $f_S(\omega\,;\delta\mathbf{u})$ on $\mathcal{C}^c$ due to external forces is written as

$$f_S(\omega\,;\delta\mathbf{u}) = f(\omega\,;\delta\mathbf{u}) + f_{p_{\text{given}}}(\omega\,;\delta\mathbf{u}) \quad , \tag{168}$$

in which antilinear form $f(\omega\,;\delta\mathbf{u})$ is defined by Eq. (29) and antilinear form $f_{p_{\text{given}}}(\omega\,;\delta\mathbf{u})$ is defined by Eq. (30).

Variational formulation for $\omega \neq 0$. For all fixed real $\omega \neq 0$, the variational formulation is written as follows. Find $\mathbf{u}$ in $\mathcal{C}^c$ such that Eq. (165) is satisfied for all $\delta\mathbf{u}$ in $\mathcal{C}^c$.

Solution for $\omega = 0$. For $\omega = 0$, $\mathbf{u}(0)$ must be searched for in $\mathcal{C}^c_{\text{elas}} = \mathcal{C}^c \setminus \mathcal{C}_{\text{rig}} \subset \mathcal{C}^c$ in which $\mathcal{C}_{\text{rig}}$ is the set of rigid body displacements. The restriction to $\mathcal{C}^c_{\text{elas}}$ of the variational formulation defined by Eq. (165) yields

$$k^S(\mathbf{u}(0), \delta\mathbf{u}) = f_S(0; \delta\mathbf{u}) \quad , \quad \forall \delta\mathbf{u} \in \mathcal{C}^c_{\text{elas}} \quad . \tag{169}$$

Since $k^S(\mathbf{u}(0), \mathbf{u}(0)) > 0$ for all $\mathbf{u}(0) \neq 0$ in $\mathcal{C}^c_{\text{elas}}$, Eq. (169) has a unique solution $\mathbf{u}(0)$ in $\mathcal{C}^c_{\text{elas}}$.

10.2. Linear operator equation

The linear operator equation corresponding to the variational formulation defined by Eq. (165) is

$$\mathbf{A}(\omega)\,\mathbf{u} = \mathbf{f}_S(\omega) \quad , \quad \mathbf{u} \in \mathcal{C}^c \quad , \tag{170}$$

in which $\mathbf{A}(\omega)$ is the operator defined by

$$\mathbf{A}(\omega) = \mathbf{A}^S(\omega) - \omega^2\, \mathbf{A}^E(\omega/c_E) \quad , \tag{171}$$

in which the dynamic stiffness linear operator $\mathbf{A}^S(\omega)$ of the master structure is defined by Eq. (50) and is written as

$$\mathbf{A}^S(\omega) = -\omega^2 \mathbf{M}^S + i\omega\, \mathbf{D}^S(\omega) + \mathbf{K}^S \quad , \tag{172}$$

where $\mathbf{M}^S$, $\mathbf{D}^S(\omega)$ and $\mathbf{K}^S$ are the mass, damping and stiffness operators of the master structure defined by Eqs. (51-2), (51-3) and (51-4) respectively. Linear operator $\mathbf{A}^E(\omega/c_E)$ is defined by Eq. (52) and element $\mathbf{f}_S(\omega)$ is defined by Eq. (59).

Pressure in the external acoustic fluid. Pressure field $p_E|_{\Gamma_E}(\omega)$ on Γ_E due to the external acoustic fluid is given by Eq. (7). At any point $\mathbf{x}$ of unbounded external acoustic fluid Ω_E, resultant pressure field $p_E(\mathbf{x}, \omega)$ is given by Eq. (8).

10.3. Finite element discretization

We consider a finite element mesh of master structure Ω_S. The finite element mesh of surface Γ_E is the trace of the finite element mesh of Ω_S (see Fig. 5).

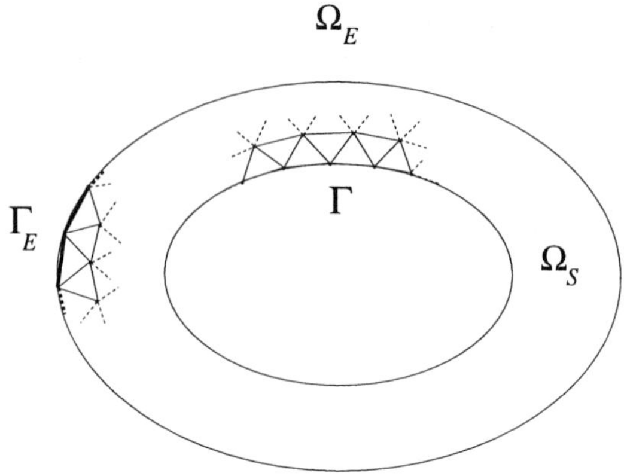

Fig. 5. Finite element mesh of the master structure

We use the finite element method to construct the discretization of the various sesquilinear forms of the problem. For the master structure, we introduce subspace $\mathcal{C}_{n_S} \subset \mathcal{C}^c$ of finite dimension n_S. Let $\mathbf{U} = (U_1, \ldots, U_{n_S})$ be the complex vector of the DOFs which are the values of $\mathbf{u}$ at the nodes of the finite element mesh of domain Ω_S. Since the finite element method uses a real basis for constructing the finite element matrices, the discretization of the variational formulation defined by Eq. (165) yields the symmetric complex matrix equation

$$[A(\omega)]\,\mathbf{U} = \mathbf{F}_S(\omega) \quad . \tag{173}$$

The complex matrix $[A(\omega)]$ is symmetric and is defined by

$$[A(\omega)] = [A^S(\omega)] - \omega^2\,[A^E(\omega/c_E)] \quad , \tag{174}$$

where $[A^S(\omega)]$ is the dynamic stiffness matrix of the master structure which is an $(n_S \times n_S)$ symmetric complex matrix (invertible for all real $\omega \neq 0$ and not invertible for $\omega = 0$ due to the presence of the rigid body modes) written as

$$[A^S(\omega)] = -\omega^2[M^S] + i\omega\,[D^S(\omega)] + [K^S] \quad . \tag{175}$$

Mass, damping and stiffness matrices $[M^S]$, $[D^S(\omega)]$ and $[K^S]$ are $(n_S \times n_S)$ symmetric real matrices. Matrix $[M^S]$ is positive definite and matrices $[D^S(\omega)]$ and $[K^S]$ are positive semidefinite. The finite element discretization of sesquilinear form $a^E(\omega/c_E\,;\,\mathbf{u}\,,\delta\mathbf{u})$ yields the $(n_S \times n_S)$ symmetric complex matrix $[A^E(\omega/c_E)]$ related to the nodal values of $\mathbf{u}$, which is written as

$$[A^E(\omega/c_E)] = -\rho_E\,[\,\Theta\,]^T\,[B_{\Gamma_E}(\omega/c_E)]\,[\,\Theta\,] \quad , \tag{176}$$

where $[B_{\Gamma_E}(\omega/c_E)]$ is the $(n_E \times n_E)$ symmetric dense complex matrix defined by Eq. (XII.164) and related to the nodal values of the normal displacement field on Γ_E. Rectangular matrix $[\Theta]$ is an $(n_E \times n_S)$ sparse real matrix defined by Eq. (69). Equations (167) and (176) show that matrix $[A^E(\omega/c_E)]$ has zero rows and zero columns corresponding to the structural nodes which are not located on interface Γ_E. Finally, the finite element discretization of antilinear form $f_{\mathrm{S}}(\omega\,;\delta\mathbf{u})$ yields the complex vector $\mathbf{F}_{\mathrm{S}}(\omega)$ in $\mathbb{C}^{n_S}$.

10.4. Symmetric reduced matrix model in the LF range

We consider the low-frequency case for which ω belongs to frequency band $\mathbb{B}_{\mathrm{LF}}$ defined by

$$\mathbb{B}_{\mathrm{LF}} =]\,0\,,\,\omega_{\mathrm{LF,final}}\,] \quad . \tag{177}$$

The methodology for constructing the symmetric reduced matrix model consists in using the Ritz-Galerkin projection of the variational formulation on a finite dimension subspace spanned by a set of structural modes of the master structure in vacuo.

1- Finite dimension subspace for the projection. We introduce subspace $\mathcal{C}^c_{N_S}$ of $\mathcal{C}^c$, of dimension $N_S \geq 1$, spanned by the finite family $\{\mathbf{u}_1,\ldots,\mathbf{u}_{N_S}\}$ of structural modes defined in Section 6.1. The projection $\mathbf{u}^{N_S}$ on $\mathcal{C}^c_{N_S}$ of $\mathbf{u}\in\mathcal{C}^c$ is written as

$$\mathbf{u}^{N_S}(\mathbf{x},\omega) = \sum_{\alpha=1}^{N_S} q^S_\alpha(\omega)\,\mathbf{u}_\alpha(\mathbf{x}) \quad , \tag{178}$$

in which $\mathbf{q}^S = (q^S_1,\ldots,q^S_{N_S})$.

2- Symmetric reduced matrix model. Using Eq. (178), the projection of the variational formulation defined by Eq. (165) yields

$$[\mathcal{A}(\omega)]\,\mathbf{q}^S = \boldsymbol{\mathcal{F}}_{\mathrm{S}}(\omega) \quad , \tag{179}$$

in which complex matrix $[\mathcal{A}(\omega)]$ is symmetric and is defined by

$$[\mathcal{A}(\omega)] = [\mathcal{A}^S(\omega)] - \omega^2[\mathcal{A}^E(\omega/c_E)] \quad , \tag{180}$$

in which $[\mathcal{A}^S(\omega)]$ is an $(N_S \times N_S)$ symmetric complex matrix written as

$$[\mathcal{A}^S(\omega)] = -\omega^2[\mathcal{M}^S] + i\omega\,[\mathcal{D}^S(\omega)] + [\mathcal{K}^S] \quad , \tag{181}$$

where $[\mathcal{M}^S]$ is an $(N_S \times N_S)$ symmetric positive-definite real matrix and $[\mathcal{D}^S(\omega)]$ and $[\mathcal{K}^S]$ are $(N_S \times N_S)$ symmetric positive-semidefinite real matrices. Generalized mass and stiffness matrices $[\mathcal{M}^S]$ and $[\mathcal{K}^S]$ are diagonal and are defined by Eqs. (89) and (90) respectively. Generalized damping matrix $[\mathcal{D}^S(\omega)]$ is dense and is defined by Eq. (91). The $(N_S \times N_S)$ symmetric complex matrix $[\mathcal{A}^E(\omega/c_E)]$ is defined by Eq. (92). Generalized force $\mathcal{F}_{\rm S}(\omega)$ is a $\mathbb{C}^{N_S}$-vector defined by Eq. (101).

3- Construction of the reduced model using the finite element approximation. The first step is the finite element calculation of the structural modes of master structure Ω_S in vacuo as explained in Step 1 of Section 6.5. The second step corresponds to calculation of the nondiagonal matrices of the reduced model. Finite element discretization leads to the construction of matrix $[\mathcal{D}^S(\omega)]$ given by Eq. (107) and to the construction of matrix $[\mathcal{A}^E(\omega/c_E)]$ given by Eqs. (111) to (113).

10.5. FRF Calculation in the LF Range

In this section, we construct the projection of the frequency response function using the symmetric reduced model presented in Section 10.4 and we introduce the quasi-static correction terms.

1- Definition of the FRF in the LF range. For all $\omega \in \mathbb{B}_{\rm LF}$, the operator-valued frequency response function $\mathbf{T}(\omega)$ is such that [18]

$$\mathbf{u}(\omega) = \mathbf{T}(\omega)\,\mathbf{f}_{\rm S}(\omega) \quad . \tag{182}$$

2- Response to an excitation vector and projection of the FRF in the LF range. For all $\omega \in \mathbb{B}_{\rm LF}$ $(\omega \neq 0)$, matrix $[\mathcal{A}(\omega)]$ is invertible and consequently,

$$\mathbf{q}^S = [\mathcal{T}(\omega)]\,\mathcal{F}_{\rm S}(\omega) \quad , \tag{183}$$

in which $[\mathcal{T}(\omega)]$ is an $(N_S \times N_S)$ symmetric complex matrix such that

$$[\mathcal{T}(\omega)] = [\mathcal{A}(\omega)]^{-1} \quad , \quad [\mathcal{T}(\omega)]^T = [\mathcal{T}(\omega)] \quad . \tag{184}$$

18 From Chapter III, we have $\mathcal{C}^c = \mathcal{C}^c_{\rm rig} \oplus \mathcal{C}^c_{\rm elas}$. For all $\omega \in \mathbb{B}_{\rm LF}$ $(\omega \neq 0)$, from Eq. (XII.25), we deduce that, for all $\mathbf{u} \neq 0 \in \mathcal{C}^c_{\rm elas}$, $\Im m\{<\mathbf{A}(\omega)\mathbf{u},\mathbf{u}>\} \neq 0$ and then $\mathbf{A}(\omega)$ is invertible on $\mathcal{C}^c_{\rm elas}$. For all $\mathbf{u} \neq 0 \in \mathcal{C}^c_{\rm rig}$, $\Re e\{<\mathbf{A}(\omega)\mathbf{u},\mathbf{u}>\} \neq 0$ and then $\mathbf{A}(\omega)$ is invertible on $\mathcal{C}^c_{\rm rig}$. Consequently, for all $\omega \in \mathbb{B}_{\rm LF}$, $\mathbf{A}(\omega)$ is continuous and invertible from $\mathcal{C}^c$ into $\mathcal{C}^{c\prime}$ and $\mathbf{T}(\omega) = \mathbf{A}(\omega)^{-1}$ is a continuous operator from $\mathcal{C}^{c\prime}$ into $\mathcal{C}^c$.

Substituting Eq. (183) into Eq. (178) yields the response of the structural-acoustic master system,

$$\mathbf{u}^{N_S}(\mathbf{x},\omega) = \sum_{\alpha=1}^{N_S}\sum_{\alpha'=1}^{N_S} [\mathcal{T}(\omega)]_{\alpha\alpha'}\, \mathcal{F}_{S,\alpha'}(\omega)\, \mathbf{u}_\alpha(\mathbf{x}) \quad . \tag{185}$$

Consequently, the projection $\mathbf{T}^{N_S}(\omega)$ of the operator-valued frequency response function $\mathbf{T}(\omega)$ is such that

$$\mathbf{u}^{N_S}(\omega) = \mathbf{T}^{N_S}(\omega)\, \mathbf{f}_{\mathrm{S}}(\omega) \quad , \tag{186}$$

where projection $\mathbf{T}^{N_S}(\omega)$ is such that

$$\mathbf{T}^{N_S}(\omega)\, \mathbf{f}_{\mathrm{S}}(\omega) = \sum_{\alpha=1}^{N_S}\sum_{\alpha'=1}^{N_S} [\mathcal{T}(\omega)]_{\alpha\alpha'}\, f_{\mathrm{S}}(\omega\,;\mathbf{u}_{\alpha'})\, \mathbf{u}_\alpha \quad . \tag{187}$$

3- Finite element discretization of the projection of the FRF. The finite element approximation of Eq. (186) is then written as

$$\mathbf{U}^{N_S} = [T^{N_S}(\omega)]\, \mathbf{F}_{\mathrm{S}}(\omega) \quad , \tag{188}$$

in which $\mathbf{U}^{N_S} = (U_1^{N_S}, \ldots, U_{n_S}^{N_S})$ is the complex vector of the DOFs which are the values of $\mathbf{u}^{N_S}$ at the nodes of the finite element mesh of domain Ω_S and $[T^{N_S}(\omega)]$ is an $(n_S \times n_S)$ symmetric complex matrix which can be written as

$$[T^{N_S}(\omega)] = \sum_{\alpha=1}^{N_S}\sum_{\alpha'=1}^{N_S} [\mathcal{T}(\omega)]_{\alpha\alpha'}\, \mathbf{U}_\alpha\, \mathbf{U}_{\alpha'}^T \quad . \tag{189}$$

4- Calculation of the pressure field on boundary Γ_E and of the resultant pressure at any point of external domain Ω_E. Finite element discretization $\mathbf{P}_{\Gamma_E}(\omega)$ of pressure field $p_E|_{\Gamma_E}(\omega)$ on Γ_E is calculated (see Section XII.12.5) by

$$\mathbf{P}_{\Gamma_E}(\omega) = -i\,\omega\,\rho_E\,[B_{\Gamma_E}(\omega/c_E)]\,[U_{\Gamma_E}]\,\mathbf{q}^S(\omega) + \mathbf{P}_{\text{given}}(\omega) \quad , \tag{190}$$

in which $[U_{\Gamma_E}]$ is defined by Eq. (113). At any point in external domain Ω_E, the finite element discretization of resultant pressure $p_E(\mathbf{x},\omega)$ is given by (see Section XII.12.5),

$$p_E(\mathbf{x},\omega) = [Z_{\text{rad}}(\mathbf{x},\omega)]\,[U_{\Gamma_E}]\,\mathbf{q}^S(\omega) + p_{\text{given}}(\mathbf{x},\omega) \quad , \tag{191}$$

in which $p_{\text{given}}(\mathbf{x},\omega)$ is given by Eq. (XII.81).

5- Introduction of quasi-static correction terms. We use the method presented in Section VI.5. Let $N_S \geq 1$ be an integer such that

$$0=\omega_1^S=\ldots=\omega_6^S<\omega_7^S\leq\ldots\leq\omega_{N_S}^S<\omega_{{}_{\rm LF,final}}<\omega_{N_S+1}^S\leq\cdots \quad , \tag{192}$$

in which $\omega_{{}_{\rm LF,final}}$ is defined by Eq. (177) and $\{\omega_\alpha^S\}_\alpha$ are the structural eigenfrequencies introduced in Section 6.1. Projection $\mathbf{T}^{N_S}(\omega)$ of operator-valued frequency response function $\mathbf{T}(\omega)$ represents the approximation corresponding to the first N_S structural modes $\{\mathbf{u}_\alpha\}_{\alpha=1,\ldots,N_S}$ whose eigenfrequencies lie inside $\mathbb{B}_{\rm LF}$. In order to accelerate convergence, we introduce an operator which includes quasi-static correction terms such that

$$\mathbf{T}_{\rm acc}^{N_S}(\omega) = \mathbf{T}^{N_S}(\omega) + \mathbf{T}_{\rm stat}^{N_S} \quad , \tag{193}$$

in which $\mathbf{T}^{N_S}(\omega)$ is given by Eq. (187) and $\mathbf{T}_{\rm stat}^{N_S}$ is defined by

$$\mathbf{T}_{\rm stat}^{N_S} = \mathbf{K}_S^{-1} - \mathbf{T}^{N_S}(0) \quad , \tag{194}$$

where $\mathbf{K}_S^{-1}$ denotes the inverse in $\mathcal{C}_{\rm elas}^c$ of operator $\mathbf{K}^S$ defined on $\mathcal{C}^c$. For N_S fixed and for all ω fixed in $\mathbb{B}_{\rm LF}$, projection $\mathbf{T}_{\rm acc}^{N_S}(\omega)$ is a more accurate approximation of $\mathbf{T}(\omega)$ than $\mathbf{T}^{N_S}(\omega)$. Operator $\mathbf{T}^{N_S}(0)$ is such that, for all square integrable functions $\mathbf{x} \mapsto \mathbf{g}(\mathbf{x})$ from Ω_S into $\mathbb{C}^3$,

$$\left(\mathbf{T}^{N_S}(0)\,\mathbf{f}\right)(\mathbf{x}) = \int_{\Omega_S} [\tau^{N_S}(\mathbf{x},\mathbf{x}')]\;\mathbf{g}(\mathbf{x}')\,d\mathbf{x}' \quad , \tag{195}$$

in which $\mathbf{f}$ is such that $<\mathbf{f}\,,\delta\mathbf{u}>= \int_{\Omega_S}\mathbf{g}(\mathbf{x})\cdot\overline{\delta\mathbf{u}(\mathbf{x})}\,d\mathbf{x}$ and where

$$[\tau^{N_S}(\mathbf{x},\mathbf{x}')] = \sum_{\alpha=7}^{N_S} \frac{1}{\mu_\alpha^S\,(\omega_\alpha^S)^2}\,\mathbf{u}_\alpha(\mathbf{x})\,\mathbf{u}_\alpha(\mathbf{x}')^T \quad . \tag{196}$$

6- Finite element discretization of the operator-valued FRF. We consider the finite element discretization introduced in Section 10.3. The finite element discretization of operator $\mathbf{K}_S^{-1}$ is the $(n_S\times n_S)$ symmetric positive-definite real matrix denoted as $[T_0]$. Using the results of Section III.7.4, this matrix is such that $\mathbf{U}_{\rm elas} = [T_0]\,\mathbf{F}$ where $\mathbf{U}_{\rm elas}$ is the unique solution of the equation $[\,K_S\,]\,\mathbf{U}_{\rm elas} = \mathbf{F}$ under the constraints $\mathbf{U}_\alpha^T\,[\,K_S\,]\,\mathbf{U}_{\rm elas} = 0$, $\alpha = 1,\ldots,6$, in which $\mathbf{U}_1,\ldots,\mathbf{U}_6$ denote the six rigid body modes. The finite element discretization of Eq. (193) is then written as

$$[T_{\rm acc}^{N_S}(\omega)] = [T^{N_S}(\omega)] + [T_{\rm stat}^{N_S}] \quad , \tag{197}$$

in which $[T^{N_S}(\omega)]$ is defined by Eq. (189) and $[T^{N_S}_{\text{stat}}]$ is such that

$$[T^{N_S}_{\text{stat}}] = [T_0] - \sum_{\alpha=7}^{N_S} \frac{1}{\mu^S_\alpha\,(\omega^S_\alpha)^2}\,\mathbf{U}_\alpha\,\mathbf{U}_\alpha^T \quad . \tag{198}$$

7- Frequency-by-frequency construction of the FRF. The numerical procedure for constructing an approximation of $\omega \mapsto \mathbf{T}(\omega)$ on low-frequency band $\mathbb{B}_{\text{LF}}$ is based on the use of projection $\mathbf{T}^{N_S}(\omega)$ or $\mathbf{T}^{N_S}_{\text{acc}}(\omega)$ if quasi-static correction terms are taken into account. In the low-frequency range, the number N_S of structural modes is small compared with the number n_S of DOFs of the finite element model, i.e., $N_S \ll n_S$. Consequently, a frequency-by-frequency construction can be used and matrix $[\mathcal{A}(\omega)]$ defined by Eq. (180) is numerically inverted frequency by frequency on $\mathbb{B}_{\text{LF}}$ in order to construct $[\mathcal{T}(\omega)]$. If quasi-static correction terms are taken into account, we have to calculate $[T^{N_S}_{\text{stat}}]$ defined by Eq. (198). It should be noted that matrix $[T_0]$ is generally not constructed explicitly as was explained in Sections IX.3.3, IX.4.3 and IX.5.4 devoted to the methods for calculating the response of the structural-acoustic master system submitted to deterministic and random excitations.

8- Approximation of matrix $[\mathcal{A}^E(\omega/c_E)]$ using spline functions. Function $\omega \mapsto [\mathcal{A}^E(\omega/c_E)]$ defined on frequency band $\mathbb{B}_{\text{LF}}$ can be approximated using spline functions to reduce the computational effort of the frequency-by-frequency method. For the details of this procedure, we refer the reader to Section 7.5.

10.6. Location of the resonant frequencies of the master structure coupled with an external acoustic fluid

For a weakly damped master structure, the coupled system exhibits peaks in LF range $\mathbb{B}_{\text{LF}}$ (resonant behavior) due to the resonances of the master structure. Once the graph of the FRF is known on $\mathbb{B}_{\text{LF}}$, the local maxima (the peaks) define the resonances and their frequency locations define the resonant frequencies denoted as $\omega_\nu^{\text{reson}}$, $\nu = 1, 2, \ldots$ (see Fig. 4). For a weakly damped master structure, resonant frequencies $\omega_\nu^{\text{reson}}$, $\nu = 1, 2, \ldots$ can be estimated by

$$\omega_\nu^{\text{reson}} \simeq \omega_\nu^{\text{conserv}} \quad , \quad \nu = 1, 2, \ldots \quad , \tag{199}$$

in which $\omega_\nu^{\text{conserv}}$, $\nu = 1, 2, \ldots$, are the solutions of the following conservative system associated with Eq. (170),

$$\Re e\,\{\mathbf{A}(\omega_\nu^{\text{conserv}})\}\,\mathbf{u}_\nu = \mathbf{0} \quad , \quad \mathbf{u}_\nu \in \mathcal{C}^c \quad , \tag{200}$$

in which $\mathbf{u}_\nu$ is a field with values in $\mathbb{R}^3$. Using the symmetric reduced matrix model, we have to solve

$$\Re e\,\{[\mathcal{A}(\omega_\nu^{\text{conserv}})]\}\ \mathbf{q}_\nu^S = \mathbf{0} \quad , \tag{201}$$

which can be rewritten as

$$\{-\omega^2\{[\mathcal{M}^S] + [\mathcal{M}^E(\omega/c_E)]\} + [\mathcal{K}^S]\}\ \mathbf{q}_\nu^S = \mathbf{0} \quad . \tag{202}$$

The $\omega_\nu^{\text{conserv}}$ terms are then the zeros of function $\omega \mapsto \det[\mathcal{A}(\omega)]$ belonging to $\mathbb{B}_{\text{LF}}$. In the particular case where $[\mathcal{M}^S] + [\mathcal{M}^E(\omega/c_E)] \simeq [\mathcal{M}^S]$ for all ω in $\mathbb{B}_{\text{LF}}$ (which is sometimes the case if the external acoustic fluid is a gas), then $\omega_\nu^{\text{conserv}} \simeq \omega_\nu^S$, i.e., the resonance frequencies are obtained for the eigenfrequencies of the master structure in vacuo. If this assumption does not hold (which is generally the case if the external acoustic fluid is a liquid), then the eigenfrequencies of the master structure are shifted. For the mathematical properties related to the problem defined by Eq. (201), we refer the reader to Lancaster, 1966 and Markus, 1986.

10.7. Bibliographical comments

For the bibliographical aspects of the external acoustic fluid and boundary integral formulation, we refer the reader to Chapter XII. For the structural-acoustic formulation in the LF range using a projection of structural modes and related applications, see for instance Soize 1996b, Soize et al., 1992 and, for slender bodies, Coupry and Soize, 1984.

11. Structure Coupled with an External and an Internal Acoustic Fluid. Case of a Zero Pressure Condition on Part of the Internal Fluid Boundary

In Section 11, we summarize the model, equations and solving methods for a master structure coupled with an external and an internal acoustic fluid for which a zero pressure condition exists on a part of the internal fluid boundary. In this case, quantity $\pi(\omega\,;\mathbf{u}\,;\varphi) = 0$ and the results presented in this section are directly deduced from Sections 2 to 9 in which all the terms related to $\pi(\omega\,;\mathbf{u}\,;\varphi)$ vanish.

11.1. Statement of the structural-acoustic problem.

We consider the problem described in Section 2 for which part $\partial\Omega$ is written $\partial\Omega = \Gamma \cup \Gamma_0 \cup \Gamma_Z$ in which Γ_0 is submitted to a zero pressure field (see Fig. 6). There is no gravity effect. Consequently, boundary Γ in Section 2 is replaced by boundary $\Gamma \cup \Gamma_0$ in this section.

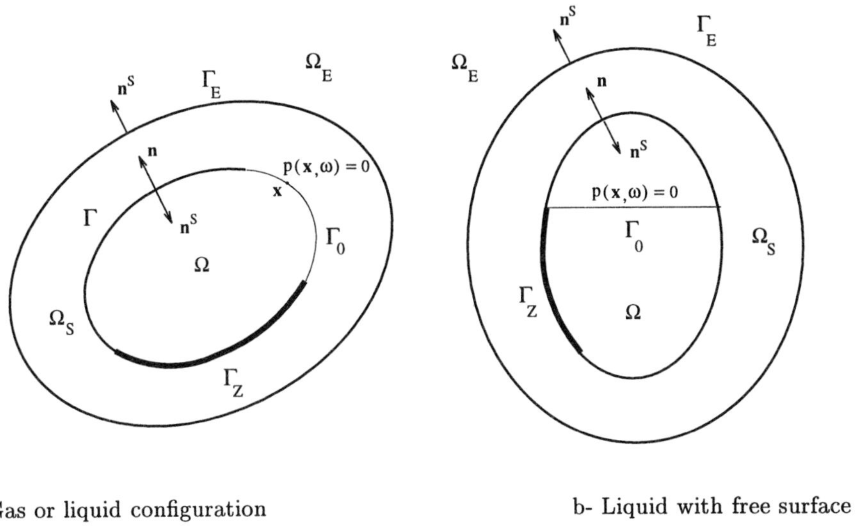

n^S
Γ_E
Ω_E
n
p(x,ω) = 0
x
Γ
n^S
Γ_0
Ω
Ω_S
Γ_Z
n^S
Γ_E
Ω_E
n
n^S
p(x,ω) = 0
Γ_0
Ω_S
Γ_Z
Ω
a- Gas or liquid configuration
b- Liquid with free surface

Fig. 6. Configurations of the structural-acoustic master system

11.2. Internal and external acoustic pressure fields.

Concerning the pressure field in internal acoustic fluid Ω, we saw in Section X.2.7-3 that $\pi(\omega\,;\mathbf{u}\,;\varphi) = 0$. Consequently, $\pi_1(\omega\,;\varphi)$, $\pi_2(\mathbf{u})$ and $\pi_Q(\omega)$ vanish in the calculation (this is achieved formally by eliminating all the terms containing $\kappa(\omega)$). Pressure field $p(\mathbf{x},\omega)$ is then given by

$$p(\mathbf{x},\omega) = \omega^2\,\rho_0\,\varphi(\mathbf{x},\omega) \quad . \tag{203}$$

Concerning the pressure field in the external acoustic fluid Ω_E, pressure field $p_E|_{\Gamma_E}(\omega)$ on Γ_E and resultant pressure field $p_E(\mathbf{x},\omega)$ at any point of unbounded external acoustic fluid Ω_E are calculated by using Eqs. (7) and (8).

11.3. Boundary value problem of the structural-acoustic master system

Since $\pi(\omega\,;\mathbf{u}\,;\varphi) = 0$ and from Section 3, we deduce the following equations of the boundary value problem. Concerning the master structure, Eqs. (9) to (11) become

$$-\omega^2 \rho_S u_i - \sigma_{ij,j}(\mathbf{u}) = g_i \quad \text{in} \quad \Omega_S \quad , \tag{204}$$

$$\sigma_{ij}(\mathbf{u})\, n_j^S = G_i - p_{\text{given}}|_{\Gamma_E}\, n_i^S - i\omega\, \mathbf{Z}_{\Gamma_E}(\omega)\{\mathbf{u}\cdot \mathbf{n}^S\}\, n_i^S \quad \text{on} \quad \Gamma_E \quad , \tag{205}$$

$$\sigma_{ij}(\mathbf{u})\, n_j^S = G_i + \omega^2 \rho_0\, \varphi\, n_i \quad \text{on} \quad \Gamma \cup \Gamma_Z \quad . \tag{206}$$

Concerning the internal acoustic fluid, $p' = p$ and Eqs. (14) to (20) become

$$-\frac{1}{c_0^2} p - i\omega\tau\rho_0 \nabla^2\varphi - \rho_0 \nabla^2\varphi = g_Q \quad \text{in} \quad \Omega \quad , \tag{207}$$

with the Dirichlet boundary condition

$$\varphi = 0 \quad \text{on} \quad \Gamma_0 \quad , \tag{208}$$

and with the Neumann boundary conditions

$$\rho_0(1+i\omega\tau)\frac{\partial\varphi}{\partial\mathbf{n}} = \rho_0\, \mathbf{u}\cdot\mathbf{n} + G_Q \quad \text{on} \quad \Gamma \quad , \tag{209}$$

$$\rho_0(1+i\omega\tau)\frac{\partial\varphi}{\partial\mathbf{n}} = \rho_0\, \mathbf{u}\cdot\mathbf{n} + \frac{\omega^2\rho_0^2}{i\omega Z}\varphi + G_Q \quad \text{on} \quad \Gamma_Z \quad , \tag{210}$$

in which $g_Q(\mathbf{x},\omega)$ is defined on Ω by

$$g_Q(\mathbf{x},\omega) = -\frac{1}{i\omega} Q(\mathbf{x},\omega) - \tau c_0^2 \frac{1}{\omega^2}\nabla^2 Q(\mathbf{x},\omega) \quad , \tag{211}$$

G_Q is defined on Γ by

$$G_Q(\mathbf{x},\omega) = \tau c_0^2 \frac{1}{\omega^2}\frac{\partial Q(\mathbf{x},\omega)}{\partial\mathbf{n}(\mathbf{x})} \quad , \tag{212}$$

and is also defined by the same equation on Γ_Z,

$$G_Q(\mathbf{x},\omega) = \tau c_0^2 \frac{1}{\omega^2}\frac{\partial Q(\mathbf{x},\omega)}{\partial\mathbf{n}(\mathbf{x})} \quad . \tag{213}$$

Eqs. (12) and (13) become

$$\frac{1}{\rho_0 c_0^2} p = \frac{\omega^2}{c_0^2}\varphi \quad \text{in} \quad \Omega \quad , \tag{214}$$

with the Dirichlet boundary condition

$$p = 0 \quad \text{on} \quad \Gamma_0 \quad . \tag{215}$$

11.4. Variational formulation of the structural-acoustic problem

The variational formulation of the boundary value problem defined by Eqs. (204) to (215) is deduced from Section 4.

Admissible function spaces. The admissible function space of displacement field $\mathbf{u}$ of master structure Ω_S is the complex vector space $\mathcal{C}^c$ defined in Section 4.1. Concerning internal acoustic fluid Ω, we introduce the complex vector spaces $\mathcal{E}^c_{0,\varphi}$ for field φ and $\mathcal{E}^c_{0,p}$ for field p such that

$$\mathcal{E}^c_{0,\varphi} = \{\ \varphi \in \mathcal{E}^c \quad ; \quad \varphi = 0 \quad \text{on} \quad \Gamma_0\}\quad , \tag{216}$$

$$\mathcal{E}^c_{0,p} = \{\ p \in H^c \quad ; \quad p = 0 \quad \text{on} \quad \Gamma_0\}\quad , \tag{217}$$

in which $\mathcal{E}^c$ is defined in Section 4.1 and and $H^c = L^2(\Omega)$. Consequently, we have $\mathcal{E}^c_{0,\varphi} \subset \mathcal{E}^c_{0,p}$.

Variational formulation in terms of $\{\mathbf{u}, \varphi, p\}$. Eqs. (204) to (206) related to the master structure yield

$$a^S(\omega; \mathbf{u}, \delta\mathbf{u}) - \omega^2 a^E(\omega/c_E; \mathbf{u}, \delta\mathbf{u}) - \omega^2 c_0(\varphi, \delta\mathbf{u}) = f_S(\omega; \delta\mathbf{u})\ ,\ \forall \delta\mathbf{u} \in \mathcal{C}^c, \tag{218}$$

in which $a^S(\omega\,;\mathbf{u}\,,\delta\mathbf{u})$, $a^E(\omega/c_E;\mathbf{u}\,,\delta\mathbf{u})$ and $c_0(\varphi, \delta\mathbf{u})$ are defined by Eqs. (24-1), (25) and (27-2) respectively. In Eq. (218), $f_S(\omega\,;\delta\mathbf{u})$ is the antilinear form on $\mathcal{C}^c$ defined by

$$f_S(\omega\,;\delta\mathbf{u}) = f(\omega\,;\delta\mathbf{u}) + f_{p_{\text{given}}}(\omega\,;\delta\mathbf{u})\quad , \tag{219}$$

in which antilinear forms $f(\omega\,;\delta\mathbf{u})$ and $f_{p_{\text{given}}}(\omega\,;\delta\mathbf{u})$ are defined by Eqs. (29) and (30) respectively. Eqs. (207) to (213) related to the internal acoustic fluid yield

$$-\overline{c_0(\delta\varphi\,,\mathbf{u})} + i\omega\, d_T(\omega; \varphi, \delta\varphi) + k(\varphi, \delta\varphi) + s_Z(\omega; \varphi, \delta\varphi) - \frac{1}{c_0^2} c_1(p, \delta\varphi)$$

$$= f_Q(\omega\,;\delta\varphi)\quad ,\quad \forall \delta\varphi \in \mathcal{E}^c_{0,\varphi}\quad , \tag{220}$$

in which sesquilinear forms $d_T(\omega; \varphi, \delta\varphi)$ and $k(\varphi, \delta\varphi)$ are defined on $\mathcal{E}^c \times \mathcal{E}^c$ by Eqs. (XI.18) and (XI.16) respectively, sesquilinear form $s_Z(\omega; \varphi, \delta\varphi)$ is defined on $\mathcal{E}^c \times \mathcal{E}^c$ by

$$s_Z(\omega\,;\varphi\,,\delta\varphi) = -\rho_0^2\omega^2 \int_{\Gamma_Z} \frac{1}{i\omega\, Z}\, \varphi\, \overline{\delta\varphi}\; ds\quad , \tag{221}$$

and sesquilinear form $c_1(p, \delta\varphi)$ is defined on $H^c \times \mathcal{E}^c$ by

$$c_1(p, \delta\varphi) = \int_\Omega p \, \overline{\delta\varphi} \, d\mathbf{x} \quad . \tag{222}$$

Substituting Eqs. (211) to (213) into Eq. (34), using Green's formula and since $G_{\mathrm{Q,z}}$ vanishes, antilinear form $f_{\mathrm{Q}}(\omega\,; \delta\varphi)$ defined on $\mathcal{E}^c_{0,\varphi}$ is written as

$$f_{\mathrm{Q}}(\omega; \delta\varphi) = -\int_\Omega \frac{Q}{i\omega} \overline{\delta\varphi} \, d\mathbf{x} + \frac{\tau c_0^2}{\omega^2} \int_\Omega \boldsymbol{\nabla} Q \cdot \boldsymbol{\nabla} \overline{\delta\varphi} \, d\mathbf{x} \quad . \tag{223}$$

Finally, Eq. (214) yields

$$-\omega^2 \frac{1}{c_0^2} \overline{c_1(\delta p\,, \varphi)} + \frac{1}{\rho_0^2} m(p, \delta p) = 0 \quad , \quad \forall \delta p \in \mathcal{E}^c_{0,p} \quad , \tag{224}$$

in which sesquilinear form $m(p, \delta p)$ defined on $H^c \times H^c$ is such that

$$m(p, \delta p) = \frac{\rho_0}{c_0^2} \int_\Omega p \, \overline{\delta p} \, d\mathbf{x} \quad . \tag{225}$$

The variational formulation for $\omega \neq 0$ is then as follows. For all fixed real $\omega \neq 0$, find $\mathbf{u}$ in $\mathcal{C}^c$, φ in $\mathcal{E}^c_{0,\varphi}$ and p in $\mathcal{E}^c_{0,p}$ such that, for all $\delta\mathbf{u}$ in $\mathcal{C}^c$, $\delta\varphi$ in $\mathcal{E}^c_{0,\varphi}$ and δp in $\mathcal{E}^c_{0,p}$, Eqs. (218), (220) and (224) are satisfied.

Solution for $\omega = 0$. Eqs. (38) to (44) are modified as follows,

$$k^S(\mathbf{u}(0)\,, \delta\mathbf{u}) = f_{\mathrm{S}}(0; \delta\mathbf{u}) \quad , \quad \forall \delta\mathbf{u} \in \mathcal{C}^c_{\mathrm{elas}} \quad , \tag{226}$$

which has a unique solution $\mathbf{u}(0)$ in $\mathcal{C}^c_{\mathrm{elas}}$ and

$$k(\varphi(0), \delta\varphi) = c_0(\delta\varphi\,, \mathbf{u}(0)) \quad , \quad \forall \delta\varphi \in \mathcal{E}^c_{0,\varphi} \quad , \tag{227}$$

which has a unique solution in $\mathcal{E}^c_{0,\varphi}$ because $k(\varphi(0), \varphi(0)) > 0$ for all $\varphi(0) \neq 0$ in $\mathcal{E}^c_{0,\varphi}$. Finally, Eq. (44) is replaced by

$$p(\mathbf{x}, 0) = 0 \quad , \quad \forall \mathbf{x} \in \Omega \quad . \tag{228}$$

Symmetric variational formulation for $\omega \neq 0$. For all fixed real $\omega \neq 0$, find $\mathbf{u}$ in $\mathcal{C}^c$, φ in $\mathcal{E}^c_{0,\varphi}$ and p in $\mathcal{E}^c_{0,p}$ such that,

$$a^S(\omega;\mathbf{u},\delta\mathbf{u}) - \omega^2 a^E(\omega/c_E;\mathbf{u},\delta\mathbf{u}) - \omega^2 c_0(\varphi,\delta\mathbf{u}) = f_S(\omega;\delta\mathbf{u})\,,\ \forall\delta\mathbf{u}\in\mathcal{C}^c\,. \quad (229)$$

$$-\omega^2\overline{c_0(\delta\varphi,\mathbf{u})} + i\omega^3\, d_T(\omega;\varphi,\delta\varphi) + \omega^2 k(\varphi,\delta\varphi) + \omega^2 s_z(\omega;\varphi,\delta\varphi) - \omega^2\frac{1}{c_0^2}c_1(p,\delta\varphi) = \omega^2 f_Q(\omega\,;\delta\varphi)\ ,\ \forall\delta\varphi\in\mathcal{E}^c_{0,\varphi}\,, \quad (230)$$

$$-\omega^2\frac{1}{c_0^2}\,\overline{c_1(\delta p\,,\varphi)} + \frac{1}{\rho_0^2}m(p,\delta p) = 0\quad ,\quad \forall\delta p\in\mathcal{E}^c_{0,p}\ . \quad (231)$$

Linear operator equation for $\omega \neq 0$. Considering Eqs. (229) to (231), Eqs. (48) and (49) are modified as follows,

$$\mathbf{A}(\omega)\begin{bmatrix}\mathbf{u}\\ \varphi\\ p\end{bmatrix} = \begin{bmatrix}\mathbf{f}_S(\omega)\\ \omega^2\,\mathbf{f}_Q(\omega)\\ \mathbf{0}\end{bmatrix}\quad ,\quad \{\mathbf{u},\varphi,p\}\in\mathcal{C}^c\times\mathcal{E}^c_{0,\varphi}\times\mathcal{E}^c_{0,p}\quad , \quad (232)$$

in which $\mathbf{A}(\omega)$ is the restriction to $\mathcal{C}^c\times\mathcal{E}^c_{0,\varphi}\times\mathcal{E}^c_{0,p}$ of the linear operator [19]

$$\mathbf{A}(\omega) = \begin{bmatrix}\mathbf{A}^S(\omega)-\omega^2\mathbf{A}^E(\omega/c_E) & -\omega^2\mathbf{C}_0 & \mathbf{0}\\ -\omega^2\,{}^t\mathbf{C}_0 & i\omega^3\mathbf{D}_T(\omega)+\omega^2\mathbf{K}+\omega^2\mathbf{S}_z(\omega) & -\frac{\omega^2}{c_0^2}\mathbf{C}_1\\ \mathbf{0} & -\frac{\omega^2}{c_0^2}\,{}^t\mathbf{C}_1 & \frac{1}{\rho_0^2}\mathbf{M}\end{bmatrix}. \quad (233)$$

Linear operators $\mathbf{A}^S(\omega)$, $\mathbf{A}^E(\omega/c_E)$, $\mathbf{C}_0$, $\mathbf{C}_1$ and $\mathbf{M}$ are defined in Section 4.3 and operators $\mathbf{K}$, $\mathbf{D}_T(\omega)$ and $\mathbf{S}_z(\omega)$ are defined by Eqs. (XI.17), (XI.19) and (XI.126) respectively. Elements $\mathbf{f}_S(\omega)$ and $\mathbf{f}_Q(\omega)$ are defined by Eqs. (59) and (60).

11.5. Finite element discretization

We consider the finite element mesh of master structure Ω_S and the finite element mesh of internal acoustic fluid Ω introduced in Section 5. The finite element discretization of the symmetric variational formulation defined by Eqs. (229) to (231) yields

$$[A(\omega)]\begin{bmatrix}\mathbf{U}\\ \mathbf{\Phi}\\ \mathbf{P}\end{bmatrix} = \begin{bmatrix}\mathbf{F}_S(\omega)\\ \omega^2\,\mathbf{F}_Q(\omega)\\ \mathbf{0}\end{bmatrix}\quad , \quad (234)$$

---

[19] Linear operator $\mathbf{A}(\omega)$ is continuous from $\mathcal{C}^c\times\mathcal{E}^c\times H^c$ into $\mathcal{C}^{c\prime}\times\mathcal{E}^{c\prime}\times H^c$.

with the constraints

$$\Phi_j = 0 \text{ and } P_j = 0 \quad \text{for nodes } j \text{ belonging to } \Gamma_0 \quad , \tag{235}$$

corresponding to the finite element discretization of the Dirichlet conditions defined by Eqs. (208) and (215). Complex matrix $[A(\omega)]$ is defined by

$$[A(\omega)] = \begin{bmatrix} [A^S(\omega)] - \omega^2 [A^E(\omega/c_E)] & -\omega^2 [C_0] & [0] \\ -\omega^2 [C_0]^T & i\omega^3 [D_T(\omega)] + \omega^2 [K] + \omega^2 [S_Z(\omega)] & -\frac{\omega^2}{c_0^2} [C_1] \\ [0] & -\frac{\omega^2}{c_0^2} [C_1]^T & \frac{1}{\rho_0^2} [M] \end{bmatrix} . \tag{236}$$

The matrices appearing in Eqs. (234) and (236) are defined in Section 5.

11.6. Symmetric reduced matrix model in the LF range

The reduced model is directly deduced from Section 6 in which the acoustic modes corresponding to a zero pressure on Γ_0 are defined in Section XI.11.6,

$$[\mathcal{A}(\omega)] \begin{bmatrix} \mathbf{q}^S \\ \mathbf{q} \\ \mathbf{r} \end{bmatrix} = \begin{bmatrix} \mathcal{F}_S(\omega) \\ \omega^2 \mathcal{F}_Q(\omega) \\ \mathbf{0} \end{bmatrix} \quad , \tag{237}$$

in which the complex matrix $[\mathcal{A}(\omega)]$ is symmetric and is defined by

$$[\mathcal{A}(\omega)] = \begin{bmatrix} [\mathcal{A}^S(\omega)] - \omega^2 [\mathcal{A}^E(\omega/c_E)] & -\omega^2 [\mathcal{C}_0] & [0] \\ -\omega^2 [\mathcal{C}_0]^T & i\omega^3 [\mathcal{D}_T(\omega)] + \omega^2 [\mathcal{K}] + \omega^2 [\mathcal{S}_Z(\omega)] & -\frac{\omega^2}{c_0^2} [\mathcal{C}_1] \\ [0] & -\frac{\omega^2}{c_0^2} [\mathcal{C}_1]^T & \frac{1}{\rho_0^2} [\mathcal{M}] \end{bmatrix} . \tag{238}$$

The matrices appearing in Eqs. (237) and (238) are defined in Section 6.4. For the construction of the reduced model using the finite element approximation, we refer the reader to Section 6.5.

11.7. Frequency-by-frequency construction of the FRF in the LF range and location of the resonant frequencies of the coupled system

The construction of the FRF is explained in Section 7 and must be adapted as follows. The elimination of $\mathbf{r}$ in Eq. (237) yields

$$[\widetilde{\mathcal{A}}(\omega)] \begin{bmatrix} \mathbf{q}^S \\ \mathbf{q} \end{bmatrix} = \begin{bmatrix} \mathcal{F}_S(\omega) \\ \omega^2 \mathcal{F}_Q(\omega) \end{bmatrix} \quad , \tag{239}$$

in which complex matrix $[\widetilde{\mathcal{A}}(\omega)]$ is symmetric and is defined by

$$[\widetilde{\mathcal{A}}(\omega)] = \begin{bmatrix} [\mathcal{A}^S(\omega)] - \omega^2[\mathcal{A}^E(\omega/c_E)] & -\omega^2[\mathcal{C}_0] \\ -\omega^2\,[\mathcal{C}_0]^T & i\omega^3[\mathcal{D}_T(\omega)] + \omega^2[\mathcal{K}] + \omega^2[\mathcal{S}_Z(\omega)] - \omega^4[\mathcal{M}] \end{bmatrix} . \tag{240}$$

The finite element discretization of the projection of the FRF is performed as explained in Section 7.3. Frequency-by-frequency construction of the FRF in the LF range is performed as explained in Sections 7.4 and 7.5. The location of the resonant frequencies of the coupled system is carried out as explained in Section 8.

11.8. Structural-acoustic modes of the master structure coupled with an internal acoustic fluid

The structural-acoustic modes defined in Section 9 are the solutions of

$$\begin{bmatrix} \mathbf{K}^S & \mathbf{0} & \mathbf{0} \\ \mathbf{0} & \mathbf{0} & \mathbf{0} \\ \mathbf{0} & \mathbf{0} & \frac{1}{\rho_0^2}\mathbf{M} \end{bmatrix} \begin{bmatrix} \mathbf{u} \\ \varphi \\ p \end{bmatrix} = \omega^2 \begin{bmatrix} \mathbf{M}^S & \mathbf{C}_0 & \mathbf{0} \\ {}^t\mathbf{C}_0 & -\mathbf{K} & \frac{1}{c_0^2}\mathbf{C}_1 \\ \mathbf{0} & \frac{1}{c_0^2}{}^t\mathbf{C}_1 & \mathbf{0} \end{bmatrix} \begin{bmatrix} \mathbf{u} \\ \varphi \\ p \end{bmatrix} . \tag{241}$$

The reduced spectral problem defined by Eq. (143) is then replaced by

$$\begin{bmatrix} [\mathcal{K}^S] & [0] & [0] \\ [0] & [0] & [0] \\ [0] & [0] & \frac{1}{\rho_0^2}[\mathcal{M}] \end{bmatrix} \begin{bmatrix} \mathbf{q}^S \\ \mathbf{q} \\ \mathbf{r} \end{bmatrix} = \omega^2 \begin{bmatrix} [\mathcal{M}^S] & [\mathcal{C}_0)] & [0] \\ [\mathcal{C}_0]^T & -[\mathcal{K}] & \frac{1}{c_0^2}[\mathcal{C}_1] \\ [0] & \frac{1}{c_0^2}[\mathcal{C}_1]^T & [0] \end{bmatrix} \begin{bmatrix} \mathbf{q}^S \\ \mathbf{q} \\ \mathbf{r} \end{bmatrix} , \tag{242}$$

in which the matrices appearing in Eq. (242) are defined in Section 9.2 using the acoustic modes of Section XI.11.6. Eliminating $\mathbf{q}$ in Eq.(242) yields the symmetric generalized eigenvalue problem with the positive-definite "mass matrix",

$$\begin{bmatrix} [\mathcal{K}_1] & [0] \\ [0] & [\mathcal{K}_2] \end{bmatrix} \begin{bmatrix} \mathbf{q}^S \\ \mathbf{r} \end{bmatrix} = \omega^2 \begin{bmatrix} [\mathcal{M}_1] & [\mathcal{M}_c] \\ [\mathcal{M}_c]^T & [\mathcal{M}_2] \end{bmatrix} \begin{bmatrix} \mathbf{q}^S \\ \mathbf{r} \end{bmatrix} , \tag{243}$$

in which

$$[\mathcal{K}_1]_{\alpha\beta} = \mu_\alpha^S\, \lambda_\alpha^S\, \delta_{\alpha\beta} \quad , \tag{244}$$

$$[\mathcal{K}_2]_{\alpha\beta} = \frac{\mu_\alpha}{\rho_0^2}\, \delta_{\alpha\beta} \quad , \tag{245}$$

$$[\mathcal{M}_1]_{\alpha\beta} = \mu_\alpha^S\, \delta_{\alpha\beta} + \sum_{\gamma=1}^{N} \frac{1}{\mu_\gamma \lambda_\gamma} [\mathcal{C}_0]_{\alpha\gamma}\, [\mathcal{C}_0]_{\beta\gamma} \quad , \tag{246}$$

$$[\mathcal{M}_2]_{\alpha\beta} = \frac{1}{\rho_0^2}\frac{\mu_\alpha}{\lambda_\alpha}\,\delta_{\alpha\beta} \quad , \tag{247}$$

$$[\mathcal{M}_c]_{\alpha\beta} = \frac{1}{\rho_0\lambda_\alpha}[\mathcal{C}_0]_{\alpha\beta} \quad . \tag{248}$$

The structural-acoustic eigenfrequencies constitute a sequence of real numbers $\{\omega_j\}_j$ such that

$$0 = \omega_1 = \ldots = \omega_6 < \omega_7 \leq \omega_8 < \ldots \quad . \tag{249}$$

It should be noted that the solution $\omega = 0$ corresponds to the six rigid body modes of the master structure. The approximation of the structural-acoustic mode associated with ω_j is given by

$$\mathbf{u}_j^{N_S}(\mathbf{x}) = \sum_{\alpha=1}^{N_S} q_{j,\alpha}^S\,\mathbf{u}_\alpha(\mathbf{x}) \quad , \tag{250}$$

$$p_{N,j}(\mathbf{x}) = \sum_{\alpha=1}^{N} r_{j,\alpha}\,\varphi_\alpha(\mathbf{x}) \quad , \tag{251}$$

in which $\{\mathbf{q}_j^S, \mathbf{r}_j\}$ is the eigenvector in $\mathbb{R}^{N_S}\times\mathbb{R}^N$ associated with the eigenfrequency ω_j. Introducing $\mathbf{q}_j^S = (q_{j,1}^S, \ldots q_{j,N_S}^S)$ and $\mathbf{r}_j = (r_{j,1}, \ldots r_{j,N})$, we have

$$\begin{bmatrix} [\mathcal{K}_1] & [0] \\ [0] & [\mathcal{K}_2] \end{bmatrix} \begin{bmatrix} \mathbf{q}_j^S \\ \mathbf{r}_j \end{bmatrix} = \omega_j^2 \begin{bmatrix} [\mathcal{M}_1] & [\mathcal{M}_c] \\ [\mathcal{M}_c]^T & [\mathcal{M}_2] \end{bmatrix} \begin{bmatrix} \mathbf{q}_j^S \\ \mathbf{r}_j \end{bmatrix} \quad . \tag{252}$$

Let $[\,U\,]$ and $[\,\Phi\,]$ be the matrices of the structural modes and acoustic modes defined by Eqs. (104) and (XI.142). The finite element discretizations of Eqs. (250) and (251) allowing calculation of the jth structural-acoustic mode are

$$\mathbf{U}_j^{N_S} = \sum_{\alpha=1}^{N_S} q_{j,\alpha}^S\,\mathbf{U}_\alpha \quad , \tag{253}$$

$$\mathbf{P}_j^N = \sum_{\alpha=1}^{N} r_{j,\alpha}\,\mathbf{\Phi}_\alpha \quad . \tag{254}$$

12. Case of an Axisymmetric Structural-Acoustic Master System

We refer the reader to Section III.8.2 for the master structure, and to Section XI.12 for the internal acoustic fluid. Let n be the circumferential wave number. From Section XI.12, we deduce that, for $n = 0$, the pressure field is given by

$$p_0(r, z\,; \omega) = \omega^2 \rho_0 \varphi_0(r, z\,; \omega) + \pi(\omega) \quad , \tag{255}$$

in which $\pi(\omega)$ is not equal to zero, and for all $n \geq 1$, the pressure field is given by

$$p_n^{\pm}(r, z\,; \omega) = \omega^2 \rho_0 \varphi_n^{\pm}(r, z\,; \omega) \quad , \tag{256}$$

i.e. $\pi(\omega)$ is equal to zero and then leads to a formulation of the type studied in Section 11.

13. Response to Deterministic and Random Excitations

In the structural-acoustic problem defined in Section 2, the excitations are (1)- for the master structure, surface force field $\mathbf{G}$ and body force field $\mathbf{g}$, (2)- for the internal fluid, acoustic source density Q and (3)- for the external fluid, acoustic source density Q_E and an incident plane wave described by ψ_{inc}. For calculation of the reponse of this structural-acoustic master system submitted to deterministic or random excitations, we refer the reader to the general methodology presented in Chapter IX. Below, we give some elements of the methodology for two typical cases.

Case of an external turbulent boundary layer. The excitation due to the random wall pressure $p_{\text{turb}}(\mathbf{y}, t)$ induced on surface Γ_E by an external turbulent boundary layer corresponds to the case of a random surface force field $\mathbf{G}(\mathbf{y}, t) = -p_{\text{turb}}(\mathbf{y}, t)\, \mathbf{n}^S(\mathbf{y})$ applied on surface Γ_E of the structure (see Fig. 1). If we assume that second-order centered random field $\{p_{\text{turb}}(\mathbf{y}, t), \mathbf{y} \in \Gamma_E, t \in \mathbb{R}\}$ is mean-square stationary with respect to time, the second term on the right-hand side of Eq. (IX.91) shows that the required data is the matrix-valued cross-spectral density function of random field $\{\mathbf{G}(\mathbf{y}, t), \mathbf{y} \in \Gamma_E, t \in \mathbb{R}\}$ which is written as $S_{p_{\text{turb}}}(\mathbf{y}, \mathbf{y}', \omega)$ because

$$[S_{\mathbf{G}}(\mathbf{y}, \mathbf{y}'\omega)] = S_{p_{\text{turb}}}(\mathbf{y}, \mathbf{y}', \omega)\, \mathbf{n}^S(\mathbf{y})\, \mathbf{n}^S(\mathbf{y}')^T \quad . \tag{257}$$

Models describing function $S_{p_{\text{turb}}}(\mathbf{y}, \mathbf{y}', \omega)$ can be found in Bakewell, 1968; Batchelor, 1967; Blake, 1984 and 1986; Chase, 1980 and 1987; Corcos, 1963 and 1964; Dowling and Ffowcs-Williams, 1983; Ffowcs-Williams, 1965 and 1982; Hinze, 1959; Lauchle, 1977; Willmarth, 1975.

Case of an external random field induced by continuously distributed multipole sources. The excitation due to an external random field due to continuously distributed multipole sources located in a specified bounded domain K_Q included in external fluid Ω_E corresponds to the case of a random acoustic source density $Q_E(\mathbf{x}, t)$ defined in K_Q. If we assume that second-order centered random field $\{Q_E(\mathbf{x}, t), \mathbf{x} \in K_Q, t \in \mathbb{R}\}$ is mean-square stationary with respect to time, the cross-spectral density function $S_{p_{\text{given}}}(\mathbf{y}, \mathbf{y}', \omega)$ of time-stationary random field $\{p_{\text{given}}(\mathbf{y}, t), \mathbf{y} \in \Gamma_E, t \in \mathbb{R}\}$ can easily be deduced from the cross-spectral density function $S_{Q_E}(\mathbf{x}, \mathbf{x}', \omega)$ of time-stationary random field Q_E using the results of Section XII.6.4. Consequently, this case corresponds to a random surface force field $\mathbf{G}(\mathbf{y}, t) = -p_{\text{given}}(\mathbf{y}, t)\, n^S(\mathbf{y})$ applied on surface Γ_E of the structure (see Fig. 1). The second term in the right-hand side of Eq. (IX.91) shows that the required data is the matrix-valued cross-spectral density function of random field $\{\mathbf{G}(\mathbf{y}, t), \mathbf{y} \in \Gamma_E, t \in \mathbb{R}\}$ which is written as

$$[S_{\mathbf{G}}(\mathbf{y}, \mathbf{y}'\omega)] = S_{p_{\text{given}}}(\mathbf{y}, \mathbf{y}', \omega)\, \mathbf{n}^S(\mathbf{y})\, \mathbf{n}^S(\mathbf{y}')^T \quad . \tag{258}$$

CHAPTER XIV

Structural-Acoustic Master System in the MF Range

1. Introduction

In this chapter, we present the solution method for solving the vibration problem in the MF range of a master structure containing an internal acoustic fluid and surrounded by an external acoustic fluid. Modeling of the master structure is developed in Chapters V and VII, modeling of the internal fluid in Chapters X and XI and modeling of the external fluid in Chapter XII.

In Section 2, we state a general structural-acoustic problem in the frequency domain with prescribed internal acoustic source density, mechanical forces, external acoustic source density and external acoustic plane wave. A part of the internal fluid-structure interface has acoustical properties modeled by a given wall acoustic impedance.

In Section 3, we establish the boundary value problem of the structural-acoustic master system. The external acoustic fluid is represented by the boundary impedance operator defined on the external fluid-structure interface. This boundary operator was introduced and constructed in Chapter XII. The master structure is described by the local equations introduced in Chapter V. The internal acoustic fluid is described by the local equations of Section XI.3 expressed in terms of field ψ.

In Section 4, we construct the variational formulation of the boundary value problem.

In Section 5, we present the finite element discretization of the variational formulation established in Section 4.

Section 6 is devoted to the method for calculating the frequency response function in the medium-frequency range, based on the frequency transform technique presented in Chapter VII.

In Section 7, we consider the particular case of a master structure coupled with an external acoustic fluid (without internal acoustic fluid) which is a simplification of the general case presented in Sections 2 to 6. This section is introduced in order to have a self-contained presentation of this standard case.

In Section 8, we consider the general structural-acoustic problem introduced in Section 2 for which a zero pressure condition is imposed on a part of the boundary of the internal acoustic domain.

Finally, in Section 9 we discuss the case of axisymmetric structural-acoustic master systems and in Section 10 the response to deterministic and random excitations.

2. Statement of the Structural-Acoustic Problem in the Frequency Domain

2.1. Geometry and mechanical assumptions

The physical space $\mathbb{R}^3$ is referred to a cartesian reference system and we denote the generic point of $\mathbb{R}^3$ as $\mathbf{x} = (x_1, x_2, x_3)$. We study the linear vibrations of a structural-acoustic master system around a static equilibrium state taken as a natural state at rest.

Master structure Ω_S. The master structure at equilibrium occupies a three-dimensional bounded domain Ω_S of $\mathbb{R}^3$ with a sufficiently smooth boundary $\partial\Omega_S = \Gamma \cup \Gamma_Z \cup \Gamma_E$ with $\Gamma \cap \Gamma_Z \cap \Gamma_E = \emptyset$. The outward unit normal to $\partial\Omega_S$ is denoted as $\mathbf{n}^S = (n_1^S, n_2^S, n_3^S)$ (see Fig. 1). The displacement field in Ω_S is denoted as $\mathbf{u}(\mathbf{x}, \omega) = (u_1(\mathbf{x}, \omega), u_2(\mathbf{x}, \omega), u_3(\mathbf{x}, \omega))$. The master structure is assumed to be free, i.e. is not fixed on any part of boundary $\partial\Omega_S$. A surface force field $\mathbf{G}(\mathbf{x}, \omega) = (G_1(\mathbf{x}, \omega), G_2(\mathbf{x}, \omega), G_3(\mathbf{x}, \omega))$ is given on $\partial\Omega_S$ and a body force field $\mathbf{g}(\mathbf{x}, \omega) = (g_1(\mathbf{x}, \omega), g_2(\mathbf{x}, \omega), g_3(\mathbf{x}, \omega))$ is given in Ω_S. The master structure is a dissipative medium whose viscoelastic constitutive equation is defined in Section IV.5.2.

Internal acoustic fluid Ω. Let Ω be the internal bounded domain filled with a dissipative acoustic fluid (gas or liquid) as described in Section X.3. The boundary $\partial\Omega$ of Ω is $\Gamma \cup \Gamma_Z$. The outward unit normal to $\partial\Omega$ is denoted as $\mathbf{n} = (n_1, n_2, n_3)$ and we have $\mathbf{n} = -\mathbf{n}^S$ on $\partial\Omega$ (see Fig. 1). Part Γ_Z of the boundary has acoustical properties modeled by a wall acoustic impedance $Z(\mathbf{x}, \omega)$ satisfying the hypotheses defined by Eqs. (X.18) to (X.25). We denote the pressure field in Ω as $p(\mathbf{x}, \omega)$. We assume that there is no Dirichlet boundary condition on any part of $\partial\Omega$. An acoustic source density $Q(\mathbf{x}, \omega)$ is given inside Ω and satisfies the hypotheses defined by Eq. (X.8).

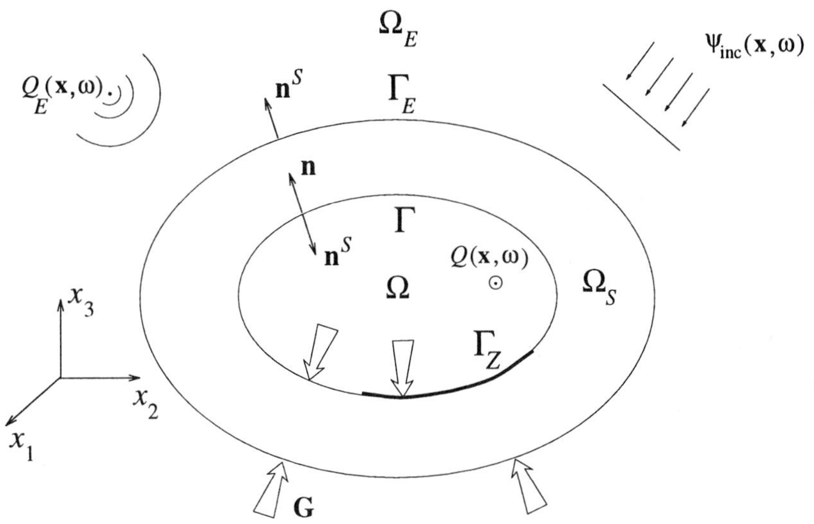

Fig. 1. Configuration of the structural-acoustic master system

External acoustic fluid Ω_E. The master structure is surrounded by an external inviscid acoustic fluid (gas or liquid) as described in Chapter XII. The fluid occupies the unbounded open three-dimensional domain Ω_E whose boundary $\partial\Omega_E$ is Γ_E. We introduce the bounded open domain Ω_i defined by $\Omega_i = \mathbb{R}^3 \backslash (\Omega_E \cup \Gamma_E)$. The boundary $\partial\Omega_i$ of Ω_i is then Γ_E. The outward unit normal to $\partial\Omega_i$ is $\mathbf{n}^S$ defined above (see Fig. 1). We denote the pressure field in Ω_E as $p_E(\mathbf{x}, \omega)$. We assume that there is no Dirichlet boundary condition on any part of Γ_E. An acoustic source density $Q_E(\mathbf{x}, \omega)$ and an incident plane wave defined by the velocity potential $\psi_{\text{inc}}(\mathbf{x}, \omega) = \psi_0(\omega) \exp\{-i\, \mathbf{k} \cdot \mathbf{x}\}$ are given inside Ω_E. As seen in Section XII.6.4, this acoustic source density and incident plane wave induce a pressure field $p_{\text{given}}(\omega)$ on Γ_E defined by Eq. (XII.78).

2.2. Structural-acoustic problem to be solved

Definition of the FRF of the structural-acoustic master system. The problem consists in calculating the operator-valued FRF of the structural-acoustic master system in the medium-frequency range, whose input is

$\{\mathbf{f}_S(\omega), \mathbf{f}_Q(\omega)\}$ and whose output is $\{\mathbf{u}(\omega), \psi(\omega)\}$ in which $\mathbf{u}(\omega)$ is the displacement field of the master structure and $\psi(\omega)$ is the field describing the internal fluid. Concerning the input, $\mathbf{f}_S(\omega)$ represents the force vector induced by $\mathbf{g}$ in Ω_S, $\mathbf{G}$ on $\partial\Omega_S$, p_{given} on Γ_E, and $\mathbf{f}_Q(\omega)$ represents Q in Ω. The main objective of this chapter is the calculation of this FRF in the medium-frequency range.

Calculation of pressure field $p(\mathbf{x}, \omega)$ in internal acoustic fluid Ω. Pressure field $p(\mathbf{x}, \omega)$ is given by Eqs. (X.85),

$$p(\mathbf{x}, \omega) = -i\omega\, \rho_0\, \psi(\mathbf{x}, \omega) + \pi(\omega\,; \mathbf{u}\,; \psi) \quad , \tag{1}$$

in which $\pi(\omega\,; \mathbf{u}\,; \psi)$ is calculated as a function of $\mathbf{u}$ and ψ as follows. Integrating Eq. (X.86) in Ω, using Green's formula and the boundary conditions defined by Eqs. (X.88) and (X.29) yields

$$\pi(\omega\,; \mathbf{u}\,; \psi) = \kappa(\omega)\, \left\{ i\omega\, \rho_0\, \pi_1(\omega\,; \psi) - \pi_2(\mathbf{u}) + \pi_Q(\omega) \right\} \quad , \tag{2}$$

where $\pi_1(\omega\,; \psi)$, $\pi_2(\mathbf{u})$ and $\pi_Q(\omega)$ are given by

$$\pi_1(\omega\,; \psi) = \int_{\Gamma_Z} \frac{\psi(\mathbf{x}, \omega)}{i\omega\, Z(\mathbf{x}, \omega)}\, ds(\mathbf{x}) \quad , \tag{3}$$

$$\pi_2(\mathbf{u}) = \int_{\Gamma \cup \Gamma_Z} \mathbf{u}(\mathbf{x}, \omega) \cdot \mathbf{n}(\mathbf{x})\, ds(\mathbf{x}) \quad , \tag{4}$$

$$\pi_Q(\omega) = \frac{1}{\rho_0} \int_\Omega \frac{Q(\mathbf{x}, \omega)}{i\omega}\, d\mathbf{x} \quad , \tag{5}$$

in which $|\Omega|$ is the volume of domain Ω,

$$|\Omega| = \int_\Omega d\mathbf{x} \quad , \tag{6}$$

and $\kappa(\omega)$ is the dimensionless complex number defined by

$$\kappa(\omega) = \frac{\rho_0\, c_0^2}{|\Omega|} \times \left\{ 1 + \frac{\rho_0\, c_0^2}{|\Omega|} \int_{\Gamma_Z} \frac{ds(\mathbf{x})}{i\omega\, Z(\mathbf{x}, \omega)} \right\}^{-1} \quad . \tag{7}$$

Real constant $\kappa(0)$ is such that

$$0 < \kappa(0) < +\infty \quad , \quad \lim_{\alpha_{\text{min}} \to +\infty} \kappa(0) = \frac{\rho_0 c_0^2}{|\Omega|} \quad . \tag{8}$$

It should be noted that the general expression of $\pi(\omega\,;\mathbf{u}\,;\psi)$ defined by Eq. (2) contains the term $\kappa(\omega)\,(i\omega/c_0^2)\int_\Omega \psi\, d\mathbf{x}$ which does not appear due to Eq. (X.80).

Calculation of pressure field $p_E|_{\Gamma_E}(\omega)$ on Γ_E and of resultant pressure field $p_E(\mathbf{x},\omega)$ at any point of unbounded external acoustic fluid Ω_E. Once the trace on Γ_E of structural displacement field $\mathbf{u}(\omega)$ is known, pressure field $p_E|_{\Gamma_E}(\omega)$ is given by Eqs. (XII.76) and (XII.77),

$$p_E|_{\Gamma_E}(\omega) = i\omega\, \mathbf{Z}_{\Gamma_E}(\omega)\{\mathbf{u}(\omega)\cdot \mathbf{n}^S\} + p_{\text{given}}|_{\Gamma_E}(\omega) \quad , \tag{9}$$

in which $p_{\text{given}}|_{\Gamma_E}(\omega)$ is defined by Eq. (XII.78) and where $\mathbf{Z}_{\Gamma_E}(\omega)$ is the acoustic impedance boundary operator whose construction is given in Sections XII.10 and XII.12. At any point $\mathbf{x}$ of unbounded external acoustic fluid Ω_E, resultant pressure field $p_E(\mathbf{x},\omega)$ is calculated using Eqs. (XII.79) and (XII.80),

$$p_E(\mathbf{x},\omega) = i\omega\, \mathbf{Z}_{\text{rad}}(\mathbf{x},\omega)\{\mathbf{u}(\omega)\cdot \mathbf{n}^S\} + p_{\text{given}}(\mathbf{x},\omega) \quad , \tag{10}$$

in which $p_{\text{given}}(\mathbf{x},\omega)$ is given by Eq. (XII.81) and $\mathbf{Z}_{\text{rad}}(\mathbf{x},\omega)$ is the radiation impedance operator whose construction is given in Sections XII.11 and XII.12.

3. Boundary Value Problem of the Structural-Acoustic Master System

In this section, we present the boundary value problem of the structural-acoustic master system consisting of the master structure coupled with the internal fluid and the external fluid. In this formulation, the velocity potential field of the external fluid is eliminated in terms of the structural displacement field using the acoustic impedance boundary operator. The unknown fields chosen will be displacement field $\mathbf{u}(\omega)$ of the structure and field $\psi(\omega)$ of the internal fluid.

Equations for the master structure. The equation of the master structure occupying domain Ω_S is written as

$$-\omega^2\,\rho_s\,u_i - \sigma_{ij,j}(\mathbf{u}) = g_i \quad \text{in} \quad \Omega_S \quad , \tag{11}$$

in which $\rho_s(\mathbf{x})$ is the mass density of the master structure. In the MF range, the constitutive equation is defined by Eqs. (IV-46) to (IV-50) where tensors $a_{ijkh}(\omega)$ and $b_{ijkh}(\omega)$ depend on ω (see Section IV.5.2). Using Eq. (9), the boundary condition $\sigma_{ij}(\mathbf{u})n_j^S = G_i - p_E|_{\Gamma_E}\, n_i^S$ on Γ_E can be written as

$$\sigma_{ij}(\mathbf{u})\, n_j^S = G_i - p_{\text{given}}|_{\Gamma_E}\, n_i^S - i\omega\, \mathbf{Z}_{\Gamma_E}(\omega)\{\mathbf{u}\cdot \mathbf{n}^S\}\, n_i^S \quad \text{on} \quad \Gamma_E \quad . \tag{12}$$

Using Eqs. (1) and (2) and since $\mathbf{n}^S = -\mathbf{n}$, the boundary condition $\sigma_{ij}(\mathbf{u})n_j^S = G_i - p(\omega)\, n_i^S = G_i + p(\omega)\, n_i$ on $\Gamma \cup \Gamma_Z$ can be written as

$$\begin{aligned} &\sigma_{ij}(\mathbf{u})\, n_j^S = G_i + \kappa\, \pi_{\mathrm{Q}}\, n_i \\ &-i\,\omega\rho_0\, \psi\, n_i + i\,\omega\rho_0\, \kappa\, \pi_1(\omega;\psi)n_i - \kappa\, \pi_2(\mathbf{u})n_i \quad \text{on} \quad \Gamma\cup\Gamma_Z \quad , \end{aligned} \tag{13}$$

in which $\pi_1(\omega\,;\varphi)$, $\pi_2(\mathbf{u})$, $\pi_{\mathrm{Q}}(\omega)$ and $\kappa(\omega)$ are given by Eqs. (3), (4), (5) and (7) respectively.

Equations for the internal acoustic fluid. Field $\psi(\mathbf{x},\omega)$ satisfies the constraint equation

$$\int_\Omega \psi\, d\mathbf{x} = 0 \quad . \tag{14}$$

Eq. (X.87) related to internal acoustic fluid Ω can be rewritten in terms of $\psi(\omega)$ as

$$\begin{aligned} &-\omega^2 \frac{\rho_0}{c_0^2}\, \psi(\mathbf{x},\omega) - i\omega\, \tau\, \rho_0 \nabla^2\psi(\mathbf{x},\omega) - \rho_0\, \nabla^2\psi(\mathbf{x},\omega) = \\ &\qquad i\omega\, g_{\mathrm{Q}} - \frac{\omega^2\rho_0\kappa}{c_0^2}\, \pi_1(\omega;\psi) - \frac{i\omega\kappa}{c_0^2}\, \pi_2(\mathbf{u}) \quad \text{in} \quad \Omega \quad , \end{aligned} \tag{15}$$

in which g_{Q} is defined by Eq. (18) below. The Neumann boundary condition on Γ is given by Eq. (X.88),

$$\rho_0(1+i\omega\tau)\frac{\partial\psi}{\partial\mathbf{n}} = i\omega\, \rho_0\, \mathbf{u}\cdot\mathbf{n} + i\omega\, G_{\mathrm{Q}} \quad \text{on} \quad \Gamma\, , \tag{16}$$

in which G_{Q} is defined by Eq. (19) below and the Neumann boundary condition on Γ_Z with wall acoustic impedance is given by Eq. (X.89),

$$\begin{aligned} &\rho_0(1+i\omega\tau)\frac{\partial\psi}{\partial\mathbf{n}} = i\omega\, \rho_0\, \mathbf{u}\cdot\mathbf{n} + \frac{\omega^2\rho_0^2}{i\omega Z}\psi \\ &- \frac{\omega^2\rho_0^2\kappa}{i\omega Z}\, \pi_1(\omega;\psi) - \frac{i\omega\kappa\rho_0}{i\omega Z}\, \pi_2(\mathbf{u}) + i\omega\, G_{\mathrm{Q}} + i\omega\, G_{\mathrm{Q,Z}} \quad \text{on}\ \Gamma_Z\, , \end{aligned} \tag{17}$$

in which $G_{Q,Z}$ is defined by Eq. (20) below. Function g_Q is defined in Ω by

$$g_Q(\mathbf{x},\omega) = \frac{1}{c_0^2}\,\kappa(\omega)\,\pi_Q(\omega) - \frac{1}{i\omega}Q(\mathbf{x},\omega) - \tau c_0^2\,\frac{1}{\omega^2}\boldsymbol{\nabla}^2 Q(\mathbf{x},\omega) \quad , \tag{18}$$

G_Q is defined on $\Gamma \cup \Gamma_Z$ by

$$G_Q(\mathbf{x},\omega) = \tau c_0^2 \frac{1}{\omega^2}\frac{\partial Q(\mathbf{x},\omega)}{\partial \mathbf{n}(\mathbf{x})} \quad , \tag{19}$$

and $G_{Q,Z}$ is defined on Γ_Z by

$$G_{Q,Z}(\mathbf{x},\omega) = \frac{\rho_0}{i\omega Z(\mathbf{x},\omega)}\kappa(\omega)\,\pi_Q(\omega) \quad . \tag{20}$$

4. Variational Formulation of the Structural-Acoustic Problem

In this section we construct the variational formulation of the boundary value problem defined by Eqs. (11) to (20) in terms of $\mathbf{u}(\omega)$ and $\psi(\omega)$

4.1. Admissible function spaces for the structural-acoustic problem

The admissible function space of displacement field $\mathbf{u}$ of master structure Ω_S is the complex vector space $\mathcal{C}^c$ of sufficiently differentiable functions [1] defined on Ω_S with values in $\mathbb{C}^3$. The admissible function space of ψ related to internal acoustic fluid Ω is the complex vector space $\mathcal{E}^c_\psi$ such that

$$\mathcal{E}^c_\psi = \{\ \psi \in \mathcal{E}^c \quad ; \quad \int_\Omega \psi\, d\mathbf{x} = 0\ \} \quad , \tag{21}$$

in which $\mathcal{E}^c$ is the complex vector space of the sufficiently differentiable functions [2] defined on Ω with values in $\mathbb{C}$.

4.2. Variational formulation in terms of $\{\mathbf{u},\psi\}$

Variational equation in $\delta\mathbf{u}$ related to the master structure. Let $\mathbf{u}$ be in $\mathcal{C}^c$ and ψ in $\mathcal{E}^c_\psi$. Multiplying Eq. (11) by $\overline{\delta\mathbf{u}} \in \mathcal{C}^c$, and integrating over domain Ω_S, using Green's formula and Eqs. (12) and (13) yields, for all $\delta\mathbf{u} \in \mathcal{C}^c$,

$$a^S(\omega;\mathbf{u},\delta\mathbf{u}) - \omega^2 a^E(\omega/c_E;\mathbf{u},\delta\mathbf{u}) + \kappa\, j(\mathbf{u},\delta\mathbf{u}) + i\omega\, c(\omega;\psi,\delta\mathbf{u}) = f_S(\omega;\delta\mathbf{u}), \tag{22}$$

[1] Vector space $\mathcal{C}^c$ is the Sobolev space $(H^1(\Omega_S))^3$.

[2] Vector space $\mathcal{E}^c$ is the Sobolev space $H^1(\Omega)$.

in which $a^S(\omega\,;\mathbf{u}\,,\delta\mathbf{u})$ is the dynamic stiffness sesquilinear form of the master structure defined on $\mathcal{C}^c \times \mathcal{C}^c$ by

$$a^S(\omega\,;\mathbf{u}\,,\delta\mathbf{u}) = -\omega^2 m^S(\mathbf{u}\,,\delta\mathbf{u}) + i\omega\, d^S(\omega\,;\mathbf{u}\,,\delta\mathbf{u}) + k^S(\omega\,;\mathbf{u}\,,\delta\mathbf{u}) \quad , \quad (23-1)$$

where mass, damping and stiffness structural sesquilinear forms m^S, d^S and k^S are defined on $\mathcal{C}^c \times \mathcal{C}^c$ and are such that

$$m^S(\mathbf{u}\,,\delta\mathbf{u}) = \int_{\Omega_S} \rho_S\, \mathbf{u}\cdot \overline{\delta\mathbf{u}}\; d\mathbf{x} \quad , \qquad (23-2)$$

$$d^S(\omega\,;\mathbf{u}\,,\delta\mathbf{u}) = \int_{\Omega_S} b_{ijkh}(\omega)\, \varepsilon_{kh}(\mathbf{u})\, \varepsilon_{ij}(\overline{\delta\mathbf{u}})\, d\mathbf{x} \quad , \qquad (23-3)$$

$$k^S(\omega\,;\mathbf{u}\,,\delta\mathbf{u}) = \int_{\Omega_S} a_{ijkh}(\omega)\, \varepsilon_{kh}(\mathbf{u})\, \varepsilon_{ij}(\overline{\delta\mathbf{u}})\, d\mathbf{x} \quad . \qquad (23-4)$$

In Eq. (22), $a^E(\omega/c_E; \mathbf{u}\,,\delta\mathbf{u})$ is the sesquilinear form defined on $\mathcal{C}^c \times \mathcal{C}^c$ such that [3]

$$-\omega^2 a^E(\omega/c_E;\, \mathbf{u}\,,\delta\mathbf{u}) = i\omega \int_{\Gamma_E} (\overline{\delta\mathbf{u}}\cdot \mathbf{n}^S)\;\; \mathbf{Z}_{\Gamma_E}(\omega)\{\mathbf{u}\cdot \mathbf{n}^S\}\; ds$$

$$= \omega^2 \rho_E \int_{\Gamma_E} (\overline{\delta\mathbf{u}}\cdot \mathbf{n}^S)\;\; \mathbf{B}_{\Gamma_E}(\omega/c_E)\{\mathbf{u}\cdot \mathbf{n}^S\}\; ds \quad , (24)$$

in which $\mathbf{B}_{\Gamma_E}(\omega/c_E)$ is related to operator $\mathbf{Z}_{\Gamma_E}(\omega)$ by Eq. (XII.11). Sesquilinear form $\jmath(\mathbf{u}, \delta\mathbf{u})$ defined on $\mathcal{C}^c \times \mathcal{C}^c$ is such that

$$\jmath(\mathbf{u}, \delta\mathbf{u}) = \pi_2(\mathbf{u})\, \pi_2(\overline{\delta\mathbf{u}}) \quad . \qquad (25)$$

Coupling sesquilinear form $c(\omega\,;\psi, \delta\mathbf{u})$ defined on $\mathcal{E}^c \times \mathcal{C}^c$ is such that

$$c(\omega\,;\psi, \delta\mathbf{u}) = c_O(\psi, \delta\mathbf{u}) + \kappa(\omega)\, c_Z(\omega\,;\psi, \delta\mathbf{u}) \quad , \qquad (26)$$

$$c_O(\psi, \delta\mathbf{u}) = \rho_0 \int_{\Gamma\cup\Gamma_Z} \psi\; \mathbf{n}\cdot \overline{\delta\mathbf{u}}\, ds \quad , \qquad (27)$$

$$c_Z(\omega\,;\psi, \delta\mathbf{u}) = -\rho_0\, \pi_1(\omega; \psi)\, \pi_2(\overline{\delta\mathbf{u}}) \quad , \qquad (28)$$

3 From Section XII.4, linear operator $\mathbf{B}_{\Gamma_E}(\omega/c_E)$ is continuous from C'_{Γ_E} into C_{Γ_E} where $C_{\Gamma_E} = H^{1/2}(\Gamma_E)$ and C'_{Γ_E} is the dual space of C_{Γ_E} (see footnote 5 of Chapter XII). Since the vector space of the trace on Γ_E of functions of $\mathcal{C}^c$ is C_{Γ_E}, the right-hand side of Eq. (24) is defined and sesquilinear form $a^E(\omega/c_E;\mathbf{u}\,,\delta\mathbf{u})$ is continuous on $\mathcal{C}^c \times \mathcal{C}^c$.

in which sesquilinear forms $c_0(\psi, \delta\mathbf{u})$ and $c_Z(\psi, \delta\mathbf{u})$ are defined on $\mathcal{E}^c \times \mathcal{C}^c$ and linear forms $\pi_1(\omega; \psi)$ and $\pi_2(\mathbf{u})$ are defined on $\mathcal{E}^c$ and $\mathcal{C}^c$ respectively. Finally, $f_S(\omega\,; \delta\mathbf{u})$ in the right-hand side of Eq. (22) is the antilinear form on $\mathcal{C}^c$ defined by

$$f_S(\omega\,; \delta\mathbf{u}) = f(\omega\,; \delta\mathbf{u}) + f_{p_{\text{given}}}(\omega\,; \delta\mathbf{u}) + \kappa(\omega)\, \pi_{\text{Q}}(\omega)\, \pi_2(\overline{\delta\mathbf{u}}) \quad , \tag{29}$$

in which antilinear form $f(\omega\,; \delta\mathbf{u})$ is defined by

$$f(\omega\,; \delta\mathbf{u}) = \int_{\Omega_S} \mathbf{g}(\mathbf{x}, \omega) \cdot \overline{\delta\mathbf{u}(\mathbf{x})}\; d\mathbf{x} + \int_{\partial\Omega_S} \mathbf{G}(\mathbf{x}, \omega) \cdot \overline{\delta\mathbf{u}(\mathbf{x})}\; ds(\mathbf{x}) \quad , \tag{30}$$

and antilinear form $f_{p_{\text{given}}}(\omega\,; \delta\mathbf{u})$ is defined by

$$f_{p_{\text{given}}}(\omega\,; \delta\mathbf{u}) = -\int_{\Gamma_E} p_{\text{given}}|_{\Gamma_E}(\mathbf{x}, \omega)\; \mathbf{n}^S(\mathbf{x}) \cdot \overline{\delta\mathbf{u}}(\mathbf{x})\, ds(\mathbf{x}) \quad . \tag{31}$$

Variational equation in $\delta\psi$ related to the internal acoustic fluid. Let $\mathbf{u}$ be in $\mathcal{C}^c$ and ψ in $\mathcal{E}^c_\psi$. Multiplying Eq. (15) by $\overline{\delta\psi} \in \mathcal{E}^c_\psi$, integrating over domain Ω, using Green's formula, taking into account the Neumann boundary conditions defined by Eqs. (16) and (17), and since $\int_\Omega \delta\psi\, d\mathbf{x} = 0$, we obtain

$$i\omega\, \widetilde{c}(\omega; \mathbf{u}, \delta\psi) - a^F(\omega\,; \psi\,, \delta\psi) = -i\omega f_{\text{Q}}(\omega; \delta\psi) \quad , \quad \forall\, \delta\psi \in \mathcal{E}^c_\psi, \tag{32}$$

in which sesquilinear form $a^F(\omega\,; \psi\,, \delta\psi)$ defined on $\mathcal{E}^c \times \mathcal{E}^c$ is such that

$$a^F(\omega\,; \psi, \delta\psi) = -\omega^2 m(\psi, \delta\psi) + i\omega\, d_\tau(\omega; \psi, \delta\psi) + k(\psi, \delta\psi) + s_z(\omega; \psi, \delta\psi)\,. \tag{33}$$

Sesquilinear form $k(\psi, \delta\psi)$ is defined on $\mathcal{E}^c \times \mathcal{E}^c$ by

$$k(\psi\,, \delta\psi) = \rho_0 \int_\Omega \boldsymbol{\nabla}\psi \cdot \boldsymbol{\nabla}\overline{\delta\psi}\; d\mathbf{x} \quad , \tag{34}$$

and is positive semidefinite. The restriction of $k(\psi\,, \delta\psi)$ to the admissible function space $\mathcal{E}^c_\psi \times \mathcal{E}^c_\psi$ is positive-definite. Positive-semidefinite sesquilinear form $d_\tau(\omega\,; \psi\,, \delta\psi)$ defined on $\mathcal{E}^c \times \mathcal{E}^c$ is such that

$$d_\tau(\omega\,; \psi\,, \delta\psi) = \tau(\omega)\, k(\psi\,, \delta\psi) \quad . \tag{35}$$

Sesquilinear form $d_T(\omega\,;\psi\,,\delta\psi)$ has the same properties as $k(\psi\,,\delta\psi)$ and is consequently positive definite on $\mathcal{E}_\psi^c \times \mathcal{E}_\psi^c$. Positive-definite sesquilinear form $m(\psi\,,\delta\psi)$ on $\mathcal{E}^c \times \mathcal{E}^c$ is such that [4]

$$m(\psi\,,\delta\psi) = \frac{\rho_0}{c_0^2}\int_\Omega \psi\,\overline{\delta\psi}\;d\mathbf{x} \quad . \tag{36}$$

Sesquilinear form $s_Z(\omega\,;\psi\,,\delta\psi)$ defined on $\mathcal{E}^c \times \mathcal{E}^c$ is such that [5]

$$s_Z(\omega\,;\psi\,,\delta\psi) = -\rho_0^2\omega^2\int_{\Gamma_Z}\frac{1}{i\omega Z}\,\psi\,\overline{\delta\psi}\;ds + \omega^2\kappa(\omega)\,\rho_0^2\,\pi_1(\omega;\psi)\,\pi_1(\omega;\overline{\delta\psi}), \tag{37}$$

and satisfies [6] $\omega^{-1}\Im m\{s_Z(\omega;\psi,\psi)\} \geq 0$. In Eq. (32), coupling sesquilinear form $\widetilde{c}(\omega\,;\mathbf{u},\delta\psi)$ defined on $\mathcal{C}^c \times \mathcal{E}^c$, is such that

$$\widetilde{c}(\omega\,;\mathbf{u},\delta\psi) = \overline{c_0(\delta\psi\,,\mathbf{u}) + \kappa(\omega)\,c_Z(\omega\,;\overline{\delta\psi},\overline{\mathbf{u}})} \quad . \tag{38}$$

Finally, $f_Q(\omega\,;\delta\psi)$ in the right-hand side of Eq. (32) is the antilinear form on $\mathcal{E}^c$ defined by

$$f_Q(\omega\,;\delta\psi) = \int_\Omega g_Q\,\overline{\delta\psi}\,d\mathbf{x} + \int_{\Gamma\cup\Gamma_Z} G_Q\,\overline{\delta\psi}\,ds + \int_{\Gamma_Z} G_{Q,Z}\,\overline{\delta\psi}\,ds \quad . \tag{39}$$

Using Eqs. (18) to (20) and Green's formula for the term containing $\boldsymbol{\nabla}^2 Q$, we obtain

$$f_Q(\omega;\delta\psi) = -\int_\Omega \frac{Q}{i\omega}\overline{\delta\psi}\,d\mathbf{x} + \frac{\tau c_0^2}{\omega^2}\int_\Omega \boldsymbol{\nabla}Q\cdot\boldsymbol{\nabla}\overline{\delta\psi}\,d\mathbf{x} + \rho_0\,\kappa(\omega)\,\pi_Q(\omega)\,\pi_1(\omega\,;\overline{\delta\psi}). \tag{40}$$

Variational formulation. For all fixed real $\omega \neq 0$, the variational formulation is written as follows. Find $\mathbf{u}$ in $\mathcal{C}^c$ and ψ in $\mathcal{E}_\psi^c$ such that, for all $\delta\mathbf{u}$ in $\mathcal{C}^c$ and $\delta\psi$ in $\mathcal{E}_\psi^c$, Eqs. (22) and (32) are satisfied.

---

4 Sesquilinear form $m(\psi\,,\delta\psi)$ is continuous on $H^c \times H^c$ with $H^c = L^2(\Omega)$ and is consequently also continuous on $\mathcal{E}^c \times \mathcal{E}^c$.

5 Sesquilinear form $s_Z(\omega\,;\psi\,,\delta\psi)$ is continuous on $H^c \times H^c$ with $H^c = L^2(\Omega)$ and is consequently also continuous on $\mathcal{E}^c \times \mathcal{E}^c$.

6 For all ω fixed in $\mathbb{R}$, since function $\mathbf{x} \mapsto \{i\omega Z(\mathbf{x},\omega)\}^{-1}$ is bounded on Γ_Z (see Eq. (X.18)), we deduce that $s_Z(\omega\,;\psi\,,\delta\psi)$ is a continuous sesquilinear form on $\mathcal{E}^c \times \mathcal{E}^c$. Due to the hypotheses on Z (see Eqs. (X.18) to (X.25)), it can be proved that $\omega^{-1}\Re e\{-is_Z(\omega;\psi,\psi)\} = \omega^{-1}\Im m\{s_Z(\omega;\psi,\psi)\} \geq 0$.

4.3. Linear operator equation

The linear operator equation corresponding to the variational formulation defined by Eqs. (22) and (32) is

$$\mathbf{A}(\omega)\begin{bmatrix}\mathbf{u}\\ \psi\end{bmatrix} = \begin{bmatrix}\mathbf{f}_{\mathrm{S}}(\omega)\\ -i\omega\,\mathbf{f}_{\mathrm{Q}}(\omega)\end{bmatrix} \quad , \quad \{\mathbf{u},\psi\}\in \mathcal{C}^{c}\times\mathcal{E}_{\psi}^{c} \quad , \tag{41}$$

in which $\mathbf{A}(\omega)$ is the linear operator [7]

$$\mathbf{A}(\omega) = \begin{bmatrix} \mathbf{A}^{S}(\omega) - \omega^2\mathbf{A}^{E}(\omega/c_E) + \kappa(\omega)\,\mathbf{J} & i\omega\,\mathbf{C}(\omega) \\ i\omega\,{}^t\overline{\mathbf{C}}(\omega) & -\mathbf{A}^{F}(\omega) \end{bmatrix} . \tag{42}$$

The dynamic stiffness linear operator $\mathbf{A}^S(\omega)$ of the master structure is such that [8]

$$<\mathbf{A}^{S}(\omega)\,\mathbf{u}\,,\delta\mathbf{u}>= a^{S}(\omega\,;\mathbf{u}\,,\delta\mathbf{u}) \quad , \tag{43}$$

and can be written as

$$\mathbf{A}^{S}(\omega) = -\omega^2\mathbf{M}^{S} + i\omega\,\mathbf{D}^{S}(\omega) + \mathbf{K}^{S}(\omega) \quad , \tag{44 – 1}$$

where $\mathbf{M}^S$, $\mathbf{D}^S(\omega)$ and $\mathbf{K}^S(\omega)$ are the mass, damping and stiffness operators of the master structure defined by

$$<\mathbf{M}^{S}\,\mathbf{u}\,,\delta\mathbf{u}>= m^{S}(\mathbf{u}\,,\delta\mathbf{u}) \quad , \tag{44 – 2}$$

$$<\mathbf{D}^{S}(\omega)\,\mathbf{u}\,,\delta\mathbf{u}> = d^{S}(\omega\,;\mathbf{u}\,,\delta\mathbf{u}) \quad , \tag{44 – 3}$$

$$<\mathbf{K}^{S}(\omega)\,\mathbf{u}\,,\delta\mathbf{u}> = k^{S}(\omega\,;\mathbf{u}\,,\delta\mathbf{u}) \quad . \tag{44 – 4}$$

Linear operator $\mathbf{A}^E(\omega/c_E)$ is such that [9]

$$<\mathbf{A}^{E}(\omega/c_E)\,\mathbf{u}\,,\delta\mathbf{u}>= a^{E}(\omega/c_E\,;\,\mathbf{u}\,,\delta\mathbf{u}) \quad . \tag{45}$$

Linear operator $\mathbf{A}(\omega)$ defined by Eq. (42) is continuous from $\mathcal{C}^c\times\mathcal{E}^c$ into $\mathcal{C}^{c\prime}\times\mathcal{E}^{c\prime}$. The restriction to $\mathcal{C}^c\times\mathcal{E}^c_\psi$ of this operator is continuous from $\mathcal{C}^c\times\mathcal{E}^c_\psi$ into $\mathcal{C}^{c\prime}\times\mathcal{E}^{c\prime}_\psi$.

Sesquilinear form $a^S(\omega;\mathbf{u}\,,\delta\mathbf{u})$ is continuous on $\mathcal{C}^c\times\mathcal{C}^c$, $\mathbf{A}^S(\omega)$ is a continuous operator from $\mathcal{C}^c$ into $\mathcal{C}^{c\prime}$ and the angle brackets in Eq. (43) denote the antiduality product between $\mathcal{C}^{c\prime}$ and $\mathcal{C}^c$. For all real $\omega\neq 0$, linear operator $\mathbf{A}^S(\omega)$ is invertible. For $\omega=0$, the restriction to $\mathcal{C}^c_S$ of continuous linear operator $\mathbf{A}^S(0)$ is invertible (see Eq. (V.56) for the definition of $\mathcal{C}^c_S$).

For all ω fixed in $\mathbb{R}$, sesquilinear form $a^E(\omega/c_0\,;\mathbf{u},\delta\mathbf{u})$ is continuous on $\mathcal{C}^c\times\mathcal{C}^c$ (see footnote 3) and then $\mathbf{A}^E(\omega/c_E)$ is a continuous linear operator from $\mathcal{C}^c$ into $\mathcal{C}^{c\prime}$. The angle brackets in Eq. (45) denote the antiduality product between $\mathcal{C}^{c\prime}$ and $\mathcal{C}^c$.

Linear operator $\mathbf{J}$ is such that [10]

$$<\mathbf{J}\,\mathbf{u}\,,\delta\mathbf{u}>= j(\mathbf{u},\delta\mathbf{u}) \quad . \tag{46}$$

Linear operator $\mathbf{A}^F(\omega)$ related to the internal fluid is such that [11]

$$<\mathbf{A}^F(\omega)\,\psi\,,\delta\psi>= a^F(\omega\,;\psi\,,\delta\psi) \quad , \tag{47}$$

and can be written as

$$\mathbf{A}^F(\omega) = -\omega^2\mathbf{M} + i\omega\mathbf{D}_\tau(\omega) + \mathbf{K} + \mathbf{S}_z(\omega) \quad , \tag{48}$$

in which linear operator $\mathbf{M}$ is such that [12]

$$<\mathbf{M}\,\psi\,,\delta\psi>= m(\psi\,,\delta\psi) \quad , \tag{49}$$

linear operator $\mathbf{K}$ is such that [13]

$$<\mathbf{K}\,\psi\,,\delta\psi>= k(\psi\,,\delta\psi) \quad , \tag{50}$$

and linear operator $\mathbf{D}_\tau(\omega)$ is such that

$$<\mathbf{D}_\tau(\omega)\,\psi\,,\delta\psi>= d_\tau(\omega\,;\psi\,,\delta\psi) \quad , \tag{51}$$

and can then be written as $\mathbf{D}_\tau(\omega) = \tau(\omega)\,\mathbf{K}$. Linear operator $\mathbf{S}_z(\omega)$ is such that [14]

$$<\mathbf{S}_z(\omega)\,\psi\,,\delta\psi>= s_z(\omega\,;\psi\,,\delta\psi) \quad . \tag{52}$$

---

10 Sesquilinear form $j(\mathbf{u},\delta\mathbf{u})$ is continuous on $\mathcal{C}^c\times\mathcal{C}^c$ and then $\mathbf{J}$ is a continuous linear operator from $\mathcal{C}^c$ into $\mathcal{C}^{c\prime}$. The angle brackets in Eq. (46) denote the antiduality product between $\mathcal{C}^{c\prime}$ and $\mathcal{C}^c$.

11 Sesquilinear form $a^F(\omega;\psi\,,\delta\psi)$ is continuous on $\mathcal{E}^c\times\mathcal{E}^c$, $\mathbf{A}^F(\omega)$ is a continuous operator from $\mathcal{E}^c$ into $\mathcal{E}^{c\prime}$ and the angle brackets in Eq. (47) denote the antiduality product between $\mathcal{E}^{c\prime}$ and $\mathcal{E}^c$. For all real $\omega\neq0$, linear operator $\mathbf{A}^F(\omega)$ is invertible. For $\omega=0$, the restriction to $\mathcal{E}^c_\psi$ of continuous linear operator $\mathbf{A}^F(0)$ is invertible.

12 Sesquilinear form $m(\psi,\delta\psi)$ is continuous on $H^c\times H^c$ with $H^c=L^2(\Omega)$ and is consequently also continuous on $\mathcal{E}^c\times\mathcal{E}^c$. Then $\mathbf{M}$ is a continuous operator from $\mathcal{E}^c$ into $\mathcal{E}^{c\prime}$ and the angle brackets in Eq. (49) denote the antiduality product between $\mathcal{E}^{c\prime}$ and $\mathcal{E}^c$.

13 Sesquilinear form $k(\psi\,,\delta\psi)$ is continuous on $\mathcal{E}^c\times\mathcal{E}^c$, $\mathbf{K}$ is a continuous operator from $\mathcal{E}^c$ into $\mathcal{E}^{c\prime}$ and the angle brackets in Eq. (50) denote the antiduality product between $\mathcal{E}^{c\prime}$ and $\mathcal{E}^c$. Continuous linear operator $\mathbf{K}$ is positive semidefinite and Hermitian from $\mathcal{E}^c$ into $\mathcal{E}^{c\prime}$. This operator is coercive on $\mathcal{E}^c_\psi$.

14 Sesquilinear form $s_z(\omega;\psi,\delta\psi)$ is continuous on $\mathcal{E}^c\times\mathcal{E}^c$ (see footnote 6) and then defines a continuous operator $\mathbf{S}_z(\omega)$ from $\mathcal{E}^c$ into $\mathcal{E}^{c\prime}$. The angle brackets in Eq. (52) denote the antiduality product between $\mathcal{E}^{c\prime}$ and $\mathcal{E}^c$.

In Eq. (42), coupling operator $\mathbf{C}(\omega)$ is such that [15]

$$<\mathbf{C}(\omega)\,\psi\,,\delta\mathbf{u}>= c(\omega\,;\psi,\delta\mathbf{u}) \quad , \tag{53}$$

and can be written as

$$\mathbf{C}(\omega) = \mathbf{C}_0 + \kappa(\omega)\,\mathbf{C}_Z(\omega) \quad , \tag{54}$$

in which operators $\mathbf{C}_0$ and $\mathbf{C}_Z(\omega)$ are defined by

$$<\mathbf{C}_0\,\psi\,,\delta\mathbf{u}>= c_0(\psi,\delta\mathbf{u}) \quad , \tag{55}$$

$$<\mathbf{C}_Z(\omega)\,\psi\,,\delta\mathbf{u}>= c_Z(\omega\,;\psi,\delta\mathbf{u}) \quad . \tag{56}$$

Coupling operator ${}^t\overline{\mathbf{C}}(\omega)$ is such that [16]

$$<{}^t\overline{\mathbf{C}}(\omega)\,\mathbf{u}\,,\delta\psi>= \widetilde{c}(\omega\,;\mathbf{u},\delta\psi) \quad , \tag{57}$$

and we have $<{}^t\overline{\mathbf{C}}(\omega)\,\mathbf{u}\,,\delta\psi>=<\mathbf{u}\,,\overline{\mathbf{C}}(\omega)\,\delta\psi>$. Finally, elements $\mathbf{f}_{\mathrm{S}}(\omega)$ and $\mathbf{f}_{\mathrm{Q}}(\omega)$ are such that [17]

$$<\mathbf{f}_{\mathrm{S}}(\omega)\,,\delta\mathbf{u}>= f_S(\omega\,;\delta\mathbf{u}) \quad , \tag{58}$$

$$<\mathbf{f}_{\mathrm{Q}}(\omega)\,,\delta\psi>= f_{\mathrm{Q}}(\omega\,;\delta\psi) \quad . \tag{59}$$

Once $\{\mathbf{u},\psi\}$ is known, internal pressure field $p(\mathbf{x},\omega)$ in Ω can be deduced from the variational formulation of Eq. (1) which is such that, for all δp in H^c,

$$\int_\Omega p\,\overline{\delta p}\,d\mathbf{x} = -i\omega\,\rho_0 \int_\Omega \psi\,\overline{\delta p}\,d\mathbf{x} + \pi(\omega\,;\mathbf{u}\,;\psi)\int_\Omega \overline{\delta p}\,d\mathbf{x} \quad . \tag{60}$$

---

15 Sesquilinear form $c(\omega\,;\psi,\delta\mathbf{u})$ is continuous on $\mathcal{E}^c \times \mathcal{C}^c$ and then $\mathbf{C}(\omega)$ is a continuous linear operator from $\mathcal{E}^c$ into $\mathcal{C}^{c\,\prime}$. The angle brackets in Eqs. (53), (55) and (56) denote the antiduality product between $\mathcal{C}^{c\,\prime}$ and $\mathcal{C}^c$.

16 Sesquilinear form $\widetilde{c}(\omega\,;\delta\psi,\mathbf{u})$ is continuous on $\mathcal{E}^c \times \mathcal{C}^c$ and then ${}^t\overline{\mathbf{C}}(\omega)$ is a continuous linear operator from $\mathcal{C}^c$ into $\mathcal{E}^{c\,\prime}$. The angle brackets in Eq. (57) denote the antiduality product between $\mathcal{E}^{c\,\prime}$ and $\mathcal{E}^c$.

17 Antilinear forms $f_S(\omega;\delta\mathbf{u})$ and $f_{\mathrm{Q}}(\omega;\delta\psi)$ are continuous on $\mathcal{C}^c$ and $\mathcal{E}^c$ respectively, and define elements $\mathbf{f}_{\mathrm{S}}(\omega)$ in $\mathcal{C}^{c\,\prime}$ and $\mathbf{f}_{\mathrm{Q}}(\omega)$ in $\mathcal{E}^{c\,\prime}$ respectively. The angle brackets in Eq. (58) denote the antiduality product between $\mathcal{C}^{c\,\prime}$ and $\mathcal{C}^c$. The angle brackets in Eq. (59) denote the antiduality product between $\mathcal{E}^{c\,\prime}$ and $\mathcal{E}^c$.

We then deduce that internal pressure field $p(\mathbf{x},\omega)$ is the unique solution of the linear operator equation

$$\frac{c_0^2}{\rho_0}\mathbf{M}p = -i\omega\,\rho_0\,{}^t\mathbf{C}_1\psi + \pi(\omega\,;\mathbf{u}\,;\psi)\,\mathbf{l} \quad , \tag{61}$$

in which $\mathbf{l}$ is the antilinear form defined on H^c such that

$$<\mathbf{l},\delta p>= \int_\Omega \overline{\delta p}\,d\mathbf{x} \quad , \tag{62}$$

and linear operator ${}^t\mathbf{C}_1$ is such that [18]

$$<{}^t\mathbf{C}_1\psi\,,\delta p>= \int_\Omega \psi\,\overline{\delta p}\,d\mathbf{x} \quad . \tag{63}$$

Pressure field $p_E|_{\Gamma_E}(\omega)$ on Γ_E due to the external acoustic fluid is given by Eq. (9) and at any point $\mathbf{x}$ of unbounded external acoustic fluid Ω_E, resultant pressure field $p_E(\mathbf{x},\omega)$ is given by Eq. (10).

5. Finite Element Discretization

We consider a finite element mesh of master structure Ω_S and a finite element mesh of internal acoustic fluid Ω. We assume that the two finite element meshes are compatible on interface $\Gamma \cup \Gamma_Z$. The finite element mesh of surface Γ_E is the trace of the mesh of Ω_S (see Fig. 2).

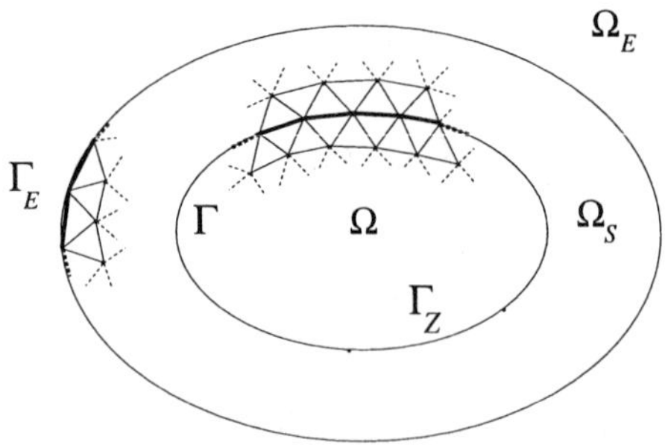

Fig. 2. Example of structure and internal fluid finite element meshes

18 Sesquilinear form $(\psi,\delta p)\mapsto\int_\Omega \psi\,\overline{\delta p}\,d\mathbf{x}$ is continuous on $\mathcal{E}^c\times H^c$ and then ${}^t\mathbf{C}_1$ is a continuous linear operator from $\mathcal{E}^c$ into $H^{c\,\prime}=H^c$. The angle brackets in Eq. (63) denote the inner product in H^c.

We used the finite element method to construct the discretization of the various sesquilinear forms of the problem. For the master structure, we introduce subspace $\mathcal{C}_{n_S} \subset \mathcal{C}^c$ of finite dimension n_S. Let $\mathbf{U} = (U_1, \ldots, U_{n_S})$ be the complex vector of the DOFs which are the values of $\mathbf{u}$ at the nodes of the finite element mesh of domain Ω_S. For the internal acoustic fluid, we introduce subspace $\mathcal{E}_n \subset \mathcal{E}^c$ of finite dimension n. Let $\mathbf{\Psi} = (\Psi_1, \ldots, \Psi_n)$ be the complex vectors of the DOFs which are the values of ψ at the nodes of the finite element mesh of domain Ω. Since the finite element method uses a real basis for constructing the finite element matrices, the finite element discretization of the variational formulation defined by Eqs. (22) and (32) yields the symmetric complex matrix equation (corresponding to the operator equation defined by Eq. (41),

$$[\mathbb{A}(\omega)] \begin{bmatrix} \mathbf{U} \\ \mathbf{\Psi} \end{bmatrix} = \begin{bmatrix} \mathbf{F}_S(\omega) \\ -i\omega\, \mathbf{F}_Q(\omega) \end{bmatrix} \quad , \tag{64}$$

with the constraint

$$\mathbf{L}^T \mathbf{\Psi} = 0 \quad . \tag{65}$$

Eq. (65) corresponds to the finite element discretization of linear constraint $\int_\Omega \psi \, d\mathbf{x} = 0$ in admissible function space $\mathcal{E}_\psi^c$ defined by Eq. (21), in which $\mathbf{L}$ is a vector in $\mathbb{R}^n$. Complex matrix $[\mathbb{A}(\omega)]$ is symmetric and is defined by

$$[\mathbb{A}(\omega)] = \begin{bmatrix} [A^S(\omega)] - \omega^2 [A^E(\omega/c_E)] + \kappa(\omega)[J] & i\omega [C(\omega)] \\ i\omega\, [C(\omega)]^T & -[A^F(\omega)] \end{bmatrix} \quad , \tag{66}$$

where $[A^S(\omega)]$ is the dynamic stiffness matrix of the master structure which is an $(n_S \times n_S)$ symmetric complex matrix (invertible for all real $\omega \neq 0$ and not invertible for $\omega = 0$ due to the rigid body modes) written as

$$[A^S(\omega)] = -\omega^2 [M^S] + i\omega\, [D^S(\omega)] + [K^S(\omega)] \quad . \tag{67}$$

Mass, damping and stiffness matrices $[M^S]$, $[D^S(\omega)]$ and $[K^S(\omega)]$ are $(n_S \times n_S)$ symmetric real matrices. Matrix $[M^S]$ is positive definite and matrices $[D^S(\omega)]$ and $[K^S(\omega)]$ are positive semidefinite (due to the presence of the rigid body modes). The finite element discretization of sesquilinear form $a^E(\omega/c_E\,; \mathbf{u}, \delta\mathbf{u})$ yields the $(n_S \times n_S)$ symmetric complex matrix $[A^E(\omega/c_E)]$ related to the nodal values of $\mathbf{u}$, written as

$$[A^E(\omega/c_E)] = -\rho_E\, [\Theta]^T\, [B_{\Gamma_E}(\omega/c_E)]\, [\Theta] \quad , \tag{68}$$

where $[B_{\Gamma_E}(\omega/c_E)]$ is the $(n_E \times n_E)$ symmetric dense complex matrix defined by Eq. (XII.164) and related to the nodal values of the normal displacement field on Γ_E. Rectangular matrix $[\,\Theta\,]$ is an $(n_E \times n_S)$ sparse real matrix corresponding to the finite element discretization $\overline{\delta \mathbf{W}}^T\,[\,\Theta\,]\,\mathbf{U}$ of the sesquilinar form

$$(\mathbf{u}, \delta w) \mapsto \theta(\mathbf{u}, \delta w) = \int_{\Gamma_E} \overline{\delta w}\; \mathbf{u} \cdot \mathbf{n}^S \, ds \tag{69}$$

defined on $\mathcal{C}^c \times \mathcal{C}_{\Gamma_E}$. Eqs. (68) and (69) show that matrix $[A^E(\omega/c_E)]$ has zero rows and zero columns corresponding to the structural nodes which are not located on interface Γ_E. Matrix $[\,J\,]$ is an $(n_S \times n_S)$ symmetric real matrix which can be written as

$$[\,J\,] = \mathbf{\Pi}_2\, \mathbf{\Pi}_2^T \quad , \tag{70}$$

in which $\mathbf{\Pi}_2$ is a vector in $\mathbb{R}^{n_S}$ such that $\mathbf{\Pi}_2^T$ corresponds to the finite element discretization of the linear form $\pi_2(\mathbf{u})$ defined by Eq. (4). Matrix $[\,J\,]$ has zero rows and zero columns corresponding to the structural nodes which are not located on interface $\Gamma \cup \Gamma_Z$. Matrix $[A^F(\omega)]$ corresponding to the finite element discretization of sesquilinear form $a^F(\omega\,;\psi\,,\delta\psi)$ is an $(n\times n)$ symmetric complex matrix (invertible for all real $\omega \neq 0$) written as

$$[A^F(\omega)] = -\omega^2\,[\,M\,] + i\omega\,[D_\tau(\omega)] + [\,K\,] + [S_z(\omega)] \quad , \tag{71}$$

in which matrices $[\,M\,]$, $[\,K\,]$, $[D_\tau(\omega)]$ and $[S_z(\omega)]$ correspond to the finite element discretization of the sesquilinear forms defined by Eqs. (36), (34), (35) and (37). The $(n\times n)$ real matrix $[\,M\,]$ is symmetric positive definite, the $(n\times n)$ real matrices $[\,K\,]$ and $[D_\tau(\omega)]$ are symmetric positive semidefinite, with rank equal to $n-1$ and the null space of which is constituted by the vector $c\,\mathbf{1}$ where c is any real constant and $\mathbf{1} = (1,\ldots,1) \in \mathbb{R}^n$. The $(n\times n)$ complex matrix $[S_z(\omega)]$ is symmetric and such that $\omega^{-1}\Im m[S_z(\omega)]$ is positive semidefinite (see footnote 6). Coupling rectangular matrix $[C(\omega)]$ corresponding to the finite element discretization of coupling sesquilinear form $c(\omega\,;\psi, \delta\mathbf{u})$ is an $(n_S \times n)$ complex matrix which is written as

$$[C(\omega)] = [C_0] + \kappa(\omega)\,[C_Z(\omega)] \quad , \tag{72}$$

in which $[C_0]$ is the $(n_S\times n)$ real matrix corresponding to the finite element discretization of sesquilinear form $c_0(\psi\,, \delta\mathbf{u})$ and $[C_Z(\omega)]$ is the $(n_S\times n)$ complex matrix corresponding to the finite element discretization of sesquilinear form $c_Z(\omega\,;\psi\,,\delta\mathbf{u})$ and such that

$$[C_Z(\omega)] = -\rho_0\,\mathbf{\Pi}_2\,\mathbf{\Pi}_1(\omega)^T \quad , \tag{73}$$

where $\boldsymbol{\Pi}_1(\omega)$ is a vector in $\mathbb{C}^n$ such that $\boldsymbol{\Pi}_1(\omega)^T$ corresponds to the finite element discretization of the linear form $\pi_1(\omega\,;\varphi)$. Matrix $[C(\omega)]$ has zero rows and zero columns corresponding to the structural and fluid nodes which are not located on interface $\Gamma \cup \Gamma_Z$. Finally, the finite element discretization of antilinear forms $f_S(\omega\,;\delta\mathbf{u})$ and $f_Q(\omega\,;\delta\psi)$ yields the complex vectors $\mathbf{F}_S(\omega)$ and $\mathbf{F}_Q(\omega)$ in $\mathbb{C}^{n_S}$ and $\mathbb{C}^n$ respectively.

Calculation of the pressure inside Ω. Let $\mathbf{P} = (p_1, \ldots, p_n)$ be the complex vector of the DOFs which are the values of p at the nodes of the finite element mesh of domain Ω. Then the finite element discretization of Eq. (60) is written as

$$\frac{c_0^2}{\rho_0}[\,M\,]\mathbf{P} = -i\omega\, c_0^2[\,M\,]\boldsymbol{\Psi} + \kappa(\omega)\big\{i\omega\,\rho_0\boldsymbol{\Pi}_1(\omega)^T\boldsymbol{\Psi} - \boldsymbol{\Pi}_2^T\mathbf{U} + \pi_Q(\omega)\big\}\ \mathbf{L}, \quad (74)$$

in which $\mathbf{L} \in \mathbb{R}^n$ is the finite element discretization of antilinear form $\mathbf{l}$ defined by Eq. (62) (which is also used in Eq. (65)) and where $\pi_Q(\omega)$ is defined by Eq. (5). It should be noted that the construction of $\mathbf{P}$ requires solving a linear matrix equation with nondiagonal matrix $(c_0^2/\rho_0)\,[\,M\,]$. An alternative discretization of Eq. (60) can be used to recover p locally element by element. Since $p(\omega)$ belongs to H^c, the pressure can be taken, for instance, as a constant over each element. In this case, no linear matrix equation of order n has to be solved.

Calculation of the pressure on boundary Γ_E and of the resultant pressure at any point of external domain Ω_E. The finite element discretization $\mathbf{P}_{\Gamma_E}(\omega)$ of pressure field $p_E|_{\Gamma_E}(\omega)$ on Γ_E is given by Eq. (XII.173) in which $\mathbf{V}$ is the vector in $\mathbb{C}^{n_E}$ such that $\mathbf{V} = [\,\Theta\,]\,\mathbf{U}$. At any point of external domain Ω_E, the finite element approximation of resultant pressure $p_E(\mathbf{x}, \omega)$ is given by Eq. (XII.174) in which $\mathbf{V}$ is defined above.

6. FRF Calculation in the MF Range

Let $\mathbb{B}_{\text{MF}}$ be the MF broad band defined by

$$\mathbb{B}_{\text{MF}} = [\omega_{\text{MF,init}}\,,\,\omega_{\text{MF,final}}]\quad . \quad (75)$$

The MF narrow band $\mathbb{B}_\nu$ is defined by

$$\mathbb{B}_\nu = [\,\Omega_\nu - \Delta\omega/2\,,\,\Omega_\nu + \Delta\omega/2\,]\quad , \quad (76)$$

where $\Omega_\nu > 0$ is the center frequency of band $\mathbb{B}_\nu$ and $\Delta\omega$ is its bandwidth. We use the finite element discretization introduced in Section 5 to construct the FRF in the MF range by the method described in Chapter VII. For all ω in $\mathbb{B}_{\text{MF}}$, we have to construct the FRF $[T(\omega)] = [\mathbb{A}(\omega)]^{-1}$ where $[\mathbb{A}(\omega)]$ is defined by Eq. (66). We then have to solve for the class of MF narrow band excitations $\theta_\nu(\omega)\,\mathbf{B}$ introduced in Section VII.4,

$$[\mathbb{A}(\omega)]\,\mathbf{Y}(\omega) = \theta_\nu(\omega)\,\mathbf{B} \quad , \quad \omega \in \mathbb{B}_\nu \quad \text{with} \quad \mathbf{Y}(\omega) = \begin{bmatrix} \mathbf{U}(\omega) \\ \mathbf{\Psi}(\omega) \end{bmatrix} \quad , \tag{77}$$

with the constraint

$$\mathbf{L}^T\mathbf{\Psi}(\omega) = 0 \quad , \quad \omega \in \mathbb{B}_\nu \quad . \tag{78}$$

Symmetric complex matrix $[\mathbb{A}(\omega)]$ can be written as

$$[\mathbb{A}(\omega)] = -\omega^2[\mathbb{M}(\omega)] + i\omega\,[\mathbb{D}(\omega)] + [\mathbb{K}(\omega)] \quad , \tag{79}$$

where $[\mathbb{M}(\omega)]$, $[\mathbb{D}(\omega)]$ and $[\mathbb{K}(\omega)]$ are the symmetric complex matrices such that

$$[\mathbb{M}(\omega)] = \begin{bmatrix} [M^S] + [A^E(\omega/c_E)] & [\,0\,] \\ [\,0\,] & -[\,M\,] \end{bmatrix} \quad , \tag{80}$$

$$[\mathbb{D}(\omega)] = \begin{bmatrix} [D^S(\omega)] & [C(\omega)] \\ [C(\omega)]^T & -[D_\tau(\omega)] - [S_Z^{\mathrm{I}}(\omega)] \end{bmatrix} \quad , \tag{81}$$

$$[\mathbb{K}(\omega)] = \begin{bmatrix} [K^S(\omega)] + \kappa(\omega)[J] & [\,0\,] \\ [\,0\,] & -[\,K\,] - [S_Z^{\mathrm{R}}(\omega)] \end{bmatrix} \quad , \tag{82}$$

in which we have written

$$[S_Z(\omega)] = [S_Z^{\mathrm{R}}(\omega)] + i\omega\,[S_Z^{\mathrm{I}}(\omega)] \quad , \tag{83}$$

where real matrices $[S_Z^{\mathrm{R}}(\omega)]$ and $[S_Z^{\mathrm{I}}(\omega)]$ are such that

$$[S_Z^{\mathrm{R}}(\omega)] = \Re e\,[S_Z(\omega)] \quad , \quad \omega\,[S_Z^{\mathrm{I}}(\omega)] = \Im m\,[S_Z(\omega)] \quad . \tag{84}$$

For construction of the frequency response function, we take $\theta_\nu(\omega) = 1$ for $\omega \in \mathbb{B}_\nu$ and $\mathbf{B}$ represents the vectors of the canonical basis of $\mathbb{R}^{n_S+n}$. Concerning the methods for calculating the response to various types of excitations, we refer the reader to Chapter IX.

6.1. Direct use of the MF method

The MF method presented in Chapter VII uses an approximation of Eq. (77) due to the frequency-dependent matrices $[\mathbb{M}(\omega)]$, $[\mathbb{D}(\omega)]$ and $[\mathbb{K}(\omega)]$ consisting in replacing Eq. (77) by

$$\left(-\omega^2[\mathbb{M}_\nu] + i\omega\,[\mathbb{D}_\nu] + [\mathbb{K}_\nu]\right)\mathbf{Y}_\nu(\omega) = \theta_\nu(\omega)\,\mathbf{B} \quad , \quad \omega \in \mathbb{B}_\nu \quad , \tag{85}$$

in which the frequency-independent complex matrices $[\mathbb{M}_\nu]$, $[\mathbb{D}_\nu]$ and $[\mathbb{K}_\nu]$ are such that

$$[\mathbb{M}_\nu] = [\mathbb{M}(\Omega_\nu)] \quad , \quad [\mathbb{D}_\nu] = [\mathbb{D}(\Omega_\nu)] \quad , \quad [\mathbb{K}_\nu] = [\mathbb{K}(\Omega_\nu)] \quad . \tag{86}$$

By inspection of Eqs. (80) to (82), it can be seen that $\omega \mapsto [\mathbb{M}(\omega)]$, $[\mathbb{D}(\omega)]$ and $[\mathbb{K}(\omega)]$ can be considered as slowly varying functions on MF narrow band $\mathbb{B}_\nu$ (from Section XIII.7.5, it is known that function $\omega \mapsto [A^E(\omega/c_E)]$ has a nonresonant behavior and varies slowly on $\mathbb{B}_\nu$; functions $\omega \mapsto [D^S(\omega)]$ and $[K^S(\omega)]$ are related to the viscoelastic coefficients of the master structure and vary slowly; function $\omega \mapsto [D_\tau(\omega)]$ is due to the acoustic dissipation inside the internal fluid and varies slowly on $\mathbb{B}_\nu$; finally, functions $\omega \mapsto [S_{\mathrm{Z}}^{\mathrm{R}}(\omega)]$, $[S_{\mathrm{Z}}^{\mathrm{I}}(\omega)]$ and $\kappa(\omega)$ are related to the wall acoustic impedance $Z(\mathbf{x},\omega)$ and vary slowly on $\mathbb{B}_\nu$). We then apply the frequency transform technique presented in Section VII.6 in order to obtain the associated LF equation which is written in the time domain as

$$[\mathbb{M}_\nu]\,\ddot{\mathbf{Y}}_0(t) + [\widetilde{\mathbb{D}}_\nu]\,\dot{\mathbf{Y}}_0(t) + [\widetilde{\mathbb{K}}_\nu]\mathbf{Y}_0(t) = \theta_0(t)\,\mathbf{B} \quad , \quad \forall t > t_{\mathrm{i}} \; , \tag{87}$$

$$\mathbf{Y}_0(t_{\mathrm{i}}) = \mathbf{0} \quad , \quad \dot{\mathbf{Y}}_0(t_{\mathrm{i}}) = \mathbf{0} \quad , \tag{88}$$

with the constraint

$$\mathbf{L}^T \mathbf{\Psi}_0(t) = 0 \quad , \quad \forall t > t_{\mathrm{i}} \quad , \tag{89}$$

in which $\mathbf{Y}_0(t)$ is such that

$$\mathbf{Y}_0(t) = \begin{bmatrix} \mathbf{U}_0(t) \\ \mathbf{\Psi}_0(t) \end{bmatrix} \quad , \tag{90}$$

and where symmetric complex matrices $[\widetilde{\mathbb{D}}_\nu]$ and $[\widetilde{\mathbb{K}}_\nu]$ are such that

$$[\widetilde{\mathbb{D}}_\nu] = [\mathbb{D}_\nu] + 2\,i\,\Omega_\nu\,[\mathbb{M}_\nu] \quad , \tag{91}$$

$$[\widetilde{\mathbb{K}}_\nu] = -\Omega_\nu^2\,[\mathbb{M}_\nu] + i\,\Omega_\nu\,[\mathbb{D}_\nu] + [\mathbb{K}_\nu] \quad . \tag{92}$$

Once $\mathbf{Y}_0$ is obtained by solving Eqs. (87) to (89) in the time domain, the expression of $\mathbf{Y}_\nu(\omega)$ in MF narrow band $\mathbb{B}_\nu$ is given by

$$\mathbf{Y}_\nu(\omega) \simeq \mathbf{1}_{\mathbb{B}_\nu}(\omega)\, \Delta t \sum_{m=m_\mathrm{i}}^{m_\mathrm{f}} \mathbf{Y}_0(m\,\Delta t)\, e^{-im\,\Delta t\,(\omega-\Omega_\nu)} \quad , \quad \forall \omega \in \mathbb{R} \quad , \tag{93}$$

in which $\mathbf{1}_{\mathbb{B}_\nu}(\omega) = 1$ if $\omega \in \mathbb{B}_\nu$ and $= 0$ if $\omega \notin \mathbb{B}_\nu$ and where integers m_i and m_f are defined in Section VII.6.4.

6.2. Special procedure

From Eq. (37), we deduce that matrix $[S_z(\omega)]$ can be rewritten as

$$[S_z(\omega)] = [S_1(\omega)] + [S_2(\omega)] \quad , \tag{94}$$

in which $[S_1(\omega)]$ is an $(n{\times}n)$ symmetric sparse complex matrix corresponding to the finite element discretization of the sesquilinear form

$$(\psi, \delta\psi) \mapsto -\rho_0^2 \omega^2 \int_{\Gamma_Z} \frac{1}{i\omega\, Z}\, \psi\, \overline{\delta\psi}\, ds \quad , \tag{95}$$

and $[S_2(\omega)]$ is a partially dense $(n \times n)$ symmetric complex matrix corresponding to the finite element discretization of the sesquilinear form

$$(\psi, \delta\psi) \mapsto \omega^2\, \kappa(\omega)\, \rho_0^2\, \pi_1(\omega\,; \psi)\, \pi_1(\omega\,; \overline{\delta\psi}) \quad . \tag{96}$$

Matrix $[S_2(\omega)]$ can then be written as

$$[S_2(\omega)] = \omega^2\, \kappa(\omega)\, \rho_0^2\, \mathbf{\Pi}_1(\omega)\, \mathbf{\Pi}_1(\omega)^T \quad , \tag{97}$$

where $\mathbf{\Pi}_1(\omega)$ is a complex vector such that $\mathbf{\Pi}_1(\omega)^T$ corresponds to the finite element discretization of linear form $\psi \mapsto \pi_1(\omega\,;\psi)$. The right-hand side of Eq. (97) shows that the block submatrix corresponding to the fluid nodes which are located on interface Γ_Z is dense and not sparse (the other elements of this matrix are zero). Similarly, matrix $[C(\omega)]$ defined by Eq. (72) is the sum of sparse real matrix $[C_0]$ with partially dense matrix $[C_Z(\omega)]$ defined by Eq. (73). Finally, matrix $[\,J\,]$ defined by Eq. (70) is partially dense. Consequently, matrix $[\mathbb{A}(\omega)]$ defined by Eq. (79) appears as the sum of a sparse matrix with a partially dense matrix. This property can be put to advantage by using the following algebraic result.

Algebraic result. Let $[A]$ be an $(N \times N)$ invertible complex matrix, $\mathbf{\Gamma}$ be a vector in $\mathbb{C}^N$, κ be a complex number and $\mathbf{F}$ be a vector in $\mathbb{C}^N$. If the linear matrix equation

$$([A] + \kappa\,\mathbf{\Gamma}\mathbf{\Gamma}^T)\,\mathbf{X} = \mathbf{F} \tag{98}$$

has a unique solution $\mathbf{X}$, then this solution can be written as

$$\mathbf{X} = \mathbf{V} - \frac{\kappa\,\mathbf{\Gamma}^T\mathbf{V}}{1 + \kappa\,\mathbf{\Gamma}^T\mathbf{W}}\,\mathbf{W} \quad , \tag{99}$$

in which $\mathbf{V}$ and $\mathbf{W}$ are the solutions of the linear matrix equations

$$[A]\,\mathbf{V} = \mathbf{F} \quad , \tag{100}$$

$$[A]\,\mathbf{W} = \mathbf{\Gamma} \quad . \tag{101}$$

It should be noted that Eqs. (100) and (101) correspond to the same matrix equation with two different second members.

Application to the FRF calculation in the MF range. Complex matrix $[S_1(\omega)]$ appearing in Eq. (94) is written as

$$[S_1(\omega)] = [S_1^{\mathrm{R}}(\omega)] + i\omega\,[S_1^{\mathrm{I}}(\omega)] \quad , \tag{102}$$

in which symmetric sparse real matrices $[S_1^{\mathrm{R}}(\omega)]$ and $[S_1^{\mathrm{I}}(\omega)]$ are such that

$$[S_1^{\mathrm{R}}(\omega)] = \Re e\,[S_1(\omega)] \quad , \quad \omega\,[S_1^{\mathrm{I}}(\omega)] = \Im m\,[S_1(\omega)] \quad . \tag{103}$$

The approximation of Eq. (77) defined by Eq. (85) is then replaced by the new approximation

$$([A_\nu(\omega)] + \kappa(\omega)\,\mathbf{\Gamma}(\omega)\,\mathbf{\Gamma}(\omega)^T)\,\mathbf{Y}_\nu(\omega) = \theta_\nu(\omega)\,\mathbf{B} \quad , \quad \omega \in \mathbb{B}_\nu \quad , \tag{104}$$

with the constraint

$$\mathbf{L}^T\mathbf{\Psi}_\nu(\omega) = 0 \quad , \quad \omega \in \mathbb{B}_\nu \quad , \tag{105}$$

in which $\mathbf{Y}_\nu(\omega) = (\mathbf{U}_\nu(\omega), \mathbf{\Psi}_\nu(\omega))$. In Eq. (104), symmetric sparse complex matrix $[A_\nu(\omega)]$ is defined by

$$[A_\nu(\omega)] = -\omega^2\,[\mathbb{M}_\nu] + i\omega\,[D_\nu] + [K_\nu] \quad , \tag{106}$$

in which frequency-independent matrix $[\mathbb{M}_\nu]$ is defined by Eqs. (86) and (80). In Eq. (106), frequency-independent matrices $[D_\nu]$ and $[K_\nu]$ are such that

$$[D_\nu] = \begin{bmatrix} [D^S(\Omega_\nu)] & [C_0] \\ [C_0]^T & -[D_\tau(\Omega_\nu)] - [S_1^{\mathrm{I}}(\Omega_\nu)] \end{bmatrix} \quad , \tag{107}$$

$$[K_\nu] = \begin{bmatrix} [K^S(\Omega_\nu)] & [\,0\,] \\ [\,0\,] & -[\,K\,] - [S_1^{\mathrm{R}}(\Omega_\nu)] \end{bmatrix} \quad , \tag{108}$$

in which $[C_0]$ is the matrix used in Eq. (72). In Eq. (104), $\mathbf{\Gamma}(\omega)$ is the complex vector such that

$$\mathbf{\Gamma}(\omega) = \begin{bmatrix} \mathbf{\Pi}_2 \\ -i\omega\,\rho_0\,\mathbf{\Pi}_1(\omega) \end{bmatrix} \quad , \tag{109}$$

in which $\mathbf{\Pi}_1(\omega)$ and $\mathbf{\Pi}_2$ are the vectors used in Eqs. (73) and (70). Using Eqs. (98) and (99), the solution of Eqs. (104) and (105) can then be written as

$$\mathbf{Y}_\nu(\omega) = \mathbf{V}_\nu(\omega) - \frac{\kappa(\omega)\mathbf{\Gamma}(\omega)^T\mathbf{V}_\nu(\omega)}{1+\kappa(\omega)\mathbf{\Gamma}(\omega)^T\mathbf{W}_\nu(\omega)}\,\mathbf{W}_\nu(\omega) \quad , \tag{110}$$

in which $\mathbf{V}(\omega)$ and $\mathbf{W}(\omega)$ are the solutions of the linear matrix equations,

$$[A_\nu(\omega)]\,\mathbf{V}_\nu(\omega) = \theta_\nu(\omega)\,\mathbf{B} \quad , \quad \omega \in \mathbb{B}_\nu \quad , \tag{111}$$

with the constraint

$$\mathbf{L}^T\mathbf{V}_\Psi(\omega) = 0 \quad , \quad \omega \in \mathbb{B}_\nu \quad , \tag{112}$$

and

$$[A_\nu(\omega)]\,\mathbf{W}_\nu(\omega) = \mathbf{\Gamma}(\omega) \quad , \quad \omega \in \mathbb{B}_\nu \quad , \tag{113}$$

with the constraint

$$\mathbf{L}^T\mathbf{W}_\Psi(\omega) = 0 \quad , \quad \omega \in \mathbb{B}_\nu \quad , \tag{114}$$

in which we have introduced the notation

$$\mathbf{V}_\nu(\omega) = \begin{bmatrix} \mathbf{V}_U(\omega) \\ \mathbf{V}_\Psi(\omega) \end{bmatrix} \quad , \quad \mathbf{W}_\nu(\omega) = \begin{bmatrix} \mathbf{W}_U(\omega) \\ \mathbf{W}_\Psi(\omega) \end{bmatrix} \quad . \tag{115}$$

Eqs. (110) and (112) and Eqs. (113) and (114) are solved using the direct MF method presented in Section 6.1.

7. Case of a Master Structure Coupled with an External Acoustic Fluid

In this section, we summarize the model, equations and solving methods for a master structure coupled with an external acoustic fluid and without an internal acoustic fluid. The following results are deduced directly from Sections 2 to 6.

7.1. Variational formulation

For all fixed real ω, the variational equation related to the master structure is

$$a^S(\omega; \mathbf{u}, \delta\mathbf{u}) - \omega^2 a^E(\omega/c_E; \mathbf{u}, \delta\mathbf{u}) = f_S(\omega; \delta\mathbf{u}) \quad , \quad \forall \delta\mathbf{u} \in \mathcal{C}^c \, , \tag{116}$$

in which $a^S(\omega\,; \mathbf{u}\,, \delta\mathbf{u})$ is the dynamic stiffness sesquilinear form of the master structure defined on $\mathcal{C}^c \times \mathcal{C}^c$ by

$$a^S(\omega\,; \mathbf{u}\,, \delta\mathbf{u}) = -\omega^2 m^S(\mathbf{u}\,, \delta\mathbf{u}) + i\omega\, d^S(\omega\,; \mathbf{u}\,, \delta\mathbf{u}) + k^S(\omega\,; \mathbf{u}\,, \delta\mathbf{u}) \quad , \tag{117}$$

where mass, damping and stiffness structural sesquilinear forms m^S, d^S and k^S are defined by Eqs. (23-2), (23-3) and (23-4) respectively. In Eq. (116), $a^E(\omega/c_E; \mathbf{u}\,, \delta\mathbf{u})$ is the sesquilinear form defined on $\mathcal{C}^c \times \mathcal{C}^c$, related to the presence of the external acoustic fluid, such that

$$-\omega^2 a^E(\omega/c_E\,; \mathbf{u}\,, \delta\mathbf{u}) = \omega^2 \rho_E \int_{\Gamma_E} (\overline{\delta\mathbf{u}} \cdot \mathbf{n}^S)\; \mathbf{B}_{\Gamma_E}(\omega/c_E)\{\mathbf{u} \cdot \mathbf{n}^S\}\, ds \quad . \tag{118}$$

Antilinear form $f_S(\omega\,; \delta\mathbf{u})$ on $\mathcal{C}^c$ due to external forces is written as

$$f_S(\omega\,; \delta\mathbf{u}) = f(\omega\,; \delta\mathbf{u}) + f_{p_{\text{given}}}(\omega\,; \delta\mathbf{u}) \quad , \tag{119}$$

in which antilinear form $f(\omega\,; \delta\mathbf{u})$ is defined by Eq. (30) and antilinear form $f_{p_{\text{given}}}(\omega\,; \delta\mathbf{u})$ is defined by Eq. (31). For all fixed real $\omega \neq 0$, the variational formulation is written as follows. Find $\mathbf{u}$ in $\mathcal{C}^c$ such that Eq. (116) is satisfied, for all $\delta\mathbf{u}$ in $\mathcal{C}^c$.

7.2. Linear operator equation

The linear operator equation corresponding to the variational formulation defined by Eq. (116) is

$$\mathbf{A}(\omega)\,\mathbf{u} = \mathbf{f}_S(\omega) \quad , \quad \mathbf{u} \in \mathcal{C}^c \quad , \tag{120}$$

in which $\mathbf{A}(\omega)$ is the operator defined by

$$\mathbf{A}(\omega) = \mathbf{A}^S(\omega) - \omega^2\,\mathbf{A}^E(\omega/c_E) \quad , \tag{121}$$

in which the dynamic stiffness linear operator $\mathbf{A}^S(\omega)$ of the master structure is written as

$$\mathbf{A}^S(\omega) = -\omega^2\mathbf{M}^S + i\omega\,\mathbf{D}^S(\omega) + \mathbf{K}^S(\omega) \quad , \tag{122}$$

where $\mathbf{M}^S$, $\mathbf{D}^S(\omega)$ and $\mathbf{K}^S(\omega)$ are the mass, damping and stiffness operators of the master structure defined by Eqs. (44-2), (44-3) and (44-4) respectively. Linear operator $\mathbf{A}^E(\omega/c_E)$ is defined by Eq. (45) and element $\mathbf{f}_S(\omega)$ is defined by Eq. (58).

Pressure in the external acoustic fluid. Pressure field $p_E|_{\Gamma_E}(\omega)$ on Γ_E due to the external acoustic fluid is given by Eq. (9). At any point $\mathbf{x}$ of unbounded external acoustic fluid Ω_E, resultant pressure field $p_E(\mathbf{x},\omega)$ is given by Eq. (10).

7.3. Finite Element Discretization

We consider a finite element mesh of master structure Ω_S. The finite element mesh of surface Γ_E is the trace of the finite element mesh of Ω_S (see Fig. 3).

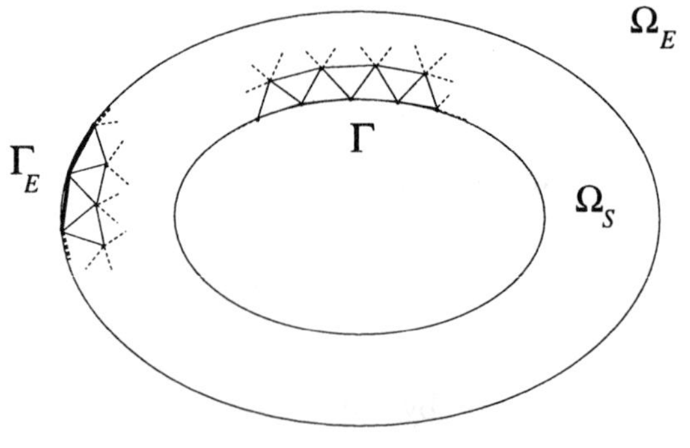

Fig. 3. Finite element mesh of the master structure

We used the finite element method to construct the discretization of the various sesquilinear forms of the problem. For the master structure, we introduce subspace $\mathcal{C}_{n_S} \subset \mathcal{C}^c$ of finite dimension n_S. Let $\mathbf{U} = (U_1, \ldots, U_{n_S})$ be the complex vector of the DOFs which are the values of $\mathbf{u}$ at the nodes of the finite element mesh of domain Ω_S. Since the finite element method uses a real basis for constructing the finite element matrices, the discretization of the variational formulation defined by Eq. (116) yields the symmetric complex matrix equation

$$[\mathbb{A}(\omega)]\,\mathbf{U} = \mathbf{F}_{\mathrm{S}}(\omega) \quad . \tag{123}$$

The complex matrix $[\mathbb{A}(\omega)]$ is symmetric and is defined by

$$[\mathbb{A}(\omega)] = [A^S(\omega)] - \omega^2\,[A^E(\omega/c_E)] \quad , \tag{124}$$

where $[A^S(\omega)]$ is the dynamic stiffness matrix of the master structure which is an $(n_S \times n_S)$ symmetric complex matrix (invertible for all real $\omega \neq 0$ and not invertible for $\omega = 0$ due to the presence of the rigid body modes) written as

$$[A^S(\omega)] = -\omega^2[M^S] + i\omega\,[D^S(\omega)] + [K^S(\omega)] \quad , \tag{125}$$

where $[M^S]$, $[D^S(\omega)]$ and $[K^S(\omega)]$ are the mass, damping and stiffness $(n_S \times n_S)$ symmetric real matrices. Matrix $[M^S]$ is positive definite and matrices $[D^S(\omega)]$ and $[K^S(\omega)]$ are positive semidefinite. The discretization of sesquilinear form $a^E(\omega/c_E\,;\,\mathbf{u}\,,\delta\mathbf{u})$ yields the $(n_S \times n_S)$ symmetric complex matrix $[A^E(\omega/c_E)]$ related to the nodal values of the vector-valued displacement field of the structure, such that

$$[A^E(\omega/c_E)] = -\rho_E\,[\,\Theta\,]^T\,[B_{\Gamma_E}(\omega/c_E)]\,[\,\Theta\,] \quad , \tag{126}$$

where $[B_{\Gamma_E}(\omega/c_E)]$ is the $(n_E \times n_E)$ symmetric dense complex matrix defined by Eq. (XII.164) and related to the nodal values of the normal displacement field on Γ_E. Rectangular matrix $[\,\Theta\,]$ is an $(n_E \times n_S)$ sparse real matrix defined by Eq. (69). Eqs. (118) and (126) show that matrix $[A^E(\omega/c_E)]$ has zero rows and zero columns corresponding to the structural nodes which are not located on interface Γ_E. Finally, the finite element discretization of antilinear form $f_S(\omega\,;\delta\mathbf{u})$ yields the complex vector $\mathbf{F}_{\mathrm{S}}(\omega)$ in $\mathbb{C}^{n_S}$.

Calculation of the pressure on boundary Γ_E and of the resultant pressure at any point of external domain Ω_E. Finite element discretization $\mathbf{P}_{\Gamma_E}(\omega)$ of pressure field $p_E|_{\Gamma_E}(\omega)$ on Γ_E is given by Eq. (XII.173) in which $\mathbf{V}$ is the vector in $\mathbb{C}^{n_E}$ such that $\mathbf{V} = [\,\Theta\,]\,\mathbf{U}$. At any point of external domain Ω_E, the finite element approximation of resultant pressure $p_E(\mathbf{x},\omega)$ is given by Eq. (XII.174) in which $\mathbf{V}$ is defined above.

7.4. FRF calculation in the MF range

Let $\mathbb{B}_{\text{MF}}$ be the MF broad band defined by

$$\mathbb{B}_{\text{MF}} = [\omega_{\text{MF,init}} \, , \, \omega_{\text{MF,final}}] \quad . \tag{127}$$

The MF narrow band $\mathbb{B}_\nu$ is defined by Eq. (VII.9),

$$\mathbb{B}_\nu = [\,\Omega_\nu - \Delta\omega/2 \, , \, \Omega_\nu + \Delta\omega/2\,] \quad , \tag{128}$$

where $\Omega_\nu > 0$ is the center frequency of band $\mathbb{B}_\nu$ and $\Delta\omega$ is its bandwidth. We use the finite element discretization introduced in Section 7.3. For all ω in $\mathbb{B}_{\text{MF}}$, we have to construct the FRF $[T(\omega)] = [\mathbb{A}(\omega)]^{-1}$ where $[\mathbb{A}(\omega)]$ is defined by Eq. (124). We then have to solve for the class of MF narrow band excitations $\theta_\nu(\omega)\,\mathbf{B}$ introduced in Section VII.4,

$$[\mathbb{A}(\omega)]\,\mathbf{U}(\omega) = \theta_\nu(\omega)\,\mathbf{B} \quad , \quad \omega \in \mathbb{B}_\nu \quad . \tag{129}$$

Symmetric complex matrix $[\mathbb{A}(\omega)]$ can be written as

$$[\mathbb{A}(\omega)] = -\omega^2[\mathbb{M}(\omega)] + i\omega\,[D^S(\omega)] + [K^S(\omega)] \quad , \tag{130}$$

where $[\mathbb{M}(\omega)]$ is the symmetric complex matrix such that

$$[\mathbb{M}(\omega)] = [M^S] + [A^E(\omega/c_E)] \quad . \tag{131}$$

For construction of the frequency response function, we take $\theta_\nu(\omega) = 1$ for $\omega \in \mathbb{B}_\nu$ and $\mathbf{B}$ represents the vectors of the canonical basis of $\mathbb{R}^{n_S}$. Concerning the methods for calculating the response to various types of excitations, we refer the reader to Chapter IX. The MF method presented in Chapter VII uses an approximation of Eq. (129) due to the frequency-dependent matrices $[\mathbb{M}(\omega)]$, $[D^S(\omega)]$ and $[K^S(\omega)]$ consisting in replacing Eq. (129) by

$$\left(-\omega^2[\mathbb{M}_\nu] + i\omega\,[D^S_\nu] + [K^S_\nu]\right)\mathbf{U}_\nu(\omega) = \theta_\nu(\omega)\,\mathbf{B} \quad , \quad \omega \in \mathbb{B}_\nu \quad , \tag{132}$$

in which the frequency independent complex matrices $[\mathbb{M}_\nu]$, $[D^S_\nu]$ and $[K^S_\nu]$ are such that

$$[\mathbb{M}_\nu] = [\mathbb{M}(\Omega_\nu)] \quad , \quad [D^S_\nu] = [D^S(\Omega_\nu)] \quad , \quad [K^S_\nu] = [K^S(\Omega_\nu)] \quad . \tag{133}$$

Functions $\omega \mapsto [\mathbb{M}(\omega)]$, $[D^S(\omega)]$ and $[K^S(\omega)]$ can be considered as slowly varying functions on MF narrow band $\mathbb{B}_\nu$ (function $\omega \mapsto [A^E(\omega/c_E)]$ has a nonresonant behavior and varies slowly on $\mathbb{B}_\nu$; functions $\omega \mapsto [D^S(\omega)]$ and $[K^S(\omega)]$ are related to the viscoelastic coefficients of the master structure and vary slowly). We then apply the frequency transform technique presented in Section VII.6 to obtain the associated LF equation which is written in the time domain as

$$[\mathbb{M}_\nu]\, \ddot{\mathbb{U}}_0(t) + [\widetilde{D}_\nu^S]\, \dot{\mathbb{U}}_0(t) + [\widetilde{K}_\nu^S]\mathbb{U}_0(t) = \theta_0(t)\, \mathbf{B} \quad , \quad \forall t > t_{\rm i} \ , \tag{134}$$

$$\mathbb{U}_0(t_{\rm i}) = \mathbf{0} \quad , \quad \dot{\mathbb{U}}_0(t_{\rm i}) = \mathbf{0} \quad , \tag{135}$$

and where complex symmetric matrices $[\widetilde{D}_\nu^S]$ and $[\widetilde{K}_\nu^S]$ are such that

$$[\widetilde{D}_\nu^S] = [D_\nu^S] + 2\, i\, \Omega_\nu\, [\mathbb{M}_\nu] \quad , \tag{136}$$

$$[\widetilde{K}_\nu^S] = -\Omega_\nu^2\, [\mathbb{M}_\nu] + i\, \Omega_\nu\, [D_\nu^S] + [K_\nu^S] \quad . \tag{137}$$

Once $\mathbb{U}_0$ is obtained by solving Eqs. (134) and (135) in the time domain, the expression of $\mathbb{U}_\nu(\omega)$ in MF narrow band $\mathbb{B}_\nu$ is given by

$$\mathbb{U}_\nu(\omega) \simeq \mathbf{1}_{\mathbb{B}_\nu}(\omega)\ \Delta t \sum_{m=m_{\rm i}}^{m_{\rm f}} \mathbb{U}_0(m\,\Delta t)\ e^{-im\,\Delta t\,(\omega - \Omega_\nu)} \quad , \quad \forall \omega \in \mathbb{R} \quad , \tag{138}$$

in which $\mathbf{1}_{\mathbb{B}_\nu}(\omega) = 1$ if $\omega \in \mathbb{B}_\nu$ and $= 0$ if $\omega \notin \mathbb{B}_\nu$ and where integers $m_{\rm i}$ and $m_{\rm f}$ are defined in Section VII.6.4.

8. Structure Coupled with an External and an Internal Acoustic Fluid. Case of a Zero Pressure Condition on Part of the Internal Fluid Boundary

In this section, we summarize the model, equations and solving methods for a master structure coupled with an external and an internal acoustic fluid for which a zero pressure condition exists on a part of the internal fluid boundary. In this case, quantity $\pi(\omega\,;\mathbf{u}\,;\psi) = 0$ and the results presented in this section are directly deduced from Sections 2 to 6 in which all the terms related to $\pi(\omega\,;\mathbf{u}\,;\psi)$ vanish.

8.1. Statement of the structural-acoustic problem

We consider the problem described in Section 2 for which boundary $\partial\Omega$ of Ω is written $\partial\Omega = \Gamma \cup \Gamma_0 \cup \Gamma_Z$ in which Γ_0 is submitted to a zero pressure field (see Fig.4).

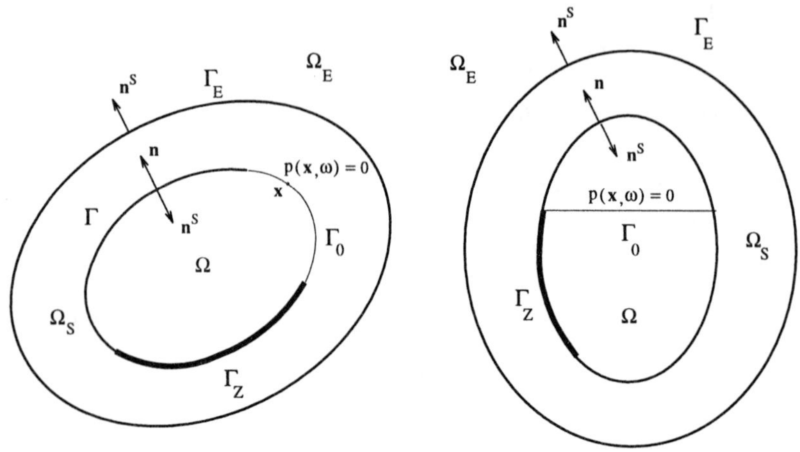

a- Gas or liquid configuration b- Liquid with free surface

Fig. 4. Configurations of the structural-acoustic master system

There is no gravity effect. Consequently, boundary Γ in Section 2 is replaced by boundary $\Gamma \cup \Gamma_0$ in this section.

8.2. Internal and external acoustic pressure fields

Concerning the pressure field in internal acoustic fluid Ω, we saw in Sections X.2.6-3 that $\pi(\omega\,;\mathbf{u}\,;\psi) = 0$. Consequently, $\pi_1(\omega\,;\psi)$, $\pi_2(\mathbf{u})$ and $\pi_Q(\omega)$ do not appear in the calculation (this is achieved formally by eliminating all the terms containing $\kappa(\omega)$). At any point $\mathbf{x}$ in $\Omega \cup \partial\Omega$, pressure field $p(\mathbf{x},\omega)$ is then given by

$$p(\mathbf{x},\omega) = -i\omega\,\rho_0\,\psi(\mathbf{x},\omega) \quad . \tag{139}$$

Pressure field $p_E|_{\Gamma_E}(\omega)$ on Γ_E and resultant pressure field $p_E(\mathbf{x},\omega)$ at any point of unbounded external acoustic fluid Ω_E are calculated using Eqs. (9) and (10).

8.3. Boundary value problem of the structural-acoustic master system

We use the fact that $\pi(\omega\,;\mathbf{u}\,;\psi) = 0$ and from Section 3, we deduce the equations of the boundary value problem. Concerning the master structure, Eqs. (11) to (13) become

$$-\omega^2\,\rho_S\,u_i - \sigma_{ij,j}(\mathbf{u}) = g_i \quad \text{in} \quad \Omega_S \quad , \tag{140}$$

$$\sigma_{ij}(\mathbf{u})\,n_j^S = G_i - p_{\text{given}}|_{\Gamma_E}\,n_i^S - i\omega\,\mathbf{Z}_{\Gamma_E}(\omega)\{\mathbf{u}\cdot\mathbf{n}^S\}\,n_i^S \quad \text{on} \quad \Gamma_E \quad , \tag{141}$$

$$\sigma_{ij}(\mathbf{u})\,n_j^S = G_i - i\omega\,\rho_0\,\psi\,n_i \quad \text{on} \quad \Gamma \cup \Gamma_Z \quad . \tag{142}$$

Concerning the internal acoustic fluid, Eqs. (15) to (20) become

$$-\omega^2 \frac{\rho_0}{c_0^2}\,\psi(\mathbf{x},\omega) - i\omega\,\tau\,\rho_0 \boldsymbol{\nabla}^2\psi(\mathbf{x},\omega) - \rho_0\,\boldsymbol{\nabla}^2\psi(\mathbf{x},\omega) = i\omega\, g_{\mathrm{Q}} \quad \text{in } \Omega \quad , \tag{143}$$

with the Dirichlet boundary condition

$$\psi = 0 \quad \text{on } \Gamma_0 \quad , \tag{144}$$

and the Neumann boundary conditions

$$\rho_0(1+i\omega\tau)\frac{\partial\psi}{\partial\mathbf{n}} = i\omega\,\rho_0\,\mathbf{u}\cdot\mathbf{n} + i\omega\, G_{\mathrm{Q}} \quad \text{on} \quad \Gamma \quad , \tag{145}$$

$$\rho_0(1+i\omega\tau)\frac{\partial\psi}{\partial\mathbf{n}} = i\omega\,\rho_0\,\mathbf{u}\cdot\mathbf{n} + \frac{\omega^2\rho_0^2}{i\omega Z}\psi + i\omega\, G_{\mathrm{Q}} \quad \text{on } \Gamma_Z \quad , \tag{146}$$

in which g_{Q} is defined on Ω by

$$g_{\mathrm{Q}}(\mathbf{x},\omega) = -\frac{1}{i\omega}Q(\mathbf{x},\omega) - \tau c_0^2\,\frac{1}{\omega^2}\boldsymbol{\nabla}^2 Q(\mathbf{x},\omega) \quad , \tag{147}$$

G_{Q} is defined on $\Gamma\cup\Gamma_Z$ by

$$G_{\mathrm{Q}}(\mathbf{x},\omega) = \tau c_0^2 \frac{1}{\omega^2}\frac{\partial Q(\mathbf{x},\omega)}{\partial\mathbf{n}(\mathbf{x})} \quad . \tag{148}$$

8.4. Variational formulation of the structural-acoustic problem

The variational formulation of the boundary value problem defined by Eqs. (140) to (148) is deduced from Section 4.

Admissible function spaces. The admissible function space of displacement field $\mathbf{u}$ of master structure Ω_S is the complex vector space $\mathcal{C}^c$ defined in Section 4.1. Concerning internal acoustic fluid Ω, we introduce the complex vector spaces $\mathcal{E}^c_{0,\psi}$ such that

$$\mathcal{E}^c_{0,\psi} = \{\, \psi \in \mathcal{E}^c \quad ; \quad \psi = 0 \quad \text{on} \quad \Gamma_0\} \quad , \tag{149}$$

in which $\mathcal{E}^c$ is defined in Section 4.1 and $H^c = L^2(\Omega)$.

Variational formulation in terms of $\{\mathbf{u},\psi\}$. Eqs. (140) to (142) related to the master structure yield

$$a^S(\omega;\mathbf{u},\delta\mathbf{u}) - \omega^2 a^E(\omega/c_E;\mathbf{u},\delta\mathbf{u}) + i\omega\, c_0(\psi,\delta\mathbf{u}) = f_S(\omega;\delta\mathbf{u}) \ , \ \forall\delta\mathbf{u}\in\mathcal{C}^c, \tag{150}$$

in which $a^S(\omega\,;\mathbf{u}\,,\delta\mathbf{u})$, $a^E(\omega/c_E;\mathbf{u},\delta\mathbf{u})$ and $c_0(\psi,\delta\mathbf{u})$ are defined by Eqs. (23), (24) and (27) respectively. In Eq. (150), $f_S(\omega;\delta\mathbf{u})$ is the antilinear form on $\mathcal{C}^c$ defined by

$$f_S(\omega\,;\delta\mathbf{u}) = f(\omega\,;\delta\mathbf{u}) + f_{p_{\text{given}}}(\omega\,;\delta\mathbf{u}) \quad , \tag{151}$$

in which antilinear forms $f(\omega\,;\delta\mathbf{u})$ and $f_{p_{\text{given}}}(\omega\,;\delta\mathbf{u})$ are defined by Eqs. (30) and (31) respectively. Eqs. (143) to (148) related to the internal acoustic fluid yield

$$i\omega\,\overline{c_0(\delta\psi\,,\mathbf{u})} - a^F(\omega\,;\psi\,,\delta\psi) = -i\omega f_Q(\omega;\delta\psi) \quad , \quad \forall\delta\psi\in\mathcal{E}^c_{0,\psi} \quad , \tag{152}$$

in which the sesquilinear form $a^F(\omega\,;\psi\,,\delta\psi)$ is given by

$$a^F(\omega\,;\psi,\delta\psi) = -\omega^2 m(\psi,\delta\psi) + i\omega\, d_\tau(\omega;\psi,\delta\psi) + k(\psi,\delta\psi) + s_1(\omega;\psi,\delta\psi)\,, \tag{153}$$

where $k(\psi,\delta\psi)$, $d_\tau(\omega\,;\psi\,,\delta\psi)$ and $m(\psi,\delta\psi)$ are defined by Eqs. (34), (35) and (36) respectively and where sesquilinear form $s_1(\omega\,;\psi\,,\delta\psi)$ defined on $\mathcal{E}^c\times\mathcal{E}^c$ is deduced from Eq. (37) and is such that

$$s_1(\omega\,;\psi\,,\delta\psi) = -\rho_0^2\omega^2\int_{\Gamma_Z}\frac{1}{i\omega Z}\,\psi\,\overline{\delta\psi}\,ds \quad . \tag{154}$$

In Eq. (152), $f_Q(\omega\,;\delta\psi)$ is the antilinear form defined on $\mathcal{E}^c$ deduced from Eq. (40) and is such that

$$f_Q(\omega;\delta\psi) = -\int_\Omega\frac{Q}{i\omega}\overline{\delta\psi}\,d\mathbf{x} + \frac{\tau c_0^2}{\omega^2}\int_\Omega\boldsymbol{\nabla}Q\cdot\boldsymbol{\nabla}\overline{\delta\psi}\,d\mathbf{x} \quad . \tag{155}$$

For all fixed real $\omega\neq 0$, the variational formulation is written as follows. Find $\mathbf{u}$ in $\mathcal{C}^c$ and ψ in $\mathcal{E}^c_{0,\psi}$ such that, for all $\delta\mathbf{u}$ in $\mathcal{C}^c$ and $\delta\psi$ in $\mathcal{E}^c_{0,\psi}$, Eqs. (150) and (152) are satisfied.

Linear operator equation. Considering Eqs. (150) and (152), Eqs. (41) and (42) are modified as follows,

$$\mathbf{A}(\omega)\begin{bmatrix}\mathbf{u}\\ \psi\end{bmatrix} = \begin{bmatrix}\mathbf{f}_S(\omega)\\ -i\omega\,\mathbf{f}_Q(\omega)\end{bmatrix} \quad , \quad \{\mathbf{u},\psi\}\in\mathcal{C}^c\times\mathcal{E}^c_{0,\psi} \quad , \tag{156}$$

in which $\mathbf{A}(\omega)$ is the restriction to $\mathcal{C}^c\times\mathcal{E}^c_{0,\psi}$ of the linear operator

$$\mathbf{A}(\omega) = \begin{bmatrix}\mathbf{A}^S(\omega) - \omega^2\mathbf{A}^E(\omega/c_E) & i\omega\,\mathbf{C}_0\\ i\omega\,{}^t\mathbf{C}_0 & -\mathbf{A}^F(\omega)\end{bmatrix} . \tag{157}$$

Linear operators $\mathbf{A}^S(\omega)$, $\mathbf{A}^E(\omega/c_E)$ and $\mathbf{C}_0$ are defined by Eqs. (43), (45) and (55) respectively. Linear operator $\mathbf{A}^F(\omega)$ defined by sesquilinear form $a^F(\omega\,;\psi,\delta\psi)$ can be written as

$$\mathbf{A}^F(\omega) = -\omega^2\mathbf{M} + i\omega\mathbf{D}_T(\omega) + \mathbf{K} + \mathbf{S}_1(\omega) \quad , \tag{158}$$

in which operators $\mathbf{M}$, $\mathbf{D}_T(\omega)$ and $\mathbf{K}$ are defined by Eqs. (49), (51) and (50) respectively. Linear operator $\mathbf{S}_1(\omega)$ is such that [19]

$$< \mathbf{S}_1(\omega)\,\psi\,,\delta\psi >= s_1(\omega\,;\psi\,,\delta\psi) \quad , \tag{159}$$

in which sesquilinear form $s_1(\omega\,;\psi\,,\delta\psi)$ is defined by Eq. (154). Finally, elements $\mathbf{f}_S(\omega)$ and $\mathbf{f}_Q(\omega)$ are defined by Eqs. (58) and (59).

8.5. Finite element discretization

We consider the finite element mesh of master structure Ω_S and the finite element mesh of internal acoustic fluid Ω introduced in Section 5. Eq. (64) becomes

$$[\mathbb{A}(\omega)]\begin{bmatrix}\mathbf{U}\\ \mathbf{\Psi}\end{bmatrix} = \begin{bmatrix}\mathbf{F}_S(\omega)\\ -i\omega\,\mathbf{F}_Q(\omega)\end{bmatrix} \quad , \tag{160}$$

with the constraint

$$\Psi_j = 0 \quad \text{for nodes} \quad j \quad \text{belonging to} \quad \Gamma_0 \quad . \tag{161}$$

corresponding to the finite element discretization of the Dirichlet conditions defined by Eq. (144). Complex matrix $[\mathbb{A}(\omega)]$ is symmetric and is defined by

$$[\mathbb{A}(\omega)] = \begin{bmatrix}[A^S(\omega)] - \omega^2[A^E(\omega/c_E)] & i\omega[C_0]\\ i\omega\,[C_0]^T & -[A^F(\omega)]\end{bmatrix} \quad , \tag{162}$$

in which matrices $[A^S(\omega)]$, $[A^E(\omega/c_E)]$, $[C_0]$ are defined by Eqs. (67), (68), (72) respectively. Matrix $[A^F(\omega)]$ is defined by

$$[A^F(\omega)] = -\omega^2\,[\,M\,] + i\omega\,[D_T(\omega)] + [\,K\,] + [S_1(\omega)] \quad , \tag{163}$$

in which matrices $[\,M\,]$, $[\,K\,]$ and $[D_T(\omega)]$ are defined in Section 5 and matrix $[S_1(\omega)]$ corresponds to the finite element discretization of the sesquilinear form defined by Eq. (154). Finally, vector $\mathbf{F}_S(\omega)$ is defined in Section 5

[19] Sesquilinear form $s_1(\omega\,;\psi\,,\delta\psi)$ is continuous on $\mathcal{E}^c\times\mathcal{E}^c$ (see also footnote 6) and defines a continuous operator $\mathbf{S}_1(\omega)$ from $\mathcal{E}^c$ into $\mathcal{E}^{c\,\prime}$. The angle brackets in Eq. (159) denote the antiduality product between $\mathcal{E}^{c\,\prime}$ and $\mathcal{E}^c$.

and vector $\mathbf{F}_Q(\omega)$ correspond to the finite element discretization of the antilinear form defined by Eq. (155).

Calculation of the pressure inside Ω. Let $\mathbf{P} = (p_1, \ldots, p_n)$ be the complex vector of the DOFs which are the values of p at the nodes of the finite element mesh of domain Ω. From Eq. (139), we deduce that

$$\mathbf{P} = -i\omega\, \rho_0 \boldsymbol{\Psi} \quad . \tag{164}$$

Calculation of the pressure on boundary Γ_E and of the resultant pressure at any point of external domain Ω_E. The finite element discretization $\mathbf{P}_{\Gamma_E}(\omega)$ of pressure field $p_E|_{\Gamma_E}(\omega)$ on Γ_E is given by Eq. (XII.173) in which $\mathbf{V}$ is the vector in $\mathbb{C}^{n_E}$ such that $\mathbf{V} = [\,\Theta\,]\,\mathbf{U}$. At any point of external domain Ω_E, the finite element approximation of resultant pressure $p_E(\mathbf{x}, \omega)$ is given by Eq. (XII.174) in which $\mathbf{V}$ is defined above.

8.6. FRF calculation in the MF cange

Let $\mathbb{B}_{\mathrm{MF}}$ be the MF broad band defined by

$$\mathbb{B}_{\mathrm{MF}} = [\,\omega_{\mathrm{MF,init}}\,,\,\omega_{\mathrm{MF,final}}\,] \quad . \tag{165}$$

The MF narrow band $\mathbb{B}_\nu$ is defined by

$$\mathbb{B}_\nu = [\,\Omega_\nu - \Delta\omega/2\,,\,\Omega_\nu + \Delta\omega/2\,] \quad , \tag{166}$$

where $\Omega_\nu > 0$ is the center frequency of band $\mathbb{B}_\nu$ and $\Delta\omega$ is its bandwidth. We consider the finite element discretization introduced in Section 8.5. For all ω in $\mathbb{B}_{\mathrm{MF}}$, we have to construct the FRF $[T(\omega)] = [\mathbb{A}(\omega)]^{-1}$ where $[\mathbb{A}(\omega)]$ is defined by Eq. (162). We then have to solve for the class of MF narrow band excitations $\theta_\nu(\omega)\,\mathbf{B}$ introduced in Section VII.4,

$$[\mathbb{A}(\omega)]\,\mathbf{Y}(\omega) = \theta_\nu(\omega)\,\mathbf{B} \quad , \quad \omega \in \mathbb{B}_\nu \quad \text{with} \quad \mathbf{Y}(\omega) = \begin{bmatrix} \mathbf{U}(\omega) \\ \boldsymbol{\Psi}(\omega) \end{bmatrix} \quad , \tag{167}$$

with the constraint

$$\Psi_j = 0 \quad \text{for nodes} \quad j \quad \text{belonging to} \quad \Gamma_0 \quad . \tag{168}$$

Symmetric complex matrix $[\mathbb{A}(\omega)]$ can be written as

$$[\mathbb{A}(\omega)] = -\omega^2 [\mathbb{M}(\omega)] + i\omega\,[\mathbb{D}(\omega)] + [\mathbb{K}(\omega)] \quad , \tag{169}$$

where $[\mathbb{M}(\omega)]$, $[\mathbb{D}(\omega)]$ and $[\mathbb{K}(\omega)]$ are the symmetric complex matrices such that

$$[\mathbb{M}(\omega)] = \begin{bmatrix} [M^S] + [A^E(\omega/c_E)] & [\,0\,] \\ [\,0\,] & -[\,M\,] \end{bmatrix} \quad , \tag{170}$$

$$[\mathbb{D}(\omega)] = \begin{bmatrix} [D^S(\omega)] & [C_0] \\ [C_0]^T & -[D_T(\omega)] - [\overset{\text{I}}{S_1}(\omega)] \end{bmatrix} \quad , \tag{171}$$

$$[\mathbb{K}(\omega)] = \begin{bmatrix} [K^S(\omega)] & [\,0\,] \\ [\,0\,] & -[\,K\,] - [\overset{\text{R}}{S_1}(\omega)] \end{bmatrix} \quad , \tag{172}$$

in which we have written

$$[S_1(\omega)] = [\overset{\text{R}}{S_1}(\omega)] + i\omega\, [\overset{\text{I}}{S_1}(\omega)] \quad , \tag{173}$$

where the real matrices $[\overset{\text{R}}{S_1}(\omega)]$ and $[\overset{\text{I}}{S_1}(\omega)]$ are such that

$$[\overset{\text{R}}{S_1}(\omega)] = \Re e\,[S_1(\omega)] \quad , \quad \omega\, [\overset{\text{I}}{S_1}(\omega)] = \Im m\,[S_1(\omega)] \quad . \tag{174}$$

For construction of the frequency response function, we take $\theta_\nu(\omega) = 1$ for $\omega \in \mathbb{B}_\nu$ and $\mathbf{B}$ represents the vectors of the canonical basis of $\mathbb{R}^{n_S+n}$. For the methods for calculating the response to various types of excitations, we refer the reader to Chapter IX. The MF method presented in Chapter VII uses an approximation of Eq. (167) due to the frequency-dependent matrices $[\mathbb{M}(\omega)]$, $[\mathbb{D}(\omega)]$ and $[\mathbb{K}(\omega)]$ consisting in replacing Eq. (167) by

$$\left(-\omega^2[\mathbb{M}_\nu] + i\omega\,[\mathbb{D}_\nu] + [\mathbb{K}_\nu]\right)\mathbf{Y}_\nu(\omega) = \theta_\nu(\omega)\,\mathbf{B} \quad , \quad \omega \in \mathbb{B}_\nu \quad , \tag{175}$$

in which the frequency-independent complex matrices $[\mathbb{M}_\nu]$, $[\mathbb{D}_\nu]$ and $[\mathbb{K}_\nu]$ are such that

$$[\mathbb{M}_\nu] = [\mathbb{M}(\Omega_\nu)] \quad , \quad [\mathbb{D}_\nu] = [\mathbb{D}(\Omega_\nu)] \quad , \quad [\mathbb{K}_\nu] = [\mathbb{K}(\Omega_\nu)] \quad . \tag{176}$$

Functions $\omega \mapsto [\mathbb{M}(\omega)]$, $[\mathbb{D}(\omega)]$ and $[\mathbb{K}(\omega)]$ can be considered as slowly varying functions on MF narrow band $\mathbb{B}_\nu$ (function $\omega \mapsto [A^E(\omega/c_E)]$ has a nonresonant behavior and varies slowly on $\mathbb{B}_\nu$; functions $\omega \mapsto [D^S(\omega)]$ and $[K^S(\omega)]$ are related to the viscoelastic coefficients of the master structure and vary slowly; function $\omega \mapsto [D_T(\omega)]$ is due to the acoustic dissipation inside the internal fluid and varies slowly on $\mathbb{B}_\nu$; finally, functions

$\omega \mapsto [S_1^{\rm R}(\omega)]$ and $[S_1^{\rm I}(\omega)]$ are related to wall acoustic impedance $Z(\mathbf{x},\omega)$ and vary slowly on $\mathbb{B}_\nu$). We then apply the frequency transform technique presented in Section VII.6 in order to obtain the associated LF equation which is written in the time domain as

$$[\mathbb{M}_\nu]\,\ddot{\mathbf{Y}}_0(t) + [\widetilde{\mathbb{D}}_\nu]\,\dot{\mathbf{Y}}_0(t) + [\widetilde{\mathbb{K}}_\nu]\mathbf{Y}_0(t) = \theta_0(t)\,\mathbf{B} \quad , \quad \forall t > t_{\rm i} \ , \tag{177}$$

$$\mathbf{Y}_0(t_{\rm i}) = \mathbf{0} \quad , \quad \dot{\mathbf{Y}}_0(t_{\rm i}) = \mathbf{0} \quad , \tag{178}$$

with the constraint

$$[\mathbf{\Psi}_0(t)]_j = 0 \quad , \quad \forall t > t_{\rm i} \quad \text{and for nodes } j \text{ belonging to } \Gamma_0 \quad , \tag{179}$$

in which $\mathbf{Y}_0(t)$ is such that

$$\mathbf{Y}_0(t) = \begin{bmatrix} \mathbf{U}_0(t) \\ \mathbf{\Psi}_0(t) \end{bmatrix} \quad , \tag{180}$$

and where symmetric complex matrices $[\widetilde{\mathbb{D}}_\nu]$ and $[\widetilde{\mathbb{K}}_\nu]$ are such that

$$[\widetilde{\mathbb{D}}_\nu] = [\mathbb{D}_\nu] + 2\,i\,\Omega_\nu\,[\mathbb{M}_\nu] \quad , \tag{181}$$

$$[\widetilde{\mathbb{K}}_\nu] = -\Omega_\nu^2\,[\mathbb{M}_\nu] + i\,\Omega_\nu\,[\mathbb{D}_\nu] + [\mathbb{K}_\nu] \quad . \tag{182}$$

Once $\mathbf{Y}_0$ is obtained by solving Eqs. (177) to (179) in the time domain, the expression of $\mathbf{Y}_\nu(\omega)$ in MF narrow band $\mathbb{B}_\nu$ is given by

$$\mathbf{Y}_\nu(\omega) \simeq \mathbf{1}_{\mathbb{B}_\nu}(\omega)\ \Delta t \sum_{m=m_{\rm i}}^{m_{\rm f}} \mathbf{Y}_0(m\,\Delta t)\ e^{-im\,\Delta t\,(\omega-\Omega_\nu)} \quad , \quad \forall \omega \in \mathbb{R} \quad , \tag{183}$$

in which $\mathbf{1}_{\mathbb{B}_\nu}(\omega) = 1$ if $\omega \in \mathbb{B}_\nu$ and $= 0$ if $\omega \notin \mathbb{B}_\nu$ and where integers $m_{\rm i}$ and $m_{\rm f}$ are defined in Section VII.6.4.

9. Case of an Axisymmetric Structural-Acoustic Master System

We refer the reader to Section III.8.2 for the master structure. Concerning the internal acoustic fluid, Ω is an axisymmetric domain, with Γ and Γ_Z an axisymmetric part of boundary $\partial\Omega$. Wall acoustic impedance $Z(\mathbf{x},\omega)$ on Γ_Z is assumed to have the same axisymmetry property. Consequently, using Section III.8.2, $p(\mathbf{x},\omega)$ and $\psi(\mathbf{x},\omega)$ can be written as

$$p(r,\theta,z) = p_0(r,z) + \sum_{n=1}^{+\infty} \{p_n^+(r,z)\,\cos n\theta + p_n^-(r,z)\,\sin n\theta\} \quad , \tag{184}$$

$$\psi(r,\theta,z) = \psi_0(r,z) + \sum_{n=1}^{+\infty} \{\psi_n^+(r,z)\,\cos n\theta + \psi_n^-(r,z)\,\sin n\theta\} \quad . \tag{185}$$

From Eq. (1), we deduce that

$$p_0(r,z) = -i\omega\,\rho_0\,\psi_0(r,z) + \pi(\omega) \quad , \tag{186}$$

in which $\pi(\omega)$ is not equal to zero, while for all $n \geq 1$,

$$p_n^{\pm}(r,z) = -i\omega\,\rho_0\,\psi_n^{\pm}(r,z) \quad . \tag{187}$$

These equations show that $\pi(\omega) = 0$ for $n \geq 1$. Consequently, in the boundary value problem, for $n \geq 1$, we set $\pi(\omega) = 0$ which leads to a formulation of the type presented in Section 8.

10. Response to Deterministic and Random Excitations

In the structural-acoustic problem defined in Section 2, the excitations are (1)- for the master structure, surface force field $\mathbf{G}$ and body force field $\mathbf{g}$, (2)- for the internal fluid, acoustic source density Q and (3)- for the external fluid, acoustic source density Q_E and an incident plane wave described by ψ_{inc}. For calculation of the reponse of this structural-acoustic master system submitted to these deterministic or random excitations, we refer the reader to the general methodology presented in Chapter IX. In addition, Section XIII.13 gives some elements of the methodology for the case of an external turbulent boundary layer and the case of an external random field due to continuously distributed multipole sources located in a specified bounded domain included in the external fluid.

11. Bibliographical Comments

The MF structural-acoustic formulation presented in this chapter is a generalization of a formulation which can be found in Soize et al., 1986b and 1992; and, for slender bodies, in Chabas and Soize, 1986. Related applications in the MF range can be found in Soize, 1996b; Soize et al., 1989. Concerning the construction of reduced structural-acoustic models in the MF range, the method presented in Chapter VIII for the reduced model in structural dynamics in the MF range can *a priori* be used. For the case of external structural-acoustic problems the method is efficient (Soize, 1997b). For the case of internal structural-acoustic problems research is in progress.

CHAPTER XV

Fuzzy Structure Theory

1. Introduction

In this chapter, we present the fuzzy structure theory introduced by Soize, 1986. The objective of this theory is to predict the medium-frequency local dynamical response of a master structure coupled with a large number of complex secondary subsystems such as equipment units or secondary structures attached to the master structure. These subsystems are called fuzzy substructures due to their structural complexity and because the details of them are unknown, or are not accurately known (the terminology "fuzzy" has nothing to do with the mathematical theory concerning fuzzy sets and fuzzy logic). This fuzzy structure theory is stated as an inverse problem and introduces a random boundary impedance operator to model the effects of the fuzzy substructures on the master structure. As an inverse problem, the random boundary impedance operator is constructed using the concept of a type I or type II homogeneous fuzzy impedance law which depends on a set of parameters called the mean coefficients and the deviation coefficients of the law. An appropriate method is proposed for identifying these coefficients. The theoretical basis of the fuzzy structure theory and the construction of the type I fuzzy law can be found in Soize, 1986 and the corresponding validation in Chabas, Desanti and Soize, 1986. The theoretical basis of the type II fuzzy law and its validation in the MF range are given in Soize 1993b. The identification method for the homogeneous fuzzy impedance law coefficients and its validation can be found in Soize, 1996a. We emphasize the fact that this fuzzy structure theory does not correspond to a classical structural-dynamics problem with random uncertainties, based on the use of the local dynamical response of the dynamical subsystems. In effect, the fuzzy structure theory does not use any local dynamical response of the fuzzy substructures. Since 1991, much research has been published concerning the particular problem of a master structure coupled with a large number of simple linear oscillators. We can mention Pierce, Sparrow and Russel, 1993; Maidanik, 1995; Russell and Sparrow, 1995; Strasberg and Feit, 1996; Weaver, 1997, and concerning a complex experiment, Photiadis, Bucaro and Houston, 1997.

In Section 2, we introduce the concept of fuzzy structure and we show the important role played by the presence of the fuzzy substructures in the dynamical response of the master structure.

In Section 3, we state the fuzzy structure theory as an inverse problem in structural dynamics and we introduce the basic requirements for the construction of the fuzzy structure theory.

In Section 4, we construct the random equation of the master structure coupled with the fuzzy substructures in terms of the displacement field of the master structure.

In Section 5, we present the construction of a homogeneous model of a fuzzy substructure, that is to say the construction of a random boundary impedance operator representing the effects of a fuzzy substructure on the master structure. This construction introduces the concept of a homogeneous fuzzy impedance law.

Section 6 is devoted to the construction of type I and type II homogeneous fuzzy impedance laws which depend on mean coefficients and deviation coefficients.

In Section 7, we present a recursive method for solving the random equation of the master structure coupled with the fuzzy substructures, expressed in terms of the random displacement field of the master structure.

Section 8 deals with the Ritz-Galerkin approximation and finite element discretization of the solving method presented in Section 7.

In Section 9, we present an identification method for the very important mean coefficient used in the type I or type II homogeneous fuzzy impedance law. This method leads to solving a nonlinear constraint optimization problem.

Finally, Section 10 is devoted to application of the fuzzy structure theory to the case of a real structure consisting of a master structure coupled with a very large number of simple oscillators.

2. Statement of the Problem

2.1. Concept of fuzzy structure

A *fuzzy structure* is defined as a *master structure* (or primary structure) coupled with *fuzzy substructures* which are defined as complex subsystems constituted by a great number of secondary subsystems such as equipment units and secondary structures attached to the master structure (see Fig. 1).

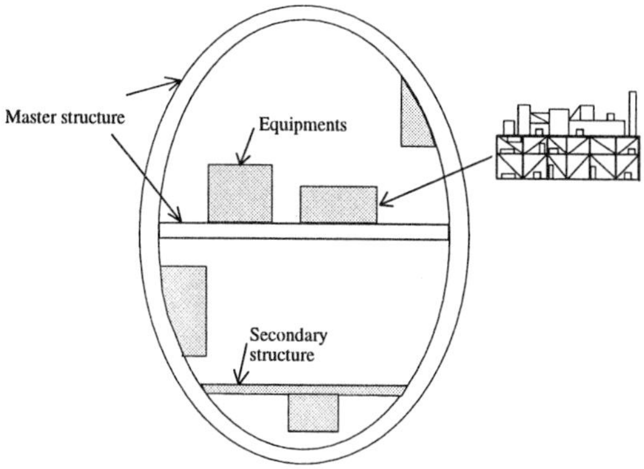

Fig. 1. Fuzzy structure defined as a master structure coupled with fuzzy substructures (equipment and secondary structures)

More precisely, a fuzzy substructure is the part of the structure that is not accessible to conventional modeling due to the structural complexity of secondary subsystems and because details of it are unknown, or are not accurately known. This explains the choice of the word "fuzzy" proposed by Soize, 1986, in introducing the *fuzzy structure theory*. It should be noted that the terminology "fuzzy" has nothing to do with the mathematical theory concerning fuzzy sets and fuzzy logic.

2.2. Effects of fuzzy substructures on the dynamical response of the master structure

As explained in Chapter I, we recall that the dynamical response of a weakly damped master structure (without fuzzy substructures) exhibits LF, MF and HF frequency ranges (see Fig. 2). It should be noted that for a very simply shaped homogeneous structure such as a rectangular plate of constant thickness, the MF frequency range does not generally exist. Let us now consider a fuzzy structure, i.e. a master structure coupled with fuzzy substructures. Since each fuzzy substructure is constituted by discrete or continuous weakly damped bounded structures, a fuzzy substructure is a resonant mechanical system which has a countable number of eigenfrequencies. In addition, since a fuzzy substructure is made of a large number of secondary subsystems, the modal density of such a substructure is high in the MF frequency band of interest. Below the first eigenfrequency of the fuzzy substructures, the effects of the fuzzy substructures on the master structure are mainly due to added-mass effects. In this case, the morphology of the dynamical response of the master structure is not altered by the presence of fuzzy substructures. Above the first eigenfrequency of the fuzzy substructures, experimental results show that the presence of fuzzy substructures induces an "apparent strong damping" in the master structure for the MF range (see Fig. 3 or Fig. 4) and possibly in the LF range (see Fig. 3), and that this strong damping cannot be explained by the small mechanical viscous damping existing in the master structure.

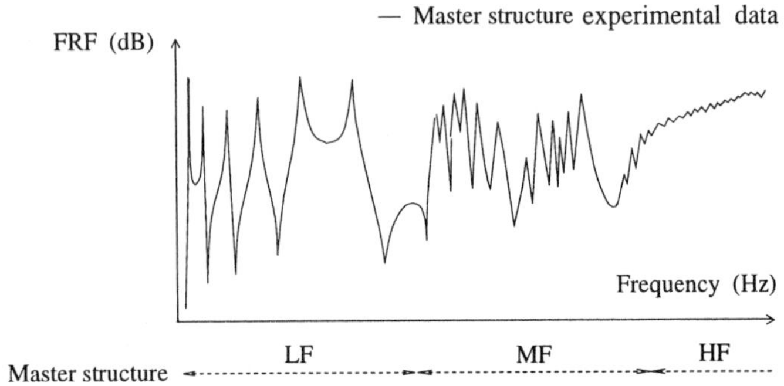

Fig. 2. LF, MF and HF dynamical response of a master structure without fuzzy substructures. (Qualitative drawing.)

This apparent strong damping is explained by the fact that a fuzzy substructure has a high modal density above its first eigenfrequency. Consequently, a significant amount of mechanical energy is transferred from the master structure to all the dynamical subsystems constituting the fuzzy substructure. This apparent strong damping is therefore due to the net transmitted power flowing from the master structure to the fuzzy substructures. This phenomenon, illustrated in Fig. 3 and Fig. 4, shows that the resonance morphology is strongly attenuated for the master structure. It is then of prime importance to introduce an appropriate model of this phenomenon. That is why the fuzzy structure theory was introduced, in order to explain the above experimental results and to improve the model predicting the local dynamical responses of the master structure in the LF and MF ranges (Soize, 1986).

First case. This case is illustrated by Fig. 3. It corresponds to the situation where the first eigenfrequency of the fuzzy substructures lies in the LF range of the master structure uncoupled with the fuzzy substructures. Fig. 3 shows a small shift of the first LF resonances. The dotted line represents the response of the master structure uncoupled with the fuzzy substructures and the solid line represents the response of the master structure coupled with the fuzzy substructures which induce an added-mass effect in the LF range. In this case, the LF range of the master structure coupled with the fuzzy substructures is narrower than the LF range of the master structure uncoupled with the fuzzy substructures. The presence of fuzzy substructures induces an "apparent strong damping" in the master structure for the MF range.

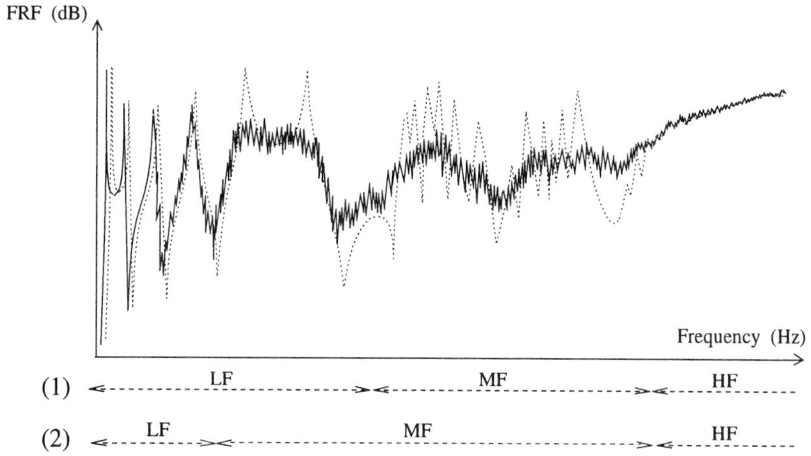

Fig. 3. Effects of fuzzy substructures on the master structure - First case. (Qualitative drawing.)
...... Response of the master structure uncoupled with the fuzzy substructures.
——— Response of the master structure coupled with the fuzzy substructures.
(1) Dynamical behavior of the uncoupled master structure.
(2) Dynamical behavior of the coupled master structure (fuzzy structure).

Second case. This case is illustrated by Fig. 4. It corresponds to the situation where the first eigenfrequency of the fuzzy substructures lies in the MF range of the master structure uncoupled with the fuzzy substructures. Fig. 4 shows a small shift of the LF resonances and small modifications of the dynamical response at the beginning of the MF range. The dotted line represents the response of the master structure uncoupled with the fuzzy substructures and the solid line represents the response of the master structure coupled with the fuzzy substructures which only induce an added-mass effect below the first eigenfrequency of the fuzzy substructures. In this case, the LF range of the master structure coupled with the fuzzy substructures coincides with the LF range of the master structure uncoupled with the fuzzy substructures. In addition, the MF range of the master structure coupled with the fuzzy substructures coincides with the MF range of the master structure uncoupled with the fuzzy substructures.

The presence of fuzzy substructures induces an "apparent strong damping" in the master structure in the MF range.

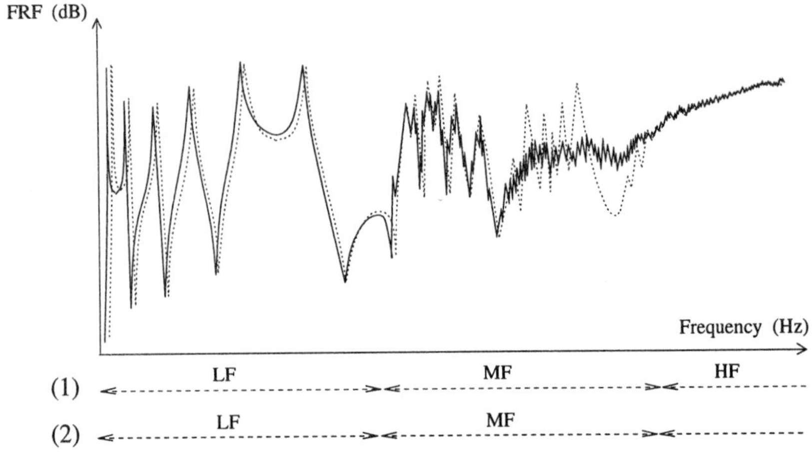

Fig. 4. Effects of fuzzy substructures on the master structure - Second case. (Qualitative drawing.)
...... Response of the master structure uncoupled with the fuzzy substructures.
—— Response of the master structure coupled with the fuzzy substructures.
(1) Dynamical behavior of the uncoupled master structure.
(2) Dynamical behavior of the coupled master structure (fuzzy structure).

3. Fuzzy Structure Theory Stated as an Inverse Problem

By construction, the fuzzy structure theory belongs to the class of inverse problems and is not a classical structural-dynamics problem with random uncertainties. In order to clarify this fundamental difference, the first section below is devoted to a short review of structural-dynamics problem with random uncertainties.

3.1. Structural-dynamics problem with random uncertainties stated as a direct problem

Here we recall the definition of a classical structural-dynamics problem with random uncertainties. Let us consider the boundary value problem for a structure made of two substructures, studied in Section V.7 (see Fig. 5). To model the uncertainties in this structural-dynamics problem, we assume that the dynamic-stiffness operator of substructure Ω_2 depends on an $\mathbb{R}^m$-valued parameter $\mathbf{y}$ related to the uncertainties and belonging to a subset $\mathcal{L}$ of $\mathbb{R}^m$,

$$\mathbf{y} \in \mathcal{L} \subset \mathbb{R}^m \quad . \tag{1}$$

For instance, the components of $\mathbf{y}$ are mechanical coefficients such as the elastic and damping coefficients of the constitutive equation of the viscoelastic material, the mass density and geometrical parameters (except

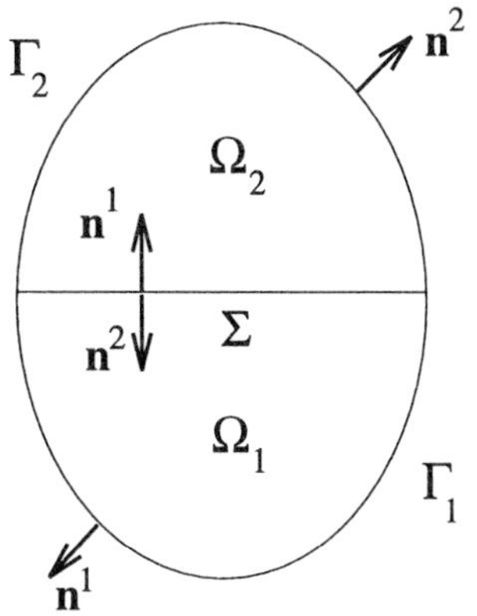

Fig. 5. Structure decomposed into two substructures

interface Σ which is assumed to be known). This operator is then written as $\mathbf{A}^2(\omega\,;\mathbf{y})$. Consequently, for all real ω, mapping $\mathbf{y} \mapsto \mathbf{A}^2(\omega\,;\mathbf{y})$ defined on $\mathcal{L}$ is known. The boundary value problem consists in finding $(\mathbf{u}^1, \mathbf{u}^2)$ in $\mathcal{C}^c_\Omega$ such that, for all $(\delta\mathbf{u}^1, \delta\mathbf{u}^2) \in \mathcal{C}^c_\Omega$ (see Equation (V.89)),

$$<\mathbf{A}^1(\omega)\,\mathbf{u}^1, \delta\mathbf{u}^1> + <\mathbf{A}^2(\omega\,;\mathbf{y})\,\mathbf{u}^2, \delta\mathbf{u}^2> = <\mathbf{f}^1(\omega)\,, \delta\mathbf{u}^1> \quad , \qquad (2)$$

in which the admissible function space $\mathcal{C}^c_\Omega$ is defined by Eq. (V.85). It is proved in Section V.7.3 that the problem defined by Eq. (2) is equivalent to a problem posed in subdomain Ω_1 by introducing boundary dynamic stiffness operator $\mathbf{A}_\Sigma(\omega)$ defined by Eqs. (V.101) and (V.106) (see Fig. 6). As seen in Section V.7.3, Eq. (2) is equivalent to the following operator equation

$$(\mathbf{A}^1(\omega) + \mathbf{A}_\Sigma(\omega\,;\mathbf{y}))\,\mathbf{u}^1(\omega) = \mathbf{f}^1(\omega) \quad , \qquad (3)$$

in which $\mathbf{u}^1(\omega) \in \mathcal{C}^c_{\Omega_1}$ where $\mathcal{C}^c_{\Omega_1}$ is defined in Section V.7.2. In terms of impedance operators, Eq. (3) can be rewritten as

$$i\omega\,(\mathbf{Z}^1(\omega) + \mathbf{Z}_\Sigma(\omega\,;\mathbf{y}))\,\mathbf{u}^1(\omega) = \mathbf{f}^1(\omega) \quad , \qquad (4)$$

in which impedance operator $\mathbf{Z}^1(\omega)$ related to domain Ω_1 is defined by

$$i\omega\,\mathbf{Z}^1(\omega) = \mathbf{A}^1(\omega) \quad , \qquad (5)$$

and boundary impedance operator $\mathbf{Z}_\Sigma(\omega\,;\mathbf{y})$ related to interface Σ is defined by

$$i\omega\,\mathbf{Z}_\Sigma(\omega\,;\mathbf{y}) = \mathbf{A}_\Sigma(\omega\,;\mathbf{y}) \quad . \tag{6}$$

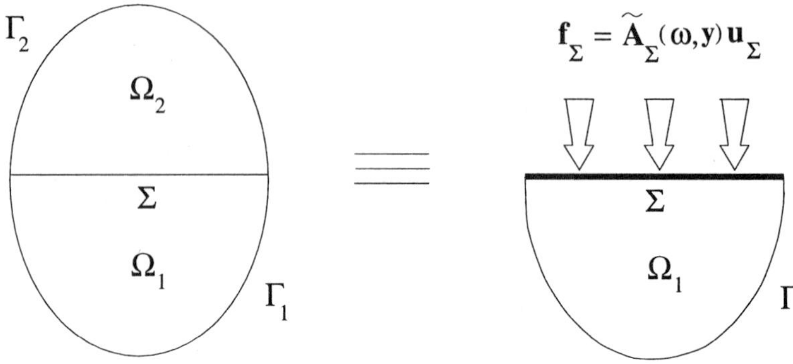

Fig. 6. Structure $\Omega_1 \cup \Omega_2$ equivalent to substructure Ω_1 with a boundary impedance operator on interface Σ

Concerning the corresponding finite element discretization, we refer the reader to Sections V.7.4 and V.7.5. In the context of random uncertainties, a probabilistic model of parameter $\mathbf{y}$ must be introduced. To do so, parameter $\mathbf{y}$ is modeled by an $\mathbb{R}^m$-valued random variable denoted as $\mathbf{Y}$. The probability distribution [1] of random variable $\mathbf{Y}$ is denoted as $P_\mathbf{Y}(d\mathbf{y})$ and, since parameter $\mathbf{y}$ belongs to $\mathcal{L}$, the support [2] of $P_\mathbf{Y}(d\mathbf{y})$ is $\mathcal{L} \subset \mathbb{R}^m$. Consequently, $\mathbf{Z}_\Sigma(\omega\,;\mathbf{Y})$ is a random variable [3] with operator values. Equation (4) is then rewritten as

$$i\omega\,(\mathbf{Z}^1(\omega) + \mathbf{Z}_\Sigma(\omega\,;\mathbf{Y}))\,\mathbf{U}^1(\omega) = \mathbf{f}^1(\omega) \quad , \tag{7}$$

in which $\mathbf{Z}_\Sigma(\omega\,;\mathbf{Y})$ is a random operator and the unknown $\mathbf{u}^1(\omega)$ is a random variable denoted as $\mathbf{U}^1(\omega)$ with values in admissible function space

---

[1] A probability distribution $P_\mathbf{Y}(d\mathbf{y})$ of an $\mathbb{R}^m$-valued random variable is a positive bounded measure on $\mathbb{R}^m$ such that $\int_{\mathbb{R}^m} P_\mathbf{Y}(d\mathbf{y})=1$. If this probability distribution has a density with respect to the Lebesgue measure $d\mathbf{y}$ on $\mathbb{R}^m$, then $P_\mathbf{Y}(d\mathbf{y})=p_\mathbf{Y}(\mathbf{y})d\mathbf{y}$ and the probability density function $p_\mathbf{Y}(\mathbf{y})$ is a positive valued function defined on $\mathbb{R}^m$ which is integrable and such that $\int_{\mathbb{R}^m} p_\mathbf{Y}(\mathbf{y})d\mathbf{y}=1$. Concerning probability theory and random variables, see Guikhman and Skorokhod, 1979; Halmos, 1976; Kree and Soize, 1986.

[2] If $\mathcal{L}\subset\mathbb{R}^m$ is the support of probability distribution $P_\mathbf{Y}(d\mathbf{y})$, then $\int_\mathcal{L} P_\mathbf{Y}(d\mathbf{y})=1$ and for any part $\mathcal{L}_0\subset\mathbb{R}^m$ such that $\mathcal{L}_0\cap\mathcal{L}=\emptyset$, we have $\int_{\mathcal{L}_0} P_\mathbf{Y}(d\mathbf{y})=0$.

[3] We assume that mapping $\mathbf{y}\mapsto\mathbf{Z}_\Sigma(\omega;\mathbf{y})$ is measurable and consequently $\mathbf{Z}_\Sigma(\omega;\mathbf{Y})$ is a random variable.

$\mathcal{C}^c_{\Omega_1}$. Equation (7) corresponds to a structural-dynamics problem with random uncertainties for which appropriate methods exist for constructing the random solution. This problem is considered as a direct problem because, (1)- the uncertain vector-valued parameter is identified, (2)- its probabilistic model is known, (3)- mapping $\mathbf{y} \mapsto \mathbf{Z}_\Sigma(\omega\,;\mathbf{y})$ is known.

3.2. Fuzzy structure theory stated as an inverse problem

In the above problem, if mapping $\mathbf{y} \mapsto \mathbf{Z}_\Sigma(\omega\,;\mathbf{y})$ is unknown, i.e. cannot be constructed due to the complexity of mechanical subsystem Ω_2, which is partially unknown from a geometrical and mechanical point of view, then the above direct approach cannot be applied. If substructure Ω_2 is a fuzzy substructure, then by the definition given in Section 2.1, mapping $\mathbf{y} \mapsto \mathbf{A}^2(\omega\,;\mathbf{y})$ cannot be constructed and therefore mapping $\mathbf{y} \mapsto \mathbf{Z}_\Sigma(\omega\,;\mathbf{y})$ which is deduced from $\mathbf{A}^2(\omega\,;\mathbf{y})$ cannot be constructed either and is thus unknown. The purpose of the proposed fuzzy structure theory is to solve this problem as an inverse problem. Consequently, since $\mathbf{A}^2(\omega\,;\mathbf{y})$ cannot be constructed, the stated inverse problem consists in directly constructing a random boundary impedance operator $\mathbf{Z}_{\text{fuz}}(\omega)$ related to interface Σ to represent the effects of the fuzzy substructure on the master structure. The objective of the fuzzy structure theory is to directly construct $\mathbf{Z}_{\text{fuz}}(\omega)$ which depends on mechanical parameters (interface Σ is assumed to be known) and to propose a procedure for identifying these parameters.

3.3. Basic requirements of the construction of the fuzzy-structure-theory

(1)- The objective of the fuzzy structure theory is to predict the effects of fuzzy substructures on the local dynamical response of the master structure, not to predict the dynamical response of fuzzy substructures. Consequently, the fuzzy structure theory is used to predict the frequency response functions in the master structure coupled with fuzzy substructures. More precisely, this construction gives the real and imaginary parts of the frequency response function, i.e. the modulus and the phases. Therefore, such an approach allows vibration propagations inside the master structure to be predicted and allows an extension to the case of structural-acoustic fuzzy systems (fuzzy structure coupled with external and internal acoustic fluids, see Chapter I).

(2)- Since a fuzzy substructure is made of a large number of secondary subsystems inaccessible to conventional modeling, a statistical approach is used and a *probabilistic* mechanical model is proposed for describing the boundary impedance operator which models the effects of the fuzzy substructures on the master structure. Consequently, the model is based on the construction of a *random boundary impedance operator* $\mathbf{Z}_{\text{fuz}}(\omega)$ of the fuzzy substructures. The model of the fuzzy structure is then written as

$$i\omega\left(\mathbf{Z}(\omega) + \mathbf{Z}_{\text{fuz}}(\omega)\right)\mathbf{U}(\omega) = \mathbf{f}(\omega) \quad , \tag{8}$$

in which $\mathbf{Z}(\omega)$ is the impedance operator of the master structure, $\mathbf{Z}_{\text{fuz}}(\omega)$ is the random boundary operator introduced in Section 3.2 to represent the effects of the fuzzy substructures on the master structure, $\mathbf{U}(\omega)$ is the random displacement field of the master structure and $\mathbf{f}(\omega)$ represents the external forces applied to the master structure. Modeling of the fuzzy structure corresponding to Eq. (8) does not introduce any additional unknown field with respect to the model of the master structure.

(3)- A requirement of the fuzzy structure theory is that it must construct random operator $\mathbf{Z}_{\text{fuz}}(\omega)$ in such a way that Eq. (8) can be solved using an efficient solver, especially in the MF range. This requirement is consistent with the construction of an inverse problem.

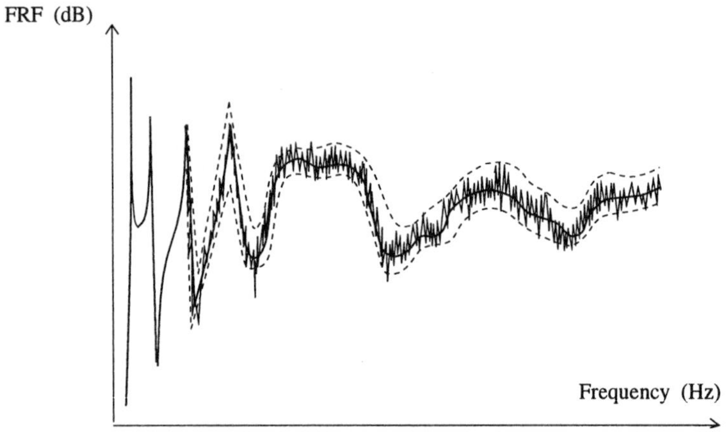

Fig. 7. FRF in the LF and MF ranges. (Qualitative drawing.)
—— Response of the coupled master structure (experimental results).
—— Mean response of the coupled master structure (fuzzy structure theory).
- - - Upper and lower values associated with a confidence interval (fuzzy structure theory).

(4)- Random displacement field $\mathbf{U}(\omega)$ of the master structure, as the solution of Eq. (8), is described by its probabilistic quantities such as mean value, second-order moments, etc. For instance, Fig. 7 shows the mean value and the upper and lower values associated with a confidence interval of a component $U_j(\mathbf{x}, \omega)$ of random displacement field $\mathbf{U}(\omega)$ at a point $\mathbf{x}$ located in the master structure obtained by the fuzzy structure theory, with respect to the fuzzy structure experimental result in the LF and MF ranges.

4. Random Equation of the Master Structure Coupled With Fuzzy Substructures in Terms of the Displacement Field of the Master Structure

The geometry of the fuzzy structure and the notations are defined in Fig. 8.

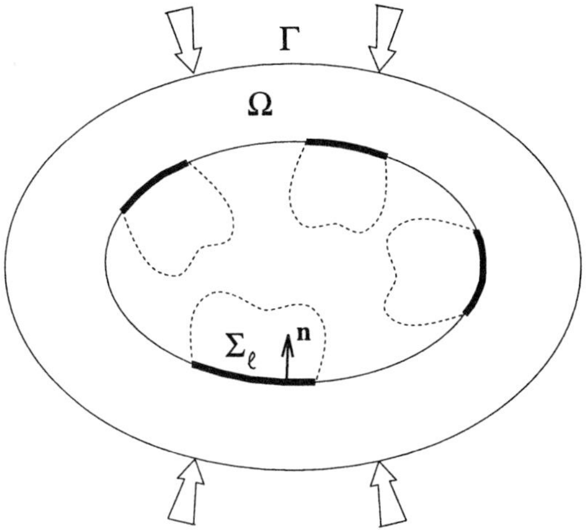

Fig. 8. Geometrical configuration of a fuzzy structure

The master structure occupies the open bounded domain Ω of $\mathbb{R}^3$ referred to a Cartesian reference system $\{\mathbf{i}, \mathbf{j}, \mathbf{k}\}$. The boundary of domain Ω is written as $\partial\Omega = \Gamma \cup \Sigma$ such that $\Gamma \cap \Sigma = \emptyset$. This master structure is coupled with L fuzzy substructures. Fuzzy substructure ℓ is attached to the master structure by boundary Σ_ℓ and there is no mechanical connection between any two fuzzy substructures. The part $\Sigma = \cup_{\ell=1}^{L} \Sigma_\ell$ is the common boundary between the fuzzy substructures and the master structure. The outward unit normal to $\partial\Omega$ is denoted $\mathbf{n}$ and the two-dimensional surface area element is denoted ds. We then have $\int_{\Sigma_\ell} ds(\mathbf{x}) = |\Sigma_\ell|$, where $|\Sigma_\ell|$ is the area of surface Σ_ℓ. We are interested in the dynamical response of the master structure coupled with the fuzzy substructures in the MF broad band defined by

$$\mathbb{B}_{\text{MF}} = [\omega_{\text{MF,init}} \, , \, \omega_{\text{MF,final}}] \subset \mathbb{R}^+ \quad , \quad \text{with} \quad \omega_{\text{MF,init}} > 0 \quad . \tag{9}$$

Of course, as explained in Section 2.2, the master structure uncoupled with the fuzzy substructures can have an LF dynamical response for ω in $\mathbb{B}_{\text{MF}}$ (see Fig. 3). Consequently, we need an LF model and an MF model for the master structure.

Modeling the master structure. The master structure is assumed to be a free structure submitted to a given body force field $\mathbf{g}(\mathbf{x}, \omega)$ defined in Ω and surface force field $\mathbf{G}(\mathbf{x}, \omega)$ defined on Γ. These forces are assumed to be in equilibrium for each ω in $\mathbb{B}_{\text{MF}}$ (see Eq. (V.49)). The model of the master structure is described in Chapter V. The admissible function space [4] $\mathcal{C}^c$ of the displacement fields $\mathbf{u}(\mathbf{x}, \omega) = (u_1(\mathbf{x}, \omega), u_2(\mathbf{x}, \omega), u_3(\mathbf{x}, \omega))$ is defined in Section V.3. When the master structure is not coupled with the fuzzy substructures, its operator equation on $\mathcal{C}^c$ is given by (see Eqs. (V.43) and (V.47)),

$$i\omega\, \mathbf{Z}(\omega)\, \mathbf{u}(\omega) = \mathbf{f}(\omega) \quad , \tag{10}$$

in which $\mathbf{Z}(\omega)$ is the impedance operator on $\mathcal{C}^c$ of the master structure defined by

$$i\omega\, \mathbf{Z}(\omega) = -\omega^2 \mathbf{M} + i\omega\, \mathbf{D}(\omega) + \mathbf{K}(\omega) \quad , \tag{11}$$

$\mathbf{u}(\omega) \in \mathcal{C}^c$ is the displacement field of the master structure and $\mathbf{f}(\omega)$ represents the given forces and is defined by Eqs. (V.38) and (V.39). We recall that in the LF range, stiffness operator $\mathbf{K}(\omega)$ is independent of ω but in the MF range, $\mathbf{K}(\omega)$ depends on ω. From Section V.5.3 (footnote 15), for all $\omega \in \mathbb{B}_{\text{MF}}$, since $\omega \neq 0$, Eq. (10) has a unique solution in $\mathcal{C}^c$.

Modeling the fuzzy substructures. As explained in Section 3.3-(2), the effects on the master structure of fuzzy substructure ℓ attached to the master structure by boundary Σ_ℓ are represented by a random boundary impedance operator $\mathbf{Z}^{\ell}_{\text{fuz}}(\omega)$ defined on $\mathcal{C}^c$. The effects of all the fuzzy substructures on the master structure are then represented by a random boundary impedance operator, denoted as $\mathbf{Z}_{\text{fuz}}(\omega)$, defined on $\mathcal{C}^c$, and such that

$$\mathbf{Z}_{\text{fuz}}(\omega) = \sum_{\ell=1}^{L} \mathbf{Z}^{\ell}_{\text{fuz}}(\omega) \quad . \tag{12}$$

Random equation of the fuzzy structure in terms of the master structure displacement field. We deduce that the random equation of the master structure coupled with the L fuzzy substructures is written as

$$i\omega\, (\mathbf{Z}(\omega) + \mathbf{Z}_{\text{fuz}}(\omega))\, \mathbf{U}(\omega) = \mathbf{f}(\omega) \quad , \tag{13}$$

[4] Space $\mathcal{C}^c$ is the Sobolev space $(H^1(\Omega))^3$.

in which $\mathbf{Z}_{\text{fuz}}(\omega)$ is given by Eq. (12) and $\mathbf{U}(\omega)$ is the master structure random displacement field with values in $\mathcal{C}^c$. The construction of random operator $\mathbf{Z}_{\text{fuz}}(\omega)$ in the present context of the fuzzy structure theory stated as an inverse problem is described in Section 5.

5. Homogeneous Model of a Fuzzy Substructure

As explained in Section 3.2, the effects of fuzzy substructure ℓ on the master structure are represented by a random boundary impedance operator $\mathbf{Z}^{\ell}_{\text{fuz}}(\omega)$ which has to be constructed in the context of an inverse problem.

5.1. Integral representation of the random boundary impedance operator of a fuzzy substructure

For this construction, we use the following integral representation of random boundary impedance operator $\mathbf{Z}^{\ell}_{\text{fuz}}(\omega)$. For all deterministic functions $\mathbf{u}$ and $\delta\mathbf{u}$ in $\mathcal{C}^c$, this operator is written as [5]

$$< \mathbf{Z}^{\ell}_{\text{fuz}}(\omega)\,\mathbf{u}\,,\delta\mathbf{u} >= \int_{\Sigma_\ell}\int_{\Sigma_\ell}\{[\mathbb{Z}^{\ell}(\omega\,;\mathbf{x},\mathbf{x}')]\,\mathbf{u}(\mathbf{x}')\}\cdot\overline{\delta\mathbf{u}(\mathbf{x})}\,ds(\mathbf{x}')\,ds(\mathbf{x})\,, \quad (14)$$

in which, for all $\mathbf{x}$ and $\mathbf{x}'$ in Σ_ℓ, the kernel $[\mathbb{Z}^{\ell}(\omega\,;\mathbf{x},\mathbf{x}')]$ of the integral representation is a random (3×3) complex matrix which has the following algebraic symmetry property,

$$[\mathbb{Z}^{\ell}(\omega\,;\mathbf{x},\mathbf{x}')] = [\mathbb{Z}^{\ell}(\omega\,;\mathbf{x}',\mathbf{x})]^T \quad , \quad (15)$$

and the following property which corresponds to the fact that this kernel must be a random (3×3) real matrix in the time domain,

$$[\mathbb{Z}^{\ell}(-\omega\,;\mathbf{x},\mathbf{x}')] = \overline{[\mathbb{Z}^{\ell}(\omega\,;\mathbf{x},\mathbf{x}')]} \quad . \quad (16)$$

In addition, for all ω in $\mathbb{B}_{\text{MF}}$ and for all deterministic functions $\mathbf{u}$ in $\mathcal{C}^c$, random kernel $[\mathbb{Z}^{\ell}(\omega\,;\mathbf{x},\mathbf{x}')]$ must be such that random operator $\mathbf{Z}^{\ell}_{\text{fuz}}(\omega)$ defined by Eq. (14) satisfies

$$E\{|< \mathbf{Z}^{\ell}_{\text{fuz}}(\omega)\,\mathbf{u}\,,\mathbf{u} >|^2\} < +\infty \quad , \quad (17)$$

---

[5] In Eq. (14), the brackets denote the antiduality product between $\mathcal{C}^{c\,\prime}$ and $\mathcal{C}^c$ in which $\mathcal{C}^{c\,\prime}$ is the antidual space of $\mathcal{C}^c$.

and random operator $\mathbf{Z}_{\text{fuz}}(\omega)$ defined by Eqs. (12) to (17) is such that, for all deterministic elements[6] $\delta\mathbf{f}$,

$$E\{|<\delta\mathbf{f}\,,(\mathbf{Z}(\omega)+\mathbf{Z}_{\text{fuz}}(\omega))^{-1}\delta\mathbf{f}>|^2\}<+\infty \quad . \tag{18}$$

5.2. Choice of an algebraic representation of random kernel $[\mathbb{Z}^\ell(\omega\,;\mathbf{x},\mathbf{x}')]$

This choice is mainly guided by the fact that we want to construct a model which is defined by a small number of significant mechanical parameters to satisfy the constraints defined by Eqs. (15) to (18). It is assumed (see Fig. 9) that there exists an orthogonal (3×3) real matrix $[\Phi^\ell(\mathbf{x})]^T$ for transition from canonical basis $\{\mathbf{i},\mathbf{j},\mathbf{k}\}$ to a local orthonormal basis $\{\mathbf{e}_1^\ell(\mathbf{x}),\mathbf{e}_2^\ell(\mathbf{x}),\mathbf{e}_3^\ell(\mathbf{x})\}$ related to boundary Σ_ℓ at point $\mathbf{x}\in\Sigma_\ell$ such that[7]

$$[\mathbb{Z}^\ell(\mathbf{x},\mathbf{x}',\omega)]=[\Phi^\ell(\mathbf{x})]\,[\mathbb{Z}^\ell_{\text{diag}}(\mathbf{x},\mathbf{x}',\omega)]\,[\Phi^\ell(\mathbf{x}')]^T \quad , \tag{19}$$

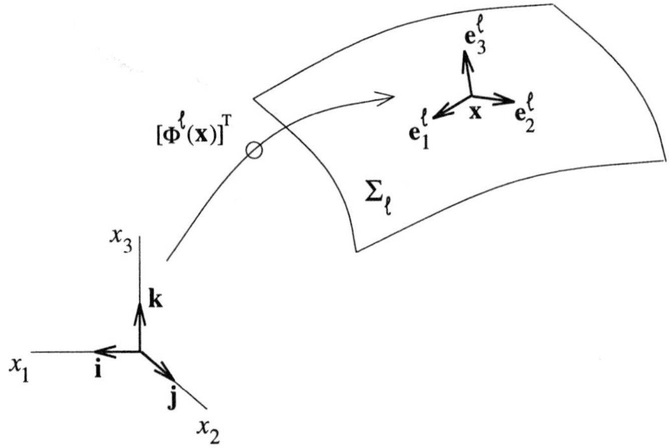

Fig. 9. Local basis related to boundary Σ_ℓ

in which $[Z^\ell_{\text{diag}}(\mathbf{x},\mathbf{x}',\omega)]$ is a random (3×3) complex diagonal matrix whose elements are written as

$$[\mathbb{Z}^\ell_{\text{diag}}(\mathbf{x},\mathbf{x}',\omega)]_{jj'}=\mathbb{Z}^\ell_j(\omega)\,\delta_{\Sigma_\ell}(\mathbf{x}-\mathbf{x}')\,\delta_{jj'} \quad , \tag{20}$$

in which $\mathbb{Z}^\ell_1(\omega)$, $\mathbb{Z}^\ell_2(\omega)$ and $\mathbb{Z}^\ell_3(\omega)$ are second-order random variables with values in $\mathbb{C}$. Local vector basis $\mathbf{e}_3^\ell(\mathbf{x})$ coincides with normal $\mathbf{n}(\mathbf{x})$ introduced

---

[6] Element $\delta\mathbf{f}$ belongs to the antidual space $\mathcal{C}^{c\,\prime}$ of $\mathcal{C}^c$. In Eqs. (17) and (18), the brackets denote the antiduality product between $\mathcal{C}^{c\,\prime}$ and $\mathcal{C}^c$.

[7] It is assumed that mapping $\mathbf{x}\mapsto[\Phi^\ell(\mathbf{x})]$ is a bounded function on Σ_ℓ.

in Section 4. In Eq. (20), $\delta_{jj'}$ is the Kronecker symbol and $\delta_{\Sigma_\ell}(\mathbf{x}-\mathbf{x}')$ is the Dirac function related to boundary Σ_ℓ which is such that, for all complex functions $\varphi(\mathbf{x})$ continuous on Σ_ℓ and $\mathbf{x}' \in \Sigma_\ell$,

$$\int_{\Sigma_\ell} \varphi(\mathbf{x})\, \delta_{\Sigma_\ell}(\mathbf{x}-\mathbf{x}')\, ds(\mathbf{x}) = \varphi(\mathbf{x}') \quad . \tag{21}$$

Substituting Eq. (19) with Eq. (20) in Eq. (14) and using Eq. (21) yields

$$<\mathbf{Z}^\ell_{\text{fuz}}(\omega)\,\mathbf{u}\,,\delta\mathbf{u}>= \int_{\Sigma_\ell} \{[\Phi^\ell(\mathbf{x}')]\,[\mathbb{Z}^\ell(\omega)]\,[\Phi^\ell(\mathbf{x}')]^T\,\mathbf{u}(\mathbf{x}')\}\cdot\overline{\delta\mathbf{u}(\mathbf{x}')}\,ds(\mathbf{x}')\,, \tag{22}$$

in which second-order random complex matrix $[\mathbb{Z}^\ell(\omega)]$ is diagonal and is defined by

$$[\mathbb{Z}^\ell(\omega)]_{jj'} = \mathbb{Z}^\ell_j(\omega)\,\delta_{jj'} \quad . \tag{23}$$

For fuzzy substructure ℓ and for local direction j, since $\mathbb{Z}^\ell_j(\omega)$ is a second-order random variable with values in $\mathbb{C}$ then Eq. (17) is satisfied. In addition, we assume that

$$\mathbb{Z}^\ell_j(-\omega) = \overline{\mathbb{Z}^\ell_j(\omega)} \quad . \tag{24}$$

Consequently, Eq. (16) is satisfied. It should be noted that Eqs. (19) and (20) imply that Eq. (15) is satisfied.

Terminology. Since for $j = 1, 2, 3$, random impedance $\mathbb{Z}^\ell_j(\omega)$ is independent of $\mathbf{x}$ and $\mathbf{x}'$, the model defined by Eqs. (19) and (20) is called a *homogeneous model* of fuzzy substructure ℓ. The construction of $\mathbb{Z}^\ell_j(\omega)$ is described in the next section introducing a parametric model. $\mathbb{Z}^\ell_j(\omega)$ is called the *homogeneous fuzzy impedance law* or the *homogeneous probabilistic impedance law* of fuzzy substructure ℓ related to local direction j at a given frequency.

5.3. Definition of a resonant frequency of fuzzy substructure ℓ for local direction j

As explained in Section 2.2, fuzzy substructure ℓ is a dynamical system having a countable number of resonant frequencies. A resonant frequency in local direction j is associated with a displacement field on Σ_ℓ, which has a nonzero contribution in local direction j.

6. Construction of a Homogeneous Fuzzy Impedance Law for a Fuzzy Substructure and for a Local Direction

In this section, we construct homogeneous fuzzy impedance law $\mathbb{Z}_j^\ell(\omega)$ for fuzzy substructure ℓ and for local direction j at a given frequency ω. Two types of homogeneous fuzzy impedance law are constructed, defined as *type I* and *type II* (type I is a particular case of type II). For additional details concerning this construction, we refer the reader to Soize, 1986 for the type I construction and Soize, 1993b for the type II construction. All the fuzzy impedance law coefficients introduced depend on local direction j at each point $\mathbf{x}$ of Σ_ℓ but are independent of this point $\mathbf{x}$. In order to simplify the notations, index j and superscript ℓ are omitted in all of Section 6. In particular, $\mathbb{Z}_j^\ell(\omega)$ is written as $\mathbb{Z}(\omega)$.

6.1. Choice of representation for random variable $\mathbb{Z}(\omega)$

The choice of the representation of second-order complex-valued random variable $\mathbb{Z}(\omega)$ is guided by the requirement introduced in Section 3.3-(3) which leads to introducing a linear random dependence of the random fluctuation part around its mean value,

$$\mathbb{Z}(\omega) = \underline{Z}(\omega) + \sum_{k=1}^{d} Z_k(\omega)\, \mathbb{X}_k \quad , \tag{25}$$

in which d is a positive integer, $\underline{Z}(\omega)$ is the mean value of random variable $\mathbb{Z}(\omega)$,

$$\underline{Z}(\omega) = E\{\mathbb{Z}(\omega)\} \quad , \tag{26}$$

$Z_1(\omega), \ldots, Z_d(\omega)$ are complex numbers and $\mathbb{X}_1, \ldots, \mathbb{X}_d$ are mutually independent, second-order, real-valued random variables. Each random variable $\mathbb{X}_k$ has a uniform probability distribution

$$P_{\mathbb{X}_k}(dx) = p_{\mathbb{X}_k}(x)\, dx \quad , \tag{27}$$

where the probability density function $p_{\mathbb{X}_k}(x)$ with respect to dx_k is written as (see Fig. 10)

$$p_{\mathbb{X}_k}(x) = \frac{1}{2\sqrt{3}}\, \mathbb{1}_{[-\sqrt{3},\sqrt{3}]}(x) \quad , \tag{28}$$

in which $\mathbb{1}_B(x)$ is the indicator function of set B (see Mathematical Notations in the appendix). The support $\mathcal{S}_{\mathbb{X}_k}$ of this probability distribution is then $\mathcal{S}_{\mathbb{X}_k} = [-\sqrt{3}, \sqrt{3}]$. Consequently, $\mathbb{X}_k$ is a second-order centered random variable and its variance is equal to 1 (normalized random variable),

$$E\{\mathbb{X}_k\} = 0 \quad , \quad E\{\mathbb{X}_k^2\} = 1 \quad . \tag{29}$$

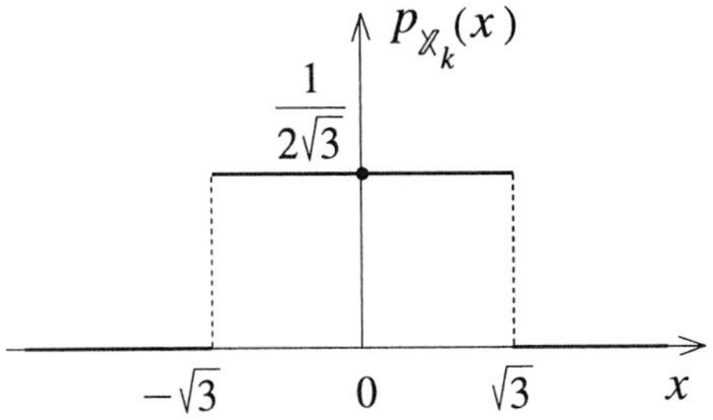

Fig. 10. Probability density function $p_{\mathbb{X}_k}(x)$

From Eqs. (25) and (29), we deduce that $\mathbb{Z}(\omega)$ defined by Eq. (25) is effectively a second-order random variable. The problem related to the construction of fuzzy impedance law $\mathbb{Z}(\omega)$ then consists in constructing deterministic quantities $\underline{Z}(\omega)$ and $Z_k(\omega)$. It should be noted that the algebraic representation defined by Eq. (25) is legitimate in the context of the construction of an inverse problem.

6.2. Underlying deterministic model for the construction of $\mathbb{Z}(\omega)$

The underlying deterministic model of random variable $\mathbb{Z}(\omega)$ is denoted as $\widetilde{z}(\omega)$ and depends on parameters Ω_c, $\mu(\omega)$, $\xi(\omega)$, $\Omega_p(\omega)$ and $\alpha(\omega)$ defined below.

Expression for the dynamic stiffness below a cutoff frequency. We introduce a positive *deterministic cutoff frequency* of fuzzy substructure ℓ related to local direction j and denoted as Ω_c. This cutoff frequency is such that for $\omega < \Omega_c$, dynamic stiffness $i\omega\,\widetilde{z}(\omega)$ behaves like a pure mass (see Section 2.2) and can be written as

$$i\omega\widetilde{z}(\omega) = -\omega^2\,\mu(\omega) \quad , \tag{30}$$

in which $\mu(\omega)$ is a positive parameter which has the dimension of a mass per unit area (kg/m^2) and which is related to fuzzy substructure ℓ for local direction j.

Expression for the dynamic stiffness above cutoff frequency Ω_c. Above the cutoff frequency, i.e. for ω fixed in $[\Omega_c, +\infty[$, dynamic stiffness $i\omega\,\widetilde{z}(\omega)$

behaves like a weakly damped resonant system (see Section 2.2) described with four positive parameters $\mu(\omega)$, $\xi(\omega)$, $\Omega_p(\omega)$ and $\alpha(\omega)$ and is written as

$$i\omega\widetilde{z}(\omega) = a(\omega) - \alpha(\omega)\frac{a(\omega)^2}{b(\omega)} \quad , \tag{31}$$

in which $\alpha(\omega)$ is a dimensionless parameter belonging to interval $[0\,,1]$ and where complex numbers $a(\omega)$ and $b(\omega)$ are defined by

$$a(\omega) = 2i\omega\,\xi(\omega)\,\mu(\omega)\,\Omega_p(\omega) + \mu(\omega)\,\Omega_p(\omega)^2 \quad , \tag{32}$$

$$b(\omega) = -\omega^2\,\mu(\omega) + 2i\omega\,\xi(\omega)\,\mu(\omega)\,\Omega_p(\omega) + \mu(\omega)\,\Omega_p(\omega)^2 \quad . \tag{33}$$

When $\alpha(\omega) = 1$, Eqs. (31) to (33) yield

$$i\omega\widetilde{z}(\omega) = \frac{-\omega^2\,\mu(\omega)(2i\omega\,\xi(\omega)\,\Omega_p(\omega) + \Omega_p(\omega)^2)}{\Omega_p(\omega)^2 - \omega^2 + 2i\omega\,\xi(\omega)\,\Omega_p(\omega)} \quad . \tag{34}$$

The underlying deterministic model defined by Eq. (34) is used to construct homogeneous fuzzy impedance $\mathbb{Z}(\omega)$, and is called the *type I homogeneous fuzzy impedance law.* It corresponds to the dynamic stiffness $i\omega\widetilde{z}(\omega)$ of the elementary dynamic model constituted by a linear oscillator excited by its support. For $0 \leq \alpha(\omega) < 1$, the underlying deterministic model defined by Eqs. (31) to (33) is used to construct homogeneous fuzzy impedance $\mathbb{Z}(\omega)$, and is called the *type II homogeneous fuzzy impedance law.* For $\alpha(\omega) = 1$, the type II law gives the type I law. Dynamic stiffness $i\omega\widetilde{z}(\omega)$ defined by Eq. (31) can be rewritten as $i\omega\widetilde{z}(\omega) = a(\omega) - \alpha(\omega)a(\omega)\,b(\omega)^{-1}\,a(\omega)$. This algebraic expression is deduced from the general expression of a reduced boundary-dynamic-stiffness-matrix model defined by Eq. (V.110-3). Concerning the proof that $\alpha(\omega)$ can be chosen as a real number belonging to the interval $[0\,,1]$, we refer the reader to Soize, 1993b. From a modeling point of view, if fuzzy substructure ℓ is composed of a large number of independent dynamical subsystems, then the type I fuzzy impedance law should be used (see Fig. 11). The type II fuzzy impedance law should be used for other cases (see Fig. 12).

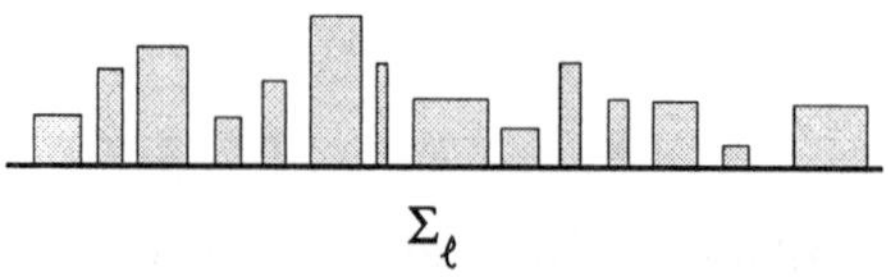

Fig. 11. Fuzzy substructure corresponding to type I fuzzy impedance law

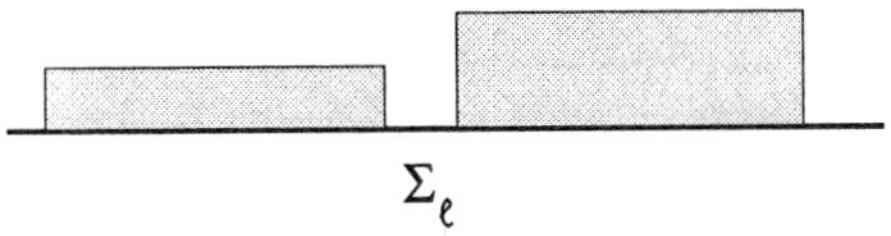

Fig. 12. Fuzzy substructure corresponding to type II fuzzy impedance law

Dimensionless parameter $\alpha(\omega)$ is called the *equivalent coupling factor* because it allows the type of attachment between the master structure and the secondary subsystems constituting fuzzy substructure ℓ to be modeled for local direction j. The case $\alpha(\omega) = 1$ corresponds to discrete attachments and the case $0 \leq \alpha(\omega) < 1$ corresponds to continuous attachments. In Eqs. (31) to (33), dimensionless parameter $\xi(\omega)$ is related to the rate of internal damping of fuzzy substructure ℓ for local direction j and is assumed to be small ($0 < \xi(\omega) \ll 1$). Finally, at $\omega \geq \Omega_c$, $\Omega_p(\omega)$ represents the eigenfrequency of a weakly damped resonant system whose dynamic stiffness is $i\omega\, \widetilde{z}(\omega)$. The dependence of $\Omega_p(\omega)$ on ω is introduced for the following reason. For ω fixed, ω does not necessarily coincide with a resonant frequency of fuzzy substructure ℓ for local direction j (see Section 5.3). We therefore introduce the probability that eigenfrequency Ω_p of the underlying deterministic model of the fuzzy impedance law is in a neighborhood of ω.

6.3. Construction of a probabilistic model of the parameters

For fuzzy substructure ℓ and for local direction j, we introduce six parameters which are Ω_c, $\mu(\omega)$, $\xi(\omega)$, $n(\omega)$, $\alpha(\omega)$ and $\Omega_p(\omega)$. Cutoff frequency Ω_c is considered as a deterministic parameter. The four parameters $\mu(\omega) > 0$, $0 < \xi(\omega) \ll 1$, $n(\omega) > 0$ and $0 \leq \alpha(\omega) \leq 1$ are modeled by random variables whose mean values are called the *mean coefficients* and whose deviations (or "spread" or "variability") are controlled by given deterministic parameters $\lambda_1(\omega)$, $\lambda_2(\omega)$, $\lambda_3(\omega)$ and $\lambda_4(\omega)$ which are called the *deviation coefficients*. Parameter $\Omega_p(\omega)$ is modeled by a random variable whose probability distribution is constructed using modal density parameter $n(\omega)$. In conclusion, four random parameters $\mu(\omega)$, $\xi(\omega)$, $n(\omega)$ and $\alpha(\omega)$ remain and consequently, in the representation of random variable $\mathbb{Z}(\omega)$ defined by Eq. (25), we have

$$d = 4 \quad . \tag{35}$$

Introduction of the modal density related to fuzzy substructure ℓ for local direction j. The modal density of fuzzy substructure ℓ for local direction j is denoted as $n(\omega)$, which means that the number of resonant frequencies of fuzzy substructure ℓ for local direction j (see Section 5.3) in a neighborhood $\delta\omega$ of ω is $n(\omega)\,\delta\omega$. The distance, denoted as $2\varepsilon(\omega)$, between two resonant frequencies of fuzzy substructure ℓ for local direction j in the neigborhood $\delta\omega$ of ω is then given by

$$2\varepsilon(\omega) = \frac{\delta\omega}{n(\omega)\,\delta\omega} = \frac{1}{n(\omega)} \quad . \tag{36}$$

Fuzzy impedance law mean coefficients as the mean values of the random parameters. The mean values of random parameters $\mu(\omega), \xi(\omega), n(\omega), \alpha(\omega)$ are defined by

$$\underline{\mu}(\omega) = E\{\mu(\omega)\}\, , \; \underline{\xi}(\omega) = E\{\xi(\omega)\}\, , \; \underline{n}(\omega) = E\{n(\omega)\}\, , \; \underline{\alpha}(\omega) = E\{\alpha(\omega)\}, \tag{37}$$

and are called the *fuzzy impedance law mean coefficients* (or simply, the *mean coefficients*). It is assumed that these mean coefficients satisfy the following property

$$\underline{\mu}(-\omega) = \underline{\mu}(\omega)\;,\quad \underline{\xi}(-\omega) = \underline{\xi}(\omega)\;,\quad \underline{n}(-\omega) = \underline{n}(\omega)\;,\quad \underline{\alpha}(-\omega) = \underline{\alpha}(\omega)\,. \tag{38}$$

(1)- The mean value $\underline{\mu}(\omega) > 0$ of random parameter $\mu(\omega)$ which has the dimension of a mass per unit area (kg/m^2) is described by a dimensionless mean coefficient $\underline{\nu}(\omega) > 0$ which is such that

$$\underline{\mu}(\omega) = \underline{\nu}(\omega)\,\frac{m^{\ell}_{\text{ref}}}{|\Sigma_\ell|} \quad , \tag{39}$$

in which $|\Sigma_\ell|$ is the area of surface Σ_ℓ and m^{ℓ}_{ref} is a reference mass related to fuzzy substructure ℓ. For instance, if the total mass of fuzzy substructure ℓ can be estimated, then m^{ℓ}_{ref} can be chosen as this total mass. Mean coefficient $\underline{\nu}(\omega)$ is called the *mean coefficient of the participating mass* of fuzzy substructure ℓ for local direction j.

(2)- Dimensionless mean coefficient $\underline{\xi}(\omega)$ is the *mean rate of internal damping* of fuzzy substructure ℓ for local direction j and is such that $0 < \underline{\xi}(\omega) \ll 1$. For instance, typical values of $\underline{\xi}(\omega)$ are 0.001 or 0.01.

(3)- Mean coefficient $\underline{n}(\omega) > 0$ is the *mean modal density* (measured in seconds) related to fuzzy substructure ℓ for local direction j.

(4)- Dimensionless mean coefficient $\underline{\alpha}(\omega)$ is the *mean equivalent coupling factor* of fuzzy substructure ℓ for local direction j and is such that $0 \leq \underline{\alpha}(\omega) \leq 1$. For $\underline{\alpha}(\omega) = 1$, the model corresponds to the type I homogeneous fuzzy impedance law and for $0 \leq \underline{\alpha}(\omega) < 1$, the model corresponds to the type II homogeneous fuzzy impedance law.

Deviation coefficients and construction of the probability distributions of random variables $\mu(\omega)$, $\xi(\omega)$, $n(\omega)$, $\alpha(\omega)$. To control the deviation of random variables $\mu(\omega)$, $\xi(\omega)$, $n(\omega)$ and $\alpha(\omega)$, we introduce the $\mathbb{R}^4$-valued function

$$\omega \mapsto \boldsymbol{\lambda}(\omega) = (\lambda_1(\omega), \lambda_2(\omega), \lambda_3(\omega), \lambda_4(\omega)) \quad , \tag{40}$$

defined on $\mathbb{B}_{\mathrm{MF}}$ and such that

$$\lambda_k(\omega) \in [0,1[\quad , \quad \forall k \in \{1,2,3\} \quad , \tag{41}$$

and for $k = 4$,

$$0 \leq \lambda_4(\omega) \leq 1 \qquad \text{as} \quad 0 \leq \underline{\alpha}(\omega) \leq \frac{1}{2} \quad , \tag{42}$$

$$0 \leq \lambda_4(\omega) \leq \frac{1}{\underline{\alpha}(\omega)} - 1 \quad \text{as} \quad \frac{1}{2} \leq \underline{\alpha}(\omega) \leq 1 \quad . \tag{43}$$

In addition, it is assumed that the $\boldsymbol{\lambda}(\omega)$ terms satisfy the property

$$\boldsymbol{\lambda}(-\omega) = \boldsymbol{\lambda}(\omega) \quad . \tag{44}$$

Parameters $\lambda_1(\omega), \lambda_2(\omega), \lambda_3(\omega)$ and $\lambda_4(\omega)$ are called the *fuzzy impedance law deviation coefficients* (or simply, the *deviation coefficients*). Random variables $\mu(\omega)$, $\xi(\omega)$, $n(\omega)$ and $\alpha(\omega)$ are then defined by

$$\mu(\omega) = \underline{\mu}(\omega)\left(1 + \frac{\lambda_1(\omega)}{\sqrt{3}}\mathbb{X}_1\right) \quad , \tag{45}$$

$$\xi(\omega) = \underline{\xi}(\omega)\left(1 + \frac{\lambda_2(\omega)}{\sqrt{3}}\mathbb{X}_2\right) \quad , \tag{46}$$

$$n(\omega) = \underline{n}(\omega)\left(1 + \frac{\lambda_3(\omega)}{\sqrt{3}}\mathbb{X}_3\right) \quad , \tag{47}$$

$$\alpha(\omega) = \underline{\alpha}(\omega)\left(1 + \frac{\lambda_4(\omega)}{\sqrt{3}}\mathbb{X}_4\right) \quad , \tag{48}$$

in which independent normalized second-order random variables $\mathbb{X}_1$, $\mathbb{X}_2$, $\mathbb{X}_3$ and $\mathbb{X}_4$ are defined in Section 6.1. From Eqs. (27), (28) and (41) to (48), we deduce that supports $S_{\mu(\omega)}$, $S_{\xi(\omega)}$, $S_{n(\omega)}$ and $S_{\alpha(\omega)}$ of the probability density functions of random variables $\mu(\omega)$, $\xi(\omega)$, $n(\omega)$ and $\alpha(\omega)$, are such that

$$\mathcal{S}_{\mu(\omega)} = [\underline{\mu}(\omega)(1-\lambda_1(\omega))\,,\,\underline{\mu}(\omega)(1+\lambda_1(\omega))] \subset\ ]0\,,2\,\underline{\mu}(\omega)[\quad , \tag{49}$$
$$\mathcal{S}_{\xi(\omega)} = [\underline{\xi}(\omega)(1-\lambda_2(\omega))\,,\,\underline{\xi}(\omega)(1+\lambda_2(\omega))] \subset\ ]0\,,2\,\underline{\xi}(\omega)[\quad , \tag{50}$$
$$\mathcal{S}_{n(\omega)} = [\underline{n}(\omega)(1-\lambda_3(\omega))\,,\,\underline{n}(\omega)(1+\lambda_3(\omega))] \subset\ ]0\,,2\,\underline{n}(\omega)[\quad , \tag{51}$$
$$\mathcal{S}_{\alpha(\omega)} = [\underline{\alpha}(\omega)(1-\lambda_4(\omega))\,,\,\underline{\alpha}(\omega)(1+\lambda_4(\omega))] \subset\ [0\,,1] \quad . \tag{52}$$

Eqs. (45) to (48) show that if $\boldsymbol{\lambda} = \mathbf{0}$ (i.e. if the deviation coefficients are equal to zero), then random variables $\mu(\omega)$, $\xi(\omega)$, $n(\omega)$ and $\alpha(\omega)$ are equal (almost surely) to $\underline{\mu}(\omega)$, $\underline{\xi}(\omega)$, $\underline{n}(\omega)$ and $\underline{\alpha}(\omega)$ respectively.

Construction of the probability distribution of random variable $\Omega_p(\omega)$

The probability distribution of random variable $\Omega_p(\omega)$ for $\omega \geq \Omega_c$ is constructed using random variable $n(\omega)$ defined by Eq. (47). From Eqs. (36) and (47), we deduce that random variable $\varepsilon(\omega)$ is defined by $\varepsilon(\omega) = b(\omega; X_3)$ in which

$$b(\omega; x_3) = \frac{1}{2\,\underline{n}(\omega)\,(1 + 3^{-1/2}\,\lambda_3(\omega)\,x_3)} \quad . \tag{53}$$

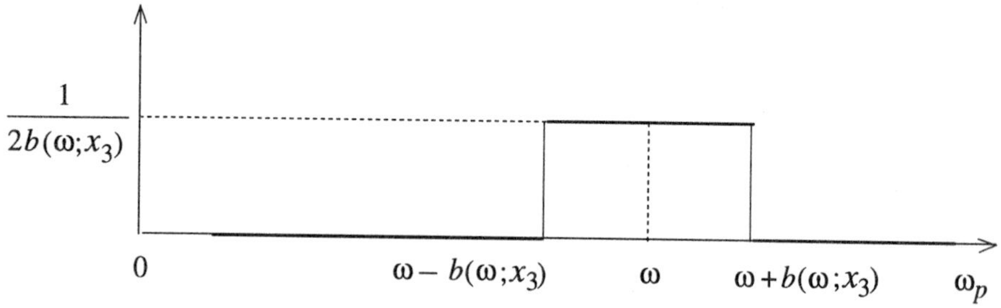

Fig. 13. Conditional probability density function $p_{\Omega_p(\omega)|X_3}(\omega_p;\omega|x_3)$

By construction, the conditional probability distribution of random variable $\Omega_p(\omega)$ given $X_3 = x_3$, denoted as $P_{\Omega_p(\omega)|X_3}(d\omega_p;\omega|x_3)$, is defined by the conditional probability density function,

$$P_{\Omega_p(\omega)|X_3}(d\omega_p;\omega|x_3) = p_{\Omega_p(\omega)|X_3}(\omega_p;\omega|x_3)\ d\omega_p \quad , \tag{54}$$

which is written as (see Fig. 13),

$$p_{\Omega_p(\omega)|X_3}(\omega_p;\omega|x_3) = \frac{1}{2\,b(\omega;x_3)}\ \mathbb{1}_{[\omega-b(\omega;x_3)\,,\,\omega+b(\omega;x_3)]}(\omega_p) \quad . \tag{55}$$

It should be noted that the introduction of this conditional probability distribution of random variable $\Omega_p(\omega)$ given $\mathbb{X}_3 = x_3$ is natural. In effect, since the distance between two resonant frequencies of fuzzy substructure ℓ for local direction j (see Section 5.3) in the neighborhood of ω is $2\,b(\omega; x_3)$, the probability for a resonant frequency to be inside an interval whose length is $2\,b(\omega; x_3)$, is equal to 1. From Eqs. (27) and (54), we deduce that the probability distribution $P_{\Omega_p(\omega),\mathbb{X}_3}(d\omega_p, dx_3; \omega)$ of $\mathbb{R}^2$-valued random variable $(\Omega_p(\omega), \mathbb{X}_3)$ has a density $p_{\Omega_p(\omega),\mathbb{X}_3}(\omega_p, x_3; \omega)$ with respect to $d\omega_p\, dx_3$ given by

$$p_{\Omega_p(\omega),\mathbb{X}_3}(\omega_p, x_3; \omega) = p_{\Omega_p(\omega)|\,\mathbb{X}_3}(\omega_p; \omega|x_3)\; p_{\mathbb{X}_3}(x_3) \quad . \tag{56}$$

From Eq. (56), we deduce that the probability distribution $P_{\Omega_p(\omega)}(d\omega_p; \omega)$ of real valued random variable $\Omega_p(\omega)$ has a density $p_{\Omega_p(\omega)}(\omega_p; \omega)$ with respect to $d\omega_p$ given by

$$p_{\Omega_p(\omega)}(\omega_p; \omega) = \int_{\mathbb{R}} p_{\Omega_p(\omega)|\,\mathbb{X}_3}(\omega_p; \omega|x_3)\; p_{\mathbb{X}_3}(x_3)\; dx_3 \quad . \tag{57}$$

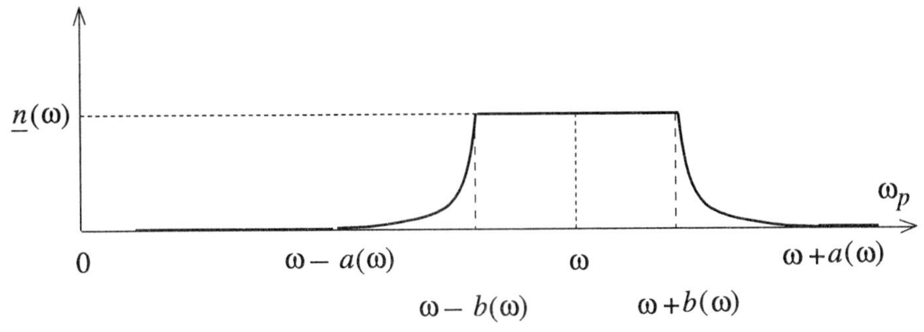

Fig. 14. Probability density function $p_{\Omega_p(\omega)}(\omega_p;\omega)$

Substituting Eqs. (28) and (55) in Eq. (57) yields (see Fig. 14)

$$p_{\Omega_p(\omega)}(\omega_p; \omega) = h(\omega_p; \omega)\; \mathbb{1}_{[\omega - a(\omega)\,,\,\omega + a(\omega)]}(\omega_p) \quad , \tag{58}$$

in which

$$h(\omega_p; \omega) = \underline{n}(\omega) \quad \text{if} \quad \omega - b(\omega) \leq \omega_p \leq \omega + b(\omega) \quad , \tag{59}$$

and

$$h(\omega_p; \omega) = \frac{1}{16\,\lambda_3(\omega)\,\underline{n}(\omega)\,(\omega - \omega_p)^2} - \underline{n}(\omega)\,\frac{(\lambda_3(\omega) - 1)^2}{4\,\lambda_3(\omega)} \quad , \tag{60}$$

if $\omega - a(\omega) \leq \omega_p \leq \omega - b(\omega)$ or if $\omega + b(\omega) \leq \omega_p \leq \omega + a(\omega)$, where $a(\omega)$ and $b(\omega)$ are defined by

$$a(\omega) = \frac{1}{2\,\underline{n}(\omega)\,(1 - \lambda_3(\omega))} \quad , \quad b(\omega) = \frac{1}{2\,\underline{n}(\omega)\,(1 + \lambda_3(\omega))} \quad . \tag{61}$$

Probability density function of random vector $\{\mathbb{X}_1, \mathbb{X}_2, \mathbb{X}_3, \mathbb{X}_4, \Omega_p(\omega)\}$

Let $\mathbb{X}_1, \mathbb{X}_2, \mathbb{X}_3, \mathbb{X}_4$ be the independent random variables defined in Section 6.1. Using Eq. (56), we then deduce that the probability distribution of $\mathbb{R}^5$-valued random vector $(\mathbb{X}_1, \mathbb{X}_2, \mathbb{X}_3, \mathbb{X}_4, \Omega_p(\omega))$ is written as

$$\begin{aligned} &P(dx_1, dx_2, dx_3, dx_4, d\omega_p; \omega) \\ &\qquad = p(x_1, x_2, x_3, x_4, \omega_p; \omega)\, dx_1\, dx_2\, dx_3\, dx_4\, d\omega_p\,, \end{aligned} \tag{62}$$

in which probability distribution $p(x_1, x_2, x_3, x_4, \omega_p; \omega)$ is such that

$$\begin{aligned} &p(x_1, x_2, x_3, x_4, \omega_p; \omega) \\ &\qquad = p_{\mathbb{X}_1}(x_1)\, p_{\mathbb{X}_2}(x_2)\, p_{\mathbb{X}_3}(x_3)\, p_{\mathbb{X}_4}(x_4)\, p_{\Omega_p(\omega)|\mathbb{X}_3}(\omega_p; \omega|x_3)\,, \end{aligned} \tag{63}$$

in which conditional probability density function $p_{\Omega_p(\omega)|\mathbb{X}_3}(\omega_p; \omega|x_3)$ is defined by Eq. (55). It should be noted that random variables $\mathbb{X}_3$ and $\Omega_p(\omega)$ are not independent.

6.4. Underlying probabilistic model for the construction of $\mathbb{Z}(\omega; \boldsymbol{\lambda})$

The substitution of the expressions of random variables $\mu(\omega)$, $\xi(\omega)$ and $\alpha(\omega)$ given by Eqs. (45), (46) and (48) into $\widetilde{z}(\omega)$ given by Eqs. (30) and (31) defines the complex-valued random variable denoted as $\widetilde{\mathbb{Z}}(\omega; \boldsymbol{\lambda})$,

$$\widetilde{\mathbb{Z}}(\omega; \boldsymbol{\lambda}) = \widetilde{z}(\omega; \mathbb{X}_1, \mathbb{X}_2, \mathbb{X}_4, \Omega_p(\omega)) \quad , \tag{64}$$

in which dispersion parameter $\boldsymbol{\lambda}(\omega)$ defined by Eq. (40) is introduced on the left-hand side of Eq. (64) due to the presence of $\boldsymbol{\lambda}(\omega)$ in Eqs. (45) to (48) (it should be noted that random variable $\Omega_p(\omega)$ depends on $\lambda_3(\omega)$). For the sake of brevity, $\boldsymbol{\lambda}(\omega)$ is omitted from the right-hand side of Eq. (64). Random variable $\widetilde{\mathbb{Z}}(\omega; \boldsymbol{\lambda})$ is called the underlying probabilistic model for the construction of random variable $\mathbb{Z}(\omega)$ which is then rewritten as $\mathbb{Z}(\omega; \boldsymbol{\lambda})$. This random variable is described by Eq. (64) and by the probability distribution of random vector $(\mathbb{X}_1, \mathbb{X}_2, \mathbb{X}_3, \mathbb{X}_4, \Omega_p(\omega))$ given by Eqs. (62) and (63). It can be satisfied that $\widetilde{\mathbb{Z}}(\omega; \boldsymbol{\lambda})$ is a second-order random variable,

$$\begin{aligned} E\{|\widetilde{\mathbb{Z}}(\omega; \boldsymbol{\lambda})|^2\} = \int_{\mathbb{R}}\int_{\mathbb{R}}\int_{\mathbb{R}}\int_{\mathbb{R}}\int_{\mathbb{R}} |\widetilde{z}(\omega; x_1, x_2, x_4, \omega_p)|^2 \times \\ p(x_1, x_2, x_3, x_4, \omega_p; \omega)\, dx_1\, dx_2\, dx_3\, dx_4\, d\omega_p < +\infty \quad , \end{aligned} \tag{65}$$

in which $p(x_1, x_2, x_3, x_4, \omega_p; \omega)$ is given by Eq. (63).

6.5. Construction of random variable $\mathbb{Z}(\omega; \boldsymbol{\lambda})$

By construction, mean value $\underline{Z}(\omega; \boldsymbol{\lambda})$ of random variable $\mathbb{Z}(\omega; \boldsymbol{\lambda})$ defined by Eq. (25), is defined as the mean value of random variable $\widetilde{\mathbb{Z}}(\omega; \boldsymbol{\lambda})$ defined by Eq. (64),

$$\begin{aligned}\underline{Z}(\omega; \boldsymbol{\lambda}) &= E\{\widetilde{\mathbb{Z}}(\omega; \boldsymbol{\lambda})\} \\ &= \int_{\mathbb{R}}\int_{\mathbb{R}}\int_{\mathbb{R}}\int_{\mathbb{R}}\int_{\mathbb{R}} \widetilde{z}(\omega; x_1, x_2, x_4, \omega_p) \times \\ &\qquad p(x_1, x_2, x_3, x_4, \omega_p; \omega)\, dx_1\, dx_2\, dx_3\, dx_4\, d\omega_p \quad ,\end{aligned} \tag{66}$$

in which $p(x_1, x_2, x_3, x_4, \omega_p; \omega)$ is given by Eq. (63). Deterministic complex values $Z_1(\omega; \boldsymbol{\lambda})$, $Z_2(\omega; \boldsymbol{\lambda})$, $Z_3(\omega; \boldsymbol{\lambda})$, $Z_4(\omega; \boldsymbol{\lambda})$ appearing in Eq. (25), are constructed as the solution in Z_1, Z_2, Z_3, Z_4 of the following optimization problem,

$$\min_{\substack{Z_k \\ \{k=1,\ldots,4\}}} \; E\Big\{\Big|\widetilde{\mathbb{Z}}(\omega; \boldsymbol{\lambda}) - \underline{Z}(\omega; \boldsymbol{\lambda}) - \sum_{k=1}^{4} Z_k\, \mathbb{X}_k\Big|^2\Big\} \quad . \tag{67}$$

From Eq. (67), since random variables $\mathbb{X}_1$, $\mathbb{X}_2$, $\mathbb{X}_3$, $\mathbb{X}_4$ are independent and using Eq. (29), we deduce that the solution of Eq. (67) is written, for all $k = 1, \ldots, 4$, as

$$\begin{aligned}Z_k(\omega; \boldsymbol{\lambda}) &= E\{\mathbb{X}_k\, \widetilde{\mathbb{Z}}(\omega; \boldsymbol{\lambda})\} \\ &= \int_{\mathbb{R}}\int_{\mathbb{R}}\int_{\mathbb{R}}\int_{\mathbb{R}}\int_{\mathbb{R}} x_k\, \widetilde{z}(\omega; x_1, x_2, x_4, \omega_p) \times \\ &\qquad p(x_1, x_2, x_3, x_4, \omega_p; \omega)\, dx_1\, dx_2\, dx_3\, dx_4\, d\omega_p \; ,\end{aligned} \tag{68}$$

in which $p(x_1, x_2, x_3, x_4, \omega_p; \omega)$ is given by Eq. (63). The construction of second-order centered random variable $\sum_{k=1}^{4} Z_k(\omega; \boldsymbol{\lambda})\, \mathbb{X}_k$ using the above procedure is called the stochastic linearization method of second-order centered random variable $\widetilde{\mathbb{Z}}(\omega; \boldsymbol{\lambda}) - \underline{Z}(\omega; \boldsymbol{\lambda})$.

6.6. Expressions of the type I and type II homogeneous fuzzy impedance laws of fuzzy substructure ℓ for local direction j

The type II homogeneous fuzzy impedance law $\mathbb{Z}(\omega; \boldsymbol{\lambda})$ of fuzzy substructure ℓ for local direction j is the complex-valued second-order random variable defined by Eq. (25) with $d = 4$,

$$\mathbb{Z}(\omega; \boldsymbol{\lambda}) = \underline{Z}(\omega; \boldsymbol{\lambda}) + \sum_{k=1}^{4} Z_k(\omega; \boldsymbol{\lambda})\, \mathbb{X}_k \quad . \tag{69}$$

Introducing the real part and the imaginary part of complex numbers $i\omega\,\underline{Z}(\omega;\boldsymbol{\lambda})$ and $i\omega\,Z_k(\omega;\boldsymbol{\lambda})$, we can write

$$i\omega\,\underline{Z}(\omega;\boldsymbol{\lambda}) = -\omega^2\,\underline{R}(\omega;\boldsymbol{\lambda}) + i\omega\,\underline{I}(\omega;\boldsymbol{\lambda}) \quad , \tag{70}$$

$$i\omega\,Z_k(\omega;\boldsymbol{\lambda}) = -\omega^2\,R_k(\omega;\boldsymbol{\lambda}) + i\omega\,I_k(\omega;\boldsymbol{\lambda}) \quad , \tag{71}$$

in which $\underline{R}(\omega;\boldsymbol{\lambda})$, $\underline{I}(\omega;\boldsymbol{\lambda})$, $R_k(\omega;\boldsymbol{\lambda})$ and $I_k(\omega;\boldsymbol{\lambda})$ are real numbers deduced from Eqs. (66), (68), (70) and (71). For the sake of brevity, in Eqs. (72) to (104), we use the notations $\underline{\mu} = \underline{\mu}(\omega)$, $\underline{\xi} = \underline{\xi}(\omega)$, $\underline{n} = \underline{n}(\omega)$, $\underline{\alpha} = \underline{\alpha}(\omega)$ and $\lambda_k = \lambda_k(\omega)$ for $k = 1, 2, 3, 4$.

Case below the cutoff frequency. For $\omega \in]0, \Omega_c[$, for the real parts we have

$$\underline{R}(\omega;\boldsymbol{\lambda}) = \underline{\mu} \quad , \tag{72}$$

$$R_1(\omega;\boldsymbol{\lambda}) = \underline{\mu}\,\frac{\lambda_1}{\sqrt{3}} \quad , \tag{73}$$

$$R_k(\omega;\boldsymbol{\lambda}) = 0 \quad \text{for} \quad k \in \{2,3,4\} \quad , \tag{74}$$

and for the imaginary parts, we have

$$\underline{I}(\omega;\boldsymbol{\lambda}) = 0 \quad , \tag{75}$$

$$I_k(\omega;\boldsymbol{\lambda}) = 0 \quad \text{for} \quad k \in \{1,2,3,4\} \quad . \tag{76}$$

Case above the cutoff frequency. For $\omega \in [\Omega_c, +\infty[$, for the real parts we have

$$\underline{R}(\omega;\boldsymbol{\lambda}) = \omega\,\underline{\mu}\,\underline{n}\;\rho^R(\omega;\boldsymbol{\lambda}) \quad , \tag{77}$$

$$R_1(\omega;\boldsymbol{\lambda}) = \omega\,\underline{\mu}\,\underline{n}\,\frac{\lambda_1}{\sqrt{3}}\;\rho^R(\omega;\boldsymbol{\lambda}) \quad , \tag{78}$$

$$R_2(\omega;\boldsymbol{\lambda}) = \omega\,\underline{\mu}\,\underline{n}\,\frac{\sqrt{3}}{\lambda_2}\;\underline{\alpha}\;\underline{J}_5(\omega;\boldsymbol{\lambda}) \quad , \tag{79}$$

$$R_3(\omega;\boldsymbol{\lambda}) = \omega\,\underline{\mu}\,\underline{n}\,\frac{\sqrt{3}}{\lambda_3}\;\underline{\alpha}\;\underline{J}_4(\omega;\boldsymbol{\lambda}) \quad , \tag{80}$$

$$R_4(\omega;\boldsymbol{\lambda}) = \omega\,\underline{\mu}\,\underline{n}\,\frac{\lambda_4}{\sqrt{3}}\;\underline{\alpha}\;(\kappa_0(\omega;\boldsymbol{\lambda}) + \underline{J}_3(\omega;\boldsymbol{\lambda})) \quad , \tag{81}$$

and for the imaginary parts we have

$$\underline{I}(\omega;\boldsymbol{\lambda}) = \frac{\pi}{2}\,\omega^2\underline{\mu}\,\underline{n}\;\rho^I(\omega;\boldsymbol{\lambda}) \quad , \tag{82}$$

$$I_1(\omega;\boldsymbol{\lambda}) = \frac{\pi}{2}\,\omega^2\underline{\mu}\,\underline{n}\;\frac{\lambda_1}{\sqrt{3}}\;\rho^I(\omega;\boldsymbol{\lambda}) \quad , \tag{83}$$

$$I_2(\omega;\boldsymbol{\lambda}) = \frac{\pi}{2}\,\omega^2\underline{\mu}\,\underline{n}\;\frac{\sqrt{3}}{\lambda_2}\;\left[\kappa_1(\omega)\,\frac{\lambda_2^2}{3}\,(1-\underline{\alpha}) + \underline{\alpha}\;\underline{J}_2(\omega;\boldsymbol{\lambda})\right] \quad , \tag{84}$$

$$I_3(\omega;\boldsymbol{\lambda}) = \frac{\pi}{2}\,\omega^2\underline{\mu}\,\underline{n}\;\frac{\sqrt{3}}{\lambda_3}\;\underline{\alpha}\;\underline{J}_1(\omega;\boldsymbol{\lambda}) \quad , \tag{85}$$

$$I_4(\omega;\boldsymbol{\lambda}) = \frac{\pi}{2}\,\omega^2\underline{\mu}\,\underline{n}\;\frac{\lambda_4}{\sqrt{3}}\;\underline{\alpha}\;(\underline{J}_0(\omega;\boldsymbol{\lambda}) - \kappa_1(\omega)) \quad . \tag{86}$$

In the above equations, $\rho^R(\omega;\boldsymbol{\lambda})$ is defined by

$$\rho^R(\omega;\boldsymbol{\lambda}) = \kappa_0(\omega;\boldsymbol{\lambda})\,(\underline{\alpha}-1) + \underline{\alpha}\;\underline{J}_3(\omega;\boldsymbol{\lambda}) \quad , \tag{87}$$

$\kappa_0(\omega;\boldsymbol{\lambda})$ is given by

$$\kappa_0(\omega;\boldsymbol{\lambda}) = \frac{1}{\omega\,\underline{n}} + \frac{1}{12\,\omega^3\,\underline{n}^3\,(1-\lambda_3^2)} \quad , \tag{88}$$

$\rho^I(\omega;\boldsymbol{\lambda})$ is defined by

$$\rho^I(\omega;\boldsymbol{\lambda}) = \kappa_1(\omega)\,(1-\underline{\alpha}) + \underline{\alpha}\;\underline{J}_0(\omega;\boldsymbol{\lambda}) \quad , \tag{89}$$

$\kappa_1(\omega)$ is given by

$$\kappa_1(\omega) = \frac{4\,\underline{\xi}}{\pi\,\omega\,\underline{n}} \quad , \tag{90}$$

and for $\kappa \in \{0,1,2,3,4,5\}$,

$$\underline{J}_\kappa(\omega;\boldsymbol{\lambda}) = \frac{1}{4}\int_{-1}^{1} dy_2 \int_{-1}^{1} dy_3\; J_\kappa(\omega\,;\,(1+\lambda_2\,y_2)\,\underline{\xi}\,,\,\lambda_3\,y_3) \quad , \tag{91}$$

with

$$J_0(\omega;x,y) = \frac{1+y}{\pi\sqrt{1-x^2}}[\arctan Y_+(\omega;x,y) - \arctan Y_-(\omega;x,y)] \quad , \tag{92}$$

$$J_1(\omega;x,y) = y\,J_0(\omega;x,y) \quad , \tag{93}$$

$$J_2(\omega; x, y) = \frac{x - \underline{\xi}}{\underline{\xi}} \, J_0(\omega; x, y) \quad , \tag{94}$$

$$J_3(\omega; x, y) = \frac{1}{\omega \, \underline{n}} - \frac{1 + y}{4\sqrt{1 - x^2}} \ln\Big[\frac{N(\omega; x, y)}{D(\omega; x, y)}\Big] \quad , \tag{95}$$

$$J_4(\omega; x, y) = y \, J_3(\omega; x, y) \quad , \tag{96}$$

$$J_5(\omega; x, y) = \frac{x - \underline{\xi}}{\underline{\xi}} \, J_3(\omega; x, y) \quad . \tag{97}$$

In Eq. (92), $Y_\pm(\omega; x, y)$ is defined by

$$Y_\pm(\omega; x, y) = \frac{\theta_\pm(\omega; y) + x^2}{x\sqrt{1 - x^2}} \quad , \tag{98}$$

in which

$$\theta_\pm(\omega; y) = \frac{1}{2} \left[(1 \pm \tau(\omega; y))^2 - 1\right] \quad , \tag{99}$$

with

$$\tau(\omega; y) = \frac{1}{2 \, \omega \, \underline{n} \, (1 + y)} \quad . \tag{100}$$

In Eq. (95), $N(\omega; x, y)$ and $D(\omega; x, y)$ are defined by

$$N(\omega; x, y) = [U_+(\omega; y) + W_+(\omega; x, y)][U_-(\omega; y) - W_-(\omega; x, y)] \quad , \tag{101}$$

$$D(\omega; x, y) = [U_+(\omega; y) - W_+(\omega; x, y)][U_-(\omega; y) + W_-(\omega; x, y)] \quad , \tag{102}$$

in which

$$U_\pm(\omega; y) = 2\,[\theta_\pm(\omega; y) + 1] \quad , \tag{103}$$

$$W_\pm(\omega; x, y) = 2\sqrt{1 - x^2}\,[1 \pm \tau(\omega; y)] \quad . \tag{104}$$

Definition. Eqs. (69) to (104) define the type II homogeneous fuzzy impedance law of fuzzy substructure ℓ for local direction j. For $\underline{\alpha}(\omega) = 1$ and $\lambda_4(\omega) = 0$, Eqs. (69) to (104) define the type I homogeneous fuzzy impedance law of fuzzy substructure ℓ for local direction j which then appears as a particular case of the type II homogeneous fuzzy impedance law of fuzzy substructure ℓ for local direction j. The parameters $\underline{\mu}(\omega)$, $\underline{\xi}(\omega)$, $\underline{n}(\omega)$ and $\underline{\alpha}(\omega)$ are called the fuzzy impedance law mean coefficients (or simply the *mean coefficients*). The associated parameters $\lambda_k = \lambda_k(\omega)$, $k = 1, 2, 3, 4$ are called the *fuzzy impedance law deviation coefficients* (or simply the *deviation coefficients*).

6.7. Properties of type I and type II homogeneous fuzzy impedance laws of a fuzzy substructure for local direction j

Verification of the constraints introduced in Section 5.2. From Section 6.1, we recall that $\mathbb{Z}(\omega)$ is a second-order random variable and consequently, Eq. (17) is satisfied. From Eqs. (38), (44), (70) to (104), we deduce that $\mathbb{Z}(-\omega) = \overline{\mathbb{Z}(\omega)}$ (see Eq. (24)) and consequently, Eq. (16) is satisfied.

Positivity of mean value $\underline{I}(\omega; \boldsymbol{\lambda})$ of the real part of $\mathbb{Z}(\omega; \boldsymbol{\lambda})$. Using the expressions given in Section 6.6, it can be proved that, for all real ω,

$$\underline{I}(\omega; \boldsymbol{\lambda}) \geq 0 \quad . \tag{105}$$

Limit case corresponding to $\boldsymbol{\lambda} \to \mathbf{0}$. Using the equations given in Section 6.6, it can be proved that, for all real ω and for $k = 1, \ldots, 4$,

$$R_k(\omega; \boldsymbol{\lambda}) \to 0 \quad , \quad I_k(\omega; \boldsymbol{\lambda}) \to 0 \quad \text{as} \quad \boldsymbol{\lambda} \to \mathbf{0} \quad . \tag{106}$$

Eqs. (69) and (106) show that the sequence of second-order random variables $\{\mathbb{Z}(\omega; \boldsymbol{\lambda})\}_{\boldsymbol{\lambda}}$ converges in mean square [8] to mean value $\underline{Z}(\omega; \boldsymbol{\lambda})$,

$$\mathbb{Z}(\omega; \boldsymbol{\lambda}) \to \underline{Z}(\omega; \mathbf{0}) \quad \text{as} \quad \boldsymbol{\lambda} \to \mathbf{0} \quad . \tag{107}$$

Limit of mean value $\underline{I}(\omega; \mathbf{0})$ as $\underline{\xi}(\omega) \to 0$. For $\omega\, \underline{n}(\omega)$ fixed,

$$\lim_{\underline{\xi}(\omega) \to 0} \underline{I}(\omega; \mathbf{0}) = 2\, \omega\, \underline{\mu}(\omega)\, \underline{\xi}(\omega)\, (1 - \underline{\alpha}(\omega)) + \frac{\pi}{2}\, \omega^2\, \underline{\mu}(\omega)\, \underline{n}(\omega)\, \underline{\alpha}(\omega) \quad . \tag{108}$$

For $\underline{\alpha}(\omega) = 1$ (type I law), Eq. (108) yields

$$\lim_{\underline{\xi}(\omega) \to 0} \underline{I}(\omega; \mathbf{0}) = \frac{\pi}{2}\, \omega^2\, \underline{\mu}(\omega)\, \underline{n}(\omega) \quad . \tag{109}$$

In practice, Eqs. (108) and (109) hold if

$$4\, \omega\, \underline{n}(\omega)\, \underline{\xi}(\omega)^2 \ll \left| \frac{1}{2\, \omega\, \underline{n}(\omega)} \pm 2 \right| \quad . \tag{110}$$

---

[8] This means that $\lim_{\boldsymbol{\lambda} \to \mathbf{0}} E\{|\mathbb{Z}(\omega;\boldsymbol{\lambda}) - \underline{Z}(\omega;\mathbf{0})|^2\} = 0$.

7. Solving Method for the Random Equation of the Master Structure Coupled with Fuzzy Substructures

7.1. Construction and properties of the random boundary impedance operator

The problem consists in constructing the random boundary impedance operator $\mathbf{Z}_{\text{fuz}}(\omega)$ defined by Eq. (12), related to the L fuzzy substructures. The following procedure is used. For each fuzzy substructure ℓ, random boundary impedance operator $\mathbf{Z}^{\ell}_{\text{fuz}}(\omega)$, defined by Eqs. (22) and (23), is constructed using a type I or type II homogeneous fuzzy impedance law of fuzzy substructure ℓ for local direction j. This law is given by Eq. (69) with Eqs. (70) to (104), and is rewritten, reintroducing indices ℓ and j, as

$$\mathbb{Z}^{\ell}_{j}(\omega;\boldsymbol{\lambda}^{\ell}_{j}) = \underline{Z}^{\ell}_{j}(\omega;\boldsymbol{\lambda}^{\ell}_{j}) + \sum_{k=1}^{4} Z^{\ell}_{j,k}(\omega;\boldsymbol{\lambda}^{\ell}_{j})\, \mathbb{X}^{\ell}_{j,k} \quad , \tag{111}$$

in which $\boldsymbol{\lambda}^{\ell}_{j} = (\lambda^{\ell}_{j,1}, \lambda^{\ell}_{j,2}, \lambda^{\ell}_{j,3}, \lambda^{\ell}_{j,4})$. Let $\mathcal{J}_{\ell} \subseteq \{1,2,3\}$ be the subset of the active local directions, i.e. local directions where the contribution is nonzero. Let $\boldsymbol{\Lambda}$ be the vector of length M constituted by all the deviation coefficients $\{\lambda^{\ell}_{j,1},\, \lambda^{\ell}_{j,2},\, \lambda^{\ell}_{j,3},\, \lambda^{\ell}_{j,4}\}$ for $\ell = 1,\ldots,L$ and for $j \in \mathcal{J}_{\ell}$. It should be noted that $M \leq 12\,L$. Similarly, let $\mathbf{X}$ be the random vector of length M associated with vector $\boldsymbol{\Lambda}$ and constituted by all the independent random variables $\{\mathbb{X}^{\ell}_{j,1},\, \mathbb{X}^{\ell}_{j,2},\, \mathbb{X}^{\ell}_{j,3},\, \mathbb{X}^{\ell}_{j,4}\}$. The components of $\mathbf{X}$ are renumbered as $\mathbb{X}_1,\ldots,\mathbb{X}_M$. Random variables $\mathbb{X}_1,\ldots,\mathbb{X}_M$ are independent and each $\mathbb{X}_i$ is a normalized random variable with a uniform distribution (see Section 6.1). From Eqs. (12), (22) and (111), it can easily be satisfied that random boundary impedance operator $\mathbf{Z}_{\text{fuz}}(\omega;\boldsymbol{\Lambda})$ can then be written as

$$\mathbf{Z}_{\text{fuz}}(\omega;\boldsymbol{\Lambda}) = \underline{\mathbf{Z}}_{\text{fuz}}(\omega;\boldsymbol{\Lambda}) + \mathbf{Z}_{\text{rand}}(\omega;\boldsymbol{\Lambda}) \quad , \tag{112}$$

in which random operator $\mathbf{Z}_{\text{rand}}(\omega;\boldsymbol{\Lambda})$ is defined by

$$\mathbf{Z}_{\text{rand}}(\omega;\boldsymbol{\Lambda}) = \sum_{i=1}^{M} \mathbf{Z}_i(\omega;\boldsymbol{\Lambda})\, \mathbb{X}_i \quad , \tag{113}$$

where deterministic operators $\underline{\mathbf{Z}}_{\text{fuz}}(\omega;\boldsymbol{\Lambda})$ and $\mathbf{Z}_1(\omega;\boldsymbol{\Lambda}),\ldots,\mathbf{Z}_M(\omega;\boldsymbol{\Lambda})$ are deduced from the above procedure. Since random variables $\mathbb{X}_1,\ldots,\mathbb{X}_M$ are centered, Eq. (113) shows that random operator $\mathbf{Z}_{\text{rand}}(\omega;\boldsymbol{\Lambda})$ is centered,

$$E\{\mathbf{Z}_{\text{rand}}(\omega;\boldsymbol{\Lambda}\} = 0 \quad , \tag{114}$$

and from Eq. (112), we then deduce that operator $\underline{\mathbf{Z}}_{\text{fuz}}(\omega;\boldsymbol{\Lambda})$ is the mean value of random operator $\mathbf{Z}_{\text{fuz}}(\omega;\boldsymbol{\Lambda})$,

$$\underline{\mathbf{Z}}_{\text{fuz}}(\omega;\boldsymbol{\Lambda}) = E\{\mathbf{Z}_{\text{fuz}}(\omega;\boldsymbol{\Lambda})\} \quad . \tag{115}$$

As in Eq. (70), we introduce the real part and the imaginary part of operator $\underline{\mathbf{Z}}_{\text{fuz}}(\omega;\boldsymbol{\Lambda})$ such that

$$i\omega\,\underline{\mathbf{Z}}_{\text{fuz}}(\omega;\boldsymbol{\Lambda}) = -\omega^2\,\underline{\mathbf{R}}_{\text{fuz}}(\omega;\boldsymbol{\Lambda}) + i\omega\,\underline{\mathbf{I}}_{\text{fuz}}(\omega;\boldsymbol{\Lambda}) \quad , \tag{116}$$

in which, due to Eq. (105), $\underline{\mathbf{I}}_{\text{fuz}}(\omega;\boldsymbol{\Lambda})$ is a symmetric positive-semidefinite operator on $\mathcal{C}^c$,

$$<\underline{\mathbf{I}}_{\text{fuz}}(\omega;\boldsymbol{\Lambda})\,\mathbf{u}\,,\mathbf{u}> \,\geq\, 0 \quad , \quad \forall\mathbf{u}\in\mathcal{C}^c \quad . \tag{117}$$

In addition, from Eqs. (12), (22), (24) and (116), we deduce that

$$\underline{\mathbf{R}}_{\text{fuz}}(-\omega;\boldsymbol{\Lambda}) = \underline{\mathbf{R}}_{\text{fuz}}(\omega;\boldsymbol{\Lambda}) \quad , \quad \underline{\mathbf{I}}_{\text{fuz}}(-\omega;\boldsymbol{\Lambda}) = \underline{\mathbf{I}}_{\text{fuz}}(\omega;\boldsymbol{\Lambda}) \quad . \tag{118}$$

Finally, from Eqs. (106) and (107), we deduce that, for $i = 1,\ldots,M$,

$$\mathbf{Z}_i(\omega;\boldsymbol{\Lambda}) \to \mathbf{0} \quad , \quad \text{as} \quad \boldsymbol{\Lambda}\to\mathbf{0} \quad , \tag{119}$$

and consequently, from Eq. (113),

$$\mathbf{Z}_{\text{rand}}(\omega;\boldsymbol{\Lambda}) \to \mathbf{0} \quad \text{as} \quad \boldsymbol{\Lambda}\to\mathbf{0} \quad . \tag{120}$$

7.2. Existence and uniqueness of the random equation

The random equation for the master structure coupled with the L fuzzy substructures is written as

$$i\omega\,(\mathbf{Z}(\omega) + \mathbf{Z}_{\text{fuz}}(\omega;\boldsymbol{\Lambda}))\,\mathbf{U}(\omega;\boldsymbol{\Lambda}) = \mathbf{f}(\omega) \quad , \tag{121}$$

in which $\mathbf{Z}_{\text{fuz}}(\omega;\boldsymbol{\Lambda})$ is given by Eqs. (112) and (113). Solution $\mathbf{U}(\omega;\boldsymbol{\Lambda})$ is a random variable with values in $\mathcal{C}^c$ depending on deviation vector $\boldsymbol{\Lambda}$ (it is recalled that $\boldsymbol{\Lambda}$ depends on ω). Substituting Eq. (112) in Eq. (121) yields

$$i\omega\,(\mathbf{Z}_{\text{det}}(\omega;\boldsymbol{\Lambda}) + \mathbf{Z}_{\text{rand}}(\omega;\boldsymbol{\Lambda}))\,\mathbf{U}(\omega;\boldsymbol{\Lambda}) = \mathbf{f}(\omega) \quad , \tag{122}$$

in which $\mathbf{Z}_{\text{det}}(\omega;\boldsymbol{\Lambda})$ is the deterministic operator which is written as

$$\mathbf{Z}_{\text{det}}(\omega;\boldsymbol{\Lambda}) = \mathbf{Z}(\omega) + \underline{\mathbf{Z}}_{\text{fuz}}(\omega;\boldsymbol{\Lambda}) \quad . \tag{123}$$

Invertibility of operator $i\omega\,\mathbf{Z}_{\rm det}(\omega;\mathbf{\Lambda})$ for $\omega \in \mathbb{B}_{\rm MF}$. Substituting Eqs. (11) and (116) in Eq. (123) yields

$$i\omega\mathbf{Z}_{\rm det}(\omega;\mathbf{\Lambda}) = -\omega^2\big(\mathbf{M}+\underline{\mathbf{R}}_{\rm fuz}(\omega;\mathbf{\Lambda})\big)+i\omega\big(\mathbf{D}(\omega)+\underline{\mathbf{I}}_{\rm fuz}(\omega;\mathbf{\Lambda})\big)+\mathbf{K}(\omega). \quad (124)$$

For all $\omega \in \mathbb{B}_{\rm MF}$, master structure damping operator $\mathbf{D}(\omega)$ and master structure stiffness operator $\mathbf{K}(\omega)$ are symmetric positive semidefinite on $\mathcal{C}^c$ (see Eqs. (V.36) and (V.30)) and $\underline{\mathbf{I}}_{\rm fuz}(\omega;\mathbf{\Lambda})$ is a symmetric positive-semidefinite operator on $\mathcal{C}^c$ (see Eq.(117)). We then conclude that, for all $\omega \in \mathbb{B}_{\rm MF}$,

$$<\big(\mathbf{D}(\omega)+\underline{\mathbf{I}}_{\rm fuz}(\omega;\mathbf{\Lambda})\big)\,\mathbf{u}\,,\mathbf{u}> \;\geq\; 0 \quad , \quad \forall \mathbf{u}\in\mathcal{C}^c \quad . \quad (125)$$

In addition, master structure mass operator $\mathbf{M}$ is symmetric positive definite on $\mathcal{C}^c$ (see Section V.4.1). It can then be proved [9] that, for all ω in $\mathbb{B}_{\rm MF}$, operator $i\omega\,\mathbf{Z}_{\rm det}(\omega;\mathbf{\Lambda})$ is invertible.

Definition of $\lambda_{\max}$ and introduction of random operator $\mathbf{T}(\omega;\mathbf{\Lambda})$. For all ω fixed in $\mathbb{B}_{\rm MF}$, deterministic operator $i\omega\,\mathbf{Z}_{\rm det}(\omega;\mathbf{\Lambda})$ is invertible. Let $\mathbf{T}(\omega;\mathbf{\Lambda})$ be the random operator such that

$$\mathbf{T}(\omega;\mathbf{\Lambda}) = -\mathbf{Z}_{\rm det}(\omega;\mathbf{\Lambda})^{-1}\,\mathbf{Z}_{\rm rand}(\omega;\mathbf{\Lambda}) \quad . \quad (126)$$

Substituting Eq. (113) in Eq. (126) yields

$$\mathbf{T}(\omega;\mathbf{\Lambda}) = -\sum_{i=1}^{M} \mathbb{X}_i\;\mathbf{Z}_{\rm det}(\omega;\mathbf{\Lambda})^{-1}\,\mathbf{Z}_i(\omega;\mathbf{\Lambda}) \quad , \quad (127)$$

in which, for each $i=1,\ldots,M$, operator $\mathbf{Z}_{\rm det}(\omega;\mathbf{\Lambda})^{-1}\,\mathbf{Z}_i(\omega;\mathbf{\Lambda})$ is a deterministic operator from $\mathcal{C}^c$ into $\mathcal{C}^c$. Since operator $\mathbf{Z}_{\rm det}(\omega;\mathbf{0})$ is invertible and due to Eq. (120), we deduce that

$$\mathbf{T}(\omega;\mathbf{\Lambda}) \rightarrow \mathbf{0} \quad \text{as} \quad \mathbf{\Lambda}\rightarrow\mathbf{0} \quad . \quad (128)$$

---

[9] Let us consider ω fixed in $\mathbb{B}_{\rm MF}\subset\mathbb{R}^{+*}$. Due to the construction of operator $-\omega^2\underline{\mathbf{R}}_{\rm fuz}(\omega;\mathbf{\Lambda})$, we can write $-\omega^2\underline{\mathbf{R}}_{\rm fuz}(\omega;\mathbf{\Lambda}) = -\omega^2\underline{\mathbf{M}}_{\rm fuz}(\omega;\mathbf{\Lambda})+\underline{\mathbf{K}}_{\rm fuz}(\omega;\mathbf{\Lambda})$ in which $\underline{\mathbf{M}}_{\rm fuz}(\omega;\mathbf{\Lambda})$ and $\underline{\mathbf{K}}_{\rm fuz}(\omega;\mathbf{\Lambda})$ are positive-semidefinite operators. Consequently, Eq. (124) can be rewritten as $i\omega\,\mathbf{Z}_{\rm det}(\omega;\mathbf{\Lambda}) = -\omega^2(\mathbf{M}+\underline{\mathbf{M}}_{\rm fuz}(\omega;\mathbf{\Lambda}))+i\omega(\mathbf{D}(\omega)+\underline{\mathbf{I}}_{\rm fuz}(\omega;\mathbf{\Lambda}))+\mathbf{K}(\omega)+\underline{\mathbf{K}}_{\rm fuz}(\omega;\mathbf{\Lambda})$. We recall that $\mathcal{C}^c = \mathcal{C}_{\rm rig}\oplus\mathcal{C}^c_{\rm elas}$ (see Eq. (V.52)). Operator $i\omega\,\mathbf{Z}_{\rm det}(\omega;\mathbf{\Lambda})$ is invertible on $\mathcal{C}^c_{\rm elas}$ because $\mathbf{D}(\omega)+\underline{\mathbf{I}}_{\rm fuz}(\omega;\mathbf{\Lambda})$ is positive definite. Operator $i\omega\,\mathbf{Z}_{\rm det}(\omega;\mathbf{\Lambda})$ is invertible on $\mathcal{C}_{\rm rig}$ because $\mathbf{M}+\underline{\mathbf{M}}_{\rm fuz}(\omega;\mathbf{\Lambda})$ is positive definite. We then deduce that operator $i\omega\,\mathbf{Z}_{\rm det}(\omega;\mathbf{\Lambda})$ is invertible on $\mathcal{C}^c$.

Due to Eq. (128), there exists $\lambda_{\max} > 0$ independent of ω such that, for all ω in $\mathbb{B}_{\rm MF}$ and for all $\mathbf{\Lambda}(\omega)$ satisfying $\|\mathbf{\Lambda}(\omega)\| \leq \lambda_{\max}$, the norm of random operator $\mathbf{T}(\omega;\mathbf{\Lambda})$ is less than 1. Consequently, random operator $\mathbf{1}-\mathbf{T}(\omega;\mathbf{\Lambda})$ is invertible and we have the following Neumann series expansion

$$\big(\mathbf{1}-\mathbf{T}(\omega;\mathbf{\Lambda})\big)^{-1} = \mathbf{1} + \sum_{k=1}^{+\infty} \mathbf{T}(\omega;\mathbf{\Lambda})^k \quad , \tag{129}$$

in which the series on the right-hand side of Eq. (129) is convergent.

Invertibility of operator $i\omega\,(\mathbf{Z}_{\rm det}(\omega;\mathbf{\Lambda}) + \mathbf{Z}_{\rm rand}(\omega;\mathbf{\Lambda}))$ for $\omega \in \mathbb{B}_{\rm MF}$. Let ω be fixed in $\mathbb{B}_{\rm MF}$ and let $\lambda_{\max}$ be defined as above. Let $\mathbf{\Lambda}(\omega)$ be such that $\|\mathbf{\Lambda}(\omega)\| \leq \lambda_{\max}$. Therefore, random operator $i\omega\,\mathbf{Z}_{\rm det}(\omega;\mathbf{\Lambda})\,(\mathbf{1}-\mathbf{T}(\omega;\mathbf{\Lambda}))$ is invertible as the product of two invertible operators. We then conclude that random operator

$$i\omega\,(\mathbf{Z}_{\rm det}(\omega;\mathbf{\Lambda}) + \mathbf{Z}_{\rm rand}(\omega;\mathbf{\Lambda})) = i\omega\,\mathbf{Z}_{\rm det}(\omega;\mathbf{\Lambda})\,(\mathbf{1}-\mathbf{T}(\omega;\mathbf{\Lambda})) \tag{130}$$

is invertible and its inverse is a random operator such that, for all $\delta\mathbf{f}$ [10],

$$E\{|<\delta\mathbf{f}\,,\{i\omega\,(\mathbf{Z}_{\rm det}(\omega;\mathbf{\Lambda}) + \mathbf{Z}_{\rm rand}(\omega;\mathbf{\Lambda}))\}^{-1}\,\delta\mathbf{f}>|^2\} < +\infty \quad . \tag{131}$$

Existence and uniqueness of the random equation. For all $\omega \in \mathbb{B}_{\rm MF}$ and for all $\mathbf{\Lambda}(\omega)$ such that $\|\mathbf{\Lambda}(\omega)\| \leq \lambda_{\max}$, Eq. (122) (or equivalently Eq. (121)) has a unique random solution $\mathbf{U}(\omega;\mathbf{\Lambda})$,

$$\mathbf{U}(\omega;\mathbf{\Lambda}) = \{i\omega\,(\mathbf{Z}_{\rm det}(\omega;\mathbf{\Lambda}) + \mathbf{Z}_{\rm rand}(\omega;\mathbf{\Lambda}))\}^{-1}\,\mathbf{f}(\omega) \quad , \tag{132}$$

which is a second-order random variable with values in $\mathcal{C}^c$,

$$E\left\{\int_\Omega \|\mathbf{U}(\mathbf{x},\omega;\mathbf{\Lambda})\|^2\,d\mathbf{x}\right\} < +\infty \quad . \tag{133}$$

7.3. Recursive method for the construction of the random solution

In this section, we assume that ω is fixed in $\mathbb{B}_{\rm MF}$ and that $\|\mathbf{\Lambda}(\omega)\| \leq \lambda_{\max}$ (see Section 7.2). Using Eq. (130), Eq. (122) can be rewritten as

$$i\omega\,\mathbf{Z}_{\rm det}(\omega;\mathbf{\Lambda})\,\big(\mathbf{1}-\mathbf{T}(\omega;\mathbf{\Lambda})\big)\,\mathbf{U}(\omega;\mathbf{\Lambda}) = \mathbf{f}(\omega) \quad , \tag{134}$$

---

[10] Element $\delta\mathbf{f}$ belongs to the antidual space $\mathcal{C}^{c\prime}$ of $\mathcal{C}^c$. In Eq. (131), the brackets denote the antiduality product between $\mathcal{C}^{c\prime}$ and $\mathcal{C}^c$.

in which random operator $\mathbf{T}$ is defined by Eq. (126). Since deterministic operator $i\omega\,\mathbf{Z}_{\text{det}}(\omega;\boldsymbol{\Lambda})$ is invertible and since random operator $\mathbf{1}-\mathbf{T}(\omega;\boldsymbol{\Lambda})$ is invertible (see Section 7.2), we deduce that

$$\mathbf{U}(\omega;\boldsymbol{\Lambda}) = \big(\mathbf{1} - \mathbf{T}(\omega;\boldsymbol{\Lambda})\big)^{-1}\big(i\omega\,\mathbf{Z}_{\text{det}}(\omega;\boldsymbol{\Lambda})\big)^{-1}\mathbf{f}(\omega) \quad . \tag{135}$$

Substituting Eq. (129) in Eq. (135) and using Eq. (127) yields

$$\begin{aligned}\mathbf{U}(\omega;\boldsymbol{\Lambda}) =& \mathbf{u}^{(0)}(\omega;\boldsymbol{\Lambda}) + \sum_{i_1=1}^{M} \mathbb{X}_{i_1}\mathbf{u}^{(1)}_{i_1}(\omega;\boldsymbol{\Lambda}) + \sum_{i_1=1}^{M}\sum_{i_2=1}^{M} \mathbb{X}_{i_1}\mathbb{X}_{i_2}\mathbf{u}^{(2)}_{i_1 i_2}(\omega;\boldsymbol{\Lambda}) \\ &+ \ldots + \sum_{i_1=1}^{M}\cdots\sum_{i_k=1}^{M} \mathbb{X}_{i_1}\ldots\mathbb{X}_{i_k}\mathbf{u}^{(k)}_{i_1,\ldots,i_k}(\omega;\boldsymbol{\Lambda}) + \ldots \quad ,\end{aligned} \tag{136}$$

in which deterministic displacement fields $\mathbf{u}^{(0)}(\omega;\boldsymbol{\Lambda})$, $\mathbf{u}^{(1)}_{i_1}(\omega;\boldsymbol{\Lambda}),\ldots$ belong to $\mathcal{C}^c$ and are defined by the following recursive method which involves the same deterministic operator $i\omega\,\mathbf{Z}_{\text{det}}(\omega;\boldsymbol{\Lambda})$.

Order 0.

$$i\omega\,\mathbf{Z}_{\text{det}}(\omega;\boldsymbol{\Lambda})\,\mathbf{u}^{(0)}(\omega;\boldsymbol{\Lambda}) = \mathbf{f}(\omega) \quad . \tag{137}$$

Order 1.

$$i\omega\,\mathbf{Z}_{\text{det}}(\omega;\boldsymbol{\Lambda})\,\mathbf{u}^{(1)}_{i_1}(\omega;\boldsymbol{\Lambda}) = \mathbf{f}^{(1)}_{i_1}(\omega;\boldsymbol{\Lambda}) \quad , \tag{138}$$

$$\mathbf{f}^{(1)}_{i_1}(\omega;\boldsymbol{\Lambda}) = -i\omega\,\mathbf{Z}_{i_1}(\omega;\boldsymbol{\Lambda})\,\mathbf{u}^{(0)}(\omega;\boldsymbol{\Lambda}) \quad . \tag{139}$$

Order 2.

$$i\omega\,\mathbf{Z}_{\text{det}}(\omega;\boldsymbol{\Lambda})\,\mathbf{u}^{(2)}_{i_1 i_2}(\omega;\boldsymbol{\Lambda}) = \mathbf{f}^{(2)}_{i_1 i_2}(\omega;\boldsymbol{\Lambda}) \quad , \tag{140}$$

$$\mathbf{f}^{(2)}_{i_1 i_2}(\omega;\boldsymbol{\Lambda}) = -i\omega\,\mathbf{Z}_{i_2}(\omega;\boldsymbol{\Lambda})\,\mathbf{u}^{(1)}_{i_1}(\omega;\boldsymbol{\Lambda}) \quad . \tag{141}$$

...

Order k.

$$i\omega\,\mathbf{Z}_{\text{det}}(\omega;\boldsymbol{\Lambda})\,\mathbf{u}^{(k)}_{i_1,\ldots,i_k}(\omega;\boldsymbol{\Lambda}) = \mathbf{f}^{(k)}_{i_1,\ldots,i_k}(\omega;\boldsymbol{\Lambda}) \quad , \tag{142}$$

$$\mathbf{f}^{(k)}_{i_1,\ldots,i_k}(\omega;\boldsymbol{\Lambda}) = -i\omega\,\mathbf{Z}_{i_k}(\omega;\boldsymbol{\Lambda})\,\mathbf{u}^{(k-1)}_{i_1,\ldots,i_{k-1}}(\omega;\boldsymbol{\Lambda}) \quad . \tag{143}$$

8. Ritz-Galerkin Approximation and Finite Element Discretization

We have to solve random Eq. (122) related to displacement field $\mathbf{U}(\omega; \mathbf{\Lambda})$ of the master structure coupled with the L fuzzy substructures,

$$i\omega\left(\mathbf{Z}_{\text{det}}(\omega; \mathbf{\Lambda}) + \mathbf{Z}_{\text{rand}}(\omega; \mathbf{\Lambda})\right)\mathbf{U}(\omega; \mathbf{\Lambda}) = \mathbf{f}(\omega) \quad , \tag{144}$$

in which $\mathbf{U}(\omega; \mathbf{\Lambda})$ is a random variable with values in $\mathcal{C}^c$. The construction of $\mathbf{U}(\omega; \mathbf{\Lambda})$ is carried out using the recursive method defined by Eqs. (136) to (143). Referring to Section 2.2, several cases must be distinguished.

If the LF modal response (see Fig. 3 or Fig. 4) of the master structure uncoupled with the fuzzy substructures is similar to the dynamical response of the master structure coupled with the fuzzy substructures (added-mass effects of the fuzzy substructures on the master structure), then the fuzzy structure theory is unnecessary and the LF method presented in Chapter VI must be used.

If the MF dynamical response (see Fig. 4) of the master structure uncoupled with the fuzzy substructures is similar to the dynamical response of the master structure coupled with the fuzzy substructures, then the fuzzy structure theory is unnecessary and the MF method presented in Chapter VII must be used.

If the LF dynamical response (see Fig. 3) or the MF dynamical response (see Fig. 3 or Fig. 4) of the master structure uncoupled with the fuzzy substructures is not similar to the dynamical response of the master structure coupled with the fuzzy substructures, then the fuzzy structure theory must be used. In this case, in the context of discretization of the master structure coupled with the fuzzy substructures, we have to consider the two cases according as the master structure uncoupled with the fuzzy substructures has a resonant response (LF case) or a nonresonant response (MF case). Below, we present the Ritz-Galerkin method for the master structure coupled with the fuzzy substructures using the structural modes of the master structure uncoupled with the fuzzy substructures for the LF case (see Chapter VI), and the eigenfunctions of the energy operator of the master structure uncoupled with the fuzzy substructures for the MF case (see Chapter VIII). Alternately, for the MF range, we present the finite element discretization of the master structure coupled with the fuzzy substructures.

8.1. Ritz-Galerkin approximation

Let $\mathcal{C}_N^c$ be the subspace of $\mathcal{C}^c$ of dimension $N \geq 1$, spanned by a complete countable family $\{\mathbf{b}_\alpha\}_{\alpha=1,\ldots,N}$ of functions in $\mathcal{C}^c$, where each $\mathbf{b}_\alpha(\mathbf{x})$ is a function from Ω into $\mathbb{R}^3$. The projection of second-order random variable $\mathbb{U}(\omega;\mathbf{\Lambda})$ with values in $\mathcal{C}^c$ on the space of the second-order random variable with values in $\mathcal{C}_N^c$ is denoted as $\mathbb{U}^N(\omega;\mathbf{\Lambda})$, and can be written at any point $\mathbf{x}$ of domain Ω occupied by the master structure as

$$\mathbb{U}^N(\mathbf{x},\omega;\mathbf{\Lambda}) = \sum_{\alpha=1}^{N} \mathbb{Q}_\alpha(\omega;\mathbf{\Lambda})\,\mathbf{b}_\alpha(\mathbf{x}) \quad , \tag{145}$$

in which

$$\mathbb{Q}(\omega;\mathbf{\Lambda}) = (\mathbb{Q}_1(\omega;\mathbf{\Lambda}),\ldots,\mathbb{Q}_N(\omega;\mathbf{\Lambda})) \quad , \tag{146}$$

is a second-order random variable with values in $\mathbb{C}^N$. The projection of Eq. (122) yields the random matrix equation of dimension N,

$$i\omega\left([\mathcal{Z}_{\text{det}}(\omega;\mathbf{\Lambda})] + [\mathcal{Z}_{\text{rand}}(\omega;\mathbf{\Lambda})]\right)\mathbb{Q}(\omega;\mathbf{\Lambda}) = \boldsymbol{\mathcal{F}}(\omega) \quad . \tag{147}$$

The recursive method for solving random matrix Eq. (147) is obtained by the projection of Eqs. (136) to (143),

$$\begin{aligned}\mathbb{Q}(\omega;\mathbf{\Lambda}) =& \mathbf{q}^{(0)}(\omega;\mathbf{\Lambda}) + \sum_{i_1=1}^{M} \mathbb{X}_{i_1}\mathbf{q}_{i_1}^{(1)}(\omega;\mathbf{\Lambda}) + \sum_{i_1=1}^{M}\sum_{i_2=1}^{M} \mathbb{X}_{i_1}\mathbb{X}_{i_2}\mathbf{q}_{i_1 i_2}^{(2)}(\omega;\mathbf{\Lambda}) \\ &+ \ldots + \sum_{i_1=1}^{M}\cdots\sum_{i_k=1}^{M} \mathbb{X}_{i_1}\ldots\mathbb{X}_{i_k}\mathbf{q}_{i_1,\ldots,i_k}^{(k)}(\omega;\mathbf{\Lambda}) + \ldots \quad ,\end{aligned} \tag{148}$$

in which deterministic vectors $\mathbf{q}^{(0)}(\omega;\mathbf{\Lambda})$, $\mathbf{q}_{i_1}^{(1)}(\omega;\mathbf{\Lambda}),\ldots$ belong to $\mathbb{C}^N$ and are defined by the following recursive method which involves the same deterministic invertible symmetric $(N\times N)$ complex matrix $i\omega\,[\mathcal{Z}_{\text{det}}(\omega;\mathbf{\Lambda})]$.

Order 0.

$$i\omega\,[\mathcal{Z}_{\text{det}}(\omega;\mathbf{\Lambda})]\;\mathbf{q}^{(0)}(\omega;\mathbf{\Lambda}) = \boldsymbol{\mathcal{F}}(\omega) \quad . \tag{149}$$

Order 1.

$$i\omega\,[\mathcal{Z}_{\text{det}}(\omega;\mathbf{\Lambda})]\;\mathbf{q}_{i_1}^{(1)}(\omega;\mathbf{\Lambda}) = \boldsymbol{\mathcal{F}}_{i_1}^{(1)}(\omega;\mathbf{\Lambda}) \quad , \tag{150}$$

$$\boldsymbol{\mathcal{F}}_{i_1}^{(1)}(\omega;\mathbf{\Lambda}) = -i\omega\,[\mathcal{Z}_{i_1}(\omega;\mathbf{\Lambda})]\;\mathbf{q}^{(0)}(\omega;\mathbf{\Lambda}) \quad . \tag{151}$$

Order 2.

$$i\omega\,[\mathcal{Z}_{\text{det}}(\omega;\mathbf{\Lambda})]\;\mathbf{q}_{i_1 i_2}^{(2)}(\omega;\mathbf{\Lambda}) = \boldsymbol{\mathcal{F}}_{i_1 i_2}^{(2)}(\omega;\mathbf{\Lambda}) \quad , \tag{152}$$

$$\boldsymbol{\mathcal{F}}_{i_1 i_2}^{(2)}(\omega;\mathbf{\Lambda}) = -i\omega\,[\mathcal{Z}_{i_2}(\omega;\mathbf{\Lambda})]\;\mathbf{q}_{i_1}^{(1)}(\omega;\mathbf{\Lambda}) \quad . \tag{153}$$

...

Order k.

$$i\omega\,[\mathcal{Z}_{\text{det}}(\omega;\boldsymbol{\Lambda})]\;\mathbf{q}^{(k)}_{i_1,\ldots,i_k}(\omega;\boldsymbol{\Lambda}) = \boldsymbol{\mathcal{F}}^{(k)}_{i_1,\ldots,i_k}(\omega;\boldsymbol{\Lambda}) \quad , \tag{154}$$

$$\boldsymbol{\mathcal{F}}^{(k)}_{i_1,\ldots,i_k}(\omega;\boldsymbol{\Lambda}) = -i\omega\,[\mathcal{Z}_{i_k}(\omega;\boldsymbol{\Lambda})]\;\mathbf{q}^{(k-1)}_{i_1,\ldots,i_{k-1}}(\omega;\boldsymbol{\Lambda}) \quad , \tag{155}$$

in which $[\mathcal{Z}_{\text{det}}(\omega;\boldsymbol{\Lambda})]$ and $[\mathcal{Z}_1(\omega;\boldsymbol{\Lambda})],\ldots,[\mathcal{Z}_M(\omega;\boldsymbol{\Lambda})]$ are deterministic symmetric $(N\times N)$ complex matrices corresponding to the projection of deterministic operators $\mathbf{Z}_{\text{det}}(\omega;\boldsymbol{\Lambda})$ and $\mathbf{Z}_1(\omega;\boldsymbol{\Lambda}),\ldots,\mathbf{Z}_M(\omega;\boldsymbol{\Lambda})$ respectively. Deterministic vector $\boldsymbol{\mathcal{F}}(\omega)$ in $\mathbb{C}^N$ corresponds to the projection of $\mathbf{f}(\omega)$.

Utilization of the structural modes. This case corresponds to LF modeling of the master structure. Stiffness operator $\mathbf{K}$ is then independent of ω. Using the method presented in Chapter VI, we choose as basis $\{\mathbf{b}_\alpha\}_{\alpha=1,\ldots,N}$ the elastic structural modes $\{\mathbf{u}_\alpha\}_{\alpha=1,\ldots,N}$ of the master structure uncoupled with the fuzzy substructures, as defined in Sections 6 and 7 of Chapter III. For further details concerning this case of fuzzy structure, we refer the reader to Soize, 1995b.

Utilization of the eigenfunctions of the energy operator. This case corresponds to MF modeling of the master structure. Stiffness operator $\mathbf{K}(\omega)$ then depends on ω. Using the method presented in Chapter VIII, we choose as basis $\{\mathbf{b}_\alpha\}_{\alpha=1,\ldots,N}$, the eigenfunctions $\{\mathbf{e}_\alpha\}_{\alpha=1,\ldots,N}$ of the energy operator of the master structure uncoupled with the fuzzy substructures, as defined in Sections 5.4 and 7.2 of Chapter VIII.

8.2. Finite element discretization

We consider a finite element mesh of master structure Ω. The finite element mesh of surface Σ is the trace of the finite element mesh of Ω (see Fig. 15).

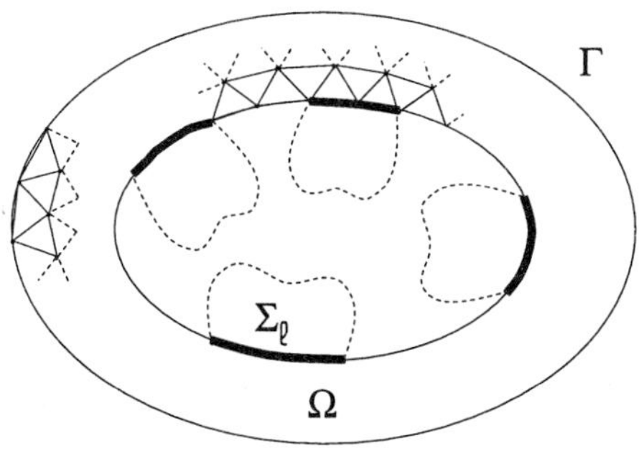

Fig. 15. Finite element mesh of the fuzzy structure

We introduce subspace $\mathcal{C}_n^c \subset \mathcal{C}^c$ of finite dimension n. Let $\mathbb{U} = (\mathbb{U}_1, \ldots, \mathbb{U}_n)$ be the complex vector of the DOFs which are the values of $\mathbb{U}$ at the nodes of the finite element mesh of domain Ω. The finite element discretization of Eq. (122) yields the random matrix equation of dimension n,

$$i\omega\left([Z_{\text{det}}(\omega;\mathbf{\Lambda})] + [Z_{\text{rand}}(\omega;\mathbf{\Lambda})]\right)\mathbb{U}(\omega;\mathbf{\Lambda}) = \mathbf{F}(\omega) \quad . \tag{156}$$

The recursive method for solving random matrix Eq. (156) is obtained by finite element discretization of Eqs. (136) to (143),

$$\begin{aligned}\mathbb{U}(\omega;\mathbf{\Lambda}) =& \mathbf{U}^{(0)}(\omega;\mathbf{\Lambda}) + \sum_{i_1=1}^{M} \mathbb{X}_{i_1}\mathbf{U}_{i_1}^{(1)}(\omega;\mathbf{\Lambda}) + \sum_{i_1=1}^{M}\sum_{i_2=1}^{M} \mathbb{X}_{i_1}\mathbb{X}_{i_2}\mathbf{U}_{i_1 i_2}^{(2)}(\omega;\mathbf{\Lambda}) \\ &+ \ldots + \sum_{i_1=1}^{M}\cdots\sum_{i_k=1}^{M} \mathbb{X}_{i_1}\ldots\mathbb{X}_{i_k}\mathbf{U}_{i_1,\ldots,i_k}^{(k)}(\omega;\mathbf{\Lambda}) + \ldots \quad ,\end{aligned} \tag{157}$$

in which deterministic vectors $\mathbf{U}^{(0)}(\omega;\mathbf{\Lambda})$, $\mathbf{U}_{i_1}^{(1)}(\omega;\mathbf{\Lambda}), \ldots$ belong to $\mathbb{C}^n$ and are defined by the following recursive method which involves the same deterministic invertible symmetric $(n \times n)$ complex matrix $i\omega\,[Z_{\text{det}}(\omega;\mathbf{\Lambda})]$.

Order 0.

$$i\omega\,[Z_{\text{det}}(\omega;\mathbf{\Lambda})]\;\mathbf{U}^{(0)}(\omega;\mathbf{\Lambda}) = \mathbf{F}(\omega) \quad . \tag{158}$$

Order 1.

$$i\omega\,[Z_{\text{det}}(\omega;\mathbf{\Lambda})]\;\mathbf{U}_{i_1}^{(1)}(\omega;\mathbf{\Lambda}) = \mathbf{F}_{i_1}^{(1)}(\omega;\mathbf{\Lambda}) \quad , \tag{159}$$

$$\mathbf{F}_{i_1}^{(1)}(\omega;\mathbf{\Lambda}) = -i\omega\,[Z_{i_1}(\omega;\mathbf{\Lambda})]\;\mathbf{U}^{(0)}(\omega;\mathbf{\Lambda}) \quad . \tag{160}$$

Order 2.

$$i\omega\,[Z_{\text{det}}(\omega;\mathbf{\Lambda})]\;\mathbf{U}_{i_1 i_2}^{(2)}(\omega;\mathbf{\Lambda}) = \mathbf{F}_{i_1 i_2}^{(2)}(\omega;\mathbf{\Lambda}) \quad , \tag{161}$$

$$\mathbf{F}_{i_1 i_2}^{(2)}(\omega;\mathbf{\Lambda}) = -i\omega\,[Z_{i_2}(\omega;\mathbf{\Lambda})]\;\mathbf{U}_{i_1}^{(1)}(\omega;\mathbf{\Lambda}) \quad . \tag{162}$$

...

Order k.

$$i\omega\,[Z_{\text{det}}(\omega;\mathbf{\Lambda})]\;\mathbf{U}_{i_1,\ldots,i_k}^{(k)}(\omega;\mathbf{\Lambda}) = \mathbf{F}_{i_1,\ldots,i_k}^{(k)}(\omega;\mathbf{\Lambda}) \quad , \tag{163}$$

$$\mathbf{F}_{i_1,\ldots,i_k}^{(k)}(\omega;\mathbf{\Lambda}) = -i\omega\,[Z_{i_k}(\omega;\mathbf{\Lambda})]\;\mathbf{U}_{i_1,\ldots,i_{k-1}}^{(k-1)}(\omega;\mathbf{\Lambda}) \quad , \tag{164}$$

in which $[Z_{\text{det}}(\omega;\mathbf{\Lambda})]$ and $[Z_1(\omega;\mathbf{\Lambda})], \ldots, [Z_M(\omega;\mathbf{\Lambda})]$ are deterministic symmetric $(n \times n)$ complex matrices corresponding to the finite element discretization of deterministic operator $\mathbf{Z}_{\text{det}}(\omega;\mathbf{\Lambda})$ and deterministic operators $\mathbf{Z}_1(\omega;\mathbf{\Lambda}), \ldots, \mathbf{Z}_M(\omega;\mathbf{\Lambda})$ respectively. Deterministic vector $\mathbf{F}(\omega)$ in $\mathbb{C}^n$

corresponds to the finite element discretization of $\mathbf{f}(\omega)$. Eqs. (158), (159), (161) and (163) are solved using the MF method presented in Chapter VII. For further details concerning the implementation of this procedure, we refer the reader to Soize, 1986; Chabas, Desanti and Soize, 1986; Soize,1993b.

9. Identification Method for the Parameters of Type I and Type II Fuzzy Impedance Laws

9.1. Statement of the problem

We consider the master structure coupled with the L fuzzy substructures described in Section 4. In Sections 5 and 6, we presented the fuzzy structure theory which allows us to construct the model of the master structure coupled with the L fuzzy substructures using a homogeneous fuzzy impedance law (type I and type II laws). Finally, in Section 7, we presented the solving method which allows us to calculate the local dynamical response of the master structure coupled with its fuzzy substructures. The fuzzy impedance law coefficients must of course be known and consequently identification methods must be developed to identify these coefficients. Among those coefficients, one, namely, the mean coefficient $\underline{\mu}_j^\ell(\omega)$ for fuzzy substructure ℓ and for local direction j, or the corresponding dimensionless mean coefficient $\underline{\nu}_j^\ell(\omega)$ of the participating mass of fuzzy substructure ℓ for local direction j, defined by Eq. (39), is of prime importance. In Section 9, we present an identification method for dimensionless mean coefficient $\underline{\boldsymbol{\nu}}(\omega) = \{\underline{\nu}_j^\ell(\omega)\}_{\ell,j}$ of dimension M in frequency band $\mathbb{B}_{\text{MF}}$ (see Soize, 1996a) .

9.2. Strategy of the identification method

Since the objective is to estimate the dimensionless mean coefficient $\underline{\boldsymbol{\nu}}(\omega) \in \mathbb{R}^M$ of the participating mass of the fuzzy substructures, it is logical to take $\boldsymbol{\Lambda} = \mathbf{0}$ as a first approximation. It should be noted that $\boldsymbol{\Lambda} = \mathbf{0}$ is used only to determine $\underline{\boldsymbol{\nu}}(\omega)$ in the context of the identification method. Once $\underline{\boldsymbol{\nu}}(\omega)$ is known, $\boldsymbol{\Lambda}$ is generally not taken equal to $\mathbf{0}$ for calculating the dynamical response of the master structure coupled with its fuzzy substructures. The strategy of the identification method is as follows. Frequency band $\mathbb{B}_{\text{MF}}$ is written as

$$\mathbb{B}_{\text{MF}} = \cup B_m \quad , \quad \cap B_m = \emptyset \quad . \tag{165}$$

For each fixed frequency band B_m, the real structure is then submitted to external given time-stationary stochastic forces. As explained above, the fuzzy structure theory allows us to calculate the local dynamical response of the master structure coupled with its fuzzy substructures. For $\mathbf{\Lambda} = \mathbf{0}$, the calculation of the local dynamical response is given by Eq. (137) which involves deterministic operator $i\omega\, \mathbf{Z}_{\text{det}}(\omega; \mathbf{0})$. In the identification method, it is assumed that $\underline{\boldsymbol{\nu}}$ is constant over frequency band B_m. Using the time-stationary solution of this problem, we deduce the *mean power flow equation* associated with Eq. (122), for the master structure coupled with the fuzzy substructures. The mean power quantities which appear in this mean power flow equation depend on $\underline{\boldsymbol{\nu}}$ and are the mean power dissipated in the master structure, the mean value of the net transmitted power flowing from the master structure to fuzzy substructures and the

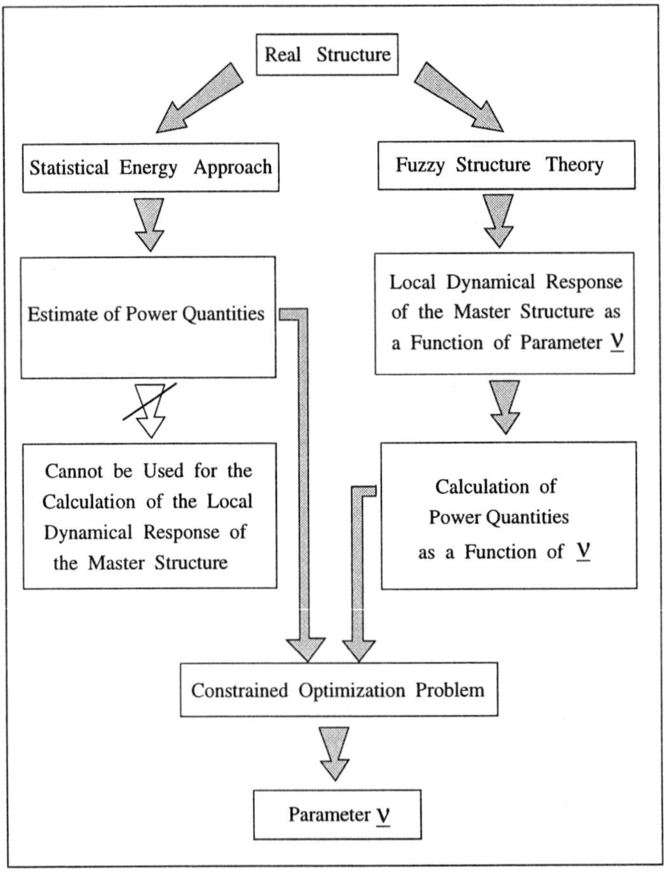

Fig. 16. Identification method of the fuzzy impedance law coefficient $\underline{\boldsymbol{\nu}}$ - Flow chart

mean input power from the external excitation forces applied to the master structure. In the context of the identification procedure, an estimate of these mean power quantities related to the real structure is obtained using a statistical energy approach based on the mean power balance equations. To do so, the bandwidth of B_m is chosen in such a way that statistical energy approaches can be used. It should be noted that these mean power quantities do not allow us to calculate the local dynamical response of the master structure. Finally, $\underline{\boldsymbol{\nu}}$ is obtained as the solution of a nonlinear constrained optimization problem. The strategy of this identification method is illustrated by the flow chart shown in Fig. 16 and is detailed below.

9.3. Equation of the master structure coupled with the fuzzy substructures for $\boldsymbol{\Lambda} = \mathbf{0}$

We consider frequency band B_m on which $\underline{\nu}_j^\ell$ is independent of ω. From Eq. (119), taking $\boldsymbol{\Lambda} = \mathbf{0}$ in Eq. (122) yields

$$i\omega\, \mathbf{Z}_{\text{det}}(\omega; \mathbf{0})\, \mathbf{U}(\omega; \mathbf{0}) = \mathbf{f}(\omega) \quad , \tag{166}$$

which coincides with Eq. (137) for $\boldsymbol{\Lambda} = \mathbf{0}$ (zero order equation of the recursive method presented in Section 7.3). Using Eq. (124), Eq. (166) can be rewritten as

$$\left(-\omega^2\left(\mathbf{M}+\underline{\mathbf{R}}_{\text{fuz}}(\omega;\mathbf{0})\right)+i\omega\left(\mathbf{D}(\omega)+\underline{\mathbf{I}}_{\text{fuz}}(\omega;\mathbf{0})\right)+\mathbf{K}(\omega)\right)\mathbf{U}(\omega;\mathbf{0}) = \mathbf{f}(\omega), \tag{167}$$

in which operators $\underline{\mathbf{R}}_{\text{fuz}}(\omega; \mathbf{0})$ and $\underline{\mathbf{I}}_{\text{fuz}}(\omega; \mathbf{0})$ are linear with respect to $\underline{\boldsymbol{\nu}}$ and can be written as

$$\underline{\mathbf{R}}_{\text{fuz}}(\omega; \mathbf{0}) = \sum_{\ell=1}^{L} \sum_{j\in\mathcal{J}_\ell} \underline{\nu}_j^\ell\, \underline{\mathcal{R}}_{\text{fuz},j}^\ell(\omega) \quad , \tag{168}$$

$$\underline{\mathbf{I}}_{\text{fuz}}(\omega; \mathbf{0}) = \sum_{\ell=1}^{L} \sum_{j\in\mathcal{J}_\ell} \underline{\nu}_j^\ell\, \underline{\mathcal{I}}_{\text{fuz},j}^\ell(\omega) \quad , \tag{169}$$

where $\mathcal{J}_\ell \subseteq \{1, 2, 3\}$ denotes the subset of the active local directions, i.e. the local directions on which the contribution is nonzero. In Eqs. (168) and (169), $\underline{\mathcal{R}}_{\text{fuz},j}^\ell(\omega)$ and $\underline{\mathcal{I}}_{\text{fuz},j}^\ell(\omega)$ are known deterministic operators independent of $\underline{\boldsymbol{\nu}}$. In effect, for fuzzy substructure ℓ and for local direction j, the mean value of the fuzzy impedance law is given by Eq. (70),

$$i\omega\, \underline{Z}_j^\ell(\omega; \mathbf{0}) = -\omega^2\, \underline{R}_j^\ell(\omega; \mathbf{0}) + i\omega\, \underline{I}_j^\ell(\omega; \mathbf{0}) \quad . \tag{170}$$

Dimensionless mean coefficient $\underline{\nu}_j^\ell$ of the participating mass is defined by Eq. (39),

$$\underline{\mu}_j^\ell = \underline{\nu}_j^\ell\, \frac{m_{\text{ref}}^\ell}{|\Sigma_\ell|} \quad . \tag{171}$$

Substituting Eq. (171) in Eqs. (72), (75), (77) and (82) yields, for $\mathbf{\Lambda} = \mathbf{0}$,

$$\underline{R}_j^\ell(\omega;\mathbf{0}) = \underline{\nu}_j^\ell\,\underline{\mathcal{R}}_j^\ell(\omega) \quad , \quad \underline{I}_j^\ell(\omega;\mathbf{0}) = \underline{\nu}_j^\ell\,\underline{\mathcal{I}}_j^\ell(\omega) \quad , \tag{172}$$

in which $\underline{\mathcal{R}}_j^\ell(\omega)$ and $\underline{\mathcal{I}}_j^\ell(\omega)$ are known real coefficients which are independent of $\underline{\nu}_j^\ell$. Using the construction introduced in Section 7.1 and from Eq. (171), we then deduce Eqs. (168) and (169). Since $\mathbf{U}(\omega;\mathbf{0})$ is a deterministic displacement field which depends on $\underline{\boldsymbol{\nu}}$, displacement field $\mathbf{U}(\omega;\mathbf{0})$ is rewritten as $\mathbf{u}(\omega;\boldsymbol{\nu})$. Using Eqs. (168) and (169), Eq. (167) can be rewritten as

$$\mathbf{A}_{\text{det}}(\omega;\underline{\boldsymbol{\nu}})\;\mathbf{u}(\omega;\underline{\boldsymbol{\nu}}) = \mathbf{f}(\omega) \quad , \tag{173}$$

in which the deterministic operator $\mathbf{A}_{\text{det}}(\omega;\boldsymbol{\nu})$ is defined by

$$\mathbf{A}_{\text{det}}(\omega;\underline{\boldsymbol{\nu}}) = -\,\omega^2\Big(\mathbf{M} + \sum_{\ell=1}^{L}\sum_{j\in\mathcal{J}_\ell}\underline{\boldsymbol{\nu}}_j^\ell\;\underline{\boldsymbol{\mathcal{R}}}_{\text{fuz},j}^\ell(\omega)\Big) + i\omega\Big(\mathbf{D}(\omega) + \sum_{\ell=1}^{L}\sum_{j\in\mathcal{J}_\ell}\underline{\boldsymbol{\nu}}_j^\ell\;\underline{\boldsymbol{\mathcal{I}}}_{\text{fuz},j}^\ell(\omega)\Big) + \mathbf{K}(\omega) \quad . \tag{174}$$

From Section 7.2, operator $i\omega\,\mathbf{Z}_{\text{det}}(\omega;\mathbf{0})$ is invertible and since $\mathbf{A}_{\text{det}}(\omega;\underline{\boldsymbol{\nu}}) = i\omega\,\mathbf{Z}_{\text{det}}(\omega;\mathbf{0})$, we deduce that for all $\omega \in B_m$, Eq. (174) has a unique solution $\mathbf{u}(\omega;\underline{\boldsymbol{\nu}})$ in $\mathcal{C}^c$ such that

$$\mathbf{u}(\omega;\underline{\boldsymbol{\nu}}) = \mathbf{T}_{\text{det}}(\omega;\underline{\boldsymbol{\nu}})\;\mathbf{f}(\omega) \quad , \tag{175}$$

in which

$$\mathbf{T}_{\text{det}}(\omega;\underline{\boldsymbol{\nu}}) = \mathbf{A}_{\text{det}}(\omega;\underline{\boldsymbol{\nu}})^{-1} \quad . \tag{176}$$

Eqs. (173) and (174) can be rewritten as

$$\mathbf{f}_{\text{cons}}(\omega;\underline{\boldsymbol{\nu}}) + \mathbf{f}_{\text{diss}}(\omega;\underline{\boldsymbol{\nu}}) + \sum_{\ell=1}^{L}\mathbf{f}_{\text{mast-fuz}}^\ell(\omega;\underline{\boldsymbol{\nu}}) = \mathbf{f}(\omega) \quad , \tag{177}$$

in which

$$\mathbf{f}_{\text{cons}}(\omega;\underline{\boldsymbol{\nu}}) = \Big\{-\omega^2\Big(\mathbf{M} + \sum_{\ell=1}^{L}\sum_{j\in\mathcal{J}_\ell}\underline{\boldsymbol{\nu}}_j^\ell\;\underline{\boldsymbol{\mathcal{R}}}_{\text{fuz},j}^\ell(\omega)\Big) + \mathbf{K}(\omega)\Big\}\,\mathbf{u}(\omega;\underline{\boldsymbol{\nu}}) \quad , \tag{178}$$

$$\mathbf{f}_{\text{diss}}(\omega;\underline{\boldsymbol{\nu}}) = \Big\{i\omega\,\mathbf{D}(\omega)\Big\}\,\mathbf{u}(\omega;\underline{\boldsymbol{\nu}}) \quad , \tag{179}$$

$$\mathbf{f}_{\text{mast-fuz}}^\ell(\omega;\underline{\boldsymbol{\nu}}) = \Big\{i\omega\sum_{j\in\mathcal{J}_\ell}\underline{\boldsymbol{\nu}}_j^\ell\;\underline{\boldsymbol{\mathcal{I}}}_{\text{fuz},j}^\ell(\omega)\Big\}\mathbf{u}(\omega;\underline{\boldsymbol{\nu}}) \quad . \tag{180}$$

The inverse Fourier transform of Eq. (177) yields the following time equation in the time domain

$$\mathbf{f}_{\text{cons}}(t;\underline{\boldsymbol{\nu}}) + \mathbf{f}_{\text{diss}}(t;\underline{\boldsymbol{\nu}}) + \sum_{\ell=1}^{L} \mathbf{f}^{\ell}_{\text{mast-fuz}}(t;\underline{\boldsymbol{\nu}}) = \mathbf{f}(t) \quad . \tag{181}$$

9.4. Mean power flow equation for the master structure coupled with the fuzzy substructures for $\boldsymbol{\Lambda} = \mathbf{0}$

In this section, we establish the mean power flow equation for the master structure coupled with the fuzzy substructures for $\boldsymbol{\Lambda} = \mathbf{0}$ and for a given frequency band B_m.

Description of the stochastic excitation force applied to the master structure. We assume that there is no surface force field applied to the master structure. The given force field $\mathbf{f}(t)$ is then due to the body force field applied to the master structure, and field $\mathbf{f}(t)$ is modeled by an $\mathbb{R}^3$-valued stochastic field denoted as $\mathbb{F}(\mathbf{x},t)$ such that, for any functions $\mathbf{a}(\mathbf{x})$ and $\mathbf{b}(\mathbf{x})$ belonging to $\mathcal{C}^c$,

$$F_{\mathbf{a}}(t) = <\mathbb{F}(t), \mathbf{a}> \quad , \quad F_{\mathbf{b}}(t) = <\mathbb{F}(t), \mathbf{b}> \quad , \tag{182}$$

are [11] limited band stationary second-order centered stochastic processes indexed by $\mathbb{R}$ with values in $\mathbb{C}$. In addition, we assume that the cross-correlation function defined by

$$R_{\mathbf{ab}}(\tau) = E\{F_{\mathbf{a}}(t+\tau)\overline{F_{\mathbf{b}}(t)}\} \quad , \tag{183}$$

is written as

$$R_{\mathbf{ab}}(\tau) = \int_{\mathbb{R}} e^{i\omega\tau} S_{\mathbf{ab}}(\omega)\, d\omega \quad , \tag{184}$$

in which $S_{\mathbf{ab}}(\omega)$ is the cross spectral density function defined by

$$S_{\mathbf{ab}}(\omega) = f_0^2 \; \mathbb{1}_{B_m \cup \widetilde{B}_m}(\omega) \int_{\Omega} \chi(\mathbf{x})\; \mathbf{b}(\mathbf{x})\cdot\overline{\mathbf{a}(\mathbf{x})}\; d\mathbf{x} \quad , \tag{185}$$

in which f_0 is a real number, $\mathbb{1}_{B_m \cup \widetilde{B}_m}(\omega)$ is the indicator function of limited band $B_m \cup \widetilde{B}_m$, where $\widetilde{B}_m$ denotes the negative frequency band which is symmetric to B_m with respect to the origin, and where $\chi(\mathbf{x})$ is a given bounded positive-valued function on Ω. This model means that stochastic

[11] The brackets denote the antiduality product between $\mathcal{C}^{c\,\prime}$ and $\mathcal{C}^c$.

field $\mathbb{F}(\mathbf{x},t)$ is stationary in time (with limited band $B_m \cup \widetilde{B}_m$), inhomogeneous and δ-correlated in space with intensity $\chi(\mathbf{x})$.

Time-stationary response of the master structure coupled with the fuzzy substructures. Let $\mathbb{U}(t;\underline{\boldsymbol{\nu}})$ be the time-stationary random field describing the time-stationary response of the master structure coupled with the fuzzy substructures whose deterministic operator-valued frequency response function is defined by Eqs. (175) and (176). The equality concerning the time-stationary stochastic fields corresponding to Eq. (181) is written as

$$\mathbb{F}_{\text{cons}}(t;\underline{\boldsymbol{\nu}}) + \mathbb{F}_{\text{diss}}(t;\underline{\boldsymbol{\nu}}) + \sum_{\ell=1}^{L} \mathbb{F}^{\ell}_{\text{mast-fuz}}(t;\underline{\boldsymbol{\nu}}) = \mathbb{F}(t) \quad . \tag{186}$$

Mean power flow equation for the master structure. For all fixed t, we introduce the real-valued random variables[12]

$$\Pi_{\text{cons}}(t;\underline{\boldsymbol{\nu}}) = <\mathbb{F}_{\text{cons}}(t;\underline{\boldsymbol{\nu}})\, , \, \partial_t \mathbb{U}(t;\underline{\boldsymbol{\nu}})> \quad , \tag{187}$$

$$\Pi_{\text{diss}}(t;\underline{\boldsymbol{\nu}}) = <\mathbb{F}_{\text{diss}}(t;\underline{\boldsymbol{\nu}})\, , \, \partial_t \mathbb{U}(t;\underline{\boldsymbol{\nu}})> \quad , \tag{188}$$

$$\Pi^{\ell}_{\text{mast-fuz}}(t;\underline{\boldsymbol{\nu}}) = <\mathbb{F}^{\ell}_{\text{mast-fuz}}(t;\underline{\boldsymbol{\nu}})\, , \, \partial_t \mathbb{U}(t;\underline{\boldsymbol{\nu}})> \quad , \tag{189}$$

$$\Pi_{\text{in}}(t;\underline{\boldsymbol{\nu}}) = <\mathbb{F}(t)\, , \, \partial_t \mathbb{U}(t;\underline{\boldsymbol{\nu}})> \quad , \tag{190}$$

which represent instantaneous random powers. For all fixed t in $\mathbb{R}$, we denote the mathematical expectation of random variable $\Pi(t;\underline{\boldsymbol{\nu}})$ as

$$<\Pi(\underline{\boldsymbol{\nu}})> = E\{\Pi(t;\underline{\boldsymbol{\nu}})\} \quad , \tag{191}$$

which is independent of time t due to the stationarity of stochastic process $\{\Pi(t;\underline{\boldsymbol{\nu}}), t \in \mathbb{R}\}$. From Eq. (186) to (191), we deduce that

$$<\Pi_{\text{cons}}(\underline{\boldsymbol{\nu}})> + <\Pi_{\text{diss}}(\underline{\boldsymbol{\nu}})> + \sum_{\ell=1}^{L} <\Pi^{\ell}_{\text{mast-fuz}}(\underline{\boldsymbol{\nu}})> = <\Pi_{\text{in}}(\underline{\boldsymbol{\nu}})> \quad . \tag{192}$$

From Eqs. (118) and (168), and since $\mathbf{K}(-\omega) = \mathbf{K}(\omega)$, we deduce that operator $\mathbf{A}_{\text{cons}}(\omega;\underline{\boldsymbol{\nu}})$ defined by

$$\mathbf{A}_{\text{cons}}(\omega;\underline{\boldsymbol{\nu}}) = -\omega^2 \Big(\mathbf{M} + \sum_{\ell=1}^{L} \sum_{j \in \mathcal{J}_\ell} \underline{\nu}^{\ell}_{j}\, \underline{\mathcal{R}}^{\ell}_{\text{fuz},j}(\omega)\Big) + \mathbf{K}(\omega) \quad , \tag{193}$$

---

12 The brackets on the right-hand side of Eqs. (187) to (191) denote the antiduality product between $\mathcal{C}^{c\,\prime}$ and $\mathcal{C}^{c}$.

is a real operator such that

$$\mathbf{A}_{\text{cons}}(-\omega;\underline{\boldsymbol{\nu}}) = \mathbf{A}_{\text{cons}}(\omega;\underline{\boldsymbol{\nu}}) \quad . \tag{194}$$

In addition, for all $\mathbf{u}$ and $\delta\mathbf{u}$ in $\mathcal{C}^c$, we have [13]

$$< \mathbf{A}_{\text{cons}}(\omega;\underline{\boldsymbol{\nu}})\, \mathbf{u}\, , \delta\mathbf{u} > = < \mathbf{u}\, , \mathbf{A}_{\text{cons}}(\omega;\underline{\boldsymbol{\nu}})\, \delta\mathbf{u} > \quad . \tag{195}$$

Since $\mathbf{U}(t,\underline{\boldsymbol{\nu}})$ is a stochastic field with values in $\mathbb{R}^3$ (and not in $\mathbb{C}^3$) and from Eqs. (193) and (194), we deduce that

$$< \Pi_{\text{cons}}(\underline{\boldsymbol{\nu}}) > = 0 \quad . \tag{196}$$

Substituting Eq. (196) in Eq. (192) yields, for $\boldsymbol{\Lambda} = \mathbf{0}$, the mean power flow equation related to frequency band B_m for the master structure coupled with the fuzzy substructures,

$$< \Pi_{\text{diss}}(\underline{\boldsymbol{\nu}}) > + \sum_{\ell=1}^{L} < \Pi^{\ell}_{\text{mast-fuz}}(\underline{\boldsymbol{\nu}}) > = < \Pi_{\text{in}}(\underline{\boldsymbol{\nu}}) > \quad , \tag{197}$$

in which $< \Pi_{\text{diss}}(\underline{\boldsymbol{\nu}}) >$ is the mean power dissipated in the master structure, $< \Pi^{\ell}_{\text{mast-fuz}}(\underline{\boldsymbol{\nu}}) >$ is the mean value of the net transmitted power flowing from the master structure to fuzzy substructure ℓ and $< \Pi_{\text{in}}(\underline{\boldsymbol{\nu}}) >$ is the mean input power from external excitation forces applied to the master structure.

9.5. Estimation of mean powers of the real structure using a statistical energy approach

The real structure is constituted by the master structure coupled with L substructures, each substructure ℓ being constituted by secondary subsystems which are complex and which have a fuzzy character. This master structure is submitted to a time-stationary stochastic excitation force defined in Section 9.4. For frequency band B_m, a statistical energy approach is used to calculate a rough estimate of the mean power $< \Pi^{\text{SEA}}_{\text{diss}} >$ dissipated in the master structure, the mean value of the net transmitted power $< \Pi^{\text{SEA},\ell}_{\text{mast-fuz}} >$ flowing from the master structure to substructure ℓ across

---

[13] The brackets on the left-hand side of Eq. (195) denote the antiduality product between $\mathcal{C}^{c\prime}$ and $\mathcal{C}^c$, and the brackets on the right-hand side of Eq. (195) denote the antiduality product between $\mathcal{C}^c$ and $\mathcal{C}^{c\prime}$.

interface Σ_ℓ and the mean input power $<\Pi_{\text{in}}^{\text{SEA}}>$ from the external excitation forces applied to the master structure. These mean power quantities are used only in the identification procedure (it should be noted that they do not allow us to calculate the local dynamical response of the master structure and *a fortiori*, the local dynamical responses of the substructures). These mean powers satisfy the mean power balance equations for the master structure and for the substructures. One of these mean power balance equations represents the mean power flow equation for the master structure coupled with the substructures,

$$<\Pi_{\text{diss}}^{\text{SEA}}> + \sum_{\ell=1}^{L} <\Pi_{\text{mast-fuz}}^{\text{SEA},\ell}> = <\Pi_{\text{in}}^{\text{SEA}}> \quad . \tag{198}$$

For instance, these mean power quantities can be estimated using the Statistical Energy Analysis method (see for instance Lyon and DeJong, 1995).

9.6. Estimate of $\underline{\boldsymbol{\nu}}$ as the solution of a nonlinear constrained optimization problem

An estimate of $\underline{\boldsymbol{\nu}} \in \mathbb{R}^M$ is obtained by solving a nonlinear constrained optimization problem based on the use of the mean power flow equation for the master structure coupled with the fuzzy substructures. We then introduce the cost function $J(\underline{\boldsymbol{\nu}})$ from $\mathbb{R}^M$ into $\mathbb{R}^+$

$$J(\underline{\boldsymbol{\nu}}) = \sum_{\ell=1}^{L} \left(<\Pi_{\text{mast-fuz}}^{\ell}(\underline{\boldsymbol{\nu}})> - <\Pi_{\text{mast-fuz}}^{\text{SEA},\ell}> \right)^2 \quad , \tag{199}$$

and the space $\mathcal{D}_M \subset \mathbb{R}^M$ of constraints such that, for all $\underline{\boldsymbol{\nu}}$ in $\mathcal{D}_M$,

$$\nu_j^\ell \geq 0 \quad , \quad \ell \in \{1, \ldots, L\} \quad , \quad j \in \mathcal{J}_\ell \quad , \tag{200}$$

$$<\Pi_{\text{diss}}(\underline{\boldsymbol{\nu}})> - <\Pi_{\text{diss}}^{\text{SEA}}> = 0 \quad , \tag{201}$$

$$<\Pi_{\text{in}}(\underline{\boldsymbol{\nu}})> - <\Pi_{\text{in}}^{\text{SEA}}> = 0 \quad . \tag{202}$$

The estimate of $\underline{\boldsymbol{\nu}}$ is then obtained by minimizing $J(\underline{\boldsymbol{\nu}})$ on $\mathcal{D}_M$, i.e. by solving the following nonlinear constrained optimization problem

$$\min_{\underline{\boldsymbol{\nu}} \in \mathcal{D}_M} J(\underline{\boldsymbol{\nu}}) \quad . \tag{203}$$

The nonlinear constrained optimization problem defined by Eq. (203) is justified as follows. Subtracting the mean power flow equation (198) associated with the statistical energy approach from the mean power flow equation (197) associated with the fuzzy structure theory approach, and using the constraints defined by Eqs. (201) and (202), we deduce that, for all $\underline{\boldsymbol{\nu}}$ in $\mathcal{D}_M$, we have

$$\sum_{\ell=1}^{L} <\Pi^{\ell}_{\text{mast-fuz}}(\underline{\boldsymbol{\nu}})>=\sum_{\ell=1}^{L} <\Pi^{\text{SEA},\ell}_{\text{mast-fuz}}> \quad , \tag{204}$$

which justifies the choice of the cost function.

9.7. Solving procedure for the nonlinear constrained optimization problem

For $\underline{\boldsymbol{\nu}}$ fixed in $\mathcal{D}_M$, we have to calculate the value of cost function $J(\underline{\boldsymbol{\nu}})$ and the constraints defined by Eqs. (201) and (202), i.e., mean power quantities $<\Pi_{\text{diss}}(\underline{\boldsymbol{\nu}})>$, $<\Pi^{\ell}_{\text{mast-fuz}}(\underline{\boldsymbol{\nu}})>$ and $<\Pi_{\text{in}}(\underline{\boldsymbol{\nu}})>$. Consequently, a finite dimension approximation of the problem is required. For this we can use the Ritz-Galerkin method presented in Section 8.1 or the finite element discretization presented in Section 8.2. For the sake of brevity, we present the Ritz-Galerkin method (the corresponding equations for the finite element discretization are similar).

Ritz-Galerkin projection. Using the notations introduced in Section 8.1, the projection of time-stationary stochastic field $\mathbf{U}(t;\underline{\boldsymbol{\nu}})$ defined in Section 9.4 is written as

$$\mathbf{U}^N(\mathbf{x},t;\underline{\boldsymbol{\nu}}) = \sum_{\alpha=1}^{N} \mathbb{Q}_\alpha(t;\underline{\boldsymbol{\nu}})\; \mathbf{b}_\alpha(\mathbf{x}) \quad . \tag{205}$$

Then, using the classical spectral analysis formula for linear filtering of stationary stochastic processes (see Chapter IX), we obtain the following approximations

$$<\Pi^N_{\text{diss}}(\underline{\boldsymbol{\nu}})>= 2\int_{B_m} \omega^2 \text{tr}\{[\mathcal{T}_{\text{det}}(\omega;\underline{\boldsymbol{\nu}})]^*\,[\mathcal{D}(\omega)]\,[\mathcal{T}_{\text{det}}(\omega;\underline{\boldsymbol{\nu}})]\,[S_{\mathcal{F}}(\omega)]\}\,d\omega, \tag{206}$$

$$<\Pi^{N,\ell}_{\text{mast-fuz}}(\underline{\boldsymbol{\nu}})>= 2\sum_{j\in\mathcal{J}_\ell}\int_{B_m} \underline{\nu}^{\ell}_j\,\omega^2\; \text{tr}\{[\mathcal{T}_{\text{det}}(\omega;\underline{\boldsymbol{\nu}})]^*\,[\underline{\mathcal{I}}^{\ell}_{\text{fuz},j}(\omega)]\,[\mathcal{T}_{\text{det}}(\omega;\underline{\boldsymbol{\nu}})]\,[S_{\mathcal{F}}(\omega)]\}\,d\omega\,, \tag{207}$$

$$<\Pi^N_{\text{in}}(\underline{\boldsymbol{\nu}})>= 2\,\Re e\int_{B_m} i\omega\;\; \text{tr}\{[\mathcal{T}_{\text{det}}(\omega;\underline{\boldsymbol{\nu}})]\,[S_{\mathcal{F}}(\omega)]\}\,d\omega \quad , \tag{208}$$

in which $[\mathcal{T}_{\text{det}}(\omega;\underline{\boldsymbol{\nu}})]$ is an $(N \times N)$ complex symmetric matrix defined by

$$[\mathcal{T}_{\text{det}}(\omega;\underline{\boldsymbol{\nu}})] = [\mathcal{A}_{\text{det}}(\omega;\underline{\boldsymbol{\nu}})]^{-1} \quad , \tag{209}$$

where the $(N \times N)$ complex symmetric matrix $[\mathcal{A}_{\text{det}}(\omega;\underline{\boldsymbol{\nu}})]$ is such that

$$[\mathcal{A}_{\text{det}}(\omega;\underline{\boldsymbol{\nu}})]_{\beta\alpha} = < \mathbf{A}_{\text{det}}(\omega;\underline{\boldsymbol{\nu}})\,\mathbf{b}_\alpha , \mathbf{b}_\beta > \quad , \tag{210}$$

with $\mathbf{A}_{\text{det}}(\omega;\underline{\boldsymbol{\nu}})$ given by Eq. (174), $[\mathcal{D}(\omega)]$ and $[\underline{\mathcal{I}}^{\ell}_{\text{fuz},j}(\omega)]$ are $(N \times N)$ real positive-semidefinite symmetric matrices defined by

$$[\mathcal{D}(\omega)]_{\beta\alpha} = < \mathbf{D}(\omega)\,\mathbf{b}_\alpha , \mathbf{b}_\beta > \quad , \tag{211}$$

$$[\underline{\mathcal{I}}^{\ell}_{\text{fuz},j}(\omega)] = < \underline{\boldsymbol{\mathcal{I}}}^{\ell}_{\text{fuz},j}(\omega)\,\mathbf{b}_\alpha , \mathbf{b}_\beta > \quad , \tag{212}$$

and finally, $[S_{\boldsymbol{\mathcal{F}}}(\omega)]$ is an $(N \times N)$ complex Hermitian matrix such that $[S_{\boldsymbol{\mathcal{F}}}(\omega)]_{\beta\alpha} = S_{\mathbf{b}_\beta \mathbf{b}_\alpha}(\omega)$, i.e. using Eq. (185),

$$[S_{\boldsymbol{\mathcal{F}}}(\omega)]_{\beta\alpha} = f_0^2 \; \mathbb{1}_{B_m \cup \widetilde{B}_m}(\omega) \int_\Omega \chi(\mathbf{x}) \; < \mathbf{b}_\alpha(\mathbf{x}) , \overline{\mathbf{b}_\beta(\mathbf{x})} > \, d\mathbf{x} \quad , \tag{213}$$

which represents the matrix-valued spectral density function of second-order stationary stochastic process $\boldsymbol{\mathcal{F}}(t) = (\mathcal{F}_1(t), \ldots, \mathcal{F}_N(t))$ in which $\mathcal{F}_\alpha(t) = < \mathbb{F}(t) , \mathbf{b}_\alpha >$.

Optimization algorithm. Let $\underline{\boldsymbol{\nu}} \mapsto \mathbf{G}(\underline{\boldsymbol{\nu}}) = (G_1(\underline{\boldsymbol{\nu}}), \ldots, G_{M+2}(\underline{\boldsymbol{\nu}}))$ be the function from $\mathbb{R}^M$ into $\mathbb{R}^{M+2}$ defined by

$$\mathbf{G}(\underline{\boldsymbol{\nu}}) = \begin{bmatrix} -\underline{\boldsymbol{\nu}} \\ (<\Pi^N_{\text{diss}}(\underline{\boldsymbol{\nu}})> - <\Pi^{\text{SEA}}_{\text{diss}}>)^2 \\ (<\Pi^N_{\text{in}}(\underline{\boldsymbol{\nu}})> - <\Pi^{\text{SEA}}_{\text{in}}>)^2 \end{bmatrix} \quad . \tag{214}$$

The constrained optimization problem defined by Eq. (203) can then be rewritten as follows

$$\min_{\mathbf{G}(\boldsymbol{\nu}) \leq 0} J(\boldsymbol{\nu}) \quad , \tag{215}$$

in which $\mathbf{G}(\boldsymbol{\nu}) \leq 0$ means that $G_j(\boldsymbol{\nu}) \leq 0$ for $j \in \{1, \ldots, M+2\}$. The nonlinear constrained optimization problem defined by Eq. (215) can be solved, for instance, using the Sequential Quadratic Programming (SQP) method (see for instance Fletcher, 1980).

10. Case of a Real Structure Constituted by a Master Structure Coupled with a Very Large Number of Simple Oscillators. Application of the Fuzzy Structure Theory

Let us consider a real structure constituted by a master structure coupled with L substructures, each substructure ℓ being constituted by secondary subsystems which are complex and which have a fuzzy character. We have explained that to calculate the response of the master structure using the fuzzy structure theory, it is necessary to know the mean coefficients and the deviation coefficients of the fuzzy impedance laws used in constructing the random boundary impedance operator of each fuzzy substructure modeling a set of secondary subsystems. As explained in Section 9 for the general case, dimensionless mean coefficient $\underline{\nu}_j^\ell$ of the participating mass of fuzzy substructure ℓ for local direction j cannot be deduced by theoretical considerations and the identification method presented in Section 9 must be used. Nevertheless, we can present a particular real structure for which the fuzzy impedance law coefficients can be deduced from theoretical considerations without using the identification method. This particular real structure is a master structure coupled with a very large number of simple oscillators. The bandwith of frequency band $\mathbb{B}_{\text{MF}}$ defined by Eq. (9) is denoted as

$$\Delta\omega = \omega_{\text{MF, final}} - \omega_{\text{MF, init}} \quad . \tag{216}$$

Description of the particular real structure. We consider the general master structure introduced in Section 4 with L interfaces Σ_ℓ. On each surface Σ_ℓ are attached N_{osc}^ℓ simple linear oscillators acting in direction $\mathbf{n}$ normal to Σ_ℓ. These N_{osc}^ℓ simple linear oscillators constitute substructure ℓ. Each oscillator has a known mass denoted as m_{osc}^ℓ and a known critical damping rate denoted as ξ_{osc}^ℓ. The total mass of substructure ℓ, denoted as m_{ref}^ℓ, is such that

$$m_{\text{ref}}^\ell = N_{\text{osc}}^\ell \, m_{\text{osc}}^\ell \quad . \tag{217}$$

These N_{osc}^ℓ oscillators are randomly located on surface Σ_ℓ and have a uniform probability distribution on Σ_ℓ. The eigenfrequency of each oscillator is a random variable uniformly distributed on frequency band $\mathbb{B}_{\text{MF}}$. Consequently, the stiffness of each oscillator is a random variable.

Estimation of the fuzzy impedance law coefficients. For the real structure defined above, for each fuzzy substructure ℓ, and for frequency band $\mathbb{B}_{\text{MF}}$, there are no contributions in local directions $j = 1$ and $j = 2$ and consequently, set $\mathcal{J}_\ell$ defined in Section 7.1 is such that

$$\mathcal{J}_\ell = \{3\} \quad . \tag{218}$$

For $N^{\ell}_{\rm osc} \gg 1$, the mean coefficients $\underline{\mu}^{\ell}_{3}(\omega)$, $\underline{\xi}^{\ell}_{3}(\omega)$, $\underline{n}^{\ell}_{3}(\omega)$, $\underline{\alpha}^{\ell}_{3}(\omega)$ and the associated deviation coefficients $\lambda^{\ell}_{3,1}(\omega)$, $\lambda^{\ell}_{3,2}(\omega)$, $\lambda^{\ell}_{3,3}(\omega)$, $\lambda^{\ell}_{3,4}(\omega)$ of the fuzzy impedance law for fuzzy substructure ℓ and for local direction $j = 3$ are defined below (see Chabas, Desanti and Soize, 1986). We have

$$\underline{\alpha}^{\ell}_{3}(\omega) = 1 \quad , \quad \lambda^{\ell}_{3,4}(\omega) = 0 \quad , \tag{219}$$

which corresponds to a type I law. Mean coefficient $\underline{\mu}^{\ell}_{3}(\omega)$ of fuzzy substructure ℓ for local direction $j = 3$ is such that

$$\underline{\mu}^{\ell}_{3}(\omega) \sim m^{\ell}_{\rm osc} \frac{\sqrt{N^{\ell}_{\rm osc}}}{|\Sigma_{\ell}|} \quad , \tag{220}$$

and Eq. (39) can be rewritten as

$$\underline{\mu}^{\ell}_{3}(\omega) = \underline{\nu}^{\ell}_{3}(\omega) \frac{m^{\ell}_{\rm ref}}{|\Sigma_{\ell}|} \quad . \tag{221}$$

Then, substituting Eqs. (217) and (220) in Eq. (221) yields the dimensionless mean coefficient $\underline{\nu}^{\ell}_{3}(\omega)$ of the participating mass of fuzzy substructure ℓ for local direction $j = 3$,

$$\underline{\nu}^{\ell}_{3}(\omega) \sim \frac{1}{\sqrt{N^{\ell}_{\rm osc}}} \quad . \tag{222}$$

The dimensionless mean rate $\underline{\xi}^{\ell}_{3}(\omega)$ of internal damping of fuzzy substructure ℓ for local direction $j = 3$ is such that

$$\underline{\xi}^{\ell}_{3}(\omega) = \xi^{\ell}_{\rm osc} \quad . \tag{223}$$

The mean modal density $\underline{n}^{\ell}_{3}(\omega)$ related to fuzzy substructure ℓ for local direction $j = 3$ is such that

$$\underline{n}^{\ell}_{3}(\omega) = \frac{\sqrt{N^{\ell}_{\rm osc}}}{\Delta\omega} \quad . \tag{224}$$

Deviation coefficient $\lambda^{\ell}_{3,1}(\omega)$ associated with $\underline{\mu}^{\ell}_{3}(\omega)$, deviation coefficient $\lambda^{\ell}_{3,2}(\omega)$ associated with $\underline{\xi}^{\ell}_{3}(\omega)$ and deviation coefficient $\lambda^{\ell}_{3,3}(\omega)$ associated with $\underline{n}^{\ell}_{3}(\omega)$ are given by

$$\lambda^{\ell}_{3,k}(\omega) \sim \frac{\sqrt{3}}{\sqrt[4]{N^{\ell}_{\rm osc}}} \quad , \quad \forall\, k = 1, 2, 3 \quad . \tag{225}$$

From Eqs. (220), (224) and (217), we deduce that

$$\underline{\mu}^{\ell}_{3}(\omega)\, \underline{n}^{\ell}_{3}(\omega) \sim \frac{m^{\ell}_{\rm ref}}{\Delta\omega\, |\Sigma_{\ell}|} \quad . \tag{226}$$

For instance, the product $\underline{\mu}^{\ell}_{3}(\omega)\, \underline{n}^{\ell}_{3}(\omega)$ appears in Eq. (109).

APPENDIX

Mathematical Notations

1. Euclidean Space $\mathbb{R}^m$ and Hermitian Space $\mathbb{C}^m$

Let $\mathbb{K}$ be the set $\mathbb{R}$ of all real numbers or the set $\mathbb{C}$ of all complex numbers. The modulus of $z = a + ib \in \mathbb{C}$ is denoted as $|z| = (z\overline{z})^{1/2} = (a^2 + b^2)^{1/2}$ in which $\overline{z} = a - ib$ is the conjugate.

Let $\mathbf{u} = (u_1, \ldots, u_n)$ be a vector of $\mathbb{K}^n$. Vectors are usually denoted by boldface characters. We identify vector $\mathbf{u}$ with the $(n \times 1)$ column matrix of its u_j components. A vector $\mathbf{u}$ in $\mathbb{C}^n$ can be written as $\mathbf{u} = \mathbf{u}^R + i\,\mathbf{u}^I$ in which $\mathbf{u}^R = \Re e\,\mathbf{u}$ and $\mathbf{u}^I = \Im m\,\mathbf{u}$ are the real and imaginary parts of vector $\mathbf{u}$. The real space $\mathbb{R}^n$, called Euclidean space, is equipped with the usual scalar (or inner) product $(\mathbf{u}, \mathbf{v}) \mapsto \mathbf{u} \cdot \mathbf{v} = \sum_{j=1}^n u_j v_j$ and the associated norm $\|\mathbf{u}\| = (\mathbf{u} \cdot \mathbf{u})^{1/2}$. The complex space $\mathbb{C}^n$, called Hermitian space, is equipped with the Hermitian scalar (or inner) product $(\mathbf{u}, \mathbf{v}) \mapsto \mathbf{u} \cdot \overline{\mathbf{v}} = \sum_{j=1}^n u_j \overline{v_j}$ and the associated norm $\|\mathbf{v}\| = (\mathbf{v} \cdot \overline{\mathbf{v}})^{1/2}$, where $\overline{\mathbf{v}} = \mathbf{v}^R - i\,\mathbf{v}^I$ is the conjugate of $\mathbf{v} = \mathbf{v}^R + i\,\mathbf{v}^I$ in $\mathbb{C}^n$.

2. Vector Product

Let $\mathbf{u}$ and $\mathbf{v}$ be two vectors in $\mathbb{K}^3$ where $\mathbb{K}$ is $\mathbb{R}$ or $\mathbb{C}$. The vector product of $\mathbf{u}$ by $\mathbf{v}$ is denoted as $\mathbf{u} \times \mathbf{v}$.

3. Real and Complex Matrices

Let $\mathbb{K}$ be $\mathbb{R}$ or $\mathbb{C}$. Let $Mat_{\mathbb{K}}(m, n)$ be the space of all the $(m \times n)$ matrices $[A]$ whose components $A_{jk} = [A]_{jk}$ are in $\mathbb{K}$. A matrix is usually denoted by characters between brackets.

(1)- The determinant of square matrix $[A] \in Mat_{\mathbb{K}}(n, n)$ is denoted as $\det[A]$ and its trace as $\mathrm{tr}[A] = \sum_{j=1}^n [A]_{jj}$.

(2)- The transpose of $[A] \in Mat_{\mathbb{K}}(m,n)$ is denoted as $[A]^T \in Mat_{\mathbb{K}}(n,m)$.

(3)- The adjoint of $[A] \in Mat_{\mathbb{K}}(m,n)$ is denoted as $[A]^* = \overline{[A]}^T \in Mat_{\mathbb{K}}(n,m)$ (if $[A]$ is real then $[A]^* = [A]^T$).

(4)- The norm of matrix $[A] \in Mat_{\mathbb{K}}(m,n)$ is denoted as

$$\|A\| = \sup_{\|\mathbf{u}\|\leq 1} \|A\mathbf{u}\| \quad , \quad \mathbf{u} \in \mathbb{K}^n \quad ,$$

and is called the norm *subordinate* to vector spaces $\mathbb{K}^n$ and $\mathbb{K}^m$. In this case, we have the following inequality $\|A\mathbf{u}\| \leq \|A\|\,\|\mathbf{u}\|$, $\forall \mathbf{u} \in \mathbb{K}^n$.

(5)- The Frobenius norm, also called the Hilbert-Schmidt norm, of matrix $[A] \in Mat_{\mathbb{K}}(m,n)$ is defined by

$$\|A\|_F = \Big(\mathrm{tr}\{[A]^*[A]\}\Big)^{1/2} = \Big(\sum_{j=1}^{m}\sum_{k=1}^{n} |[A]_{jk}|^2\Big)^{1/2} \quad .$$

It should be noted that this norm is not subordinate to vector spaces $\mathbb{K}^n$ and $\mathbb{K}^m$. We have $\|A\| \leq \|A\|_F$.

Important remark. To simplify the notations, the column matrix of the components $u_1, \ldots, u_n$ of a vector $\mathbf{u} = (u_1, \ldots, u_n) \in \mathbb{K}^n$ will also be denoted as $\mathbf{u}$ instead of $[u]$, and the row matrix of its transpose will be denoted as $\mathbf{u}^T$ instead of $[u]^T$.

4. Linear and Bilinear Forms, Algebraic Duality and Algebraic Transposition on Complex Vector Spaces[1]

4.1. Linear and bilinear forms on complex vector spaces

Linear form. A *linear form* $\mathbf{x} \mapsto \ell(\mathbf{x})$ on a complex vector space X is a linear mapping from X into $\mathbb{C}$, i.e., for all λ and μ in $\mathbb{C}$ and, $\mathbf{x}_1$ and $\mathbf{x}_2$ in X,

$$\ell(\lambda\,\mathbf{x}_1 + \mu\,\mathbf{x}_2) = \lambda\,\ell(\mathbf{x}_1) + \mu\,\ell(\mathbf{x}_2) \quad .$$

Bilinear form. Let X and Y be complex vector spaces. A *bilinear form* $(\mathbf{x},\mathbf{y}) \mapsto b(\mathbf{x},\mathbf{y})$ on $X \times Y$ is a bilinear mapping from $X \times Y$ into $\mathbb{C}$, i.e., for all λ and μ in $\mathbb{C}$, $\mathbf{x}$, $\mathbf{x}_1$ and $\mathbf{x}_2$ in X and, $\mathbf{y}$, $\mathbf{y}_1$ and $\mathbf{y}_2$ in Y,

$$b(\lambda\,\mathbf{x}_1 + \mu\,\mathbf{x}_2\,,\mathbf{y}) = \lambda\,b(\mathbf{x}_1,\mathbf{y}) + \mu\,b(\mathbf{x}_2,\mathbf{y}) \quad ,$$

$$b(\mathbf{x}\,,\lambda\,\mathbf{y}_1 + \mu\,\mathbf{y}_2) = \lambda\,b(\mathbf{x},\mathbf{y}_1) + \mu\,b(\mathbf{x},\mathbf{y}_2) \quad .$$

[1] For more details concerning this section, we refer the reader to Reed and Simon (Vol. I), 1980; Dautray and Lions, 1992.

4.2. Algebraic duality of complex vector spaces

Let X and X^* be complex vector spaces. Let $(\mathbf{x}, \mathbf{x}^*) \mapsto b(\mathbf{x}, \mathbf{x}^*)$ be a bilinear form on $X \times X^*$ such that

$$\forall \mathbf{x}^* \in X^* \quad , \quad b(\mathbf{x}, \mathbf{x}^*) = 0 \quad \Longrightarrow \quad \mathbf{x} = \mathbf{0} \quad ,$$

$$\forall \mathbf{x} \in X \quad , \quad b(\mathbf{x}, \mathbf{x}^*) = 0 \quad \Longrightarrow \quad \mathbf{x}^* = \mathbf{0} \quad .$$

This bilinear form makes X and X^* *algebraically dual*. The *algebraic duality bracket* denoted as $<\mathbf{x}, \mathbf{x}^*>_{X,X^*}$, is defined by

$$\forall\, (\mathbf{x}, \mathbf{x}^*) \in X \times X^* \quad , \quad <\mathbf{x}, \mathbf{x}^*>_{X,X^*} = b(\mathbf{x}, \mathbf{x}^*) \quad .$$

We then identify X^* with the space of all the linear forms on X, and X^* is called the *algebraic dual* of X.

4.3. Algebraic transpose of a linear operator on complex vector spaces

Let X and Y be complex vector spaces and X^* and Y^* their algebraic duals. Let $\mathbf{B}$ be a linear operator from X into Y. The algebraic transpose of $\mathbf{B}$ is the linear operator ${}^t\mathbf{B}$ from Y^* into X^* such that

$$\forall \mathbf{x} \in X \quad , \quad \forall \mathbf{y}^* \in Y^* \quad , \quad <\mathbf{B}\mathbf{x}, \mathbf{y}^*>_{Y,Y^*} = <\mathbf{x}, {}^t\mathbf{B}\,\mathbf{y}^*>_{X,X^*} \quad .$$

A linear operator $\mathbf{B}$ from X into Y is transposable if ${}^t\mathbf{B}$ maps Y^* into X^* and we have the following diagram

$$\begin{array}{ccc} & \mathbf{B} & \\ X & \longrightarrow & Y \\ \vdots & & \vdots \\ & {}^t\mathbf{B} & \\ X^* & \longleftarrow & Y^* \end{array}$$

5. Linear and Bilinear Forms and Linear Operators on Complex Hilbert Spaces[2]

5.1. Complex Hilbert space and dual space

Let V be a complex Hilbert space equipped with the inner product $(\mathbf{u}\,,\mathbf{v})_V$ and the associated norm

$$\|\mathbf{u}\|_V = (\mathbf{u}\,,\mathbf{u})_V^{1/2} \quad .$$

This inner product is a positive-definite Hermitian (so-called symmetric) *sesquilinear form* on $V \times V$ which is linear with respect to the left argument $\mathbf{u}$ and antilinear (or semilinear) with respect to the right argument $\mathbf{v}$, i.e. for any complex number λ,

$$(\lambda\,\mathbf{u}\,,\mathbf{v})_V = \lambda\,(\mathbf{u}\,,\mathbf{v})_V \quad , \quad (\mathbf{u}\,,\lambda\,\mathbf{v})_V = \overline{\lambda}\,(\mathbf{u}\,,\mathbf{v})_V \quad .$$

A *linear form* $\mathbf{u} \mapsto f(\mathbf{u})$ on V is such that, for all complex number λ,

$$f(\lambda\,\mathbf{u}) = \lambda\,f(\mathbf{u}) \quad .$$

A linear form $\mathbf{u} \mapsto f(\mathbf{u})$ is continuous (or bounded) on V if

$$|f(\mathbf{u})| \leq c\,\|\mathbf{u}\|_V \quad , \quad \forall\,\mathbf{u} \in V \quad ,$$

in which c is a positive real constant. The *dual space* V' of V is the set of all the continuous linear forms on V. Any continuous linear form $\mathbf{u} \mapsto f(\mathbf{u})$ on V can be written as

$$f(\mathbf{u}) = <\mathbf{f}\,,\mathbf{u}>_{V',V} \quad , \quad \forall\,\mathbf{u} \in V \quad ,$$

with $\mathbf{f} \in V'$ and in which the angle brackets $<.\,,.>_{V',V}$ denote the *duality product* between V' and V, i.e. which is linear with respect to the left argument and linear with respect to the right argument (if there is no possible confusion in the notation, $<.\,,.>_{V',V}$ can be simply denoted as $<.\,,.>$).

5.2. Bounded operators on complex Hilbert spaces

Bounded operators from V into W. Let V and W be two complex Hilbert spaces. A bounded operator $\mathbf{B}$ from V into W is a linear mapping from V into W such that

$$\|\mathbf{B}\,\mathbf{u}\|_W \leq c\|\mathbf{u}\|_V \quad , \quad \forall\,\mathbf{u} \in V \quad ,$$

[2] For more details concerning this section, we refer the reader to Reed and Simon (Vol. I), 1980; Kato, 1966; Dautray and Lions, 1992; Yosida, 1966; Dunford and Schwartz, 1967.

in which c is a positive real constant. The set of all the bounded (continuous) operators from V into W is denoted as $\mathcal{L}(V,W)$. If $W = V$, then $\mathcal{L}(V,V)$ is denoted as $\mathcal{L}(V)$. If $\mathbf{B} \in \mathcal{L}(V,W)$, then the norm of $\mathbf{B}$ is defined by

$$\|\mathbf{B}\| = \sup_{\|\mathbf{u}\|_V \leq 1} \|\mathbf{B}\,\mathbf{u}\|_W \quad .$$

Adjoint of a bounded operator from V into W. Let V and W be two complex Hilbert spaces, and V' and W' their respective dual spaces. Let $\mathbf{B} \in \mathcal{L}(V,W)$ be a bounded operator from V into W. The *adjoint operator* (or the *transpose*) of $\mathbf{B}$ is the bounded operator ${}^t\mathbf{B} \in \mathcal{L}(W',V')$ from W' into V' such that

$$<\mathbf{v}\,,\mathbf{B}\,\mathbf{u}>_{W',W} = <{}^t\mathbf{B}\,\mathbf{v}\,,\mathbf{u}>_{V',V} \quad , \quad \forall \mathbf{u} \in V \quad , \quad \forall \mathbf{v} \in W' \quad ,$$

in which the angle brackets $< .\,, . >_{V',V}$ (or $< .\,, . >_{W',W}$) denote the duality product between V' and V (or between W' and W).

Continuous bilinear form on $V \times W$ and associated bounded operator from V into W'. Let V and W be two complex Hilbert spaces, and V' and W' their respective dual spaces. Let $(\mathbf{u},\mathbf{v}) \mapsto b(\mathbf{u},\mathbf{v})$ be a continuous bilinear form on $V \times W$, i.e. which is linear with respect to the left argument $\mathbf{u}$ and linear with respect to the right argument $\mathbf{v}$. Then there exists a bounded linear operator $\mathbf{B} \in \mathcal{L}(V,W')$ defined by

$$<\mathbf{B}\,\mathbf{u}\,,\mathbf{v}>_{W',W} = b(\mathbf{u},\mathbf{v}) \quad , \quad \forall \mathbf{u} \in V \quad , \quad \forall \mathbf{v} \in W \quad ,$$

in which the angle brackets $< .\,, . >_{W',W}$ denote the duality product between W' and W.

6. Antilinear and Sesquilinear Forms and Linear Operators on Complex Hilbert Spaces[3]

6.1. Complex Hilbert space, continuous antidual forms and antidual space

Let V be a complex Hilbert space equipped with the inner product $(\mathbf{u}\,,\mathbf{v})_V$ and the associated norm

$$\|\,\mathbf{u}\|_V = (\mathbf{u}\,,\mathbf{u})_V^{1/2} \quad .$$

[3] For more details concerning this section, we refer the reader to Reed and Simon (Vol. I), 1980; Kato, 1966; Dautray and Lions, 1992; Yosida, 1966; Dunford and Schwartz, 1967.

This inner product is a positive-definite Hermitian (so-called symmetric) *sesquilinear form* on $V \times V$ which is linear with respect to the left argument $\mathbf{u}$ and antilinear (or semilinear) with respect to the right argument $\mathbf{v}$, i.e. for any complex number λ,

$$(\lambda\,\mathbf{u}\,,\mathbf{v})_V = \lambda\,(\mathbf{u}\,,\mathbf{v})_V \quad , \quad (\mathbf{u}\,,\lambda\,\mathbf{v})_V = \overline{\lambda}\,(\mathbf{u}\,,\mathbf{v})_V \quad .$$

An *antilinear form* $\mathbf{u} \mapsto f(\mathbf{u})$ on V is such that, for any complex number λ,

$$f(\lambda\,\mathbf{u}) = \overline{\lambda}\,f(\mathbf{u}) \quad .$$

An antilinear form $\mathbf{u} \mapsto f(\mathbf{u})$ is continuous (or bounded) on V if

$$|f(\mathbf{u})| \leq c\,\|\mathbf{u}\|_V \quad , \quad \forall\,\mathbf{u} \in V \quad ,$$

in which c is a positive real constant. The *antidual space* V' of V is the set of all the continuous antilinear forms on V. Any continuous antilinear form $\mathbf{u} \mapsto f(\mathbf{u})$ on V can be written as

$$f(\mathbf{u}) = <\mathbf{f}\,,\mathbf{u}>_{V',V} \quad , \quad \forall\,\mathbf{u} \in V \quad ,$$

with $\mathbf{f} \in V'$ and in which the angle brackets $< .\,, . >_{V',V}$ denote the *anti-duality product* between V' and V, i.e. which is linear with respect to the left argument and antilinear with respect to the right argument (if there is no possible confusion in the notation, $< .\,, . >_{V',V}$ can be simply denoted as $< .\,, . >$).

6.2. Bounded operators on complex Hilbert spaces

Bounded operators from V into W. Let V and W be two complex Hilbert spaces. A bounded operator $\mathbf{B}$ from V into W is a linear mapping from V into W such that

$$\|\mathbf{B}\,\mathbf{u}\|_W \leq c\|\mathbf{u}\|_V \quad , \quad \forall\,\mathbf{u} \in V \quad ,$$

in which c is a positive real constant. The set of all the bounded (continuous) operators from V into W is denoted as $\mathcal{L}(V, W)$. If $W = V$, then $\mathcal{L}(V, V)$ is denoted as $\mathcal{L}(V)$. If $\mathbf{B} \in \mathcal{L}(V, W)$, then the norm of $\mathbf{B}$ is defined by

$$\|\mathbf{B}\| = \sup_{\|\mathbf{u}\|_V \leq 1} \|\mathbf{B}\,\mathbf{u}\|_W \quad .$$

Adjoint of a bounded operator from V into W. Let V and W be two complex Hilbert spaces, and V' and W' their respective antidual spaces.

Let $\mathbf{B} \in \mathcal{L}(V, W)$ be a bounded operator from V into W. The *adjoint operator* (or the *transpose*) of $\mathbf{B}$ is the bounded operator ${}^t\mathbf{B} \in \mathcal{L}(W', V')$ from W' into V' such that

$$<\mathbf{v}\,,\mathbf{B}\,\mathbf{u}>_{W',W} = <{}^t\mathbf{B}\,\mathbf{v}\,,\mathbf{u}>_{V',V} \quad , \quad \forall\mathbf{u} \in V \quad , \quad \forall\mathbf{v} \in W' \quad ,$$

in which the angle brackets $< .\,,. >_{V',V}$ (or $< .\,,. >_{W',W}$) denote the antiduality product between V' and V (or W' and W). If V (and W) is identified with its antidual space V' (and W'), i.e. $V = V'$ (and $W = W'$), then the adjoint operator of $\mathbf{B} \in \mathcal{L}(V, W)$ is denoted as $\mathbf{B}^*$. Operator $\mathbf{B}^* \in \mathcal{L}(W, V)$ and is such that

$$(\mathbf{B}\,\mathbf{u}\,,\mathbf{v})_W = (\mathbf{u}\,,\mathbf{B}^*\,\mathbf{v})_V \quad , \quad \forall\mathbf{u} \in V \quad , \quad \forall\mathbf{v} \in W \quad .$$

Continuous sesquilinear form on $V \times V$ and associated bounded operator from V into V'. Let V be a complex Hilbert space and V' its antidual space. Let $(\mathbf{u}, \mathbf{v}) \mapsto b(\mathbf{u}, \mathbf{v})$ be a continuous sesquilinear form on $V \times V$, i.e. which is linear with respect to the left argument $\mathbf{u}$ and antilinear with respect to the right argument $\mathbf{v}$. Then there exists a bounded linear operator $\mathbf{B} \in \mathcal{L}(V, V')$ defined by

$$<\mathbf{B}\,\mathbf{u}\,,\mathbf{v}>_{V',V} = b(\mathbf{u}, \mathbf{v}) \quad , \quad \forall\mathbf{u} \in V \quad , \quad \forall\mathbf{v} \in V \quad ,$$

in which the angle brackets $< .\,,. >_{V',V}$ denote the antiduality product between V' and V. The adjoint operator of $\mathbf{B}$ is the bounded linear operator ${}^t\mathbf{B} \in \mathcal{L}(V, V')$ such that

$$<\mathbf{B}\,\mathbf{u}\,,\mathbf{v}>_{V',V} = <\mathbf{u}\,,{}^t\mathbf{B}\,\mathbf{v}>_{V,V'} \quad , \quad \forall\mathbf{u} \in V \quad , \quad \forall\mathbf{v} \in V \quad .$$

Continuous Hermitian sesquilinear form and associated self-adjoint operator. Let H be a complex Hilbert space identified to its antidual space H' ($H = H'$). Let $(\mathbf{u}, \mathbf{v}) \mapsto m(\mathbf{u}, \mathbf{v})$ be a continuous sesquilinear form on $H \times H$. Then there exists a bounded linear operator $\mathbf{M} \in \mathcal{L}(H)$ defined by

$$(\mathbf{M}\,\mathbf{u}\,,\mathbf{v})_H = m(\mathbf{u}, \mathbf{v}) \quad , \quad \forall\mathbf{u} \in H \quad , \quad \forall\mathbf{v} \in H \quad .$$

Let $\mathbf{M}^* \in \mathcal{L}(H)$ be the adjoint operator of $\mathbf{M}$ which is such that

$$(\mathbf{M}\,\mathbf{u}\,,\mathbf{v})_H = (\mathbf{u}\,,\mathbf{M}^*\,\mathbf{v})_H \quad , \quad \forall\mathbf{u} \in H \quad , \quad \forall\mathbf{v} \in H \quad .$$

Sequilinear form $m(\mathbf{u}, \mathbf{v})$ on $H \times H$ is *Hermitian* (so-called *symmetric*) if

$$m(\mathbf{u}, \mathbf{v}) = \overline{m(\mathbf{v}, \mathbf{u})} \quad , \quad \forall\mathbf{u} \in H \quad , \quad \forall\mathbf{v} \in H \quad .$$

If continuous sesquilinear form $m(\mathbf{u},\mathbf{v})$ is Hermitian, then the associated bounded operator $\mathbf{M} \in \mathcal{L}(H)$ is *self-adjoint* (or Hermitian), i.e.

$$\mathbf{M}^* = \mathbf{M} \quad .$$

6.3. Unbounded operators on complex Hilbert spaces

Let H be a complex Hilbert space identified to its antidual space H' ($H = H'$). An *unbounded operator* $\mathbf{A}$ in H with domain $D(\mathbf{A})$ dense in H is a linear operator defined on a subset $D(\mathbf{A})$ of H (dense in H) with range into H

$$\mathbf{A} \quad : \quad D(\mathbf{A}) \hookrightarrow H \longrightarrow H \quad .$$

The notation $D(\mathbf{A}) \hookrightarrow H$ means that $D(\mathbf{A}) \subset H$ and that the embedding (injection) from $D(\mathbf{A})$ into H is dense in H. By definition, $\mathbf{A}$ is not a bounded operator ($\mathbf{A} \notin \mathcal{L}(H)$) and $\mathbf{A}\,\mathbf{u} \in H$, $\forall \mathbf{u} \in D(\mathbf{A})$.

Adjoint of an unbounded operator in H. The *adjoint* of unbounded operator $\mathbf{A}$ in H with domain $D(\mathbf{A})$ is the unbounded operator $\mathbf{A}^*$ in H with domain $D(\mathbf{A}^*)$ such that

$$(\mathbf{A}\,\mathbf{u}\,,\mathbf{v})_H = (\mathbf{u}\,,\mathbf{A}^*\,\mathbf{v})_H \quad , \quad \forall \mathbf{u} \in D(\mathbf{A}) \quad , \quad \forall \mathbf{v} \in D(\mathbf{A}^*) \quad .$$

Symmetric unbounded operator in H. Unbounded operator $\mathbf{A}$ in H with domain $D(\mathbf{A})$ is *symmetric* if

$$(\mathbf{A}\,\mathbf{u}\,,\mathbf{v})_H = (\mathbf{u}\,,\mathbf{A}\,\mathbf{v})_H \quad , \quad \forall \mathbf{u} \in D(\mathbf{A}) \quad , \quad \forall \mathbf{v} \in D(\mathbf{A}) \quad .$$

Self-adjoint unbounded operator in H. Let $\mathbf{A}$ be an unbounded operator in H with domain $D(\mathbf{A})$ and let $\mathbf{A}^*$ be its adjoint which is an unbounded operator in H with domain $D(\mathbf{A}^*)$. Unbounded operator $\mathbf{A}$ is *self-adjoint* if $D(\mathbf{A}^*) = D(\mathbf{A})$ and

$$(\mathbf{A}\,\mathbf{u}\,,\mathbf{v})_H = (\mathbf{u}\,,\mathbf{A}\,\mathbf{v})_H \quad , \quad \forall \mathbf{u} \in D(\mathbf{A}) \quad , \quad \forall \mathbf{v} \in D(\mathbf{A}) \quad ,$$

that is to say

$$\mathbf{A}^* = \mathbf{A} \quad .$$

It should be noted that for an unbounded operator $\mathbf{A}$, if $\mathbf{A}$ is self-adjoint, then $\mathbf{A}$ is symmetric, but the converse is not true. An unbounded operator $\mathbf{A}$ can be symmetric and not self-adjoint (if $D(\mathbf{A}^*) \neq D(\mathbf{A})$).

6.4. Bounded and unbounded operators on a triplet of complex Hilbert spaces

Let V and H be two complex Hilbert spaces such that the embedding (injection) $V \hookrightarrow H$ from V into H is dense and continuous. Let V' and H' be the antidual spaces and we identify H with H' ($H' = H$). We then have the triplet of complex Hilbert spaces

$$V \hookrightarrow H = H' \hookrightarrow V' \quad ,$$

the embeddings being dense and continuous. Let $(\mathbf{u}, \mathbf{v}) \mapsto a(\mathbf{u}, \mathbf{v})$ be a continuous sesquilinear form on $V \times V$ which is coercive (V-elliptic)

$$\Re e\{a(\mathbf{u}, \mathbf{u})\} \geq c \, \|\mathbf{u}\|_V^2 \quad , \quad \forall \mathbf{u} \in V \quad ,$$

and Hermitian (so-called symmetric)

$$a(\mathbf{u}, \mathbf{v}) = \overline{a(\mathbf{v}, \mathbf{u})} \quad , \quad \forall \mathbf{u} \in V \quad , \quad \forall \mathbf{v} \in V \quad ,$$

(then $\Re e\{a(\mathbf{u}, \mathbf{u})\} = a(\mathbf{u}, \mathbf{u}) \geq 0$). Let $\mathbf{A} \in \mathcal{L}(V, V')$ be the bounded linear operator associated with $a(\mathbf{u}, \mathbf{v})$ such that

$$<\mathbf{A}\,\mathbf{u} \,, \mathbf{v}>_{V',V} = a(\mathbf{u}, \mathbf{v}) \quad , \quad \forall \mathbf{u} \in V \quad , \quad \forall \mathbf{v} \in V \quad ,$$

in which the angle brackets $< .\,, . >_{V',V}$ denote the antiduality product between V' and V. Let $\mathbf{A}_H$ be the restriction of $\mathbf{A} \in \mathcal{L}(V, V')$ to H, defined on the domain $D(\mathbf{A}_H)$ such that

$$D(\mathbf{A}_H) = \{\, \mathbf{u} \in V \; ; \; \mathbf{A}\,\mathbf{u} \in H \,\} \quad .$$

Then, $\mathbf{A}_H$ is an unbounded (closed) operator in H with domain $D(\mathbf{A}_H)$ dense in H and $\mathbf{A}_H$ is self-adjoint

$$\mathbf{A}_H^* = \mathbf{A}_H \quad .$$

Generally, in order to simplify the notation the subscript H is omitted.

6.5. Case of real Hilbert spaces

If the Hilbert spaces V, W and H are real, then the terms antilinear form, antidual space, antiduality product, sesquilinear form, Hermitian sesquilinear form, are replaced respectively by linear form, dual space, duality product, bilinear form, symmetric bilinear form. For instance, if H is a real Hilbert space, a bilinear form $m(\mathbf{u}, \mathbf{v})$ on $H \times H$ is symmetric on H if $m(\mathbf{u}, \mathbf{v}) = m(\mathbf{v}, \mathbf{u})$.

7. Gradient, Divergence, Curl and Laplacian Operators

Let $\mathbf{x} \mapsto f(\mathbf{x})$ be a real- or complex-valued function defined on $\mathbb{R}^3$ and $\mathbf{x} \mapsto \mathbf{v}(\mathbf{x})$ be a function defined on $\mathbb{R}^3$ with values in $\mathbb{R}^3$ or $\mathbb{C}^3$. We denote the *gradient operator* with respect to $\mathbf{x}$ as $\boldsymbol{\nabla}$,

$$(\boldsymbol{\nabla} f)_j = \frac{\partial f}{\partial x_j} \quad , \quad j \in \{1, 2, 3\} \quad .$$

Then, $\boldsymbol{\nabla} f(\mathbf{x})$ is a vector in $\mathbb{R}^3$. The *Laplacian operator* is written as $\boldsymbol{\nabla}^2 = \boldsymbol{\nabla} \cdot \boldsymbol{\nabla}$. The *divergence* and the *curl* of function $\mathbf{x} \mapsto \mathbf{v}(\mathbf{x})$ with respect to $\mathbf{x}$ are written as $\boldsymbol{\nabla} \cdot \mathbf{v}$ and $\boldsymbol{\nabla} \times \mathbf{v}$ respectively.

8. Fourier Transform of Functions[4]

In this section, we recall some important results concerning the Fourier transform used throughout the book.

Spaces $L^1(\mathbb{R})$ and $L^2(\mathbb{R})$. Let $L^1(\mathbb{R})$ be the set of all integrable functions from $\mathbb{R}$ into $\mathbb{R}$ or $\mathbb{C}$. A function $t \mapsto h(t)$ belonging to $L^1(\mathbb{R})$ is such that

$$\|h\|_{L^1} = \int_{\mathbb{R}} |h(t)|\, dt < +\infty \quad .$$

Let $L^2(\mathbb{R})$ be the set of all square integrable functions from $\mathbb{R}$ into $\mathbb{R}$ or $\mathbb{C}$. A function $t \mapsto h(t)$ belonging to $L^2(\mathbb{R})$ is such that

$$\|h\|_{L^2}^2 = \int_{\mathbb{R}} |h(t)|^2\, dt < +\infty \quad .$$

From a mathematical point of view, these sets are defined in the context of measure and integration theory and consequently functions h are only defined almost everywhere on $\mathbb{R}$ (see for instance Halmos, 1976; Reed and Simon, 1980).

[4] For more details concerning this section, we refer the reader to Halmos, 1976; Dautray and Lions, 1992; Schwartz, 1966; Soize, 1993a.

Fourier transform in $L^1(\mathbb{R})$. The Fourier transform of a function h belonging to $L^1(\mathbb{R})$ is a continuous function $\omega \mapsto h(\omega)$ from $\mathbb{R}$ into $\mathbb{C}$ (using the customary abusive notation) such that, for all real ω,

$$h(\omega) = \int_{\mathbb{R}} e^{-i\omega t}\, h(t)\, dt \quad ,$$

and such that

$$\lim_{|\omega|\to+\infty} h(\omega) = 0 \quad .$$

Consequently, h does not generally belong to $L^1(\mathbb{R})$ but is a bounded function on $\mathbb{R}$. If h belongs to $L^1(\mathbb{R})$, then there is an inverse Fourier transform such that, for all real t in $\mathbb{R}$,

$$h(t) = \frac{1}{2\pi}\int_{\mathbb{R}} e^{i\omega t}\, h(\omega)\, d\omega \quad .$$

If h does not belong to $L^1(\mathbb{R})$, but if h is, in addition, a bounded variation function on any closed and bounded interval of $\mathbb{R}$, we then have, for all real t (the Fourier-Dirichlet theorem),

$$\lim_{a\to+\infty} \frac{1}{2\pi}\int_{-a}^{a} e^{i\omega t}\, h(\omega)\, d\omega = \frac{1}{2}(h(t_+) + h(t_-)) \quad .$$

Fourier transform in $L^2(\mathbb{R})$. The Fourier transform of a function h belonging to $L^2(\mathbb{R})$ is a function $\omega \mapsto h(\omega)$ from $\mathbb{R}$ into $\mathbb{C}$, belonging to $L^2(\mathbb{R})$, such that, for almost all real ω,

$$h(\omega) = \int_{\mathbb{R}} e^{-i\omega t}\, h(t)\, dt \quad .$$

If $t \mapsto h(t)$ belongs to $L^2(\mathbb{R})$, then $\omega \mapsto h(\omega)$ belongs to $L^2(\mathbb{R})$ and consequently its inverse Fourier transform is in $L^2(\mathbb{R})$ and such that, for almost all real t,

$$h(t) = \frac{1}{2\pi}\int_{\mathbb{R}} e^{i\omega t}\, h(\omega)\, d\omega \quad ,$$

and we have the Plancherel equality

$$\int_{\mathbb{R}} |h(t)|^2\, dt = \frac{1}{2\pi}\int_{\mathbb{R}} |h(\omega)|^2\, d\omega \quad .$$

Fourier transform of the derivatives of functions on $\mathbb{R}$. If the first and second derivatives (in the usual or generalized sense) of h with respect to t are functions belonging to $L^1(\mathbb{R})$ or $L^2(\mathbb{R})$, we then have

$$\dot{h}(\omega) = i\omega\, h(\omega) \quad , \quad \ddot{h}(\omega) = -\omega^2\, h(\omega) \quad ,$$

in which $\dot{h}(\omega)$ and $\ddot{h}(\omega)$ denote the Fourier transform of functions $t \mapsto \dot{h}(t)$ and $t \mapsto \ddot{h}(t)$ respectively.

Fourier transform in $L^2(\mathbb{R}^n)$. Let $\mathbf{t} = (t_1, \ldots, t_n)$ and $\boldsymbol{\omega} = (\omega_1, \ldots, \omega_n)$ be in $\mathbb{R}^n$. We denote the Euclidean scalar product of $\boldsymbol{\omega}$ with $\mathbf{t}$ as

$$\boldsymbol{\omega} \cdot \mathbf{t} = \omega_1 t_1 + \ldots + \omega_n t_n \quad ,$$

and the Lebesgue measure on $\mathbb{R}^n$ related to variables $\mathbf{t}$ and $\boldsymbol{\omega}$ as $d\mathbf{t} = dt_1 \ldots dt_n$ and $d\boldsymbol{\omega} = d\omega_1 \ldots d\omega_n$ respectively. Let $L^2(\mathbb{R}^n)$ be the set of all the square integrable functions from $\mathbb{R}^n$ into $\mathbb{R}$ or $\mathbb{C}$. A function $\mathbf{t} \mapsto h(\mathbf{t})$ belonging to $L^2(\mathbb{R}^n)$ is such that

$$\|h\|_{L^2}^2 = \int_{\mathbb{R}^n} |h(\mathbf{t})|^2 \, d\mathbf{t} < +\infty \quad .$$

The Fourier transform of a function h belonging to $L^2(\mathbb{R}^n)$ is a function $\boldsymbol{\omega} \mapsto h(\boldsymbol{\omega})$ from $\mathbb{R}^n$ into $\mathbb{C}$, belonging to $L^2(\mathbb{R}^n)$, such that, for almost everywhere $\boldsymbol{\omega}$ in $\mathbb{R}^n$,

$$h(\boldsymbol{\omega}) = \int_{\mathbb{R}^n} e^{-i\,\boldsymbol{\omega}\cdot\mathbf{t}}\, h(\mathbf{t})\, d\mathbf{t} \quad .$$

If $t \mapsto h(t)$ belongs to $L^2(\mathbb{R}^n)$, then $\omega \mapsto h(\omega)$ belongs to $L^2(\mathbb{R}^n)$ and consequently its inverse Fourier transform is in $L^2(\mathbb{R}^n)$ and such that, for almost all $\mathbf{t}$ in $\mathbb{R}^n$,

$$h(\mathbf{t}) = \frac{1}{(2\pi)^n} \int_{\mathbb{R}^n} e^{i\,\boldsymbol{\omega}\cdot\mathbf{t}}\, h(\boldsymbol{\omega})\, d\boldsymbol{\omega} \quad ,$$

and we have the Plancherel equality

$$\int_{\mathbb{R}^n} |h(\mathbf{t})|^2 \, d\mathbf{t} = \frac{1}{(2\pi)^n} \int_{\mathbb{R}^n} |h(\boldsymbol{\omega})|^2 \, d\boldsymbol{\omega} \quad .$$

Remarks concerning Fourier transforms. In general, we use either an abusive notation consisting in using the same symbol for a quantity and its

Fourier transform, i.e. denoting the Fourier transform as $h(\omega)$, or another symbol if convenient, i.e. denoting the Fourier transform of $t \mapsto h(t)$ as $T(\omega)$.

9. Indicator Function $\mathbb{1}_B$

Let X be any set and let B be any subset of X. The indicator function $\mathbf{x} \mapsto \mathbb{1}_B(\mathbf{x})$ from X into $\mathbb{R}$ is such that $\mathbb{1}_B(\mathbf{x}) = 1$ if $\mathbf{x} \in B$ and $\mathbb{1}_B(\mathbf{x}) = 0$ if $\mathbf{x} \notin B$.

10. Order Symbols o and O

Let f and g be arbitrary real functions of x defined in the vicinity of $x = 0$. We are interested in the behavior of function $f(x)$ if $x \to 0$. We use symbols O and o which are defined as follows. The expression

$$f = O(g)$$

means that function $|f(x)/g(x)|$ is bounded by a finite positive constant if $x \to 0$. Therefore the function $O(g)$ converges to a positive constant which is not zero. For instance, $f = O(1)$ means that f is a bounded function in the vicinity of $x = 0$. We can see that $\sin x = O(x)$ because for $x \to 0$, $(\sin x)/x = 1 + x^2/6 + \ldots$. The expression

$$f = o(g)$$

means that function $f(x)/g(x) \to 0$ if $x \to 0$. Therefore $o(g) \to 0$ if $x \to 0$. For instance, $o(1)$ means that $f(x) \to 0$ if $x \to 0$, and $1 - \cos x = o(x)$ because $(1 - \cos x)/x = x/2 - x^3/8 + \ldots$.

References

Abraham, R. and Marsden, J. E. (1978). *Foundations of Mechanics*. Benjamin/Cummings, Reading, Massachusetts.

Amini, S. and Harris, P. J. (1990). A comparison between various boundary integral formulations of the exterior acoustic problem. *Comp. Meth. Appl. Mech. Eng.*, **84**, 59–75.

Amini, S., Harris, P. J. and Wilton, D. T. (1992). *Coupled Boundary and Finite Element Methods for the Solution of the Dynamic Fluid-Structure Interaction Problem*. Lecture Notes in Eng., Vol 77. Springer, New York.

Angelini, J. J. and Hutin, P. M. (1983). Exterior Neumann problem for Helmholtz equation. Problem of irregular frequencies. *La Recherche Aérospatiale*, **3**, 43–52 (English edition).

Angelini, J. J. and Soize, C. (1989). Boundary integral / finite element mixed numerical method for 3D harmonic RCS. Part I: Formulation and numerical analysis. Tech. Rep. 5-2894RN081R (in French), ONERA, Châtillon, France.

Angelini, J. J., Soize, C. and Soudais, P. (1992). Hybrid numerical method for solving the harmonic Maxwell equations: II. Construction of the numerical approximations. *La Recherche Aérospatiale*, **4**, 45–55 (English edition).

Angelini, J. J., Soize, C. and Soudais, P. (1993). Hybrid numerical method for harmonic 3D Maxwell equations: scattering by mixed conducting and inhomogeneous-anisotropic dielectric media. *IEEE Trans. on Antennas and Propagation*, **41**(1), 66–76.

Argyris, J. and Mlejnek, H. P. (1991). *Dynamics of Structures*. North-Holland, Amsterdam.

Arnold, V. (1978). *Mathematical Methods of Classical Mechanics*. Springer, New York.

Bakewell, H. P. (1968). Turbulent wall pressure fluctuations on a body of revolution. *J. Acoust. Soc. Am.*, **43**(6).

Batchelor, G. K. (1967). *The Theory of Homogeneous Turbulence*. Cambridge University Press, Massachusetts.

Bathe, K. J. and Wilson, E. L. (1976). *Numerical Methods in Finite Element Analysis*. Prentice-Hall, New York.

Belytschko, T. and Geers, T. L. (eds). (1977). *Computational Methods for Fluid-Structure Interaction*. ASME-AMD, Vol. 26, New York.

Belytschko, T. and Hughes, T. J. R. (1983). *Computational Methods for Transient Analysis*. North-Holland, New York.

Bendat, J. S. and Piersol, A. G. (1971). *Random Data: Analysis and Measurement Procedures*. Wiley, New York.

Bendat, J. S. and Piersol, A. G. (1980). *Engineering Applications of Correlation and Spectral Analysis*. Wiley, New York.

Berger, M. and Gostiaux, B. (1987). *Differential Geometry: Manifolds, Curves, and Surfaces*. Springer-Verlag, New York.

Bermúdez, A. and Rodriguez, R. (1994). Finite element computation of the vibration modes of a fluid-solid system. *Comp. Meth. Appl. Mech. Eng.*, **119**, 355–370.

Bettess, P. (1993). *Infinite Elements*. Penshaw Press, Sunderland, U.K.

Blake, W. K. (1984). Aero-hydroacoustics for ships. Tech. Rep. DTNSRDC-84/010, David Taylor NSRDC, Bethesda, Maryland.

Blake, W. K. (1986). *Mechanics of Flow Induced Sound and Vibration*. Academic Press, New York.

Bland, D. R. (1960). *The Theory of Linear Viscoelasticity*. Pergamon, London.

Bluman, G. W. and Kumei, S. (1989). *Symmetries and Differential Equations*. Springer, New York.

Bowman, J. J., Senior, T. B. A. and Uslenghi, P. L. E. (1969). *Electromagnetic and Acoustic Scattering by Simple Shapes*. North-Holland, Amsterdam.

Brebbia, C. A. (1978). *The Boundary Element Method*. Pentech Press, London.

Brebbia, C. A. and Dominguez, J. (1992). *Boundary Elements: An Introductory Course*. McGraw-Hill, New York.

Brebbia, C. A., Telles, J. C. F. and Wrobel, L. C. (1984). *Boundary Element Techniques*. Springer-Verlag, Berlin.

Brezis, H. (1987). *Analyse Fonctionnelle. Théorie et Applications*. Masson, Paris.

Bucy, R. S. and Joseph, P. D. (1968). *Filtering for Stochastic Processes with Applications to Guidance*. Wiley, New York.

Burton, A. J. and Miller, G. F. (1971). The application of integral equation methods to the numerical solution of some exterior boundary value problems. *Proc. R. Soc. London Ser. A.*, **323**, 201–210.

Caughey, T. K. (1960). Classical normal modes in damped linear dynamic systems. *Journal of Applied Mechanics*, pp. 269–271.

Chabas, F. and Soize, C. (1986). Hydroelasticity of slender bodies in an unbounded fluid in the medium frequency range. *La Recherche Aérospatiale*, **4**, 39–51 (English edition).

Chabas, F. and Soize, C. (1987). Modeling mechanical subsystems by boundary impedance in the finite element method. *La Recherche Aérospatiale*, **5**, 59–75 (English edition).

Chabas, F., Desanti, A. and Soize, C. (1986). Probabilistic structural modeling in linear dynamic analysis of complex mechanical systems. II Numerical analysis and applications. *La Recherche Aérospatiale*, **5**, 49–67 (English edition).

Chase, D. M. (1980). Modeling the wave vector-frequency spectrum of turbulent boundary layer wall pressure. *Journal of Sound and Vibration*, **70**(1).

Chase, D. M. (1987). The character of the turbulent wall pressure spectrum at subconvective wave numbers and a suggested comprehensive model. *Journal of Sound and Vibration*, **112**, 125–147.

Chatelin, F. (1993). *Eigenvalues of Matrices*. Wiley, New York.

Chen, C. C., Ragiah, H. and Atluri, S. H. (1990). An effective method for solving the hypersingular integral equations in 3-D acoustics. *J. Acoust. Soc. Am.*, **88**, 918–937.

Chen, G. and Zhou, J. (1992). *Boundary Element Methods*. Academic Press, New York.

Chorin, A. J. and Marsden, J. E. (1993). *A Mathematical Introduction to Fluid Mechanics*. Springer-Verlag, New York.

Ciarlet, P. G. (1979). *The Finite Element Method for Elliptic Problems*. North-Holland, Amsterdam.

Ciarlet, P. G. (1988). *Mathematical Elasticity, Vol.I: Three-Dimensional Elasticity*. North-Holland, Amsterdam.

Clough, R. W. and Penzien, J. (1975). *Dynamics of Structures*. McGraw-Hill, New York.

Colton, D. L. and Kress, R. (1992). *Integral Equation Methods in Scattering Theory*. Krieger Publishing Company, Malabar, Florida.

Corcos, G. M. (1963). Resolution of pressure in turbulence. *J. Acoust. Soc. Am.*, **35**(2), 192–199.

Corcos, G. M. (1964). The structure of the turbulent pressure field in boundary layer flows. *J. Fluid Mech.*, **18**, 353–378.

Costabel, M. (1988). Boundary integral operators on Lipschitz domains: elementary results. *SIAM Journal of Mathematical Analysis*, **19**, 613–626.

Costabel, M. and Stephan, E. (1985). A direct boundary integral equation method for transmission problems. *Journal of Mathematical Analysis and Applications*, **106**, 367–413.

Coupry, G. and Soize, C. (1984). Hydroelasticity and the field radiated by a slender elastic body into an unbounded fluid. *Journal of Sound and Vibration*, **96**(2), 261–273.

Craggs, A. (1971). The transient response of a coupled plate-acoustic system using plate acoustic finite elements. *Journal of Sound and Vibration*, **15**, 509–528.

Craig, R. R. (1985). A review of time domain and frequency domain component mode synthesis method. In *Combined Experimental-Analytical Modeling of Dynamic Structural Systems* (Eds. D.R. Martinez and A.K. Miller). ASME-AMD, Vol. 67, New York.

Craig, R. R. and Bampton, M. C. C. (1968). Coupling of substructures for dynamic analysis. *AIAA Journal*, **6**, 1313–1319.

Crandall, S. H. and Mark, D. W. (1973). *Random Vibration in Mechanical Systems*. Academic Press, New York.

Cremer, L., Heckl, M. and Ungar, E. E. (1988). *Structure-Born Sound*. Springer-Verlag, Berlin.

Crighton, D. G., Dowling, A. P., Ffowcs-Williams, J. E., Heckl, M. and Leppington, F. G. (1992). *Modern Methods in Analytical Acoustics*. Springer-Verlag, Berlin.

Daniel, W. T. J. (1980). Performance of reduction methods for fluid-structure and acoustic eigenvalue problems. *Int. J. Num. Meth. Eng.*, **15**, 1585–1594.

Dautray, R. and Lions, J.-L. (1992). *Mathematical Analysis and Numerical Methods for Science and Technology*. Springer-Verlag, Berlin.

David, J. M. and Soize, C. (1994). Prediction of the high-frequency behavior of coupled fluid-structure systems by the SEA method and applications. In *Computational Methods for Fluid-Structure Interaction* (Eds. J.M. Crolet and R. Ohayon), pp. 55–77. Longman Scientific and Technical, Harlow.

Dieudonné, J. (1969). *A Treatise on Analysis*. Academic Press, New York.

Doob, J. L. (1953). *Stochastic Processes*. Wiley, New York.

Dowling, A. P. and Ffowcs Williams, J. E. (1983). *Sound and Sources of Sound*. Ellis Horwood, Chichester, U.K.

Dunford, N. and Schwartz, J. T. (1967). *Linear operators I*. Interscience, New York.

Elishakoff, I. (1983). *Probabilistic Methods in the Theory of Structures*. Wiley, New York.

Everstine, G. C. (1981). A symmetric potential formulation for fluid-structure interaction. *Journal of Sound and Vibration*, **79**(1), 157–160.

Everstine, G. C. and Yang, M. K. (eds). (1984). *Advances in Fluid-Structure Interaction*. ASME-PVP, Vol. 78, San Antonio.

Fahy, F. (1987). *Sound and Structural Vibration*. Academic Press, London.

Farhat, C. and Geradin, M. (1994). On a component mode method and its application to incompatible substructures. *Computers and Structures*, **51**(5), 459–473.

Felippa, C. A. (1985). Symmetrization of the contained compressible-fluid vibration eigenproblem. *Comm. in Appl. Num. Meth.*, **1**, 241–247.

Felippa, C. A. (1988). Symmetrization of coupled eigenproblems by eigenvector augmentation. *Comm. in Appl. Num. Meth.*, **4**, 561–563.

Felippa, C. A. and Ohayon, R. (1990). Mixed variational formulation of finite element analysis of acoustoelastic/slosh fluid-structure interaction. *Int. J. of Fluids and Structures*, **4**, 35–57.

Ffowcs-Williams, J. E. (1965). Surface pressure fluctuations induced by boundary layer flow at finite Mach number. *J. Fluid Mech.*, **22**(3), 507–519.

Ffowcs-Williams, J. E. (1982). Boundary layer pressures and the Corcos model: a development to incorporate low-wave-number constraints. *J. Fluid Mech.*, **125**, 9–25.

Finlayson, B. A. (1972). *The Method of Weighted Residuals and Variational Principles*. Academic Press, New York.

Fletcher, R. (1980). *Practical Methods of Optimization - Vol. 2, Constrained Optimization.* Wiley, New York.

Fung, Y. C. (1968). *Foundations of Solid Mechanics.* Prentice Hall, Englewood Cliffs, New Jersey.

Geers, T. L. and Felippa, C. A. (1983). Doubly asymptotic approximations for vibration analysis of submerged structures. *J. Acoust. Soc. Am.*, **73**, 1152–1159.

Geers, T. L. and Zhang, P. (1994). Doubly asymptotic approximations for internal acoustic domains: formulation and evaluation. *Journal of Applied Mechanics*, **61**, 893–906.

Géradin, M. and Rixen, D. (1994). *Mechanical Vibrations.* Wiley, Chichester, U.K.

Germain, P. (1973). *Cours de Mécanique des Milieux Continus.* Masson, Paris.

Germain, P. (1986). *Mécanique, Tomes I and II, Ecole Polytechnique.* Ellipses, Paris.

Givoli, D. and Keller, J. B. (1989). A finite element method for large domains. *Comp. Meth. Appl. Mech. Eng.*, **76**, 41–66.

Gladwell, G. M. L. (1966). A variational formulation of damped acousto-structural vibration problems. *Journal of Sound and Vibration*, **4**, 172–186.

Golub, G. H. and Van Loan, C. F. (1989). *Matrix Computations.* The Johns Hopkins University Press, Baltimore and London.

Guikhman, L. and Skorokhod, A. V. (1979). *The Theory of Stochastic Processes.* Springer-Verlag, Berlin.

Guyan, R. J. (1965). Reduction of stiffness and mass matrices. *AIAA Journal*, **3**, 380.

Hackbusch, W. (1995). *Integral Equations, Theory and Numerical Treatment.* Birkhauser Verlag, Basel.

Halmos, P. R. (1976). *Measure Theory.* Springer-Verlag, Berlin.

Harari, I. and Hughes, T. J. R. (1994). Studies of domain-based formulations for computing exterior problems of acoustics. *Int. J. Num. Meth. Eng.*, **37**(17), 2891–3014.

Hinze, J. O. (1959). *Turbulence. An Introduction to Mechanism and Theory.* McGraw-Hill, New York.

Hörmander, L. (1985). *The Analysis of Linear Partial Differential Operators.* Springer-Verlag, Berlin. (Vol. I and II (1983), Vol. III and IV (1985)).

Hughes, T. J. R. (1987). *The Finite Element Method.* Prentice Hall, Englewood Cliffs, New Jersey.

Hurty, W. C. (1965). Dynamic analysis of structural systems using component modes. *AIAA Journal*, **3**(4), 678–685.

Irons, B. M. (1970). Role of part inversion in fluid-structure problems with mixed variables. *AIAA Journal*, **7**, 568–568.

Jenkins, G. M. and Watt, D. G. (1968). *Spectral Analysis and its Applications.* Holden Day, San Francisco.

Jones, D. S. (1974). Integral equations for the exterior acoustic problem. *Quart. J. Mech. Appl. Math.*, **1**(27), 129–142.

Jones, D. S. (1986). *Acoustic and Electromagnetic Waves.* Oxford University Press, New York.

Junger, M. C. and Feit, D. (1993). *Sound, Structures and Their Interaction.* Acoust. Soc. Am. Publications on Acoustics, Woodbury, NY, U.S.A. (originally published in 1972, MIT Press, Cambridge).

Kanarachos, A. and Antoniadis, I. (1988). Symmetric variational principles and modal methods in fluid-structure interaction problems. *Journal of Sound and Vibration*, **121**(1), 77–104.

Kato, T. (1966). *Perturbation Theory for Linear Operator.* Springer-Verlag, New York.

Kehr-Candille, V. and Ohayon, R. (1992). Elastoacoustic damped vibrations. Finite element and modal reduction methods. In *New Advances in Computational Structural Mechanics* (Eds. O.C. Zienkiewicz and P. Ladevèze), Elsevier, Amsterdam.

Kellogg, O. D. (1953). *Foundations of Potential Theory.* Dover, New York.

Kirsch, A. (1989). Surface gradients and continuity properties for some integral operators in classical scattering theory. *Mathematical Methods in the Applied Sciences*, **11**, 789–804.

Kree, P. and Soize, C. (1983). *Mécanique Aléatoire.* Dunod, Paris.

Kree, P. and Soize, C. (1986). *Mathematics of Random Phenomena.* Reidel, Dordrecht.

Kress, R. (1989). *Linear Integral Equations.* Springer, New York.

Lancaster, P. (1966). *Lambda-Matrices and Vibrating Systems.* Pergamon Press, Oxford.

Landau, L. and Lifchitz, E. (1992a). *Elasticity Theory.* Pergamon Press, Oxford.

Landau, L. and Lifchitz, E. (1992b). *Fluid Mechanics.* Pergamon Press, Oxford.

Landau, L. and Lifchitz, E. (1992c). *Mechanics.* Pergamon Press, Oxford.

Lauchle, G. C. (1977). Noise generated by axisymmetric turbulent boundary layer flow. *J. Acoust. Soc., Am.*, **61**(3), 694–703.

Leissa, A. W. (1993a). *Vibrations of Plates.* Acoust. Soc. Am. Publications on Acoustics, Woodbury, NY, U.S.A. (originally published in 1969, NASA SP-160).

Leissa, A. W. (1993b). *Vibrations of Shells.* Acoust. Soc. Am. Publications on Acoustics, Woodbury, NY, U.S.A. (originally published in 1973, NASA SP-288).

Lesueur, C. (1988). *Rayonnement Acoustique des Structures.* Eyrolles, Paris.

Leung, A. Y. T. (1993). *Dynamic Stiffness and Substructures.* Springer-Verlag, New York.

Lighthill, J. (1978). *Waves in Fluids.* Cambridge University Press, MA.

Lin, Y. K. (1967). *Probabilistic Theory of Structural Dynamics.* McGraw-Hill, New York (Reprint R.E., Krieger, Melbourne, Fla., 1976).

Liu, W. K., Zhang, Y. and Ramirez, M. R. (1991). Multiple scale finite element methods. *Int. J. Num. Meth. Eng.*, **32**, 969–990.

Ludwig, W. and Falter, C. (1988). *Symmetries in Physics: Group Theory Applied to Physical Problems.* Springer Verlag, New York.

Lyon, R. H. (1975). *Statistical Energy Analysis Systems: Theory and Applications.* MIT Press, Cambridge, Massachusetts.

Lyon, R. H. and DeJong, R. G. (1995). *Theory and Application of Statistical Energy Analysis.* Butterworth-Heinemann, Boston.

MacNeal, R. (1971). A hybrid method of component mode synthesis. *Computers and Structures*, **1**, 581–601.

Maidanik, G. (1981). Extension and reformulation of statistical energy analysis with use of room acoustics concepts. *Journal of Sound and Vibration*, **78**(3), 417–423.

Maidanik, G. (1995). Power dissipation in a sprung mass attached to a master structure. *J. Acoust. Soc. Am.*, **98**(6), 3527–3533.

Maidanik, G. and Dickey, J. (1988). Response of coupled one-dimensional dynamic systems. *Journal of Sound and Vibration*, **121**, 187–194.

Maidanik, G. and Dickey, J. (1994). Flexural response matrix for ribbed panels. *J. Acoust. Soc. Am.*, **95**(3), 1245–1254.

Malvern, L. (1969). *Introduction to the Mechanics of a Continuous Medium.* Prentice Hall, Englewood Cliffs, New Jersey.

Mandel, J. (1966). *Cours de Mécanique des Milieux Continus, Tome II, Mécanique des Solides.* Gauthier-Villars, Paris.

Markus, A. S. (1986). *Introduction of the Spectral Theory of Polynomial Operator Pencils.* American Mathematical Society, Providence, Rhode Island.

Marsden, J. E. and Hughes, T. J. R. (1983). *Mathematical Foundations of Elasticity.* Prentice Hall, Englewood Cliffs, New Jersey.

Mason, W. P. (1971). *Physical Acoustics. Principles and Methods.* Volume II, Part A, Academic Press, New York.

Mathews, I. C. (1986). Numerical techniques for three-dimensional steady-state fluid-structure interaction. *J. Acoust. Soc. Am.*, **79**, 1317–1325.

Maue, A. W. (1949). Zur Formulierung eines allgemeinen Beugungsproblems durch eine Integralgleichung (Formulation of a general diffraction problem using an integral equation). *Zeitschrift der Physik*, **126**, 601–618.

Meirovitch, L. (1980). *Computational Methods in Structural Dynamics*. Sijthoff and Noordhoff, The Netherlands.

Meirovitch, L. (1990). *Dynamics and Control of Structures*. Wiley, New York.

Mikhlin, S. G. (1964). *Variational Methods in Mathematical Physics*. Pergamon, Oxford.

Mikhlin, S. G. (1970). *Mathematical Physics, an Advanced Course*. North-Holland, Amsterdam.

Min, K. W., Igusa, T. and Achenbach, J. D. (1992). Frequency window method for strongly coupled and multiply connected structural systems: multiple-mode windows. *J. Appl. Mech.*, **59**(2), 244–252.

Mitzner, K. M. (1966). Acoustic scattering from an interface between media of greatly different density. *Journal of Mathematical Physics*, **7**, 2053–2060.

Morand, H. J. P. and Ohayon, R. (1976). Investigation of variational formulations for the elasto-acoustic problem. *Proc. of the Int. Symp. on Finite Element Method in Flow Problems*, Rappallo, Italy.

Morand, H. J. P. and Ohayon, R. (1979). Substructure variational analysis for the vibrations of coupled fluid-structure systems. *Int. J. Num. Meth. Eng.*, **14**(5), 741–755.

Morand, H. J. P. and Ohayon, R. (1989). Finite element method applied to the prediction of the vibrations of liquid-propelled launch vehicles. ASME-PVP, Vol. 176, New York.

Morand, H: J. P. and Ohayon, R. (1992). *Interactions Fluides Structures*. Masson, Paris.

Morand, H. J. P. and Ohayon, R. (1995). *Fluid Structure Interaction*. Wiley, New York.

Morse, P. M. and Feshbach, H. (1953). *Methods of Theoretical Physics*. McGraw-Hill, New York.

Morse, P. M. and Ingard, K. U. (1968). *Theoretical Acoustics*. McGraw-Hill, New York.

Nedelec, J. C. (1976). Curved finite element methods for the solution of singular integral equations of surfaces in 3D. *Comp. Meth. Appl. Mech. Eng.*, **8**, 61–80.

Nedelec, J. C. (1978). Approximation par potentiel de double couche du problème de Neumann extérieur. *C. R. Acad. Sc. Paris*, **t. 286**, 16 janvier, 103–106.

Nefske, D. J. and Sung, S. H. (1988). Power flow finite element analysis of dynamic systems. ASME-NCA, Vol. 3, New York.

Novozhilov, V. V. (1964). *The Theory of Thin Shells*. Noordhoff, Groningen, The Netherlands.

Oden, J. T. and Reddy, J. N. (1983). *Variational Methods in Theoretical Mechanics*. Springer-Verlag, New York.

Ohayon, R. (1979). Symmetric variational formulation of harmonic vibrations problem by coupling primal and dual principles. Application to fluid-structure coupled systems. *La Recherche Aérospatiale*, **3**, 69–77.

Ohayon, R. (1984). Transient and modal analysis of bounded medium fluid-structure problems. *Proc. Int. Conf. on Num. Meth. for Transient and Coupled Problems*, Pineridge Press, U.K.

Ohayon, R. (1985). Fluid structure interaction analysis for cyclically symmetric bounded systems. ASME-PVP, Vol. 64, pp. 71-76.

Ohayon, R. (1986). Variational analysis of a slender fluid-structure system: the elastic-acoustic beam. A new symmetric formulation. *Int. J. for Num. Meth. in Eng.*, **22**, 637–647.

Ohayon, R. (1987). Fluid-structure modal analysis. New symmetric continuum-based formulations. Finite element applications. *Proc. Int. Conf. NUMETA 87*, Martinus Sijhoff Publ., The Netherlands.

Ohayon, R. (1989). Alternative variational formulations for static and modal analysis of structures containing fluids. ASME-PVP, Vol. 176, New York.

Ohayon, R. and Felippa, C. A. (1990). The effect of wall motion on the governing equations of contained fluids. *Journal of Applied Mechanics*, **57**, 783–785.

Ohayon, R. and Nicolas-Vullierme, B. (1981). An efficient shell element for the computation of the vibrations of fluid-structure systems of revolution. *Proc. SMIRT-6*, North-Holland, Amsterdam.

Ohayon, R. and Valid, R. (1983). Symmetric coupled variational primal-dual principle in linear elastodynamics (in French). *C.R. Acad. Sc.*, Série II, Vol. 297.

Ohayon, R. and Valid, R. (1984). True symmetric variational formulations for fluid-structure interaction in bounded domains - Finite elements results. In *Numerical Methods in Coupled Systems*, New York. Wiley.

Ohayon, R., Meidinger, N. and Berger, H. (1987). Symmetric variational formulations for the vibration of damped structural-acoustic systems. Aerospace applications. *Proc. AIAA/ASME 28th Structures, Structural Dynamics Conf.*, AIAA Publisher, Washington.

Ohayon, R., Sampaio, R. and Soize, C. (1997). Dynamic substructuring of damped structures using singular value decomposition. *Journal of Applied Mechanics*, **64**, 292–298.

Oñate, E., Periaux, J. and Samuelsson, A. (eds). (1991). *The Finite Element Method in the 1990's*. Springer-Verlag, CIMNE, Barcelona.

Panich, O. I. (1965). On the question of solvability of the exterior boundary value problems for the wave equation and Maxwell's equations. *Russian Math. Surv.*, **20**, 221–226.

Papoulis, A. (1977). *Signal Analysis*. McGraw-Hill, New York.

Parlett, B. N. (1980). *The Symmetric Eigenvalue Problem*. Prentice Hall, Englewood Cliffs, New Jersey.

Petitjean, B. (1992). Neumann's external problem for the Helmholtz equation: application to thin and volumic multiconnected domains with cyclic symmetry. *La Recherche Aérospatiale*, **5**, 15–24 (English edition).

Petyt, M. (1990). *Introduction to Finite Element Vibration Analysis*. Cambridge University Press, Massachusetts.

Photiadis, D. M., Bucaro, J. A. and Houston, B. H. (1997). The effect of internal oscillators on the acoustic response of a submerged shell. *J. Acoust. Soc. Am.*, **101**(2), 895–899.

Pierce, A. D. (1989). *Acoustics: An Introduction to its Physical Principles and Applications*. Acoust. Soc. Am. Publications on Acoustics, Woodbury, NY, U.S.A. (originally published in 1981, McGraw-Hill, New York).

Pierce, A. D., Sparrow, V. W. and Russel, D. A. (1993). Fundamental structural-acoustic idealizations for structures with fuzzy internals. *ASME Paper 93-WA/NCA-17*. (also published in the *Journal of Vibration and Acoustics*, **117**, pp. 339-348, 1995).

Pinsky, P. M. and Abboud, N. N. (1989). Two mixed variational principles for exterior fluid-structure interaction problems. *Computers and Structures*, **33**(3), 621–635.

Priestley, M. B. (1981). *Spectral Analysis and Time Series*. Academic Press, New York.

Priestley, M. B. (1988). *Nonlinear and Non-Stationary Time Series Analysis*. Academic Press, New York.

Raviart, P. A. and Thomas, J. M. (1983). *Introduction à l'Analyse Numérique des Equations aux Dérivées Partielles*. Masson, Paris.

Reed, M. and Simon, B. (1980). *Methods of Modern Mathematical Physics*. Academic Press, New York.

Roberts, J. B. and Spanos, P. D. (1990). *Random Vibration and Statistical Linearization*. Wiley, New York.

Roseau, M. (1980). *Vibrations in Mechanical Systems*. Springer-Verlag, Berlin.

Rubin, S. (1975). Improved component mode representation for structural dynamic analysis. *AIAA Journal*, **18**(8), 995–1006.

Russell, D. A. and Sparrow, V. W. (1995). Backscattering from a baffled finite plate strip with fuzzy attachments. *J. Acoust. Soc. Am.*, **98**(3), 1527–1533.

Salençon, J. (1988). *Mécanique des Milieux Continus*. Ellipses, Paris.

Sanchez-Hubert, J. and Sanchez-Palencia, E. (1989). *Vibration and Coupling of Continuous Systems. Asymptotic Methods*. Springer-Verlag, Berlin.

Sandberg, G. and Göransson, P. (1988). A symmetric finite element formulation for acoustic fluid-structure interaction analysis. *Journal of Sound and Vibration*, **123**(3), 507–515.

Schenck, H. A. (1968). Improved integral formulation for acoustic radiation problems. *J. Acoust. Soc. Am.*, **44**, 41–58.

Schwartz, L. (1965). *Méthodes Mathématiques pour les Sciences Physiques.* Hermann, Paris.

Schwartz, L. (1966). *Théorie des Distributions.* Hermann, Paris.

Soedel, W. (1993). *Vibrations of Shells and Plates.* Marcel Dekker, New York.

Soize, C. (1982a). Medium frequency linear vibrations of anisotropic elastic structures. *La Recherche Aérospatiale*, **5**, 65–87 (English edition).

Soize, C. (1982b). Medium frequency linear vibrations of anisotropic elastic structures (in French). *C. R. Acad. Sc.*, Série II, Vol. 294, pp. 895–898.

Soize, C. (1986). Probabilistic structural modeling in linear dynamic analysis of complex mechanical systems. I - Theoretical elements. *La Recherche Aérospatiale*, **5**, 23–48 (English edition).

Soize, C. (1988). Methods for classical problems of stochastic dynamics (in French). In *Collection Mathématiques Appliquées, No 11*, Paris. Techniques de l'Ingénieur, Edition Périodique TI, Fascicules A1 346, pp. 1-26, A1 347, pp. 1-22, A1 348, pp. 1-16.

Soize, C. (1993a). *Méthodes Mathématiques en Analyse du Signal* (Mathematical Methods in Signal Analysis). Masson, Paris.

Soize, C. (1993b). A model and numerical method in the medium frequency range for vibroacoustic predictions using the theory of structural fuzzy. *J. Acoust. Soc. Am.*, **94**(2), 849–865.

Soize, C. (1994). *The Fokker-Planck Equation for Stochastic Dynamical Systems and its Explicit Steady State Solutions.* World Scientific, Singapore.

Soize, C. (1995a). Coupling between an undamped linear acoustic fluid and a damped nonlinear structure - Statistical Energy Analysis considerations. *J. Acous. Soc. Am.*, **98**(1), 373–385.

Soize, C. (1995b). Vibration damping in low-frequency range due to structural complexity. A model based on the theory of fuzzy structures and model parameters estimation. *Computers and Structures*, **58**(5), 901–915.

Soize, C. (1996a). Estimation of the fuzzy substructure model parameters using the mean power flow equation of the fuzzy structure. *Proceedings of the ASME Noise Control and Acoustics Division.* Volume 1, ASME/WAM, pp. 23-30, NCA-Vol. 22 (also to appear in the *Journal of Vibration and Acoustics of ASME, 1997*).

Soize, C. (1996b). A strategy for prediction and active control in structural acoustics. *La Recherche Aérospatiale*, **3**, 189–197 (English edition).

Soize, C. (1997a). Reduced models in medium frequency range for general dissipative structural-dynamics systems. *European Journal of Mechanics, A/Solids.* (In press).

Soize, C. (1997b). Reduced models in medium frequency range for general external structural-acoustics systems. *J. Acoust. Soc. Am.* (Submitted in 1997).

Soize, C., David, J. M. and Desanti, A. (1986a). Functional reduction of stochastic fields for studying stationary random vibrations. *La Recherche Aérospatiale*, **2**, 31–44 (English edition).

Soize, C., Hutin, P. M., Desanti, A., David, J. M. and Chabas, F. (1986b). Linear dynamic analysis of mechanical systems in the medium frequency range. *Computers and Structures*, **23**(5), 605–637.

Soize, C., David, J. M. and Desanti, A. (1989). Dynamic and acoustic response of coupled structure - dense fluid axisymmetric systems excited by a random wall pressure field. *La Recherche Aérospatiale*, **5**, 1–14 (English edition).

Soize, C., Desanti, A. and David, J. M. (1992). Numerical methods in elastoacoustics for low and medium frequency ranges. *La Recherche Aérospatiale*, **5**, 25–44 (English edition).

Soong, T. T. (1973). *Random Differential Equations in Science and Engineering.* Academic Press, New York.

Strang, G. and Fix, G. J. (1973). *An Analysis of the Finite Element Method.* Prentice-Hall, Englewood

Cliffs, New Jersey.

Strasberg, M. and Feit, D. (1996). Vibration damping of large structures induced by attached small resonant structures. *J. Acoust. Soc. Am.*, **99**(1), 335–344.

Sung, S. H. and Nefske, D. J. (1986). Component mode synthesis of a vehicle structural-acoustic system model. *AIAA Journal*, **24**(6), 1021–1026.

Truesdell, C. (1960). *The Elements of Continuum Mechanics*. Springer-Verlag, Berlin.

Truesdell, C. (1984). (ed), *Mechanics of Solids, Vol III, Theory of Viscoelasticity, Plasticity, Elastic Waves and Elastic Stability*. Springer-Verlag, Berlin.

Vanmarcke, E. H. (1983). *Random Fields: Analysis and Synthesis*. The MIT Press, Cambridge, Massachusetts.

Vasudevan, R. (1991). Solution of acoustic problems using time integration and transform techniques. Tech. Rep. DTRC Report SAD-91/20e-1941, David Taylor Research Center.

Vasudevan, R. and Liu, Y. N. (1991). Application of time integration and transform techniques to scattering problems. *Structural Acoustics, ASME NCA-12/AMD-128*, pp. 35–40.

Washizu, K. (1975). *Variational Methods in Elasticity and Plasticity*. Pergamon Press, Oxford.

Weaver, R. L. (1997). Mean and mean-square responses of a prototypical master/fuzzy structure. *J. Acoust. Soc. Am.*, **101**(3), 1441–1449.

Wilcox, C. H. (1984). *Sound Propagation in Stratified Fluids*. Springer-Verlag, New York.

Willmarth, W. W. (1975). Pressure fluctuations beneath turbulent boundary layers. *Annual Review of Fluid Mechanics*, **7**, 14–19.

Yosida, K. (1966). *Functional Analysis*. Springer-Verlag, Berlin.

Zienkiewicz, O. C. and Bettess, P. (1978). Fluid-structure interaction and wave forces. An introduction to numerical treatment. *Int. J. Num. Meth. Eng.*, **13**(1), 1–17.

Zienkiewicz, O. C. and Newton, R. E. (1969). Coupled vibrations of a structure submerged in a compressible fluid. Int. Symp. Finite Element Techn., Stuttgart.

Zienkiewicz, O. C. and Taylor, R. L. (1989). *The Finite Element Method*. McGraw-Hill, New York, 4th edn. (vol. 1, 1989 and vol. 2, 1991).

Zienkiewicz, O. C., Bando, K., Bettess, P., Emson, C. and Chiam, T. C. (1985). Mapped infinite element for exterior wave problems. *Int. J. Num. Meth. Eng.*, **21**, 1229–1251.

Subject Index

Symbol Index